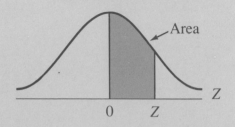

Area

0   Z   Z

AREAS UNDER THE NORMAL CURVE

z	.00	.01	.02	.03	.04	.05	.06	.07	.08	.09
2.1	.4812	.4826	.4830	.4834	.4838	.4842	.4846	.4850	.4854	.4857
2.2	.4861	.4864	.4868	.4871	.4875	.4878	.4881	.4884	.4887	.4890
2.3	.4893	.4896	.4898	.4091	.4904	.4906	.4909	.4911	.4913	.4916
2.4	.4918	.4920	.4922	.4925	.4927	.4929	.4931	.4932	.4934	.4936
2.5	.4938	.4940	.4941	.4943	.4945	.4946	.4948	.4949	.4951	.4952
2.6	.4953	.4955	.4956	.4957	.4959	.4960	.4961	.4962	.4963	.4964
2.7	.4965	.4966	.4967	.4968	.4969	.4970	.4971	.4972	.4973	.4974
2.8	.4974	.4975	.4976	.4977	.4977	.4978	.4979	.4979	.4980	.4981
2.9	.4981	.4982	.4982	.4983	.4984	.4984	.4985	.4985	.4986	.4986
3.0	.4987	.4987	.4987	.4988	.4988	.4989	.4989	.4989	.4990	.4990
3.1	.49903									
3.2	.49931									
3.3	.49952									
3.4	.49966									
3.5	.49977									
3.6	.49984									
3.7	.49989									
3.8	.49993									
3.9	.49995									
4.0	.50000									

APPLIED ELEMENTARY STATISTICS

APPLIED ELEMENTARY STATISTICS

RICHARD C. WEIMER
Frostburg State College

BROOKS/COLE PUBLISHING COMPANY
MONTEREY, CALIFORNIA

Brooks/Cole Publishing Company
A Division of Wadsworth, Inc.

Printed in the United States of America

10   9   8   7   6   5   4   3   2   1

Library of Congress Cataloging-in-Publication Data

Weimer, Richard C.
  Applied elementary statistics.

  Includes index.
  1. Statistics.   I. Title.
QA276.12.W38   1987   519.5   86–26425
ISBN 0–534–07002–7

Sponsoring Editor: *Craig Barth*
Project Development Editors: *Jody Larson and John Bergez*
Marketing Representative: *Sue Ewing*
Editorial Assistant: *Lorraine McCloud*
Production Editors: *Michael Oates and Penelope Sky*
Production Assistant: *Sara Hunsaker*
Manuscript Editor: *Susan Reiland*
Permissions Editor: *Carline Haga*
Interior Design: *Vernon T. Boes*
Cover Design: *Sharon L. Kinghan*
Cover Image: *Lisa Thompson*
Cover Photo: *Lee Hocker*
Art Coordinator: *Lisa Torri*
Interior Illustration: *Carl Brown*
Typesetting: *Syntax International, Singapore*
Cover Printing: *Phoenix Color Corporation, Long Island City, New York*
Printing and Binding: *R. R. Donnelley & Sons Company, Crawfordsville, Indiana*

Without the support, patience, and encouragement of my wife, Marlene,
and our children, Stephanie, Richard, and David,
this book would never have been completed.
I especially appreciate the joys and love given to me by my grandson, Albert Richard Winner,
who passed away on September 12, 1986, at the early age of five years.
My associations and diversions with him while writing made the task much easier.
I dedicate this text to my family with love and appreciation.

PREFACE

The material in this book was developed over a four-year period from class notes I used in teaching an introductory statistics course to students majoring in business, the social sciences, the biological sciences, and the natural sciences. The text is intended as an introduction to probability and modern statistical techniques, for students who have had at least one course in elementary algebra and who desire a strong understanding of the concepts and principles of basic probability and statistics. The logic and theory of applied statistics are developed by clear, logical exposition.

My primary purpose in writing this text was to produce an accurate, meaningful exposition of the subject that would help students understand the elementary statistical techniques used by experimental researchers.

This text differs from other works in the field by presenting statistics as a way of dealing with uncertainty and variability, showing the true roles that variability, sampling error, and sampling distributions play in the development of inferential statistics. Students who intuitively understand the underlying principles and how they are related will be better equipped for the future than those who have been conditioned to simply substitute numbers into an appropriate formula. Other distinct features include the following:

1. The text explains in an elementary fashion the logic that underlies the principles of inferential statistics, motivating readers to practice this logic themselves. Students are taught not only how to use statistical techniques but also to understand why they work.

2. Histograms, the building blocks of statistics, are used throughout to indicate basic properties of theoretical distributions.

3. Important relationships concerning hypothesis tests are pointed out whenever possible: for example, analysis of variance is presented as an extension of the two-sample t test; the relationship between t^2 and F is discussed; the chi-square test for testing more than two populations is shown to be an extension of the two-sample z test for proportions; and the relationship between z^2 and χ^2 is examined.

4. Proportions for dichotomous or binomial populations are first introduced by associating them with means for a population consisting of ones and zeroes. The properties for the sampling distribution of the sample proportion are then found by using the properties for the sampling distribution of the sample mean.

5. Assumptions underlying the various hypothesis tests are emphasized. The F test for testing the assumption of homogeneity of variances is discussed before the two-sample t test, so it is possible to test this assumption before learning how to use the two-sample t test.

6. One-way ANOVA is introduced and promoted in relation to the concept of range. Follow-up procedures are provided for testing for pairwise differences and strength of relationship after a significant F test.

Organization

The text is logically and smoothly developed. Probability is presented as a way to bridge the gap between descriptive and inferential statistics. The use of probability in inference-making is stressed throughout. The spiral approach is used to organize the material. For instance, the concepts of regression and correlation are introduced in Chapter 4 as descriptive measures. In Chapter 14 they are used to draw inferences. The concept of sum of squares is first presented in Chapter 3, then used throughout the text to simplify ideas and formulas involving variance. Sampling distributions are introduced in Chapter 8, and then used throughout the development of inferential techniques. The importance of the role of sampling distributions in inference-making is clearly demonstrated, and emphasized as a central concept underlying the development of hypothesis testing and estimation procedures.

The book is organized to allow the instructor a great deal of flexibility. Those who wish to pass quickly through Chapters 1 and 2 need teach only the concepts of histograms in Chapter 2 to effectively continue past Chapter 4. If you are not planning an intense discussion of probability, only Sections 5.6 and 5.7 need be taught as a basis for later sections (Section 5.6 is background for Chapter 12; 5.7 is important to Sections 6.3, 8.1, and 8.2). If you intend to omit binomial distributions and related concepts, all of Chapter 6 and Sections 7.3, 8.5, 9.3, part of 10.4, 11.3, and 12.1 may be skipped. After Chapter 11 has been studied, the remaining chapters can be covered in any order. Section 13.3 is optional, and may be omitted without a break in continuity.

Students often find it confusing to study normal distributions, the normal approximation to the binomial, and the central limit theorem simultaneously. Great care has therefore been given to making these three topics sequential. In addition, the central limit theorem has been developed in a way that stresses its importance in the development of large-sample statistical methods for making inferences. The sampling distribution of the sample sum is also described as approximately normal, and likened to binomial distributions being approximated by normal distributions.

Estimation for single parameters following a logical development is presented in Chapter 9, and the hypothesis-testing procedures for one-sample tests are developed in Chapter 10. The two procedures are presented side by side in Chapter 11, showing students how to make comparisons between two parameters using two populations. Both procedures are shown to produce consistent results when nondirectional tests are used.

The concept of sampling error (introduced in Chapter 8) serves as a unifying concept, explaining the natural variation in sample statistics when sampling is used. For example, in Chapter 10, if there is no evidence to suggest that the null hypothesis is false, the difference between the test statistic and the hypothesized population value is attributed to sampling error.

Students are frequently asked to classify applications according to sampling distribution and necessary assumptions. Nonparametric tests in Chapter 15 are presented as an alternative, whenever assumptions are tenuous or known to be violated. Interpreting results and identifying the associated risks are stressed throughout Chapters 8–15.

Learning Features

Each chapter begins with an introductory list of objectives and ends with a concise summary. In addition, each chapter contains:

- exercises at the end of each section
- a list of important notation
- a list of important facts and formulas
- review exercises
- an achievement test

The applications provided are realistic, yet not overly technical in nature. I have deliberately drawn upon a broad variety of areas in the natural, biological, and social sciences.

Each section is concluded by a large exercise set that is divided into two parts. Type A exercises offer additional drill and practice, and are similar to those in the text. Type B exercises are more difficult, expanding the ideas encountered in the text, asking the student to supply simple proofs of certain important facts, and often developing new ideas. Type B problems are of a level appropriate to an honors section in introductory statistics. Altogether there are more than 1,000 exercises in the text.

Acknowledgments

I appreciate the assistance of the many people who helped me prepare this book. I gratefully acknowledge the suggestions and constructive criticisms offered by hundreds of Frostburg State College students who studied statistics under my direction during the past several years. Peter Tirrell, Stewart Crall, Donna Pope, and William Byers helped check and verify the answers to the problems in the text. Shelley Drees did a superb job of typing various drafts; she had the painstaking task of deciphering my handwriting and making sense of my special notations. Special thanks go to Sandy Cochrane, who typed portions of an early draft, and to Pat Murray, for reproducing and binding versions for classroom use. I also appreciate the help and advice given to me by my colleague Kil Lee, who taught from my notes in 1984. Numerous constructive criticisms and suggestions were provided by the following reviewers: James Baker, Jefferson Community College; James Baldwin, Nassau Community College; Pat Cerrito, University of South Florida; James Daly, California Polytechnic State University; Ken Eberhard, Chabot College; Antanas Gilvydis, Malcolm X College; Raymond Guzman, Pasadena City College; Gary Itzkowitz, Glassboro State College; Keith Nelson, Beloit College; Jim Ridenhour, Austin Peay State University; Larry Ringer, Texas A & M University; Ann Thomas, University of Northern Colorado; William Tomhave, Concordia College; and John Van Druff, Fort Steilacoom Community College.

I especially appreciate the expert advice offered by Professor Susan L. Reiland, who read the entire manuscript before the final draft was prepared, and who served as copy editor for the manuscript. She offered a care for detail and overall excellence that exceeded my greatest expectations; she is a true and dedicated professional who has earned my great respect. Finally, at Brooks/Cole Publishing I want to thank Craig Barth, editor; John Bergez, technical advisor; and Michael Oates and Penelope Sky, production editors, for their encouragement and advice throughout the project.

Richard C. Weimer

CONTENTS

10 HYPOTHESIS TESTING 340

11 INFERENCES COMPARING TWO PARAMETERS 375

12 ANALYSES OF COUNT DATA 431

13 SINGLE-FACTOR ANALYSIS OF VARIANCE 467

14 LINEAR REGRESSION ANALYSIS 504

APPLIED ELEMENTARY STATISTICS

1

INTRODUCTION

Chapter Objectives

In this chapter you will learn:

- *the two meanings of the term statistics*
- *the difference between a population and a sample*
- *the difference between a statistic and a parameter*
- *the difference between descriptive and inferential statistics*
- *how induction and deduction relate to probability and statistics*
- *why it is important to study statistics*

Chapter Contents

People view statistics in several different ways. Statistics is commonly perceived as anything having to do with percentages, averages, charts, and graphs. For some, statistics is a field consisting of rules and methods for dealing with information. For others, statistics is a way of acting and thinking about worldly events—events that occur with irregularity and are governed by the laws of uncertainty.

1.1 WHAT IS STATISTICS?

As consumers of statistical information and potential users of statistical techniques, we need to understand the basic ideas and tools of statistics. Most of us are influenced daily by some aspect of statistics in the information we get from radio, television, newspapers, and magazines. For example, we may read or hear that:

1. Studies suggest that about 50% of all drownings among teenagers and adults are associated with alcohol use.
2. One-parent families now account for 26% of all U.S. families with children under 18—up from only 13% in 1970.
3. Seven out of ten Americans do not have wills.
4. The prevalence of diabetes is nearly 3 times as high in overweight people as it is in nonoverweight people.
5. More than 3000 insurance companies pay more than $8.8 billion in claims each year.
6. There is a 50% chance that the victim will never race again.
7. Children who brush their teeth with brand X toothpaste have 35% fewer cavities.
8. The median net worth of newly retired Social Security beneficiaries in 1981–1982 was between $64,700 and $68,300 for couples, and between $17,000 and $30,000 for unmarried people.
9. In 1960 it was estimated that only 1% of high school seniors had tried marijuana, whereas in 1980 it was estimated that 60% had done so.
10. Studies suggest that feelings of helplessness are correlated with a marked decrease in several immune-system cells that fight disease.

These examples indicate that statistical information is used for a variety of reasons. Among these reasons are to:

- Inform the public (all of the preceding examples).
- Provide comparisons (examples 2, 4, 8, and 9).
- Explain actions that have taken place (examples 1, 4, and 10).
- Influence decisions that will take place (examples 1 and 7).
- Justify a claim or assertion (examples 1, 7, and 10).
- Predict future outcomes (example 6).
- Estimate unknown quantities (examples 1 and 9).
- Establish a relationship or association between two factors (examples 1, 4, and 10).

Since we are consumers of statistical information, we can use statistics to study and gain an understanding of many changing events that will contribute to our understanding of the world around us. Studying statistics will enable us to give a reasonable interpretation to each of the previous examples. For instance, the figure "35%" in example 7 is open to interpretation, because we do not know the basis of the comparison. It may be difficult, if not impossible, to find a toothpaste that produces 35% fewer cavities than any other toothpaste when tested under similar conditions on independent and similar groups of children. But it should be fairly simple to find one child who uses brand X toothpaste and who has 35% fewer cavities than some other child, perhaps one who does not brush his or her teeth at all. As daily consumers of statistical information, we should be knowledgeable in both the uses and misuses of statistics. We will learn in this course how each of the numbers in the examples of this section could have been derived.

As consumers of statistical information, we are frequently confronted with graphs displaying statistical information. For example, parents spend approximately $142,700 on a child by the time the child is 17 years of age. The circle graph in Figure 1-1 indicates how this money is spent.[1] We can see at a glance that (1) food, housing, and transportation costs are approximately equal; and (2) recreation, medical care, and miscellaneous expenses share nearly equal amounts of the total budget.

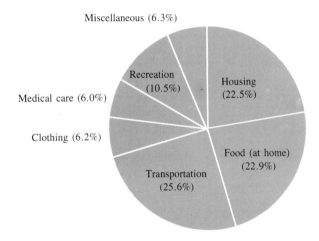

FIGURE 1-1 How $142,700 is spent to raise a child until age 17

As potential users of statistical methods and techniques, we need to be concerned with doing basic statistical research, describing the results of our scientific inquiry, making decisions based on this inquiry, and estimating unknown quantities. The following further illustrate how statistics might be used:

[1] Data source: Sandra Chofon, *The Big Book of Kids' Lists* (World Almanac Publications, 1985).

1. A television commercial claims that one product brand is superior to all other brands. If the claim is based on scientific inquiry, the statement is used to educate the viewers. Suppose we doubt the claim. In an attempt to disprove it, we might gather relevant information concerning all brands of the product, analyze the results using proper statistical procedures, and make a decision regarding the advertised claim. Frequently, advertised claims are based on insufficient information or faulty analyses of the information.

2. Suppose we want to determine the best teacher at Excel College. How should we go about making this determination? We could ask Excel students who the best teacher is, analyze the results, and make our determination. Should we survey every student? How will the survey be conducted? How will the information be analyzed? How will the best teacher be determined? Answers to these, and other questions, are a focus of statistics.

3. A life insurance company is considering offering reduced premiums to policyholders who engage in an ongoing exercise program. To aid the insurance company in making a decision, mortality information will be collected and analyzed. What type of exercise program will qualify a policyholder for a reduced premium? How much should the reduction be? What risk factors would disqualify a policyholder in an ongoing exercise program from a premium reduction? A person with a strong background in statistics could assist the life insurance company in evaluating the merits of the new program.

The next three cases serve to illustrate how statistics can be used to obtain answers to current issues of interest.

Case 1   *What role does diet play in coronary heart disease?* The role of diet in coronary heart disease has been contested for nearly two generations. The heart/diet theory maintains that reducing the level of blood cholesterol through diet will reduce one's risk of developing coronary heart disease. To partially test the relationship between blood cholesterol reduction and coronary heart disease, a study involving the drug cholestyramine, a cholesterol-lowering medication, was undertaken.[2] The study involved 3800 middle-aged men. All had blood cholesterol levels of at least 265 milligrams (per deciliter of blood), which put them in the top 5% of adult Americans in blood cholesterol values, and all were found to be free of any signs of coronary heart disease on entering the study. The participants were randomly assigned to the treatment or the control group, each of which contained 1900 men. Both groups followed a cholesterol-reducing diet for an average of 7.4 years. The treatment group also took daily doses of cholestyramine, while the control group received a placebo that was indistinguishable from cholestyramine. The study concluded that the drug group had fewer heart attacks (155 versus 187 for the control group) and fewer deaths from heart attack (30 versus 38). The difference between the two groups was judged to be statistically significant. The findings support the belief that lowering blood cholesterol by using cholestyramine in middle-aged men with cholesterol levels above 265 milligrams would be effective in reducing coronary heart disease.

---

[2] "The Latest on Diet and Your Heart," *Consumer Reports*, July 1985, pp. 423–427.

Case 2   *Is "new" Coke that good?* In the early part of 1985, Coca-Cola Company announced that it was changing its secret formula for making Coke, a formula it had used since 1886. After the "new" Coke was marketed, the staff at *Consumer Reports* attempted to answer such questions as: What does the new Coke really taste like? Is it better than "old" Coke? How does it compare with Pepsi? The research staff conducted three separate blind taste tests involving 95 staff members and 532 plastic cups. The results of the study showed no difference in preferences between Pepsi and the new Coke.[3] Both products were preferred over the old Coke by a 2-to-1 margin. All three formulations were found to consist of about 99% carbonated water and sugar, and all three were remarkably similar in sugar, each with between 6.14% and 6.22% fructose, and between 4.54% and 4.73% dextrose (corn sugar). [Miscellaneous traditions and other human factors can affect the responses in an experiment. For example, even though the results of this experiment seem to indicate that new Coke is superior in taste to old Coke, Coca-Cola Classic is currently outselling new Coke in many regions of the United States.]

Case 3   *Is tobacco smoke harmful to nonsmokers?* It has long been known that smoking is harmful to smokers. To determine if tobacco smoke is harmful to nonsmokers, a study involving lung-function tests was undertaken at the University of California at San Diego.[4] Tests were conducted on 200 middle-aged nonsmokers whose environments were relatively free of tobacco smoke and on another 200 middle-aged nonsmokers who were routinely exposed to tobacco smoke at work for 20 years or more. Both groups were also compared with smokers who do not inhale, light smokers, moderate smokers, and heavy smokers. The researchers concluded that the two groups of nonsmokers did not differ significantly in lung-test results measuring forced vital capacity and initial expiratory rate. However, they did report a statistically significant difference between the two groups in the amount of impairment in the small airways of the lungs. The nonsmokers who were passively exposed to smoke at work did not have scores that were judged to be indicative of lung disease, but their scores were similar to those for light smokers (one to ten cigarettes per day) and smokers who do not inhale. The study suggests that chronic exposure to tobacco smoke at work is deleterious to the nonsmoker and significantly reduces small-airways function.

1.2 THE LANGUAGE OF STATISTICS

As with all sciences, statistics has its own language. We begin by examining the term *statistics*. There are two meanings for *statistics*:

> **Statistics** (singular) is the science of collecting, organizing, analyzing, and interpreting information.
>
> **Statistics** (plural) are numbers calculated from a set or collection of information.

[3] "How to Tell the Real Thing in Coke," *Consumer Reports*, Aug. 1985, p. 447.
[4] "The Murky Hazards of Second-Hand Smoke," *Consumer Reports*, Feb. 1985, pp. 81–84.

As a science, statistics is concerned with describing the results of scientific inquiry, making decisions based on this inquiry, and estimating unknown quantities. The numerical characteristic that is used as an estimate serves as an example of a statistic. For example, researchers estimate that IBM's entire family of personal computers controls about 40% of the microcomputer market in the United States. The figure "40%" is an example of a **statistic**.

A basic distinction in statistics concerns the difference between a population and a sample. The totality of information or objects of concern to the statistician for a particular investigation is called a **statistical population**, or simply a **population**. Any subset of the population is called a **sample**. The collection of student GPAs at a local college could serve as a statistical population, and any sub-collection (say, the GPAs of the students in a Mathematics 101 class) could serve as a sample from this population. For the Excel College example given in Section 1.1, the population consists of the responses from the entire student body to the question "Who is the best teacher?" Since it would be extremely difficult and time-consuming to poll each student, we might instead decide to poll a representative subset of the student body. This representative subset of the population represents a sample. The information from the sample can be used to *estimate* who is the best teacher at Excel College.

EXAMPLE 1-1 A manufacturer of kerosene heaters wants to determine if its customers are satisfied with the performance of their heaters. Toward this goal, 5000 of its 200,000 customers are contacted and each is asked, "Are you satisfied with the performance of the kerosene heater you purchased?" Identify the population and the sample for this situation.

Solution The population is the hypothetical collection of responses to the question from all 200,000 customers. We have not polled the entire population, but we hope to learn something about it from the sample. The sample is the collection of 5000 responses made by the customers polled.   ■

A statistical population may be completely imaginary. For example, if a researcher is interested in the possible selling prices of 1989 automobiles, the desired information does not yet exist. Even though the information is not available, the selling prices for cars over the past several years, along with information concerning the current rate of inflation, could be used to predict the selling prices for 1989 automobiles.

A value used in statistics can be characterized as either a statistic or a parameter, depending on the extent of the information. Any numerical characteristic of a sample is called a *statistic*, while any numerical characteristic of a population is called a *parameter*. The following examples illustrate the use of the terms *statistic* and *parameter*.

EXAMPLE 1-2 In order to estimate the proportion of students at a certain college who smoke cigarettes, an administrator polled a sample of 200 students and determined the proportion of students from the sample who smoke cigarettes. Identify the parameter and the statistic.

Solution   The parameter is the proportion of all students at the administrator's college who smoke cigarettes, while the statistic is the proportion of all students in the sample of 200 who smoke cigarettes.   ■

EXAMPLE 1-3   A tip is the amount of money, above the amount of a bill, given for satisfactory service. Patrons in 1500 nightclubs were given a confidential questionnaire with their bill asking how much of a tip was being given. Calculations showed that the average tip was about 15% of the amount of the bill. Is "15%" a parameter or a statistic? Explain.

Solution   If only the 1500 establishments are under study, then the tipping information from the 1500 establishments constitutes the population and "15%" is a parameter. However, if the tipping data from the 1500 establishments form a sample from some larger population of tipping data, then "15%" is a statistic.   ■

Problem Set 1.2

1. In an attempt to reduce the number of highway accidents, the state of Maryland has implemented a campaign aimed at reducing the number of speeders and drivers under the influence of alcohol. A researcher is interested in determining the extent to which alcohol is a contributing factor in highway fatalities within the state of Maryland. The researcher secured accident information for the month of June from 5 of the 22 highway patrol offices in the state.
   a. What is the population of interest to the researcher?
   b. Describe the sample.
   c. How could the researcher use the sample information to estimate the extent to which alcohol is a contributing factor in highway fatalities within the state of Maryland?

2. The medical problem of acquired immune deficiency syndrome (AIDS) has created high levels of anxiety and concern among the public. Records reveal that 71% of all AIDS cases in the United States have been gay or bisexual men and about 18% have been intravenous drug users. Many people question the possibility of contracting AIDS through a blood transfusion. Although blood for transfusions is screened for AIDS, a medical researcher wants to study the medical records of 50 hospitals located in cities throughout the United States to determine the extent of AIDS cases verified to have been caused by blood transfusions.
   a. What is the population of interest to the researcher?
   b. Describe the sample.

   c. How could the researcher use the sample information to estimate the proportion of people who have blood transfusions aand contract AIDS?

3. A doctor recently made the claim that a tablespoon of cod liver oil daily can cure arthritis. A researcher is interested in testing the claim. Two groups, each containing 50 arthritis patients, are used. Patients in only one group are to be administered a tablespoon of cod liver oil daily for 1 year, after which all subjects in both groups will be examined for the symptoms of arthritis.
   a. What are the two populations of interest?
   b. Describe the two samples.
   c. How could the researcher use the information from the samples to test the doctor's claim?

4. From 1971 until early 1985, the National Highway Traffic Safety Administration (NHTSA) had attributed at least 207 deaths to Ford Motor Company vehicles unexpectedly backing into and over people. In addition, there had been 4597 reported injuries resulting from unexpected reverse movement of Ford vehicles. By June 1980, NHTSA had received more than 23,000 reports about Ford cars failing to engage or hold in park. Instead of a government-ordered recall of affected vehicles, an agreement was negotiated with Ford in an attempt to prevent further injuries or deaths. Ford agreed to send out warning notices, with accompanying warning stickers to be displayed on the dashboards of the vehicles, to the owners of the affected vehicles—some 23 million cars and

trucks. In mid-1981, the Center for Auto Safety checked 700 Ford vehicles in four cities to ascertain if warning stickers were being displayed on the dashboards. Only 7% of the cars observed actually displayed the warning sticker.[5]

a. Identify the population of interest to the Center for Auto Safety.

b. Describe the sample.

c. Would you say the sticker campaign was successful in reducing unexpected reverse movements in Ford vehicles? Explain.

d. Identify a population of interest to NHTSA.

5. A 6-month study was conducted to determine if stress and mood are linked to the presence of certain immune-system cells. Thirty-six people participated in the study, which involved examining blood samples taken from them at regular intervals. The blood was analyzed for changes in the number of helper and T-cells which regulate immune functions. The results showed that increased levels of stress appear to be directly tied to the decreased number of T-cells and outbreaks of herpes.[6]

a. Identify the population of interest.

b. Describe the sample.

1.3 DESCRIPTIVE AND INFERENTIAL STATISTICS

The procedures and analyses encountered in statistics fall into two general categories, descriptive and inferential, depending on the purpose of the study. **Descriptive statistics** comprises those methods used to organize and describe information that has been collected. These methods are used to analyze the information and to display information in graphical form to allow meaningful interpretations.

The methods of descriptive statistics help us describe the world around us. We use descriptive statistics when we collect such information as the average yield of wheat per acre of a particular agricultural area, or the number of people in various income categories, or the average number of points scored by a particular football team during first-quarter play. Through descriptive statistics we hope to learn how things *are*.

The following situations involve descriptive statistics:

1. A bowler wants to find his bowling average for the past 12 games.
2. A politician wants to know the exact percentage of votes cast for her in the last election.
3. Mary wants to describe the variation in her five test scores in first-quarter calculus.
4. Mr. Smith wants to determine the average weekly amount he spent on groceries the past 3 months.

Inferential statistics, on the other hand, involves the theory of probability and comprises those methods and techniques for making generalizations, predictions, or estimates about the population by using limited information. The ability to make generalizations with limited information is an important aspect of statistics. Rarely do we have the complete information needed to form an absolute truth about some worldly event. Decisions and inferences are based

---

[5] "Fords in Reverse," *Consumer Reports*, Sept. 1985, pp. 520–523.

[6] *Cumberland Times/News*, Sept. 18, 1985, p. 7.

on limited and incomplete information. The methods of inferential statistics allow us to use the available information to understand and deal with the uncertainties surrounding our daily lives. We might, for example, predict the wheat yield for the coming year on the basis of growing trends over previous years. We could estimate the increase in average income over a 5-year period, based on past knowledge of average income and other descriptive statistics. We could try to predict the total points scored for the season by a particular football team knowing the points scored in the first seven games. With inferential statistics, we state how things will be, *probably*—or sometimes only *maybe.*

The following situations (which parallel the descriptive situations given above) involve inferential statistics:

1. A bowler wants to estimate his chance of winning an upcoming tournament based on his current season average and the averages of the competing bowlers.
2. Based on an opinion poll, a politician would like to estimate her chance for reelection in the upcoming election.
3. Based on her first-quarter calculus test scores, Mary would like to predict the variation in her second-quarter calculus test scores.
4. Based on last year's grocery bills, Mr. Smith would like to predict the average weekly amount he will spend on groceries for the upcoming year.

In summary, knowledge gained from using inferential statistics provides us with an understanding of our randomly changing world based on the limited information that is available to us.

Problem Set 1.3

1. Give an example, not mentioned in the text, of each of the following concepts:
   a. Population   b. Sample
   c. Statistic   d. Parameter

2. Mr. Jackson, a candidate for mayor in a small town, wants to determine if he should campaign harder against his opponent. To determine this, he will poll 500 of the 1500 registered voters. If the results indicate he has 25% more votes than his opponent, he will not intensify his campaign efforts against the opponent.
   a. Identify the population.
   b. Identify the sample.
   c. Identify a statistic.
   d. Identify a parameter.
   e. What would Mr. Jackson do if he had 65% of the sample votes?

3. The following are prices charged for a 5-pound bag of sugar at four supermarkets:

   $1.09 $.89 $.96 $1.04

   Which of the following statements were obtained by using inferential statistics and which were obtained by using descriptive statistics? Explain.
   a. The highest price charged is $1.09.
   b. Two supermarkets charge more than $1 for a 5-pound bag of sugar.
   c. One-half of all supermarkets charge more than $.20 per pound for sugar.
   d. The prices at all supermarkets for a 5-pound bag of sugar range from $.89 to $1.09.

4. Four out of five doctors recommend preparation A. Is this conclusion drawn from a sample or a population? Explain.

5. The average cost for student textbooks last semester at a small college was determined to be $135.00, based on an enrollment of 1200 students. As a class project at the college, a statistics class polled 25 students to determine their average textbook cost last semester. It was determined to be $152.25.

a. Identify the population.
b. Identify the sample.
c. Identify two parameters.
d. Identify two statistics.
e. What might the statistics class conclude if the average book cost for the sample of 25 students is $400?

1.4 INFERENCES AND DEDUCTIONS

The study of statistics involves both induction and deduction. **Induction** involves reasoning from specific instances to the general case, whereas **deduction** involves reasoning from the general case to specific instances. (See Figure 1-2.) When we make a generalization about a population parameter based on information derived from a sample, we are using induction. The generalization is called an **inference**. When we ascribe properties of samples from a population, we are using deduction. The deductions will involve probability, the study of uncertainty. We will study probability in Chapters 5–7 and inferential statistics in Chapters 8–15.

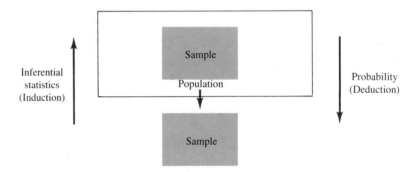

Inferential statistics (Induction)

Sample

Population

Sample

Probability (Deduction)

FIGURE 1-2 Induction versus deduction

For example, suppose 1000 automobiles were recently manufactured. Let's assume that 5% of them have a defective steering component and that a local dealer has a sample of ten of these cars. As an application of probability, we might attempt to determine the likelihood that at least two of these ten cars have defective steering components. Since the sample is a subset of the population, we would expect (deduce) that 5% of the cars in the sample would have defective steering components; that is, 5% represents the likelihood that a given car in the sample is defective. In Chapter 5, we will learn how to determine the likelihood that at least two of ten cars have defective steering components.

To illustrate inferential statistics, suppose 1000 cars were manufactured and the number of cars with defective steering systems is unknown. To estimate the percentage of cars in the population of 1000 cars with defective steering systems, we will inspect a sample of ten cars. If two of the ten cars are found to have defective steering systems, we might infer (using induction) that 20%, or 200, of the 1000 cars have defective steering systems. The proportion of defective steering systems in the sample is an example of a statistic; its value is .20. The percentage of cars in the population with defective steering systems is an example of a parameter. We will learn in Chapter 9 how to use statistics to estimate unknown parameters.

1.5 WHY STUDY STATISTICS?

There are at least four good reasons for studying statistics. By studying statistics, we are able to:

1. Learn the rules and methods for dealing with statistical information.
2. Evaluate and assess the importance of published statistical findings.
3. Better understand the empirical world around us.
4. Learn the aspects of statistical thinking as an important and essential component of a liberal arts education.

C H A P T E R   R E V I E W

IMPORTANT TERMS

For each term, provide a definition in your own words. Then check your responses against the definitions given in the chapter.

deduction	inference	sample	statistical population
descriptive statistics	inferential statistics	statistic	statistics
induction	population		

2

DESCRIPTIVE STATISTICS— ORGANIZING DATA

Chapter Objectives

In this chapter you will learn:

- *what data are*

- *that there are two general types of data*

- *that data can be classified according to the type of measurement scale used*

- *how to organize and summarize data using tables*

- *how to display data using various types of graphs*

Chapter Contents

The most basic aspect of statistics is the information involved. Without information to collect, organize, analyze, or interpret, there would be no reason to use or study statistics. The information that is used in statistics is referred to as **data**. We begin this chapter with a discussion of the various types of data.

2.1 DATA: THE BUILDING BLOCKS OF STATISTICS

Data is the plural of *datum*, a piece of information. Data can be classified into two general categories, quantitative and qualitative. **Quantitative data** refer to numerical information, such as "how much" or "how many." These quantities are measured on a numerical scale. Examples of quantitative data are weight, age, price, rank, length, and volume. **Qualitative data**, on the other hand, represent categorical or attribute information that can be classified by some criterion or quality. Examples of qualitative data are sex (male, female); color (red, green, blue); religion (Protestant, Catholic, Jewish); blood type (A, B, AB, O); favorite make of car (Ford, Chevrolet); and computer brand (IBM, Kaypro, Compaq).

Some data can be classified as either quantitative or qualitative. A numeral can thus be classified in quantitative or qualitative terms, depending on how it is used. If it is used as a label for identification purposes, it is qualitative; otherwise, it is quantitative. For example, if a serial number on a radio is used to identify the number of radios manufactured up to that point, it is a quantitative measure. But if the serial number is used only for identification purposes, it is a qualitative bit of information.

Some measurements can be made by using either quantitative or qualitative scales. For example, if the height of an individual is measured in feet and inches, then the classification is quantitative, but if height is measured as short, medium, or tall, then the classification is qualitative. In addition, height could be measured using quantitative data (feet and inches) but reported or classified using qualitative data (small, medium, or tall).

Our main purpose in analyzing data is to make meaningful interpretations. As a general rule, the amount of information contained in the data depends on the type of data. The quantitative-versus-qualitative dichotomy is not always adequate for the classification of data according to the amount of information it contains. Data can also be classified according to the type of measurement scale or procedure by which it is produced. For example, consider the numeral "4" in the following four situations:

  a. John's football-jersey numeral is 4.
  b. John is in the 4th grade.
  c. John recorded the temperature as 4° Celsius.
  d. John grew a cucumber that measured 4 inches in length.

In (a), the measurement is used for identification or classification purposes only: the football player numbered 4 identifies John. In (b), in addition to classification, the measurement 4 denotes that 4th grade is a grade level more advanced than 3rd grade and less advanced than 5th grade. In (c), the measurement 4 indicates that the temperature is higher than a temperature of 2° Celsius and lower than a temperature of 7° Celsius, and 4° is 1.5° higher than a temperature of 2.5°. The distance between 4° and 2.5° is 1.5°. However, a temperature of 4° Celsius is not "twice as warm" as a temperature of 2° Celsius. In (d), the measurement 4 indicates that the cucumber is a member of the class of all cucumbers measuring 4 inches in length; it is 1 inch longer than a 3-inch cucumber and twice as long as a cucumber measuring 2 inches in length. The situations in (a)–(d) represent four different levels of information resulting from the use of different measuring scales. The type of measurement scale determines the amount of information contained in a piece of datum.

Four types of measurement scales used in statistics are nominal, ordinal, interval, and ratio. Each of these scales will now be discussed.

Nominal Scale

Nominal scales exist for quantitative and qualitative data. A **nominal scale for quantitative data** assigns numbers to categories to distinguish one from another. Examples are numerals on basketball jerseys, postal zip codes, telephone numbers, and football scores (6 points for a touchdown, 1 point for an extra-point kick, 2 points for an extra-point run or safety, and 3 points for a field goal). A **nominal scale for qualitative data** is an unordered grouping of data into discrete categories, where each datum can go into only one group. Examples are sex (male, female), ethnic category, blood type, and religion. Nominal scales are used mainly for identification or classification purposes.

Ordinal Scale

Data measured on a nominal scale that is ordered in some fashion are referred to as **ordinal data**. An ordinal scale places measurements into categories, each category indicating a different level of some attribute that is being measured. Examples of ordinal-data scales are:

1. Letter grades: A, B, C, D, and F. These grades indicate categories of achievement, as well as levels of achievement.
2. Academic ranks: instructor, assistant professor, associate professor, and professor. A professor has higher academic rank than an instructor.
3. Residence numbers on a particular street: 421 North Street, 423 North Street, and so on. The residence at 423 North Street is located between the residences located at 421 North Street and 425 North Street.
4. Teacher ratings: poor, fair, good, and superior.
5. Grades of school: 1st, 2nd, 3rd, and so on.

The difference or distance between values measured on an ordinal scale cannot be determined. Even though we often code a letter grade of A as 4, B as 3, C as 2, D as 1, and F as 0, we should not say, for example, that an A is twice as good as a C or that the A-student knows twice as much as the C-student. All we can say is that an A grade is a better or higher grade than a C grade. An ordinal scale has no unit of distance.

Interval Scale

Data measured on an ordinal scale for which distances between values can be calculated are called **interval data**. The distance between two values is relevant. Interval data are necessarily quantitative. An interval scale does not always have a zero point, a point which indicates the absence of what we are measuring. Examples of interval data are:

1. *IQ test scores*   An IQ score of 110 is higher than an IQ score of 105 (ordinal data). Not only can we say that an IQ score of 110 is higher than an IQ score of 105, but we can also say that it is 5 points higher. However, we cannot say that a person with an IQ score of 180 is twice as smart as a person with an IQ of 90. A given difference between two IQ scores does not always have the same meaning. For example, the differences $(100 - 90)$ and $(150 - 140)$ may have different interpretations even though they both equal 10. Although a person with an IQ of 140 is more intelligent (as measured by the intelligence test) than a person with an IQ of 100, we cannot say that a person having an IQ of 150 is as much more intelligent than a person with an IQ of 140 as a person with an IQ of 100 is more intelligent than a person with an IQ of 90.

2. *Celsius temperatures*   A temperature of $80°\,C$ is $40°$ warmer than a temperature of $40°\,C$. But it is not correct to say that $80°\,C$ is twice as warm as $40°\,C$. Note also that a temperature of $0°\,C$ does not represent the absence of heat. The zero point on the Celsius temperature scale was arbitrarily set at the icing point. This point indicates that some heat is present. (Theoretically, $-273°C$ represents the absolute minimum temperature, the temperature at which the molecules of a substance approach a speed of 0.)

3. *Calendar dates*   Ronald Reagan was inaugurated as the 40th president of the United States in 1981, 192 years after George Washington (1789). We can specify the distance between these two ordered events—192 years—but year 0 (if it existed) would not represent the absence of time.

Ratio Scale

Data measured on an interval scale with a zero point meaning "none" are called **ratio data**. Because the zero point of the Celsius scale does not represent the complete absence of heat, the Celsius scale is not a ratio scale. The Kelvin temperature scale, where $0\,K$ corresponds to $-273°\,C$, is an example of a ratio temperature scale. Examples of other ratio scales are scales commonly used to measure units, such as feet, pounds, dollars, and centimeters. The results of counting objects are also ratio data. Ten apples are twice as many as five apples. A 200-pound person will always weigh twice as much as a 100-pound person, using any ratio scale (for example, a scale based on ounces, grams, or kilograms).

EXAMPLE 2-1   Suppose a group of teachers is surveyed concerning religious affiliation and 15 indicated Protestant, 21 indicated Catholic, and 7 indicated Jewish. What type of data is involved?

Solution   The response from each teacher is either "Protestant," "Catholic," or "Jewish," and these responses constitute nominal (categorical) or qualitative data, not quantitative data. On the other hand, the numbers 15, 21, and 7 result from counting the response data; these results thus represent quantitative data. Numbers resulting from operations performed on the data, such as adding, should not be confused with the data collection.   ■

Problem Set 2.1

1. Classify the following data as quantitative or qualitative:
   a. Type of religion for 50 people
   b. House numbers for dwellings on North Street
   c. Sizes of shoes worn by 25 adults
   d. License plate numbers for 20 cars
   e. Social Security numbers for 17 retired police officers
   f. Heights (in inches) of 5 basketball players
   g. Weights (in ounces) of 12 chickens
   h. Ethnic classifications of 20 employees
   i. GPAs of the members of the junior class
   j. Letter grades of 15 students in Philosophy 209

2. Classify the data in each of the three columns of the accompanying table as nominal, ordinal, interval, or ratio. In addition, classify the data as quantitative or qualitative.

   The Distribution of Vehicles Registered at Excel College

Rank	Type of vehicle	Number registered
1	Car	150
2	Truck	25
3	Motorcycle	15
4	Bicycle	10

3. Memorial Hospital maintains the following information on each patient:
   a. Social Security number
   b. Date of last admission
   c. Date of birth
   d. Insurance company
   e. Employer
   f. Home address
   g. Home telephone number
   Classify each as nominal, ordinal, interval, or ratio.

4. The accompanying figure shows a numerical scale to measure teaching effectiveness.

1	3	5
Needs improvement	Effective and competent	Truly outstanding

   a. Identify the type of measurement scale.
   b. Suppose this scale is used by 30 students to evaluate their statistics instructor. Would the results be easier to interpret than the results that would be obtained if the 30 students evaluated their instructor by submitting a written statement of free response? Explain.

5. Give an example, not given in this section, of an ordinal scale for quantitative data.

6. Can an interval scale exist for qualitative data? Explain.

2.2 ORGANIZING DATA USING TABLES

The objective of data organization is to arrange a set of data into a useful form in order to reveal essential features and to simplify certain analyses. Data that are not organized in some fashion are referred to as **raw data**. One method of arranging raw data is to construct an ordered array; that is, arrange the data from low to high (or from high to low). If the number of data is large, the data array may be difficult to manage or comprehend. As a result, tables are often used as a general approach to organizing raw data. In this section, we will discuss various types of tables used for organizing data; in Section 2.3, we will discuss graphical means for displaying raw data organized in table form. The type of data—nominal, ordinal, interval, or ratio—will determine the type of display.

Ungrouped Frequency Tables

A widely used means for reporting qualitative or quantitative data is the **frequency table**. The **frequency** of a measurement or category is the total number of times the measurement or category occurs in a collection of data. The symbol f is used to denote the frequency of a measurement. For example, here is a sample of data representing the number of free throws missed by a basketball team during the last seven games:

$$7 \quad 2 \quad 8 \quad 4 \quad 2 \quad 7 \quad 2$$

The number 7 occurs with a frequency of 2, 2 occurs with a frequency of 3, and 8 and 4 each occur with a frequency of 1. This information can be summarized as shown in Table 2-1, where x denotes the measurement and f denotes the frequency of each measurement.

Table 2-1 Frequency Table of Free-Throw Data

x	f
2	3
4	1
7	2
8	1

EXAMPLE 2-2 Construct a frequency table for the following data representing the number of absences per class period during the 1985 fall term for students enrolled in Statistics 101:

9	8	7	8	4	3
2	1	0	5	3	2
1	1	7	3	2	8
7	6	6	4	3	2
2	0	9	4	6	9
6	9	4	3	5	7
3	2	1	4	4	2

Solution As an intermediate step, we use **tally marks** to aid in determining the frequency f for each observation, with x representing the number of absences. Corresponding to each observation, we place a tally mark (|) in the tally column adjacent to its value. After all the tallies are placed, they are counted for each measurement x to determine the frequency.

Number of absences (x)	Tally	Frequency (f)
0	||	2
1	||||	4
2	⊬⊬ ||	7
3	⊬⊬ |	6
4	⊬⊬ |	6
5	||	2
6	||||	4
7	||||	4
8	|||	3
9	||||	4
		42

Note that the sum of the frequencies in a frequency table represents the size of the data collection. In this case, the sum of the frequencies (42) represents the 42 class periods for which absences were recorded.   ■

Grouped Frequency Tables

Frequency tables, such as the one shown in Table 2-1, are correctly termed **ungrouped frequency tables**: each measurement has its own corresponding frequency. A **grouped frequency table**, on the other hand, shows frequencies according to groups or classes of measurements.

For example, suppose Memorial Hospital wants to study whether its emergency room staffing is adequate. To start the study, the department manager tracks the number of people visiting the emergency room each day for a 12-day period, with the following results:

7   43   8   22   13   28   36   18   23   21   15   52

To simplify the data, the manager then constructs six groupings or classes: the first class will represent 1 to 10 patients; the second class, 11 to 20 patients; the third class, 21 to 30 patients; and so on. From this grouping, he or she prepares a grouped frequency table (Table 2-2) to show how often during the 12 days the number of patients fell into each group.

For the class 1–10, 1 is called the **lower class limit** and 10 is called the **upper class limit**. There are two measurements that fall between 1 and 10, inclusive; three measurements fall between 11 and 20, inclusive; four measurements fall between 21 and 30, inclusive; one measurement falls between 31 and 40, inclusive; and so on.

Notice that when examining a grouped frequency table without the raw data (that is, the data before statistical processing), we do not know the individ-

**Table 2-2 Grouped Frequency
Table for Emergency Room Data**

Class	Frequency (f)
1–10	2
11–20	3
21–30	4
31–40	1
41–50	1
51–60	1

ual measurements. For example, in Table 2-2 we know that two measurements fall in the class 1–10, but we do not know what the two measurements are. This would not be the case for an ungrouped frequency table, where all the measurements are known.

The following three basic guidelines should be used when constructing a grouped frequency table:

1. Each class should have the same width.
2. No two classes should overlap.
3. Each piece of data should belong to some class.

Class boundaries and **class widths** for a grouped frequency table are determined by considering the **unit** or precision of measurement. For the classes in Table 2-2, the precision of measurement is the nearest whole number (since we are counting people), so the unit of measurement is 1. The **lower class boundary** of an interval is located one-half unit below the lower class limit, and the **upper class boundary** of an interval is located one-half unit above the upper class limit. For the first class in Table 2-2, the lower class boundary is $[1 - .5(1)] = .5$ and the upper boundary is $[10 + .5(1)] = 10.5$. Note that measurements of .5 and 10.5 could not fall in the first class, but that any measurement *between* .5 and 10.5 could. Of course, 10.5 is not a possible measurement, so the class boundaries have mathematical meaning only. The width w for any class in a grouped frequency table is found by subtracting the lower class boundary from the upper class boundary. Thus, for the first class in Table 2-2, $w = 10.5 - .5 = 10$.

Consider the next two examples.

1.

**Number of Seeds
in Oranges**

Class	Frequency
3–6	5
→ 7–10	6
11–14	7
15–18	3

The precision of measurement for the classes is 1, since whole-number data are involved. For the class 7–10, if we add $(.5)(1) = .5$ to the upper class limit

10, we get the upper class boundary, 10.5. To get the lower class boundary, we subtract .5 from the lower class limit to get $7 - .5 = 6.5$. (Refer to the following illustration.) The width of the class 7–10 is then found by subtracting the lower class boundary from the upper class boundary, or $w = 10.5 - 6.5 = 4$.

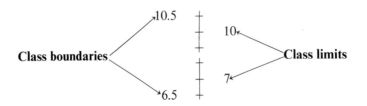

2.

Weights (in Pounds) of Newborn Babies

Class	Frequency
3.0–4.4	1
4.5–5.9	1
6.0–7.4	7
→ 7.5–8.9	8
9.0–10.4	1

The precision of measurement for the classes is .1 pound. For the class 7.5–8.9, subtracting one-half the unit from the lower class limit yields $7.5 - (.5)(.1) = 7.5 - .05 = 7.45$, the lower class boundary. The upper class boundary is found by adding one-half the unit to the upper class limit, yielding $8.9 + (.5)(.1) = 8.95$. Note that no weight can correspond to either boundary because the precision of measurement is to the nearest one-tenth of a pound.

In any grouped frequency table, the class width can also be found simply by subtracting two consecutive upper class limits or two consecutive lower class limits. For the emergency room data originally given in Table 2-2, we can compute the class width as indicated in Table 2-3. Notice, however, that the class width is *not* found by subtracting the lower class limit from the upper class limit.

Table 2-3 Computation of Class Width for Table 2-2

	Class		Frequency (f)
	1–10 $\{w = 20 - 10$		2
	11–20 $= 10$		3
$w = 31 - 21\}$	21–30		4
$= 10$	31–40		1
	41–50		1
	51–60		1

Selecting Classes for Grouped Frequency Tables

In order to construct a grouped frequency table for a given collection of data, we need to answer three questions concerning the classes:

1. How many classes should be used?
2. What should be the width of each class?
3. At what value should the first class start?

Selecting the number of classes involves several considerations. If all the data are grouped into a small number of classes, characteristics of the original data are hidden and pertinent information may be lost. On the other hand, too many classes provide too much detail and defeat the purpose of grouping—to condense the data in a meaningful manner that is easy to interpret. In addition, too many classes may produce too many empty classes and render the grouping of data meaningless.

The number of classes, denoted by c, depends on the situation and the amount of data involved. Since there is no general agreement among statisticians regarding the number of classes to use and the choice is arbitrary, we will use between 5 and 15 classes, inclusive, in this text. A useful suggestion for the number of classes is given by **Sturges' rule**, which states that the number of classes needed is approximately

$$c = 3.3(\log n) + 1$$

where $\log n$ is the logarithm of n to the base 10. The value of c is usually rounded to the nearest whole number. For example, if $n = 25$ measurements are involved, Sturges' rule would suggest using six classes, since

$$
\begin{aligned}
c &= 3.3(\log n) + 1 \\
&= 3.3(\log 25) + 1 \\
&= 3.3(1.3979) + 1 \\
&\simeq 6
\end{aligned}
$$

where "$\simeq$" means "approximately equal to." Some researchers believe that in most situations Sturges' rule provides a value for c that permits the construction of a grouped frequency table that presents a realistic picture of the raw data.

Once we agree on the number of class intervals to use, the width of each class is found by using the **range** R, which is the difference between the largest measurement U and the smallest measurement L:

$$R = U - L$$

Because c classes must cover the range, we divide the range by the number of classes to find the class width w:

$$w = \frac{R}{c}$$

Since the smallest measurement should fall in the first class, the lower limit of the first class should be near and at most as large as the smallest measurement L. So that we can have general agreement on the classes in our grouped frequency tables, we will always start the first class with the smallest measurement. (This will be especially helpful when checking our answers.) In actual practice, the first class typically begins at a point which permits the classes to be expressed in convenient intervals. But there are occasions where exceptions to the guidelines are justified (see Problem 19 at the end of this section).

EXAMPLE 2-3   Professor Smith gave a final examination consisting of 100 questions to his Introductory Accounting class. The following data represent the number of correct answers on each test. Construct a grouped frequency table with five classes to aid Professor Smith in analyzing the results.

17	15	78	21	10	32	7	65	18	87
4	22	34	42	9	9	82	79	98	4
44	64	62	77	2	81	45	37	83	44
77	13	41	16	17	13	82	37	5	54
7	67	88	41	61	22	92	16	67	85

Solution   *Step 1.*   We first determine the range R. Since the largest measurement is $U = 98$ and the smallest measurement is $L = 2$, the range is

$$R = U - L = 98 - 2 = 96$$

Step 2.   We next determine w, the width of each class. Since the number of classes is given as $c = 5$, the width is

$$w = \frac{R}{c} = \frac{96}{5} = 19.2$$

Any value close to and at least as large as 19.2 can serve as the class width. The unit of precision for the examination scores is 1. If we round the value 19.2 to the next higher unit value, we get $w = 20$, a convenient number. Typically, the value of R/c is *rounded up* to the next value determined by the precision of measurement to produce a convenient value.

Step 3.   We start at $L = 2$ and construct the first class with a width of 20. Suppose the first class extends from 2 to l, where l represents the unknown upper class limit (refer to the accompanying diagram).

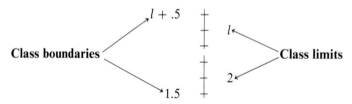

Since the unit of measurement is 1 and $.5(1) = .5$, the upper class boundary can be represented as $l + .5$. The width of the first class is found by subtracting the lower class boundary from the upper class boundary. Thus,

$$w = (l + .5) - 1.5$$

Since $w = 20$, we have

$$20 = l - 1$$

Solving this equation for l, we find that the upper class limit is $l = 21$. Thus, the first class becomes 2–21.

Step 4.   To obtain each class following the first class, we add $w = 20$ to the lower and upper limits of each preceding class. Thus,

$$
\begin{aligned}
&\qquad\qquad 2\text{–}21 \\
&\qquad\qquad 22\text{–}41 \quad (=\mathbf{21} + 20) \\
(=\mathbf{22} + 20)\quad &\qquad 42\text{–}61 \\
&\qquad\qquad 62\text{–}81 \\
&\qquad\qquad 82\text{–}101 \quad (=\mathbf{81} + 20)
\end{aligned}
$$

Step 5.   To determine the frequency of each class we use a tally column. If a piece of data falls in a class, we record a tally mark (|) in the tally column corresponding to that class. Table 2-4 contains our grouped frequency table for the 50 examination scores.

Table 2-4 Grouped Frequency Table of Final Examination Scores

Class number	Class	Tally	Frequency (f)
1	2–21	⫢⫢⫢⫢ ⫢⫢⫢⫢ ⫢⫢⫢⫢ |||	18
2	22–41	⫢⫢⫢⫢ |||	8
3	42–61	⫢⫢⫢⫢		6
4	62–81	⫢⫢⫢⫢ ⫢⫢⫢⫢	10
5	82–101	⫢⫢⫢⫢ |||	8
			50

Class Mark

The midpoint of each class is called the **class mark** and is denoted by X. When data are summarized into a grouped frequency table, some information is lost; we do not know the exact values of the measurements that lie in each class. The best we can do is to let each measurement within a given class be represented by the class mark of that class. Using class marks, instead of the raw data, makes the computations easier but results in a lack of accuracy. For a given class, the class mark is found using the formula

$$X = \frac{l_1 + l_2}{2}$$

where l_1 is the lower class limit and l_2 is the upper class limit of the class. For the first class in Table 2-5, the class mark is

$$X = \frac{l_1 + l_2}{2} = \frac{2 + 21}{2} = 11.5$$

For the second class, the class mark is

$$X = \frac{l_1 + l_2}{2} = \frac{22 + 41}{2} = 31.5$$

Notice that the class mark for class 2 can also be found by adding $w = 20$ to the class mark for class 1 ($11.5 + 20 = 31.5$). In general, each class mark for classes following the first class can be found by adding $w = 20$ to the preceding class mark. Thus, the remaining three class marks are found to be 51.5, 71.5, and 91.5. Table 2-5 displays a grouped frequency table containing the class marks.

Table 2-5 Class Marks for Grouped Frequency Table of Example 2-3

Class number	Class	Tally	f	X
1	2–21	|||| |||| |||| |||	18	11.5
2	22–41	|||| |||	8	31.5
3	42–61	||||		6	51.5
4	62–81	|||| ||||	10	71.5
5	82–101	|||| |||	8	91.5
			50	

EXAMPLE 2-4 The following data set represents the amounts of cash (in dollars) spent in a particular weekend by 25 graduate students. Construct a grouped frequency table containing five classes.

39.78	28.30	28.31	17.95	44.47
46.65	31.47	33.45	29.17	48.39
$U \rightarrow$ 82.71	43.63	41.17	47.32	52.16
25.94	50.32	35.25	35.70	17.89 $\leftarrow L$
60.20	48.14	22.78	38.22	23.25

Solution *Step 1.* Compute the range R. Since $U = 82.71$ and $L = 17.89$, the range is

$$R = U - L = 82.71 - 17.89 = 64.82$$

Step 2. Compute the class width w. Since $c = 5$, we have

$$w = \frac{R}{c} = \frac{64.82}{5} = 12.96$$

Thus, the class width is approximately equal to 13.

Step 3. We start at $L = 17.89$ and construct a class with width $w = 13$. The unit is .01 and $(.5)(.01) = .005$. Let x represent the upper boundary of the first class. Then the width is found by subtracting the lower class boundary from the upper class boundary:

$$w = (x + .005) - 17.885$$
$$13 = (x + .005) - 17.885$$
$$13 = x - 17.88$$
$$x = 30.88$$

Hence, the first class is 17.89–30.88.

Step 4. To obtain the remaining classes, we add 13 to the preceding class limits:

17.89–30.88		← First class
13	13	Add $w = 13$.
30.89–43.88		← Second class
13	13	Add $w = 13$.
43.89–56.88		← Third class

The remaining two classes are found in a similar fashion. They are 56.89–69.88 and 69.89–82.88.

Step 5. The frequencies are determined for the five classes using tallies, as shown in Table 2-6.

Step 6. Class marks are found for each class using the midpoint formula given previously. The class mark for the first class is

$$X = \frac{l_1 + l_2}{2} = \frac{17.89 + 30.88}{2} = 24.385$$

Each successive class mark is found by adding $w = 13$ to the preceding class mark. Table 2-6, the grouped frequency table for the data, also displays the class marks.

Table 2-6 Grouped Frequency Table

Class number	Class	Tally	f	X
1	17.89–30.88	ⱵⱵ |||	8	24.385
2	30.89–43.88	ⱵⱵ |||	8	37.385
3	43.89–56.88	ⱵⱵ ||	7	50.385
4	56.89–69.88			1	63.385
5	69.89–82.88			1	76.385
			25	

Relative Frequency Tables

It is sometimes useful to express each value or class in a frequency table as a fraction or a percentage of the total number of measurements. The **relative frequency** for a measurement (or class) is found by dividing the frequency, f, of the measurement (or class) by the total number of measurements, *n*. The table is then called a **relative frequency table**. For example, the relative frequency for class 1 in Table 2-6 is $n = \frac{8}{25} = .32$. Table 2-7 displays a relative frequency table for the data in Example 2-4. Note that the sum of the "relative frequency" column is 1.00.

Table 2-7 Relative Frequency Table for Data of Example 2-4

Class	Relative frequency
17.89–30.88	.32
30.89–43.88	.32
43.89–56.88	.28
56.89–69.88	.04
69.89–82.88	.04
	1.00

A relative frequency table has several advantages over a frequency table. One important advantage is that we can make meaningful comparisons between similar sets of data having the same classes but different total frequencies. For example, consider Table 2-8, which shows the starting salaries of recent mathematics graduates at two state colleges, A and B. By examining the two parts of the table, we see that each college graduates three mathematics majors making starting salaries from $22,000 to $24,999. But comparing the relative frequencies, we see that college A has $\frac{3}{20} = 15\%$ of its mathematics graduates earning $22,000 to $24,999, while college B has $\frac{3}{10} = 30\%$ of its mathematics graduates earning from $22,000 to $24,999.

Table 2-8 Frequency Tables for Starting Salaries at Two Colleges

College A		College B	
Salary class	*f*	Salary class	*f*
$10,000–12,999	0	$10,000–12,999	1
13,000–15,999	2	13,000–15,999	1
16,000–18,999	7	16,000–18,999	2
19,000–21,999	6	19,000–21,999	2
22,000–24,999	3	22,000–24,999	3
25,000–27,999	2	25,000–27,999	1

Cumulative Frequency Tables

There are many occasions where we are interested in the number of observations less than or equal to some value. A quality control engineer might want to know how many days a production process produced at most 100 defective items. A teacher might be interested in the number of students who received a score less than or equal to 70% on an examination. Or a basketball coach might be interested in the number of games in which the opponents scored at most 60 points.

The cumulative frequency for any measurement (or class) is the total of the frequency for that measurement (or class) and the frequencies of all measurements (or classes) of smaller value. Table 2-9 illustrates a cumulative frequency table for the data of Example 2-2.

Table 2-10 illustrates a cumulative frequency table for the data in Example 2-3.

Table 2-9 Cumulative Frequency Table for Absence Data of Example 2-2

Frequency table			Cumulative frequency table	
x	f		x	Cumulative frequency
0	2		0	2
1	4		1	$6 = (2 + 4)$
2	7		2	$13 = (6 + 7)$
3	6		3	$19 = (13 + 6)$
4	6		4	$25 = (19 + 6)$
5	2		5	$27 = (25 + 2)$
6	4		6	$31 = (27 + 4)$
7	4		7	$35 = (31 + 4)$
8	3		8	$38 = (35 + 3)$
9	4		9	$42 = (38 + 4)$
	42			

Table 2-10 Cumulative Frequency Table for Examination Score Data of Example 2-3

Frequency table			Cumulative frequency table	
Class	f		Class	Cumulative frequency
2–21	18		2–21	18
22–41	8		22–41	$26 = (18 + 8)$
42–61	6		42–61	$32 = (26 + 6)$
62–81	10		62–81	$42 = (32 + 10)$
82–101	8		82–101	$50 = (42 + 8)$
	50			

Cumulative Relative Frequency Tables

Cumulative frequency tables can also be constructed for tables containing relative frequencies or percentages. The procedures are identical to those used for cumulative frequency tables except that relative frequencies or percentages are used. When this is done, the table is called a **cumulative relative frequency table**.

A cumulative relative frequency table is displayed in Table 2-11 for the data in Example 2-3. It was obtained from Table 2-10 by computing cumulative relative frequencies for the cumulative frequencies.

Table 2-11 Cumulative Relative Frequency Table for Data of Example 2-3

Class	Cumulative frequency	Cumulative relative frequency
2–21	18	$\frac{18}{50} = .36$
22–41	26	$\frac{26}{50} = .52$
42–61	32	$\frac{32}{50} = .64$
62–81	42	$\frac{42}{50} = .84$
82–101	50	$\frac{50}{50} = 1.00$

Cumulative relative frequencies have many uses. One use is in scoring standardized tests such as the Scholastic Aptitude Test (SAT) and many college entrance exams. Test scores are usually given as *percentiles*. A **percentile score** tells what part of the tested population scored *lower*; for example, if 590 is said to be the 90th percentile in the mathematics portion of the SAT, it means that 90% of the scores on the mathematics portion of the test were lower than 590.

A cumulative relative frequency table that employs class boundaries can be used to determine percentiles. For example, suppose Table 2-12 reports the heights (in inches) of 200 male freshmen at a college.

Table 2-12 Heights (in Inches) of Male Freshmen

Height	f	Relative frequency	Cumulative relative frequency
59.5–62.5	2	.01	.01
62.5–65.5	12	.06	.07
65.5–68.5	24	.12	.19
68.5–71.5	46	.23	.42
71.5–74.5	62	.31	.73
74.5–77.5	36	.18	.91
77.5–80.5	16	.08	.99
80.5–83.5	2	.01	1.00
	200		

The following results are apparent from observation of the table:

1. A height of 74.5 inches is at the 73rd percentile.
2. The 50th percentile is between 71.5 inches and 74.5 inches.
3. The 19th percentile is 68.5 inches.
4. The 75th percentile is between 74.5 inches and 77.5 inches.

EXAMPLE 2-5   Suppose the 70th percentile of weights of all adult males is 175 pounds and the 85th percentile is 195 pounds. What percentage of men weigh strictly between 175 pounds and 195 pounds?

Solution By definition, 70% of adult males weigh less than 175 pounds and 85% of adult males weigh less than 195 pounds. So, $.85 - .70 = .15 = 15\%$ of adult males weigh strictly between 175 pounds and 195 pounds. ∎

Bivariate Tables

If data result from measuring one characteristic per source, the data are referred to as **univariate data**. (A **source** is anything that yields a piece of data.) For example, if the data are weights of a class of 30 statistics students, then a source is a student and a measurement is the weight of a student. Thus far in the text, we have dealt only with univariate data.

Bivariate data result when two different measurements are taken on the same source. Sometimes bivariate data are represented as ordered pairs (x, y), where x represents the first measurement and y represents the second measurement. For example, the height and weight for each of 30 statistics students represent bivariate data. Again, a source is a student. Bivariate data consist of two collections of measurements that are related and can be paired.

A **bivariate frequency table** is an arrangement of data into two categories of classification. The data used to construct the bivariate frequency table usually result from frequency counts. Each category is identified with a symbol, called a *variable*. A variable is used to represent data within a category. The categories may be discrete numbers, number intervals, or qualitative values (such as sex, hair color, or religion). For example, suppose data were collected from a sample of voters concerning their political philosophies and party affiliations. Each voter was asked to identify his or her political philosophy as liberal, conservative, or other, and his or her party affiliation as Democrat, Republican, or other. The two variables of classification are political philosophy and party affiliation. The political philosophy variable has three categories or levels of classification (liberal, conservative, other), while the party affiliation variable also has three categories or levels (Democrat, Republican, other). The data were tabulated in Table 2-13.

Table 2-13 Bivariate Frequency Table

Party affiliation	Political philosophy			Total
	Liberal	Conservative	Other	
Democrat	78	65	37	180
Republican	84	79	7	170
Other	38	46	16	100
Total	200	190	60	450

Among other things, the following information can be easily read from the table:

1. There were 78 voters who indicated they were liberal Democrats.
2. There were 79 voters who indicated they were conservative Republicans.

3. There were 450 voters polled.
4. There were 170 Republicans polled.
5. There were 60 voters who classified their political philosophy as "other."

Consider the next two examples.

EXAMPLE 2-6   The statistics grades and sex (M, F) for 32 college students are shown in the following table. Construct a frequency table for the bivariate data.

Student	Grade	Sex	Student	Grade	Sex
1	B	M	17	C	F
2	C	F	18	E	F
3	C	F	19	C	M
4	C	M	20	B	F
5	B	F	21	D	M
6	B	F	22	E	M
7	A	M	23	B	M
8	C	M	24	B	M
9	D	F	25	C	M
10	C	M	26	C	F
11	B	F	27	D	M
12	A	F	28	B	F
13	C	M	29	D	F
14	D	F	30	A	M
15	D	F	31	E	M
16	A	F	32	A	F

Solution   *Step 1.*   We use tally marks to determine the totals for each of the ten sex/grade combinations.

Sex	Grade				
	A	B	C	D	E
M	||	|||	ⅢⅠ	||	||
F	|||	Ⅲ	||||	||||	|

Step 2.   Next we find the totals for the two rows, five columns, and ten sex/grade combinations.

Sex	Grade					Total
	A	B	C	D	E	
M	2	3	6	2	2	15
F	3	5	4	4	1	17
Total	5	8	10	6	3	32

■

EXAMPLE 2-7   A study of college faculty was undertaken to study their attitudes toward collective bargaining between administration and labor unions. The results are summarized in the following table.

Faculty rank	Attitude toward collective bargaining			Total
	Favor	Oppose	Unsure	
Professor	45	8	2	55
Associate Professor	31	16	3	50
Assistant Professor	42	19	4	65
Instructor	12	4	14	30
Total	130	47	23	200

Use the table to answer each of the following:

a. What percentage of the faculty oppose collective bargaining?
b. What percentage of the faculty are associate professors?
c. The professors who favor collective bargaining form what percentage of the faculty?
d. What percentage of instructors favor collective bargaining?
e. What percentage of those faculty who oppose collective bargaining are professors?
f. What percentage of the total faculty are represented by assistant professor or above who favor collective bargaining?

Solution   **a.** $\frac{47}{200} = 23.5\%$
b. $\frac{50}{200} = 25\%$
c. $\frac{45}{200} = 22.5\%$
d. $\frac{12}{30} = 40\%$
e. $\frac{8}{47} = 17.02\%$
f. $(42 + 31 + 45)/200 = 59\%$ ■

Problem Set 2.2 _____

A

1. Using the grouped frequency table given here, construct each of the following:
a. A relative frequency table
b. A cumulative frequency table
c. A cumulative relative frequency table

Class	f
1–4	14
5–8	18
9–12	12
13–16	16
17–20	20

2. Using the accompanying ungrouped frequency table, construct each of the following:
a. A relative frequency table
b. A cumulative frequency table
c. A cumulative relative frequency table

x	f
4	1
7	3
8	6
9	4
10	2

3. The following data represent the amounts (in dollars) spent by a sample of 25 students for snacks during a final examination period.

57	28	63	38	29
64	84	88	42	36
89	77	72	39	47
72	69	68	41	52
39	72	45	52	84

Using six classes, construct a grouped frequency table.

4. Refer to the data given in Problem 3. Use Sturges' rule to construct a grouped frequency table.

responses are as follows:

4	3	6	5	6	5
5	4	5	4	5	3
6	4	4	4	6	2
9	5	4	3	3	8
11	7	8	7	4	10

Construct an ungrouped relative frequency table.

10. For the data in Problem 9, construct a cumulative relative frequency table.

11. Students in a small school were classified according to class standing and music preference. The results are recorded in the accompanying bivariate table.

Music preference	Class standing				Total
	Freshman	Sophomore	Junior	Senior	
Rock	16	11	7	6	40
Country	10	12	3	5	30
Classical	3	1	2	4	10
Jazz	23	11	2	4	40
Folk	3	0	6	1	10
Total	55	35	20	20	130

5. By using the data in Problem 3 and Sturges' rule, construct a grouped cumulative frequency table.

6. By using the data in Problem 3 and five classes, construct a grouped cumulative relative frequency table.

7. The following is a sample of prices (in cents) charged for unleaded gasoline in a particular city during a certain month.

130.9	121.9	132.9	120.8	115.9
117.9	131.9	121.9	126.9	122.8
137.9	115.9	115.9	121.9	126.9
119.9	118.9	119.8	116.9	129.9
123.9	127.9	122.8	119.9	126.9

Construct a grouped frequency table using five classes.

8. For the data in Problem 7, construct a cumulative frequency table using six classes.

9. A sample of 30 students were asked how many books they purchased for classes last semester. Their

a. What percentage of the freshmen prefer classical music?
b. What percentage of the rock music fans are sophomores?
c. What percentage of the student body prefer country music?
d. What percentage of the student body are juniors?
e. What percentage of the student body are juniors or seniors?
f. What percentage of the student body prefer country or folk music?

12. A sample of voters were polled concerning their preference among three candidates running for mayor. The results are recorded by sex in the following table.

Sex	Candidate			Total
	A	B	C	
Male	15	16	4	35
Female	5	4	1	10
Total	20	20	5	45

a. What percentage of the voters prefer candidate B?

b. What percentage of the voters are male?

c. What percentage of the males prefer candidate C?

d. What percentage of the voters who prefer candidate A are female?

13. The following data represent the monthly telephone bills (in dollars) of 25 residents of a small community:

9.80	36.05	18.50	11.48	11.15
15.12	13.47	27.81	16.66	20.35
30.22	15.49	10.80	23.83	15.35
23.48	25.81	26.83	10.77	9.98
35.87	22.02	21.07	10.96	13.38

a. What percentage of the group paid over $14.00?

b. What percentage of the group paid over $14.00 but less than $20.00?

c. Arrange the data in a grouped frequency distribution with five classes.

14. For the data in Problem 13, construct a cumulative frequency table using five classes.

15. Consider the following grouped frequency table:

Classes	f
4.5–9.4	2
9.5–14.4	3
14.5–19.4	4
19.5–24.4	1
24.5–29.4	8

a. Find w, the width of each class.

b. Find the five class marks.

c. Find the boundaries for the first class.

B

16. The following data represent the weekly amounts (in dollars) spent on food by 50 newlywed couples:

57.10	70.89	59.17	60.08	49.16
50.25	46.39	55.01	68.81	58.70
58.32	50.82	45.43	57.20	62.30

55.45	58.14	57.14	58.63	55.14
64.00	47.75	52.59	42.73	60.32
48.10	59.62	59.46	57.16	58.19
60.37	52.41	74.13	62.38	51.15
51.42	58.76	46.37	47.16	63.51
69.48	46.02	54.16	65.07	48.09
56.17	66.94	67.08	58.10	71.28

Construct a grouped frequency table having the smallest number of classes for which the class width is $w = 2.75$.

17. The miles-per-gallon calculations for 40 fuel refills of a 1980 automobile are as follows:

26.6	28.7	29.2	26.4	29.3
25.8	28.7	29.0	30.0	28.1
28.3	27.8	27.6	31.9	26.6
28.4	30.3	30.4	29.2	29.3
26.5	28.7	28.8	28.3	28.8
27.1	28.9	31.2	30.2	29.2
30.3	32.0	28.4	30.3	29.5
28.4	27.4	30.8	29.5	31.5

Construct a grouped relative frequency table with eight classes.

18. An efficiency study was conducted by a supermarket. The following data represent the lengths of time (in minutes) required to service 50 customers at checkout:

3.5	1.8	2.3	.7	5.2
.8	1.2	1.7	1.2	.2
.2	1.3	.6	.6	1.8
.7	1.5	1.3	1.4	1.1
4.0	2.5	1.9	.8	1.2
.6	2.8	2.4	.3	3.1
.4	1.4	.7	1.2	.7
.9	.9	.9	3.0	1.1
1.2	2.3	1.7	2.3	1.7
1.6	.3	1.0	1.0	.5

Construct a relative frequency table with ten classes.

19. For the data in Problem 7, construct a grouped frequency table with seven classes and with class marks that end in .9 cent.

2.3 GRAPHICAL REPRESENTATION OF DATA ———

A **graph** is a pictorial means for portraying and summarizing data. A pictorial presentation of the data often makes certain features of the data more apparent than a frequency table. One result of presenting data in graphical form is that new characteristics of the data are often discovered. Graphical displays of data have enjoyed increased use in the media; this is due, in part, to the popularity and use of computer graphics.

Graphs are of many types. Some of the more common types are the circle graph, bar graph, line graph, stem-and-leaf diagram, histogram, and ogive. We will discuss each of these in some detail.

Bar and Circle Graphs

Two of the commonly used graphical types are **bar graphs** and **circle graphs**. Both types are generally used for categorical or nominal data. Circle graphs are used only to portray parts of a total; they are very popular for displaying budget information.

Table 2-14 Recipients of Charitable Giving

Recipients	Amount (in billions of dollars)
Religion	31.0
Arts and humanities	4.1
Social services	6.9
Education	9.0
Health	9.2
Other	4.7

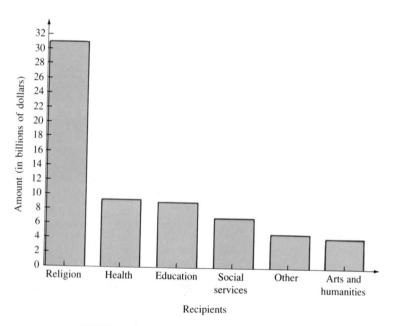

FIGURE 2-1 Recipients of 1983 charitable giving

Both types of graphs will be illustrated using the data in Table 2-14, which shows the recipients of charitable giving by U.S. citizens in 1983.[1]

Figure 2-1 represents a vertical-bar graph for the data in Table 2-14.

Table 2-15 organizes the computations necessary for constructing a circle graph for the charity-recipient data. Each entry in the "%" column was obtained by dividing the amount by the total amount (64.9) and then multiplying by 100. Entries in the "degrees" column were obtained by multiplying the entries in the "%" column by 360°, the number of degrees in a circle.

Table 2-15 Computations for Constructing Circle Graph for Data of Table 2-14

Recipient	Amount	%	Degrees	
Religion	31.0	47.8	172.1	← (.478 × 360)
Health	9.2	14.2	51.1	
Education	9.0	13.9	50.0	
Social services	6.9	10.6	←⌐ 38.2	$\left(\dfrac{6.9}{64.9} \times 100 \right)$
Arts and humanities	4.1	6.3	22.7	
Other	4.7	7.2	25.9	
Totals	64.9	100.0	360.0	

The circle graph in Figure 2-2 was constructed using a protractor, a device for measuring angles, and the information from Table 2-15. We can see from a glance at the circle graph that religion gets the biggest share—an amount approximately equal to the total amount for the remaining recipients. Arts and humanities get the smallest share.

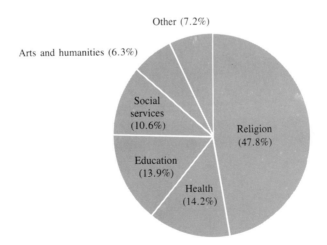

FIGURE 2-2 Recipients of 1983 charitable giving

[1] *U.S. News and World Report*, Apr. 23, 1984, p. 12.

Many computer programs for data analysis allow the computer to draw circle graphs when given percentages or raw data.

Stem-and-Leaf Diagrams

The use of a grouped frequency table has one rather obvious disadvantage: the original data are lost in the grouping process. To overcome this limitation, a **stem-and-leaf diagram** can be used. Stem-and-leaf diagrams offer a quick and novel way for displaying numerical data. If a numeral has two or more digits, then it can be split into a stem and a leaf. A **stem** is the leading digit (part) of the numeral, while the **leaf** constitutes the trailing digit(s). For example, the numeral 278 could be split two ways:

$$2 \mid 78 \qquad \text{or} \qquad 27 \mid 8$$

$$\uparrow \quad \uparrow \qquad\qquad\qquad \uparrow \quad \uparrow$$

Stem   Leaf   Stem   Leaf

A graphical display for data classified using stems and leaves is very simple to accomplish. Each piece of data serves as a leaf on some stem. For illustrative purposes, let's construct a stem-and-leaf diagram for the following collection of 25 algebra test scores:

78	67	65	87	75
64	84	82	81	68
98	59	57	79	65
85	76	89	67	94
65	71	54	59	80

Since all the scores fall between 50 and 99, let's use the tens digit in each case to serve as the stem and the units digit to serve as the leaf.

Step 1.   Place the stems in a vertical array using a vertical line segment, called the **trunk**, to separate the leaves from the stems:

```
5 |
6 |
7 |
8 |
9 |
```

Step 2.   Place each leaf to the right of its stem. Since the first score is 78, we place the leaf, 8, on its stem, 7:

```
        5 |
        6 |
Stem → 7 | 8 ← Leaf
        8 |
        9 |
```

Continuing this process for each score, we obtain the following stem-and-leaf diagram:

```
5 | 9  7  4  9
6 | 4  5  7  5  7  8  5
7 | 8  6  1  9  5
8 | 5  4  2  9  7  1  0
9 | 8  4
```

The following conclusions can be drawn by observing the stem-and-leaf diagram:

1. The largest score is 98.
2. The smallest score is 54.
3. The scores vary from 54 to 98.
4. Stem 9 has the fewest leaves.
5. Stems 6 and 8 contain the most leaves, with seven leaves on each stem.
6. The total number of leaves represents the size of the data set.

It should be noted that it does not make any difference in which order the leaves are placed on a stem. If the leaves on each stem are ordered from smallest to largest, the diagram is called an **ordered stem-and-leaf diagram**. An ordered stem-and-leaf diagram for the algebra test scores is as follows:

```
5 | 4  7  9  9
6 | 4  5  5  5  7  7  8
7 | 1  5  6  8  9
8 | 0  1  2  4  5  7  9
9 | 4  8
```

An ordered stem-and-leaf diagram is useful for ordering data and for computing position points. A **position point** is a number that has a certain percentage of the data falling below it. For example, with some simple addition, one can quickly observe that the 40% position point for the algebra scores data set is 68; that is, $(.40)(25) = 10$ scores fall below a score of 68. An ordered stem-and-leaf diagram represents a graphical display corresponding to a frequency table.

Sometimes it may be desirable to include fewer than ten leaf-values on a single stem to spread out the data, particularly if a few stems contain a large number of leaves. When this is done, we have a visual display that corresponds to a grouped frequency table. When the number of classes is increased, certain salient characteristics of the data may become more apparent. If each stem in a stem-and-leaf diagram is split into two stems, called **substems** and each having the same number of leaf-values, the resulting stem-and-leaf diagram is referred to as a **double-stem diagram**. Let's use the previous collection of 25 test scores and construct a double-stem stem-and-leaf diagram with each stem having five possible leaf-values. This is accomplished by splitting each stem into two substems, called *a* and *b*. Substem *a* will contain digits 0 through 4 as its leaves and substem *b* will contain digits 5 through 9 as its leaves. For example, substem

5a will contain leaves 0–4 and substem 5b will contain leaves 5–9. The resulting double-stem diagram is shown here:

```
5a | 4
5b | 7 9 9
6a | 4
6b | 5 5 5 7 7 8
7a | 1
7b | 5 6 8 9
8a | 0 1 2 4
8b | 5 7 9
9a | 4
9b | 8
```

The following two examples illustrate additional modifications of the above construction procedures for stem-and-leaf diagrams.

EXAMPLE 2-8   A national survey by utility regulators found that power costs vary widely across the United States. The 25 most expensive cities (rated by average cost in cents per kilowatt-hour) in 1984 are as follows:

16.5	14.3	14.3	13.9	13.8
13.1	12.8	12.1	12.0	11.8
11.6	11.4	11.3	11.3	11.2
11.2	11.1	11.1	10.8	10.8
10.8	10.8	10.7	10.6	10.6

Construct a stem-and-leaf diagram for the data.

Solution   We will ignore the decimal points in the data; each value in the final array can be converted back to its original value by multiplying by .1. Also, the numbers will be treated as three-digit numbers ranging from 106 to 165. By using double-digit stems, we get the following ordered diagram:

```
10 | 6 6 7 8 8 8 8
11 | 1 1 2 2 3 3 4 6 8
12 | 0 1 8
13 | 1 8 9
14 | 3 3
15 |
16 | 5
```

The leaf 5 on stem 16 represents 16.5 cents. We can easily determine that 20% of the average costs are above 13.1 cents.   ■

EXAMPLE 2-9   The following data represent the 1-year percentage changes in the prison popu-lations at 25 federal and state prisons as of June 30, 1984.[2]

_____

[2] *U.S. News and World Report*, Sept. 10, 1984, p. 16.

0.6	12.9	10.8	11.7	0.4
− 11.1	0.6	2.5	0.2	− 4.4
− 1.4	− 3.2	− 1.7	− 1.2	7.0
− 10.1	19.2	20.6	− 0.5	9.8
2.1	16.3	8.8	20.8	4.1

Construct a stem-and-leaf diagram for the data.

Solution   Ignoring decimals, we observe that the data range from − 111 to 208. Let's use stem values of − 1, − 0, +0, 1, and 2. To make all of the values three-digit numbers, we put a zero in front of any two-digit value. Thus, 4.1 is represented as 041. We need two stems for zero to indicate the signs of numbers. For example, the stem for the value 0.6 is +0, the stem for 7.0 is +0, and the stem for − 1.7 is − 0. The value 0.6 must be represented in the stem-and-leaf diagram as +006; 7.0 must be represented as +070; and − 1.7 must be represented as − 017. The stem-and-leaf diagram is shown here:

− 1	01	11								
− 0	05	12	14	17	32	44				
+0	02	04	06	06	21	25	41	70	88	98
1	08	17	29	63	92					
2	06	08								

Remember that each value in the diagram must be converted by multiplying by .1 before interpretations are made.   ■

Stem-and-leaf diagrams are capable of other adaptations, as indicated by the following:

1. Two similar distributions can be compared.

Leaves				*Stem*	*Leaves*				
		8	6	5	3	6	8		
9	8	7	3	6	2	7			
		8	6	7	3	5	5	7	6
5	4	3	0	8	3	3	4	5	

2. More than two distributions can be compared by arranging the diagrams in column form.

Stem	*Distribution 1* leaves	*Distribution 2* leaves	*Distribution 3* leaves
3	2 7	0	1 1
4	4 4 5	6 8	
5	7 9	3 4 7	5 6 8 9
6	3		3 4

Histograms

A **histogram** is a type of bar graph for a frequency distribution. Although histograms can also be constructed for ungrouped frequency distributions, in this section we will construct histograms for grouped frequency distributions only.

Two steps are typically followed when constructing a histogram for data measured using an interval or ratio scale:

1. The data are organized into a grouped frequency table.
2. A bar graph is constructed using the class boundaries to locate the bars and the frequencies to determine the heights of the bars.

EXAMPLE 2-10   The following grouped frequency table represents the extent of unemployment (in percentages) for 27 eastern cities for February 1984.[3]

Extent of unemployment (percentage)	Number of cities
3.7–5.1	5
5.2–6.6	12
6.7–8.1	6
8.2–9.6	1
9.7–11.1	0
11.2–12.6	1
12.7–14.1	2
	$\overline{27}$

Construct a histogram for the data.

Solution   The histogram is constructed by first locating the class boundaries along the horizontal axis and the frequencies on the vertical axis. For each class, a rectangular bar is constructed using the class boundaries to measure the width of the bar and the frequency to measure the height. Since all the classes in a grouped frequency table have the same width, the areas of the bars will be proportional to the heights of the bars—that is, to the frequency of the classes. To construct the histogram we follow these steps:

Step 1.   We first calculate the class boundaries. Note that the unit of measurement is .1 (of a percent). Thus, $(.5)(.1) = .05$ is subtracted from the lower class limit of a class to find the lower class boundary and .05 is added to the upper class limit of a class to find the upper class boundary.

---

[3] *U.S. News and World Report*, Apr. 23, 1984, p. 14.

Class	Boundaries	f
3.7–5.1	3.65–5.15	5
5.2–6.6	5.15–6.65	12
6.7–8.1	6.65–8.15	6
8.2–9.6	8.15–9.65	1
9.7–11.1	9.65–11.15	0
11.2–12.6	11.15–12.65	1
12.7–14.1	12.65–14.15	2
		27

Step 2. A bar graph is now constructed using the class boundaries and frequencies. The boundaries are located along the horizontal axis and the frequencies are located along the vertical axis, as shown in Figure 2-3. Note that the horizontal axis is "broken" to draw attention to the fact that the scale does not start at zero. We break the axis to indicate that we are not deliberately trying to distort perspective.

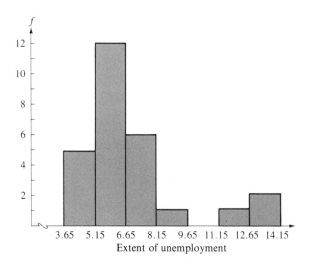

FIGURE 2-3 Bar graph for unemployment data   ■

A histogram improves our ability to compare the frequencies of corresponding classes; a class frequency can be easily compared with those of its neighboring classes. We can immediately see that the second class in the histogram in Figure 2-3 has the largest frequency and that the frequency of this class is twice the frequency of the third class. For the classes that measure unemployment between 8.15% and 11.15%, there is a rapid decline in the number of cities represented.

The shape of a histogram may change dramatically with a change in the number of intervals, n, or the width of the intervals, c. For this reason, we should be careful about drawing conclusions about the shape of the sample distribution. For example, the three histograms shown in Figure 2-4 represent a sample of 100 measurements for different values of n and c. The histogram in (a) has five classes and a class width of 9.95; the histogram in (b) has eight classes with a class width of 6.22; and the histogram in (c) has 11 classes with a class width of 4.60. Notice how the appearance changes as the number of intervals and class width vary.

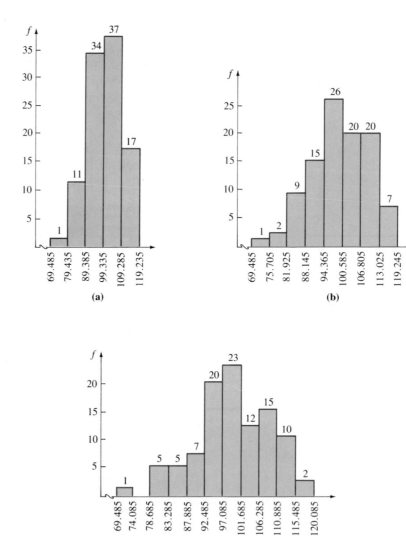

FIGURE 2-4 Effect of changes in number of intervals and class width on appearance of histogram. The horizontal axis represents lengths in inches

Relative Frequency Histograms

A **relative frequency histogram** can be constructed by changing the vertical scale of a frequency histogram. Instead of beginning with a grouped frequency table, we begin with a grouped relative frequency table. The height of the bars on a relative frequency histogram will indicate the *proportion* of the total represented by each class. The basic shape of a relative frequency histogram is similar to the shape of the corresponding frequency histogram.

For the data in Example 2-10 we have the relative frequency table shown in Table 2-16 and the corresponding relative frequency histogram depicted in Figure 2-5.

Table 2-16 Relative Frequency Table for Unemployment Data

Class	Boundaries	f	Relative frequency
3.7–5.1	3.65–5.15	5	.19
5.2–6.6	5.15–6.65	12	.44
6.7–8.1	6.65–8.15	6	.22
8.2–9.6	8.15–9.65	1	.04
9.7–11.1	9.65–11.15	0	.00
11.2–12.6	11.15–12.65	1	.04
12.7–14.1	12.65–14.15	2	.07
		27	1.00

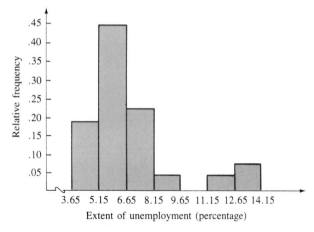

FIGURE 2-5 Relative frequency histogram for unemployment data of Example 2-10

Converting a Stem-and-Leaf Diagram to a Histogram

It is very easy to convert a stem-and-leaf diagram to a histogram, since the stem-and-leaf diagram contains the raw data. Consider the following example.

EXAMPLE 2-11   A 1984 survey of the cost (in cents) of a pound of sugar in 16 world capitals produced the following data:

32	36	16	15	32	33	26	9
31	31	26	37	49	37	50	44

Construct a stem-and-leaf diagram and a histogram for the data.

Solution   A stem-and-leaf diagram is as follows:

```
0 | 9
1 | 6 5
2 | 6 6
3 | 2 6 2 3 1 1 7 7
4 | 9 4
5 | 0
```

We obtain the corresponding histogram by converting the stems to class-interval labels and using the number of leaves as the frequency for each class. The stem 0 corresponds to the class interval 0–9, the stem 1 corresponds to the interval 10–19, the stem 3 corresponds to the interval 30–39, and so forth. The class boundaries for these class intervals are used to construct the histogram. The result is shown in Figure 2-6. Notice that the histogram has the shape of the stem-and-leaf diagram, but is turned 90° counterclockwise.

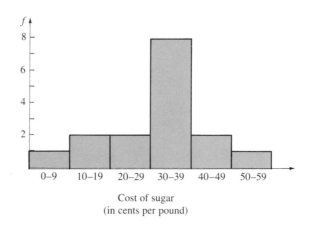

Cost of sugar
(in cents per pound)

FIGURE 2-6 Histogram of sugar prices   ■

Line Graphs or Frequency Polygons

A **line graph** or **frequency polygon** is constructed using a grouped frequency table with class marks. The line graph offers a useful alternative to the histogram; the choice of which to use is usually a personal matter. A line graph creates the impression that the frequencies change more smoothly, whereas a histogram suggests that the frequencies change abruptly. A line graph or frequency polygon can be constructed for data displayed in a grouped frequency table by identifying each class mark and its corresponding frequency, (X, f), with a point on the graph. These points are then joined by a sequence of line segments, as illustrated in the next example.

EXAMPLE 2-12   The following grouped frequency table reports the 1984 average annual incomes (to the nearest $100) of factory workers in 27 eastern cities.[4]

Average income	Number of cities
$12,500–14,300	1
14,400–16,200	5
16,300–18,100	3
18,200–20,000	7
20,100–21,900	6
22,000–23,800	1
23,900–25,700	3
25,800–27,600	1

Construct a frequency polygon for the data.

Solution   *Step 1.*   We first find the class marks, designated X.

Average income	f	X
$12,500–14,300	1	13,400
14,400–16,200	5	15,300
16,300–18,100	3	17,200
18,200–20,000	7	19,100
20,100–21,900	6	21,000
22,000–23,800	1	22,900
23,900–25,700	3	24,800
25,800–27,600	1	26,700

Step 2.   We now construct the line graph shown in Figure 2-7. The class marks are located on the horizontal axis and the frequencies are located on the vertical axis. Note that the line graph is "tied down" at both ends by connecting the first and last points to the horizontal axis at points a distance of $w = 1900$ from the nearest class marks.

The following salient characteristics of the data are shown in the line graph of Figure 2-7:

1. Most of the cities fall between the extreme ends of the scale. Only one city has factory workers earning an approximate average annual income of $13,400, and only one city has factory workers earning an approximate average annual income of $26,700.
2. The data appear to be centered at approximately $19,000.

[4] *U.S. News and World Report*, Apr. 30, 1984, p. 65.

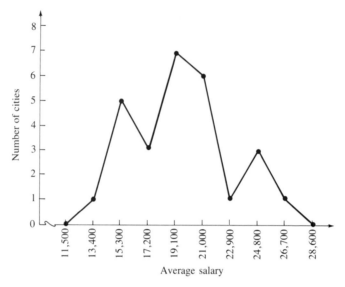

FIGURE 2-7 Average annual salaries of factory workers
(to the nearest $100)   ■

Ogives

A line graph constructed from a cumulative frequency table or a cumulative relative frequency table is called an **ogive** (pronounced ō′jĭv). Ogives offer a graphical means for interpolating or approximating the number or percentage of observations less than or equal to a specified value.

For an ogive, an *upper* class boundary and its corresponding cumulative frequency (or cumulative relative frequency) are used to locate a point. Consecutive points are then joined by line segments. Cumulative frequencies (or cumulative relative frequencies) are always located on the vertical axis. The following example illustrates the construction of an ogive.

EXAMPLE 2-13 Construct an ogive using cumulative frequencies for the data in Example 2-12.

Solution *Step 1.* We first find the cumulative frequencies.

Average income	Upper boundary	*f*	Cumulative frequency
$12,500–14,300	14,350	1	1
14,400–16,200	16,250	5	6
16,300–18,100	18,150	3	9
18,200–20,000	20,050	7	16
20,100–21,900	21,950	6	22
22,000–23,800	23,850	1	23
23,900–25,700	25,750	3	26
25,800–27,600	27,650	1	27

Step 2. We use the boundaries of the classes to locate the points on the horizontal axis and the frequencies for the vertical axis.

Step 3.   We construct the ogive (Figure 2-8). Notice that the cumulative frequency for the lower boundary of the first class is 0. We can determine at a glance the number of cities having average salaries for factory workers below any specified amount.

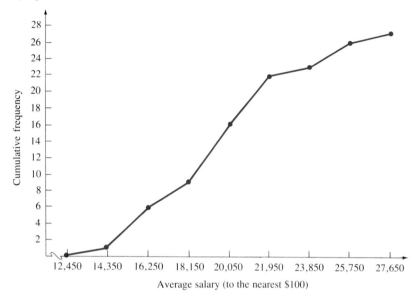

FIGURE 2-8 Annual incomes of factory workers for 27 eastern cities   ∎

Using Ogives to Determine Percentiles

A cumulative relative frequency ogive can be used to determine percentiles, as described in the next example.

EXAMPLE 2-14   For the data in Example 2-13, construct a cumulative relative frequency ogive and use it to approximate the 50th percentile (P_{50}) and the 75th percentile (P_{75}). Recall that the 75th percentile is the measurement below which 75% of the measurements fall.

Solution   *Step 1.*   We first find the relative cumulative frequencies using the cumulative frequencies.

Average income	Upper boundary	Cumulative f	Relative cumulative f
$12,500–14,300	14,350	1	.037
14,400–16,200	16,250	6	.222
16,300–18,100	18,150	9	.333
18,200–20,000	20,050	16	.593
20,100–21,900	21,950	22	.815
22,000–23,800	23,850	23	.852
23,900–25,700	25,750	26	.963
25,800–27,600	27,650	27	1.000

Step 2. We use the boundaries of the classes to locate the points on the horizontal axis and the frequencies for the vertical axis.

Step 3. We construct the ogive (Figure 2-9). We see that P_{50}, the 50th percentile, is between $18,150 and $20,050 (approximately $19,500) and that P_{75}, the 75th percentile, is a little less than $22,000. Hence, approximately 50% of the cities have factory workers earning an average annual salary less than $19,500 and 75% of the cities have factory workers earning an average salary less than $22,000. As a result, approximately 25% of the factory workers earn between $19,500 and $22,000.

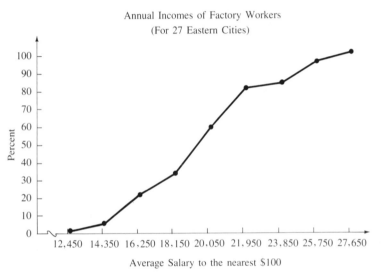

FIGURE 2-9 Ogive of annual incomes of factory workers
for 27 eastern cities   ■

Histograms, Ogives, and the Shapes of Populations

Histograms and ogives for sample data give the researcher an indication of the shape of the population from which the sample was selected. The histogram of a sample suggests the shape of the corresponding population frequency curve. A relative frequency histogram for a sample should have a shape similar to the relative frequency distribution for the population, and an ogive for a sample should have approximately the same shape as the ogive for the population. Since populations are generally represented by relative frequency curves or cumulative relative frequency curves, it is important that we understand their sample counterparts. For example, suppose an automatic bottle filler in a brewery is supposed to deliver 12 ounces of beer to each bottle. A sample of 50 bottles revealed the following contents (in ounces):

12.335	12.111	12.166	11.900	11.889
12.057	11.848	12.151	11.717	11.584
12.497	12.083	12.018	11.704	12.187
12.082	12.491	11.929	11.743	12.035
12.335	12.520	11.988	12.080	12.001

11.990	11.748	12.103	12.185	12.100
11.846	12.240	12.339	11.611	11.856
11.629	11.912	11.786	11.853	11.655
12.101	11.886	12.410	11.956	12.108
11.923	11.853	11.919	12.130	12.408

The histogram in Figure 2-10(a) illustrates the distribution of bottle contents for the sample of 50 bottles of beer. The histogram for the sample approximates the bell-shaped population (called a **normal distribution**) illustrated in Figure 2-10(b). The ogive for the sample illustrated in Figure 2-10(c) approximates the "S" shape of the cumulative relative frequency distribution for the normal distribution illustrated in Figure 2-10(d). A cumulative relative frequency distribution for a bell-shaped or normal population will always have an S-shaped appearance.

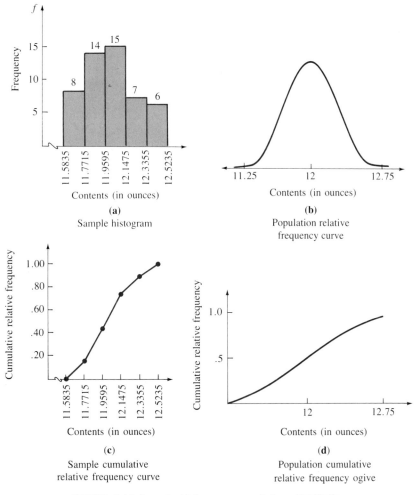

FIGURE 2-10 Sample histogram, population distribution, and ogives for beer-contents data

Frequently the data produced by a particular process or application will have a known form, such as a bell-shaped distribution. This information, as we shall later see, can be used to evaluate sample data taken from the population.

Problem Set 2.3

A

1. Consider the following sample of grades:

A	C	D	B	C
C	C	D	F	F
D	F	A	D	C
B	C	D	D	B

Construct:
a. A bar graph   **b.** A circle graph

2. Construct a frequency histogram for the data listed here. Use six bars.

17	14	16	8	31
16	14	9	17	11
25	24	28	10	48
24	12	13	43	24
32	37	33	42	11
34	16	41	21	15

3. Construct a frequency polygon for the data in Problem 2 using nine points (including the endpoints).

4. Construct a stem-and-leaf diagram for the data in Problem 2.

5. Construct a relative frequency histogram for the accompanying histogram.

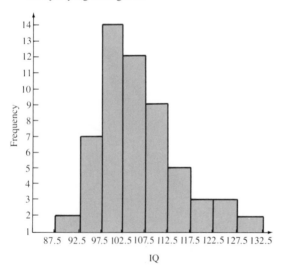

IQs of a random selection of graduating students at ABC High School

6. The number of calories burned per hour by a 130-pound woman engaging in ten activities is as shown in the table (*Shape*, July 1985, p. 23). Construct a bar graph for the data.

Activity	Calories burned per hour
Lying down, awake	72
Sitting at rest	95
Working at desk	128
Dressing or undressing	140
Walking level, 2.6 mph	190
Bicycling level, 5.5 mph	295
Tennis	384
Hiking, no load	425
Swimming, slow crawl	450
Jogging, 5.3 mph	550

7. Construct an ogive corresponding to the following grouped frequency table:

Class	f
1–4	14
5–8	18
9–12	12
13–16	16
17–20	20

8. What kinds of graphs are appropriate for qualitative data? For quantitative data?

9. Of 100 hospital patients, 30 had type O blood, 38 had type A blood, 22 had type B blood, and 10 had type AB blood. Construct a circle graph for these data.

10. The following table shows the ten most commonly performed plastic surgery operations during 1984 (data from American Society of Plastic and Reconstructive Surgeons). Construct a bar graph for the data.

Operation	Number performed
Hand surgery	160,000
Laceration repair	150,000
Tumor removal	100,000
Breast augmentation	75,000
Industrial injury	70,000
Eyelid surgery	57,000
Nose surgery	55,000
Burn repair	45,000
Reconstruction	45,000
Facelift	40,000

11. For the following English test scores, construct a stem-and-leaf diagram:

67	71	90	46	51
74	34	65	55	63
71	66	54	46	22
69	61	57	46	84

12. Construct a relative frequency histogram using 7 classes for the following 1984 EPA mileage estimates for 30 new cars:

22	31	20	27	21
29	27	35	47	29
27	23	51	41	30
34	27	35	27	27
31	38	25	27	44
35	34	32	21	19

13. For the data in Problem 12, construct a double-stem stem-and-leaf diagram.

14. According to the 1983–1984 catalog of a state college in Maryland, a student's principal expenses for 1 year are as listed in the accompanying table. Construct a circle graph showing the data.

Expense	Amount
Registration fee	$ 20.00
Activities fee	45.00
College Center fee	147.50
Athletic fee	94.00
Tuition	1030.00
Board	1130.00
Room	1065.00
Total	$3531.50

B

15. Convert the accompanying circle graph to a bar graph.

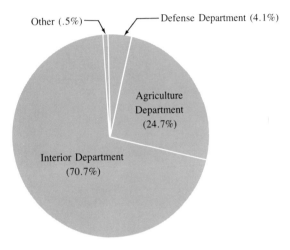

Land owned in 1975 by federal agencies
(total of 761 million acres)

16. Convert the accompanying bar graph to a circle graph.

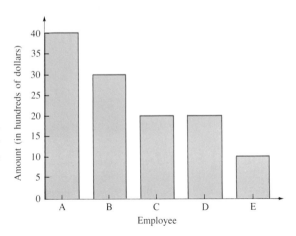

Last year's sales (in hundreds of dollars)

17. Refer to the histogram shown in Problem 5. Find each of the following:
a. The total number of IQ scores
b. The number of IQ scores falling between 97.5 and 102.5

c. The width of each bar

d. The relative frequency for the IQ scores comprising the 110 IQ bar

e. The percentage of IQ scores falling below 117.5

18. Refer to the accompanying line graph. Find each of the following:

a. The total number of families represented

b. The total number of children represented

c. The percentage of families that have four children

d. The percentage of families that have fewer than three children

19. The accompanying table expresses the time (in hours and minutes) bus drivers in five cities must work in order to purchase McDonald's Big Macs, small fries, and medium cokes for a family of four. Figures are given for the years 1979 and 1984.[5] Construct a single bar graph comparing the costs for the five cities for both years.

City	1979	1984
Chicago	0:58	1:02
Tokyo	1:29	1:19
Paris	1:52	2:18
Dusseldorf	1:41	1:47
London	2:02	2:24

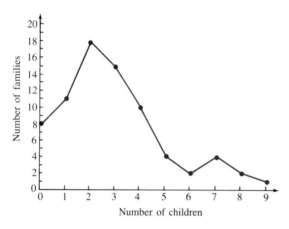

Number of children per family

CHAPTER SUMMARY

In this chapter we learned that data can be classified as quantitative or qualitative. Data can also be classified according to the type of measurement scale used—nominal, ordinal, interval, or ratio. Tables and graphs are used to organize the data so that they can be more easily used and understood. We studied grouped and ungrouped frequency tables, including relative frequency and cumulative frequency tables. In addition, we saw that data distributions can also be represented graphically by using circle graphs, bar graphs, histograms, stem-and-leaf diagrams, frequency polygons or line graphs, and ogives.

---

[5] *Reader's Digest*, Dec. 1984, pp. 107–108.

C H A P T E R   R E V I E W

IMPORTANT TERMS

For each term, provide a definition in your own words. Then check your responses against the definitions given in the chapter.

bar graph	frequency	ogive	relative frequency table
bivariate data	frequency polygon	ordinal data	source
bivariate frequency table	frequency table	percentile score	stem
circle graph	graph	position point	stem-and-leaf diagram
class boundaries	grouped frequency table	qualitative data	Sturges' rule
class limits	histogram	quantitative data	substems
class mark	interval data	range	tally marks
class width	leaf	ratio data	trunk
cumulative frequency table	line graph	raw data	ungrouped frequency table
cumulative relative frequency table	nominal data	relative frequency	unit
	nominal distribution	relative frequency histogram	univariate data
double stem diagram	nominal scale		

IMPORTANT SYMBOLS

X, class mark	f, frequency	R, range	n, total number of measurements
U, largest measurement	w, class width	l_1, lower class limit	
L, smallest measurement	c, number of classes	l_2, upper class limit	

IMPORTANT FACTS AND FORMULAS

1. For a grouped frequency table the number of classes should be between 5 and 15, inclusive.

2. The class width is found by dividing the range by the number of classes and rounding the result to the nearest unit of measurement, or $w = R/c$.

3. For a grouped frequency table, the first class always begins with the smallest measurement.

4. For a grouped frequency table, all classes have the same width.

5. Class boundaries are used to construct histograms and ogives.

6. Sturges' rule: The number of classes needed in a grouped frequency table is approximately equal to $c = 3.3(\log n) + 1$.

REVIEW PROBLEMS

1. The heights of 50 female college students (in centimeters) are as follows:

157	155	171	150	163
150	172	161	154	174
163	148	152	163	149
158	176	164	157	153
169	161	160	164	155
162	151	167	167	167
170	158	163	175	169
169	158	150	156	157
174	162	150	151	165
170	156	170	153	154

    a. Construct a grouped frequency table using ten classes.
    b. Construct a stem-and-leaf diagram.
    c. Construct an ogive using the result of part (a).
    d. Construct a histogram using the result of part (a).

2. Classify the following data as either quantitative or qualitative:
    a. The weights (in ounces) of 20 apples
    b. The colors of 10 cars
    c. The length (in centimeters) of a 12-inch ruler
    d. The religious preferences of 15 people
    e. The letter grades for a class of students
    f. The percentage grades for a class of students
    g. The sex classifications of 50 teachers
    h. The status (off/on) of 30 light switches
    i. The street addresses of 100 relatives
    j. The jersey numbers for the members of a baseball team
    k. The number π (pi)

3. Twenty people were asked to identify their religious preference. If C denotes Catholic, P denotes Protestant, J denotes Jewish, and A denotes Atheist, construct a frequency table, a circle graph, and a bar graph for the following data:

C	P	P	J	J
A	J	C	P	P
C	J	J	C	P
P	A	P	C	J

4. Find the widths of the following classes given their limits.
    a. 7–16   **b.** 3.4–7.8   **c.** 1.3–4.5
    d. 1.23–4.78   **e.** .03–.09

5. If a grouped frequency table is to contain eight classes, the smallest measurement is 14 and the largest is 94, find the width of each class.

6. Consider the following bivariate frequency table.

	Grade		
Sex	**Pass**	**Fail**	**Total**
Male	11	15	26
Female	14	10	24
Total	25	25	50

Find each of the following:
    a. The number of females who passed
    b. The percentage of males who failed
    c. The percentage of those who passed who are males

7. An experiment was conducted to determine the effect of a certain drug on serum cholesterol level (in mg/100 ml) in 30-year-old males. The following measurements were obtained:

245	185	230	210	245
165	225	160	240	225
235	120	210	190	220
140	215	195	145	185
195	285	175	260	195
170	205	225	265	

    a. Construct a stem-and-leaf diagram.
    b. Construct a grouped frequency table with ten classes.
    c. Construct a relative frequency histogram using the table in part (b).

8. The heights (to the nearest inch) of 33 students are as follows:

66	65	64	68	69
65	68	68	64	66
64	63	71	70	67
69	71	59	67	72
70	67	69	69	66
63	67	70	66	70
67	64	80		

Construct a grouped frequency histogram having eight bars.

9. Construct a double-stem stem-and-leaf diagram for the data in Problem 8.

10. The table represents U.S. average life expectancies for the years 1950 and 1983.

	1950		1983	
	Males	Females	Males	Females
Newborn	65.5	71.0	70.9	78.3
15-year-olds	68.6	73.5	72.2	79.3
25-year-olds	69.4	74.0	72.9	79.6
35-year-olds	70.2	74.5	73.7	80.0
45-year-olds	71.6	75.6	74.5	80.5
65-year-olds	77.7	80.0	79.5	83.8

a. Construct a bar graph for the average male life expectancies for 1983.

b. Construct a bar graph for the average female life expectancies for 1983.

c. Construct a bar graph comparing male and female average life expectancies for the year 1983.

d. Construct a line graph comparing average male life expectancies for the years 1950 and 1983.

11. The following data represent the weights (in pounds) for a sample of college students:

114	115	116	120	123
126	128	129	131	132
132	133	134	135	135
137	138	139	142	142

143	146	147	152	157
158	161	164	165	167
168	168	170	170	172
174	174	174	175	175
176	177	177	178	180
184	184	184	186	187
189	194	195	195	200
201	202	206	207	209

a. Construct a stem-and-leaf diagram.

b. Construct a double-stem stem-and-leaf diagram. Does this diagram reveal any characteristics of the data not revealed by the stem-and-leaf diagram in part (a)? Offer an explanation for the difference in shape.

c. By using the diagram in part (a), construct a histogram for the raw data.

12. The accompanying frequency table contains the speeds (in miles per hour) of a sample of 60 cars traveling on 14th Avenue in New York as measured by a policeman with radar. Construct a relative frequency ogive for the data.

Class	f
28–33	1
34–39	3
40–45	6
46–51	28
52–57	14
58–63	8

CHAPTER ACHIEVEMENT TEST

The following data indicate the weights (in pounds) lost by a group of women in the first 2 weeks of a daily exercise program:

1	2	12	3	15	5	12	11	3	4
3	5	0	7	17	6	17	13	2	5
5	7	1	11	3	9	9	8	18	8
10	9	4	12	1	8	8	7	11	9
15	11	8	4	5	11	3	14	12	10

Use this data set to answer Problems 1–6.

(20 points)   **1.** Construct an ungrouped frequency table.

(10 points)   **2.** Construct a stem-and-leaf diagram and corresponding histogram.

(10 points)   **3.** Construct a cumulative relative frequency table.

(10 points)   **4.** Construct a grouped frequency table containing five classes.

(20 points)   **5.** Construct an ogive using relative frequencies and the table constructed in Problem 4.

(10 points)   **6.** Construct a histogram having five bars using the table constructed in Problem 4.

(5 points)   **7.** Find the width of the class 10–20, where 10 and 20 are the class limits.

(5 points)   **8.** If 50 measurements have a smallest value of 17 and a largest value of 96, find the smallest whole-number class width for a grouped frequency table containing 11 classes.

(10 points)   **9.** Consider the following bivariate table representing the number of A and B grades for three sections of Speech 102 taught by Professor Smith:

	Section		
Grade	**I**	**II**	**III**
A	5	6	1
B	3	2	5

a. Find the percentage of A grades that were earned in section I.

b. Find the percentage of section II grades that are B's.

c. What percentage of the total number of grades are section II grades?

d. What percentage of the total number of grades are A grades?

e. What percentage of the total number of grades are the A grades made in section II?

DESCRIPTIVE STATISTICS— ANALYSES OF UNIVARIATE DATA

Chapter Objectives

In this chapter you will learn:

- *four measures of central tendency*
- *how to calculate the measures of central tendency for grouped and ungrouped data*
- *how to find percentiles for grouped data*
- *advantages and disadvantages of using each measure of central tendency*

- *the concept of skewness*
- *the concept of sum of squares*
- *three measures of dispersion*
- *how to calculate the measures of dispersion for grouped and ungrouped data*
- *Chebyshev's theorem*
- *standard scores*

Chapter Contents

In Chapter 2, we presented the methods for organizing data using tables and graphs. Those techniques represent visual means for discovering relationships, patterns, or trends within the data.

In this chapter, we want to supplement the visual interpretations made possible by tables and graphs with numerical measures of characteristics enjoyed by most collections of quantitative data. The characteristics include center, spread, and position points for a set of data. We begin by discussing the concept of center for a data set.

3.1 MEASURES OF CENTRAL TENDENCY

The first characteristic of a set of data that we want to measure is the center or central tendency. The purpose of a measure of **central tendency** is to summarize a collection of data so that we can obtain a general overview; such a measure will serve as a representative for the rest of the data. A measure of central tendency of a set of data will also provide a sense of the central value for a seemingly disorganized set of observations. Consider the following four examples:

1. Weights in pounds: 5, 6, 12, 15, and 20.
2. Grades for a test: 31, 74, 78, 79, 80, and 81.
3. Colors of cars: 3 white, 4 red, 7 black, and 1 blue.
4. Faculty ranks: 7 professors, 3 associate professors, 2 assistant professors, and 10 instructors.

In examples 1 and 2, the measurement scale used is ratio; in (3), nominal; and in (4), ordinal. Which measurement(s) would you use to describe the central value or to represent the data set for each example? There are many measures of central tendency used to locate a center of a set of data. Four of the more common measures are the mean, the median, the mode, and the midrange. The **mean** is the arithmetic average. The **median** is the middle ordered score. The **mode**, if it exists, is the most frequent score. The **midrange** is the arithmetic average of the largest and smallest measurements. To describe the center measurements for the four examples just given, we would use the mean for example 1, the median for examples 2 and 4, and the mode for example 3. We will now examine each measure of central tendency in detail and learn the reasons for the choices in the four examples.

Mean

The **mean** or arithmetic average of a set of numbers is found by adding the numbers and then dividing the sum by n, the number of measurements. For example, the following ten scores represent the number of points scored in ten basketball games by player A:

6   10   3   7   6   6   8   5   9   10

The mean is

$$\frac{6 + 10 + 3 + 7 + 6 + 6 + 8 + 5 + 9 + 10}{10} = \frac{70}{10} = 7$$

The value 7 represents, in some sense, the central or "middle" number of points scored in ten games by player A.

Means can be calculated for both samples and populations. They are computed the same way, but are denoted differently. The sample mean is denoted by $\bar{x}$ and the population mean is denoted by the Greek letter μ (pronounced mu). A formula for calculating the mean of a sample of numerical data is given by

$$\bar{x} = \frac{\sum x}{n}$$

where $\bar{x}$ denotes the sample mean, x denotes a sample measurement, $\sum x$ denotes the sum of the sample measurements, and n is the size of the sample. A formula for finding the population mean is given by

$$\mu = \frac{\sum x}{N}$$

where μ is the mean of the population and N is the size of the population.

EXAMPLE 3-1   The annual amounts (in billions of dollars) of U.S. agricultural exports from 1974 to 1983 are 21.9, 21.9, 23.0, 23.6, 29.4, 34.7, 41.2, 43.3, 39.1, and 33.7 (*U.S. News and World Report*, Sept. 5, 1983). Determine the mean μ if the data constitute a population.

Solution   The sum of the ten measurements is $\sum x = 311.8$. As a result, the population mean is

$$\mu = \frac{\sum x}{N} = \frac{311.8}{10} = 31.18$$

Thus, the average amount of agricultural exports over the 10-year period is $31.18 billion dollars.   ∎

The sample mean cannot be found for all types of data. Suppose we have recorded the color of hair for each of ten college students. The phrase "average hair color" has no meaning. The data in this situation are qualitative and the mean can be computed only for quantitative data.

Suppose we have the following sample of ages (in years) for ten college freshmen:

18   18   18   18   19   19   19   20   20   21

To find the sample mean $\bar{x}$, we add the ten ages and divide the sum by 10. Adding the ten ages can be shortened by adding the four products (4)(18), (3)(19),

(2)(20), and (1)(21). Each product can be written as fx, where f is the frequency of occurrence of an age x (see Table 3-1). The sum of the f values equals n, and the sum of the fx values equals $\sum x$.

Table 3-1 Raw Data and Frequency Table for Ages of Ten Freshmen

x	tally	x	tally	fx
18	||||	18	4	72
19	|||	19	3	57
20	||	20	2	40
21	|	21	1	21
			10	190

(a) Raw data (b) Frequency table

By applying the definition of the sample mean to the data on freshmen ages, we obtain

$$\bar{x} = \frac{\sum x}{n} = \frac{190}{10} = 19$$

But the sample mean is also equal to $\bar{x} = \sum fx / \sum f = 190/10 = 19$. Hence, to find the mean for sample data displayed in a frequency table, we use the following formula:

$$\bar{x} = \frac{\sum fx}{\sum f} \tag{3-1}$$

Mean For Grouped Data

If we are finding the mean for data that have been displayed in a grouped frequency table, we use the class marks to represent the measurements for each class. Then formula (3-1) can be used to determine the approximate sample mean, $\bar{x}_a$, since the original data are unknown and each observation is represented by its class mark.

EXAMPLE 3-2 The following data represent the number of records sold each day for a 25-day period by a music shop located in a shopping mall:

60	36	61	56	19
49	57	54	59	28
33	67	46	53	30
35	51	42	21	28
63	38	15	24	35

For convenience, the data have been displayed in the following grouped frequency table:

Number of records sold	Number of days
15–25	4
26–36	7
37–47	3
48–58	6
59–69	5

Find each of the following:

    a. $\bar{x}$, the average number of records sold per day
    b. $\bar{x}_a$, the approximate average number of records sold per day

Solution   **a.** With the help of a hand-held calculator, we determine the sum of the 25 measurements to be $\sum x = 1060$. Hence,

$$\bar{x} = \frac{\sum x}{n} = \frac{1060}{25} = 42.40$$

Thus, the average number of records sold per day is 42.40.

b. We first find the class marks, X. (Recall from Chapter 2 that a class mark is the midpoint of a class interval.) Each class mark is then multiplied by its corresponding frequency, as shown in Table 3-2.

Table 3-2 Class Marks Multiplied by Frequencies for Example 3-2

Class	f	X	fX
15–25	4	20	80
26–36	7	31	217
37–47	3	42	126
48–58	6	53	318
59–69	5	64	320

The products from Table 3-2 are now added and the sum is divided by the sum of the frequencies to obtain the approximate sample mean:

$$\bar{x}_a = \frac{\sum fX}{\sum f} = \frac{1061}{25} = 42.44$$

Note that $\bar{x}_a = 42.44$ is only an approximate value for the mean of the 25 original sample measurements; the approximation is considered good, compared with the exact value, $\bar{x} = 42.40$, obtained in part (a). ■

A Disadvantage to Using the Mean

The mean has one serious shortcoming—it is affected by extreme measurements on one end of a distribution. It depends on the value of every measurement, and extreme values can lead to the mean misrepresenting the data. To see why this is so, suppose a marathon runner has run in six of the largest marathon races in the country. He placed (in order of marathon run) in the following positions:

$$3 \quad 5 \quad 4 \quad 6 \quad 2 \quad 85$$

For the last race, in which he placed 85th, he went all out trying to win the race. He ran in first place for the first 22 miles, but developed extreme cramping and had to walk some of the last 4 miles. If the mean is used to describe the runner's ability, then the value 17.5 would be used. Since he finished in no more than sixth place in the first five races, it does not seem reasonable to use the mean to measure his overall running ability. Perhaps the median would provide a better measure. The mean is affected too much by the extreme value 85 in this example.

Median

For data measured on an interval scale, the median is the middle ordered score. For example, the median of the ordered test scores 9, 22, 37, 45, and 57 is 37. In general, the median is found by first ranking the data. If there is an odd number of scores, then the median will be the score in the middle; if there is an even number of scores, the median is the mean of the two scores occupying the middle positions. The median for a population is denoted by $\tilde{\mu}$ and the median for a sample is denoted by $\tilde{x}$.

For example, suppose in their last seven games, the Bobcats scored the following numbers of points:

$$6 \quad 10 \quad 3 \quad 21 \quad 0 \quad 35 \quad 14$$

The median number of points scored is found by first ranking the scores:

$$0 \quad 3 \quad 6 \quad 10 \quad 14 \quad 21 \quad 35$$

The median score is easily seen to be 10, since only one score occupies the middle position. If the Bobcats score 42 points their next game, then the eight ordered scores would form the following sequence:

$$0 \quad 3 \quad 6 \quad 10 \quad 14 \quad 21 \quad 35 \quad 42$$

Since there is now an even number of scores, the values 10 and 14 occupy the middle positions and the median is found to be 12, the average of 10 and 14.

Since the median is the middle value for a distribution, there may not be as many measurements above it as there are below it. As an illustration of why this may be the case, consider the following sample of five values:

$$6 \quad 6 \quad 6 \quad 7 \quad 8$$

The median value of 6 has no values below it and two values above it.

There are both advantages and disadvantages to using the median for interval data. One advantage is that the median is not affected by extreme scores on one end of the distribution. This was the reason for choosing the median to represent the "middle" measurement for the grade data illustrated in Example 3-2 at the beginning of this section. A disadvantage to using the median is that it is not easily found for a large set of data, since the measurements must first be ranked.

For large sets of data that have been organized in a frequency table (where the values of x are ranked), the median is found as follows:

> If n is odd, the median is the measurement with rank $(n + 1)/2$ and if n is even, the median is the average of the measurements with ranks $n/2$ and $(n/2) + 1$.

Notice that $(n + 1)/2$ does not represent one of the measurements, but represents the number of values that must be counted to arrive at the median. For the five ranked values 4, 8, 12, 13, and 14, the measurement with rank $(5 + 1)/2 = 3$ is 12.

EXAMPLE 3-3 Find the median for the sample data organized in Table 3-3, a frequency table representing the number of absences each class period during the 1985 spring term for an introductory philosophy class.

**Table 3-3 Absence Data
for Example 3-3**

Number of absences	Frequency
0	10
1	9
2	7
3	6
4	8

Solution As a consequence of the given rule and the fact that there are 40 measurements involved, the median is the average of the 20th and 21st measurements. Notice that since there is an even number of measurements, two measurements occupy the middle positions. We can count in either direction (from the smallest measurement to the largest measurement, or from the largest measurement to the smallest measurement) to arrive at the median value. In either case, the 20th value is 2. Since the 20th value counting from the smallest measurement is the 21st value counting from the largest measurement, we need only average the 20th and 21st values counting from the smallest value. Thus, the median for the data is 2 absences.   ■

Median For Grouped Data

When data are organized into classes using a grouped frequency table, we shall assume that the values in each class are distributed evenly throughout the class. This assumption will enable us to identify a point in a histogram such that the area of the bars that lie to the left of this point is equal to the area of the bars that lie to the right of this point. Consider the following example.

EXAMPLE 3-4   Table 3-4 represents the speeds (in miles per hour) for a sample of 35 cars traveling through a 25-miles-per-hour school zone. Find the approximate median speed.

Table 3-4 Speed data

Speed	Number of cars	Cum. f
1–5	3	3
6–10	2	5
11–15	5	10
16–20	8	18
21–25	7	25
26–30	10	35

Solution   Since $n = 35$, we want to locate the $n/2 = 35/2 = 17.5$th value. By observing the table, we notice that the 17.5th value (the approximate median) falls in the 16–20 class, because the first three classes contain a total of 10 values and the fourth class contains 8 values. Therefore, we must count $(17.5 - 10) = 7.5$ values into the 16–20 class under the assumption that the 8 values falling in this class are spread evenly throughout the class. In other words, we are seeking the measurement in the 16–20 class that is located a distance of 7.5/8 into the class. Since the width of each class is $w = 5$, to find the approximate median value, $\tilde{x}_a$, we need only add 7.5/8 of the width $w = 5$ to the lower boundary of the fourth class. Thus, the approximate value for the median is

$$\tilde{x}_a = 15.5 + \left(\frac{7.5}{8}\right)(5) = 20.1875 \qquad \blacksquare$$

A histogram for the data of Example 3-4 is given in Figure 3-1. One can easily verify that the total area of the bars below the median value $\tilde{x}_a = 20.1875$ is equal to the total area of the bars above the median value.

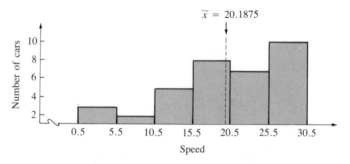

FIGURE 3-1 Histogram for data of Example 3-4

In general, if l is the lower boundary of the class in which the median value falls, f is the frequency of the class containing the median, g is the number of values that remain to be counted when we reach l (counting from the smallest value), and w is the class width, then the approximate median for the data is given by

$$\tilde{x}_a = l + \left(\frac{g}{f}\right)(w)$$

For Example 3-4, $l = 15.5$, $g = 7.5$, $f = 8$, and $w = 5$. Substituting these values into the above expression yields

$$15.5 + \left(\frac{7.5}{8}\right)(5) = 20.1875$$

the same value we obtained in the example.

The method used for finding the approximate median value for data grouped in a frequency table can also be used to find other position points, such as percentile points. A **position point** for a distribution is that value for which a specified portion of the distribution falls below it. The following example illustrates the process.

EXAMPLE 3-5   For the data used in Example 3-4, find P_{60}, the 60th percentile.

Solution   Starting with the first class in Table 3-4, we count 60% of the data—namely, $(.60)(35) = 21$ values. Thus, P_{60} must fall in the class containing the 21st measurement. This class is 21–25. The value P_{60} is located within the interval 21–25 at a distance of 2.14 from the left boundary point of the interval. [This distance was found by multiplying $(21 - 18)/7$ by 5, the width of the class.] Thus, the 60th percentile is

$$P_{60} = 20.5 + 2.14 = 22.64$$

Sixty percent of the data fall below a value of 22.64.   ■

Mode

The **mode**, if it exists, is the most frequent measurement. For example, with the measurements

$$1 \quad 1 \quad 3 \quad 3 \quad 3 \quad 2 \quad 7 \quad 8$$

the mode is 3.

The mode has the advantages of being easily found for small samples and is usually not influenced by extreme measurements on one end of an ordered set of data. This can be seen by noting the following two samples, A and B, each having a mode of 2:

A: 1,   2,   2,   2,   3,   78
B: 1,   2,   2,   2,   3,   8

The extreme measurement of 78 in sample A has no effect on the modal value.

When qualitative data are being analyzed, the mode is frequently considered to be a good measure of central location. Suppose that the blood types for a group of 12 student nurses are A, A, B, A, AB, O, O, B, O, A, B, and AB. The mode, or most frequently occurring blood type, is type A. For this data, it makes no sense to use either the mean or the median to locate a central observation.

The mode has several disadvantages as a measure of central tendency. A major disadvantage to using the mode is that there may be no mode; this situation arises when each measurement occurs with equal frequency. For example, for the measurements

1   2   3   4   5

there is no mode. As another example, the sample

2   2   3   3   4   4   5   5

has no mode. A disadvantage to using the mode with a grouped frequency distribution of data is that the value of the mode often depends on the arbitrary grouping of the data. It is for this reason that a mode for a grouped frequency distribution is often referred to as a **crude mode**.

A set of measurements can have more than one mode. For example, consider the sample of values

1   1   1   2   3   4   4   4

Both 1 and 4 are modes. In a case such as this, the collection of measurements is said to be **bimodal**.

A mode for data displayed in a frequency table is found by locating a largest frequency value, if all of the frequencies are not equal. The value of x that corresponds to the largest frequency value is then taken to be a mode. For Example 3-3, the mode is easily seen to be 0.

If the data are organized in a grouped frequency table, a mode, if it exists, can be easily identified; it corresponds to the class mark for a class having a largest frequency value. And for data displayed in a histogram, a mode is associated with a tallest bar. For the histogram shown in Figure 3-2, the modes are seen to be 20 and 40.

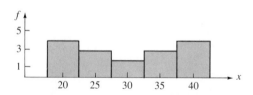

FIGURE 3-2 Histogram displaying two modes

Midrange

The **midrange** of a set of data is the average of the largest and smallest measurements. For data organized in a grouped frequency table, the midrange is approximately the average of the lower class boundary of the first class and the upper class boundary of the last class. Thus, for example, the approximate midrange for the data in Example 3-4 is

$$\frac{0.5 + 30.5}{2} = 15.5$$

EXAMPLE 3-6   Which measure of central tendency should be used to indicate the central salary of all wage earners in the United States?

Solution   The median is the preferred measure. Because of the extreme salaries on the high end of the salary scale, neither the mean nor the midrange should be used. Of course, the proper measure will depend on how the measure is to be used; to indicate our financial status on the international market, we would most likely use the mean value. One reason the modal value is not used is that there is no guarantee that a unique value will exist; there may be no value or a large number of values that occur with greatest frequency.   ■

Skewness

The shape of a histogram depends on the relative positions of the mean, median, and mode. A histogram is **symmetric** if both sides of the histogram determined by the mean are identical. As an illustration of a symmetrical distribution, consider the following data:

x	f
1	1
2	3
3	5
4	7
5	5
6	3
7	1

From the table, we can see that $\bar{x} = 4$, $\tilde{x} = 4$, and mode = 4.

The corresponding frequency histogram is shown in Figure 3-3. We can see that a symmetric histogram has its mean equal to its median. In fact, for the histogram in Figure 3-3, the mean, median, and mode are all identical. But this is not always the case for a symmetric histogram, as indicated in Figure 3-4. Here the mean and median are equal to 8, but the distribution is bimodal, with modes of 7 and 9.

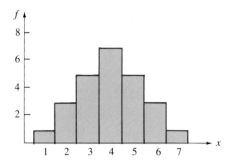

FIGURE 3-3 Symmetric frequency histogram

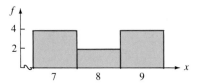

FIGURE 3-4 Symmetric bimodal frequency histogram

A histogram for which the measurements below the mean occur less frequently than the measurements above the mean is said to be **skewed to the left**. As an illustration of a distribution that is skewed to the left, consider the following data:

x	1	2	3	4	5	6	7	8	9
f	2	2	5	5	10	15	20	25	30

The mean is 6.94, the median is 7, and the mode is 9. For this distribution, there are 75 values above $\bar{x} = 6.94$ and 39 values below $\bar{x} = 6.94$. The histogram for this data set is shown in Figure 3-5. Notice that the histogram has a long tail on the left side. As such, a histogram skewed to the left is sometimes referred to as being **negatively skewed**. A negatively skewed histogram will always have its median greater than its mean.

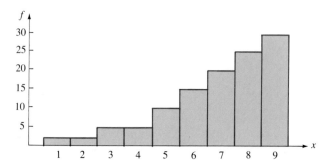

FIGURE 3-5 Distribution that is skewed to the left

A histogram for which the measurements above the mean occur less frequently than the measurements below the mean is said to be **skewed to the right**. As an illustration of a distribution that is skewed to the right, consider the following data:

x	1	2	3	4	5	6	7	8	9
f	30	25	20	15	10	5	5	2	2

The mean is 3.06, the median is 3, and the mode is 1. For this distribution, there are 75 values below the mean $\bar{x} = 3.06$ and 39 values above the mean. The histogram for this set of data is shown in Figure 3-6. Notice that the histogram has a long tail on the right end. Such a histogram is sometimes described as **positively skewed**. A positively skewed histogram will always have its mean greater than its median.

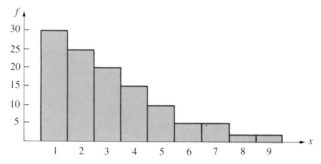

FIGURE 3-6 Distribution that is skewed to the right

Problem Set 3.1

A

1. In an effort to cut down on his coffee consumption, an office worker recorded the following numbers of cups of coffee consumed for a 20-day period:

4   5   3   6   7   1   2   3   0   5

6   5   8   4   0   2   3   7   5   6

Which measure of central tendency is most useful for this purpose? What is its numerical value?

2. A bowler has been bowling regularly for the past 5 years. Her bowling scores for the past 6 games are

201   187   162   234   208   198

For this sample, compute the value of the following statistics (if they exist):
a. Mean   **b.** Median
c. Mode   **d.** Midrange

3. If the mean income for 20 workers is $40,000, what is the total income for the 20 workers?

4. The following table contains the annual salaries (in dollars) for 25 laborers:

Annual salary	Frequency
5,500	7
6,000	5
7,000	6
8,000	4
30,000	3

a. What is the mode?   **b.** What is the mean?
c. What is the median?   **d.** What is the midrange?
e. Determine the skewness.
f. Which measure of central tendency would you use to determine the central value? Explain.

5. If 20 scores have a mean of 15 and 30 scores have a mean of 25, what is the mean of the total group of 50 scores?

6. Suppose 6 is the mean of a sample of four scores.
   a. If 5 is added to each of the scores, what is the mean of the new set of scores? [*Hint:* Try an example.]
   b. If each score is multiplied by 5, what is the mean of the new set of scores?

7. Find the mean, median, mode, and midrange for the sample data in the accompanying grouped frequency table, which shows the ages of a sample of 36 people attending an adult movie.

Class	f
8–13	2
14–19	7
20–25	13
26–31	5
32–37	9

8. The grouped frequency table shown here gives the distribution of rainfall in a certain Maryland county for the month of June over the last 29 years.

Rainfall (in inches)	Number of years
2.0–2.5	3
2.6–3.1	5
3.2–3.7	6
3.8–4.3	8
4.4–4.9	7

   a. Approximate the median amount of rainfall in the month of June.
   b. Find P_{40}, the 40th percentile, and P_{75}, the 75th percentile.
   c. Determine the skewness of the frequency histogram.

9. A teacher gave a standardized test to each of her three classes. From the data she determined the three medians and averaged them to determine the central point of her classes' ability. Can she be misled by doing this? Explain.

10. Which measure of central tendency would you use to select an accurate thermometer for purchase from a local hardware store? Explain.

11. Suppose, as owner of a men's clothing store, you are interested in restocking your shoes for the next year. Which measure of central tendency would you use to determine your shoe order? Explain.

B

12. When averaging percentages, the **geometric mean** $\bar{x}_g$ is often used. The geometric mean is defined by

$$\bar{x}_g = \sqrt[n]{x_1 x_2 x_3 \cdots x_n}$$

where $x_1, x_2, \ldots, x_n$ are positive numbers. Find $\bar{x}_g$ for the following percentages: 95, 125, 140, and 100.

13. The **harmonic mean** $\bar{x}_h$ is often used for averaging rates of travel for equal distances. It is defined as the reciprocal of the average of the data reciprocals. That is, $\bar{x}_h = n / \sum(1/x)$, where the n values of x are positive. Suppose one drives 20 miles at 30 miles per hour and 20 miles at 60 miles per hour. What is the average rate of speed for the 40-mile trip?

14. Would either $\bar{x}$ or $\bar{x}_g$ be appropriate for the data in Problem 12? Explain.

15. A race-car driver wants to average 60 miles per hour (mph) for two laps around a 1-mile track. For the first lap his time was 30 mph due to an electrical problem with the carburetion system. How fast must he travel for the second lap in order to accomplish his goal of 60 mph for the two laps?

16. Suppose a commodity was priced $2 in 1984, $4 in 1985, and $2 in 1986. The percentage change from 1984 to 1985 was 200 and the percentage change from 1985 to 1986 was 50. Find the average percentage change in price for the 3-year period. Justify your answer.

17. If a constant C is added to each measurement in a data set, show that the mean of the new set of measurements is equal to the mean of the original set of measurements plus C.

18. If each measurement in a data set is multiplied by a constant C, show that the mean of the new set of measurements is equal to C times the mean of the original set of measurements.

19. For the data in Problem 7, what is the percentile corresponding to an age of 18 years? This percentage is commonly referred to as the **percentile rank** for 18.

20. Suppose a sample consists of all the even integers between 238 and 874, inclusive. Find the mean and the median of the sample.

21. Two professional baseball players have the career records shown in the following tables. If they are equal in other playing abilities and are negotiating next season's contract, which player should receive the higher salary, based on the better batting average? Explain.

Player A			
Year	**At bat**	**Hits**	**Avg.**
1973	189	57	.302
1974	80	21	.263
1975	212	72	.340
1976	71	17	.239
1977	212	64	.302
1978	97	26	.268
1979	281	89	.317
1980	129	37	.287
1981	151	57	.377
1982	130	34	.262
Total	1552	474	.305

Player B			
Year	**At bat**	**Hits**	**Avg.**
1973	85	27	.318
1974	144	42	.292
1975	53	19	.358
1976	207	52	.251
1977	55	19	.345
1978	263	74	.281
1979	107	35	.327
1980	175	52	.297
1981	75	29	.387
1982	163	45	.276
Total	1327	394	.297

3.2 MEASURES OF DISPERSION OR VARIABILITY

Quite often the measures of central tendency alone do not adequately describe a characteristic being observed. As an example, suppose Dave and Rick each shoot 25 arrows at a target. Their scores are as follows:

	Frequency	
Score	**Dave**	**Rick**
10	2	0
9	3	0
8	4	5
7	7	8
6	2	5
5	1	4
4	1	3
3	1	0
2	2	0
1	2	0

Dave and Rick have the same average score, $\bar{x} = 6.32$. But as Figure 3-7 illustrates, Dave's performance with the bow is certainly different from Rick's performance; Dave's arrows are more spread out than Rick's arrows.

Should coach Wells of the Bobcats send Jones in as a pinch hitter? His average is .310, but in some games he strikes out every time at bat and in other games he gets a hit every time at bat. Or should he put in Smith, who has a batting average of .290 and hits once in every game he is in? The answer appears to be obvious: put Smith in; his batting performance is less variable.

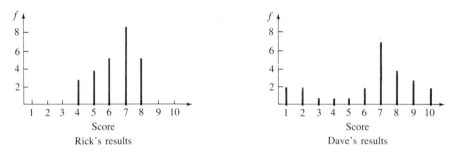

FIGURE 3-7 Variability of scores for Rick and Dave

Any collection of measurements on the same unit will vary with the precision of the measuring instrument. For example, in a box of 24 2-ounce candy bars, not all the bars will weigh exactly 2 ounces. If they do, the scale is not sensitive or precise enough. These same candy bars, weighed on a sensitive analytical scale, will not all weigh the same. The measurements will exhibit a certain degree of variability.

Variability is a very important concept in statistics. As a result, there are many measures of variability or dispersion for a collection of quantitative data. These measures include the range, variance, and standard deviation.

Range

For a distribution of measurements, the **range** is defined to be the difference between the largest measurement U and the smallest measurement L. For example, the ages in years of a family group are

30   2   1   7   4   32   10

The range is

$$R = U - L = 32 - 1 = 31$$

We used the range R in Section 2.2 to determine the width of the intervals for a grouped frequency table. Since the range is easy to determine, it is often useful for estimating other measures of variability, such as the standard deviation, which are not as easy to compute (see Problem 42 at the end of this section).

The range is not always a sensitive measure of dispersion for a collection of data. For the two sets of data pictured on number lines in Figure 3-8, which is more dispersed, A or B? The answer is clearly set A, but notice that both A and

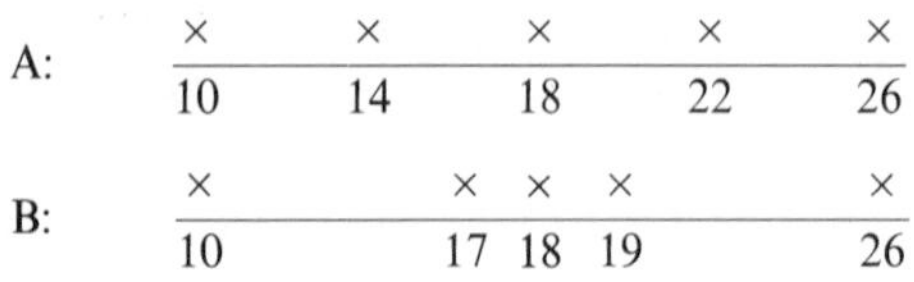

A: ×   ×   ×   ×   ×
      10   14   18   22   26

B: ×   ×   ×   ×   ×
      10   17 18 19   26

FIGURE 3-8 The range as a measure of dispersion

B have the same range. This example illustrates that the range is not a sensitive measure of dispersion. For this reason, it is not an extremely useful measure of dispersion.

Deviation Score

In statistics, the quantity $x - \bar{x}$ is called a **deviation score**. A positive deviation score for a measurement indicates that the measurement is above the mean, while a negative deviation score for a measurement indicates that the measurement is below the mean. A deviation score of 0 for a measurement means that the measurement is equal to the mean.

EXAMPLE 3-7   Compute the deviation scores for the following data representing the number of defects found by an automobile inspector on an assembly line for the last five automobiles produced:

$$1 \quad 4 \quad 6 \quad 6 \quad 8$$

Solution   The sample mean is easily found to be $\bar{x} = 5$. The deviation scores are presented in the following table:

x	$x - \bar{x}$
1	$1 - 5 = -4$
4	$4 - 5 = -1$
6	$6 - 5 = 1$
6	$6 - 5 = 1$
8	$8 - 5 = 3$

We can make the following observations:

a. The measurements 6 and 8 are above the mean and their deviation scores are positive.
b. The measurements 1 and 4 are below the mean and their deviation scores are negative.
c. The sum of the deviation scores is 0.   ■

It can be easily shown that the sum of the deviation scores for any set of numbers is 0. That is,

$$\sum(x - \bar{x}) = 0, \quad \text{for any data set} \tag{3-2}$$

Equation (3-2) has an interesting physical interpretation. The mean of a set of numbers can be geometrically described as the point on the number line that serves as the "center of gravity" for the numbers. If we imagine the number line supported by a point (fulcrum) located at the mean and 1-unit weights placed where the numbers in our sample are located on the line, then equation (3-2) implies that the weights below the mean will perfectly balance the weights above the mean. In other words, the mean serves as the "center of gravity" for the data. Consider the following diagram for the data on automobile defects:

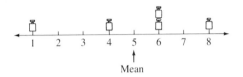

Mean

The number line with a fulcrum at 5 (the mean of the data set) would be perfectly balanced if unit weights were placed at the data values of 1, 4, 6, 6, and 8.

EXAMPLE 3-8   The following data represent the annual amounts (in billions of dollars) that the United States spent for agricultural exports from foreign countries from 1974 to 1983, respectively:

10.2   9.3   11.0   13.4   14.8   16.7   17.4   16.8

15.4   16.2

(*U.S. News and World Report*, Sept. 5, 1983). Find the deviation scores for each amount and verify that equation (3-2) holds for this data set.

Solution   The mean is found to be $\bar{x} = 14.12$. The deviation scores are contained in Table 3-5. By adding the deviation scores, we have

$$\sum(x - \bar{x}) = 0$$

Table 3-5 Data and Deviation Scores for Example 3-8

Year	Amount	Deviation score
1974	10.2	−3.92
1975	9.3	−4.82
1976	11.0	−3.12
1977	13.4	−0.72
1978	14.8	0.68
1979	16.7	2.58
1980	17.4	3.28
1981	16.8	2.68
1982	15.4	1.28
1983	16.2	2.08
		0.00

Sum of Squares

Deviation scores can be used to describe the dispersion of a given distribution of quantitative data. Recall that a deviation score represents the directed distance a measurement is from the mean of a set of data. As a result, we might think the average for all the deviation scores should provide a measure of dispersion of all the measurements about their mean. But this is not the case, since by equation (3-2), the sum of all the deviation scores is 0. The positive deviation scores are canceled by the negative deviation scores when added. To avoid this problem caused by the negative deviation scores canceling the positive ones, we can first square each deviation before adding. The resulting sum of squared deviation scores is called the **sum of squares** and is denoted by SS. As we will see later, SS is very useful in statistics for describing the dispersion of a collection of measurements about their mean.

We can compute a sum of squares for either a sample or a population. The formulas for both are as follows:

$$SS = \sum(x - \bar{x})^2 \qquad SS = \sum(x - \mu)^2 \qquad\qquad (3\text{-}3)$$

$$\text{\textit{Sample}} \qquad\qquad \text{\textit{Population}}$$

The formulas differ, but the computational procedures are the same.

To find the sum of squares (SS) for the sample of values 3, 4, 5, 6, and 7, we first find $\bar{x}$:

$$\bar{x} = \frac{3 + 4 + 5 + 6 + 7}{5} = 5$$

Then by using (3-3), we have

$$
\begin{aligned}
SS &= \sum(x - \bar{x})^2 \\
&= (3 - 5)^2 + (4 - 5)^2 + (5 - 5)^2 + (6 - 5)^2 + (7 - 5)^2 \\
&= 4 + 1 + 0 + 1 + 4 \\
&= 10
\end{aligned}
$$

In general, a sum of squares (SS) can be found as follows:

1. Determine the mean.
2. Find the deviation score for each measurement.
3. Square each of the deviation scores.
4. Find the sum of these squares.

To simplify the computations involved in calculating SS, the following computational formulas will be useful:

$$SS = \sum x^2 - \frac{(\sum x)^2}{n} \qquad SS = \sum x^2 - \frac{(\sum x)^2}{N} \qquad\qquad (3\text{-}4)$$

$$\text{\textit{Sample}} \qquad\qquad\qquad \text{\textit{Population}}$$

where $\sum x^2$ is the sum of the squares of the data, $(\sum x)^2$ is the square of the sum of the data, n is the size of the sample, and N is the size of the population. Note that $\sum x^2 \neq (\sum x)^2$. This fact can be demonstrated by observing that $2^2 + 3^2 \neq (2 + 3)^2$, or $13 \neq 25$. Both of the formulas given by (3-4) can be verified algebraically using the properties of summation found in Appendix A.

The following sample of data represent the scores made by five students on an American History test:

$$62 \qquad 80 \qquad 83 \qquad 72 \qquad 73$$

To compute SS using (3-4), we first organize the computations using the following two-column table:

x	x^2
62	3,844
80	6,400
83	6,889
72	5,184
73	5,329
370	27,646

If we use (3-4) to calculate SS, we get

$$SS = \sum x^2 - \frac{(\sum x)^2}{n} = 27{,}646 - \frac{(370)^2}{5} = 266$$

For purposes of computation, the formulas given by (3-4) are usually preferred to those given by (3-3). For one thing, the formulas of (3-4) are easier to use with a calculator, since there are fewer subtractions involved. Furthermore, the formulas in (3-4) do not require that the mean be found. If formulas (3-3) are used in situations where the mean does not terminate and rounding is involved, the calculations could lead to results lacking in precision. For example, such might be the case if we wanted to find SS for the values 0, 5, and 8. If the mean is rounded to the nearest tenth, then $\bar{x} = \frac{13}{3} \simeq 4.3$. Then by using (3-3) we have

$$SS = \sum (x - \bar{x})^2$$
$$= (0 - 4.3)^2 + (5 - 4.3)^2 + (8 - 4.3)^2$$
$$= 32.670$$

By using (3-4), we have

$$SS = \sum x^2 - \frac{(\sum x)^2}{n}$$
$$= 89 - \frac{(13)^2}{3}$$
$$= 32.667$$

To the nearest thousandth, the two answers differ by .003.

Variance

The **variance** of a population of measurements is defined to be the average of the squared deviation scores and is denoted by σ^2 (read "sigma squared"). The symbol σ is the lowercase Greek letter sigma. The variance of a population is given by:

$$\sigma^2 = \frac{SS}{N} \tag{3-5}$$

The variance of a sample is denoted by s^2 and is defined by the following formula:

$$s^2 = \frac{SS}{n-1} \tag{3-6}$$

In Chapters 8–15, we will use the sample variance s^2 to estimate an unknown population variance σ^2. If we were to compute s^2 by dividing SS by n, instead of $n-1$, we would, on the average, underestimate σ^2. For descriptive purposes only, some statisticians compute the sample variance by dividing SS by n. Of course, for large values of n, there is little difference between the values of SS/n and $SS/(n-1)$.

EXAMPLE 3-9   Suppose the American History test scores given previously (62, 80, 83, 72, and 73) constitute a population. Find the population variance σ^2.

Solution   By using (3-5), we have

$$\sigma^2 = \frac{SS}{N} = \frac{266}{5} = 53.2 \qquad\blacksquare$$

EXAMPLE 3-10   Table 3-6 shows the costs (in U.S. cents) per liter of high octane gasoline for 19 cities throughout the world (*U.S. News and World Report*, June 25, 1984). Determine the sample variance s^2.

Table 3-6 Gasoline Costs for Example 3-10

City	Cost per liter
Amsterdam	57
Brussels	53
Buenos Aires	38
Hong Kong	57
Johannesburg	48
London	56
Madrid	59
Manila	46
Mexico City	25
Montreal	47
Nairobi	57
New York	40
Oslo	65
Paris	58
Rio de Janeiro	42
Rome	76
Singapore	59
Sydney	43
Tokyo	79

Solution   We shall use formula (3-4) to compute SS. Toward this end, we first compute $\sum x$ and $\sum x^2$. With the help of a calculator, we determine that $\sum x = 1005$ and $\sum x^2 = 56{,}171$. Thus, the sum of squares is

$$SS = \sum x^2 - \frac{(\sum x)^2}{n} = 56{,}171 - \frac{(1005)^2}{19} = 3011.7895$$

By now applying (3-6), we obtain

$$s^2 = \frac{SS}{n-1} = \frac{3011.7895}{18} = 167.32$$

The sample variance of the 19 gasoline prices is 167.32 square cents.   ■

Used by itself as a descriptive measure of spread or dispersion, the variance is difficult to interpret, since the units of the variance are the squares of the units of measurement. For the gasoline price-per-liter data in Example 3-10, knowing that $s^2 = 167.32$ square cents has very little, if any, meaning by itself. We know that if the value of the variance is large, then the measurements are widely dispersed, while if the value of the variance is small, there is very little variability in the measurements. And if the variance is 0, all the measurements are equal. This is a consequence of the fact that SS is always greater than or equal to 0 and is equal to 0 only when each measurement is equal to the mean. But if we analyzed two samples of data, A and B, and found that $s_A^2 = 10$ and $s_B^2 = 5$, we would know that the measurements in sample A are more dispersed about their mean than the measurements in sample B are dispersed about their mean. For the most part and for descriptive purposes, the variance is used for comparison purposes as a relative measure of variation.

Standard Deviation

Another measure of dispersion, related to the variance, is the standard deviation. The **standard deviation** is defined to be the positive square root of the variance. The population standard deviation is denoted by σ and the sample standard deviation is denoted by s. Hence, we have the following formulas:

$$s = \sqrt{s^2} = \sqrt{\text{Sample variance}}$$
$$\sigma = \sqrt{\sigma^2} = \sqrt{\text{Population variance}}$$

For the data in Example 3-9, $\sigma = \sqrt{53.2} = 7.29$; for the data in Example 3-10, $s = \sqrt{167.32} = 12.94$.

Why do we need both the variance and standard deviation as measures of dispersion? One answer to this question involves the units of measurement. As we saw in Example 3-10, if the set of data involves measurements in cents, then the unit of variance is square cents, and the unit of standard deviation is cents. Thus, we could evaluate an expression such as $x - s$, but could not evaluate an expression such as $x - s^2$, since in the first case the units match, but in the second case, they do not.

The following two examples illustrate the use of the sample variance and the sample standard deviation for making relative comparisons.

EXAMPLE 3-11   The following data represent the average miles-per-gallon per day for 5 days for two cars driven under similar conditions:

A: 20   25   30   15   35
B: 15   27   25   23   35

a. Find the mean and range of the miles-per-gallon ratings for each car.
b. Which car seems to have obtained more consistent mileage if consistency is determined by examining the variances? Explain.

Solution   **a.** For car A, we have

$$R_A = 35 - 15 = 20$$
$$\bar{x}_A = 25$$

For car B, we have

$$R_B = 35 - 15 = 20$$
$$\bar{x}_B = 25$$

Note that both cars have the same mean and the same range of miles-per-gallon ratings.

b. We calculate the variance for car A, s_A^2:

x	$x - \bar{x}$	$(x - \bar{x})^2$
20	-5	25
25	0	0
30	5	25
15	-10	100
35	10	100
		SS = 250

As a result of (3-6), we have

$$s_A^2 = \frac{SS}{n-1} = \frac{250}{4} = 62.5$$

The variance for the gas mileages for car A is 62.5 square miles. We next calculate the variance for car B, s_B^2:

x	$x - \bar{x}$	$(x - \bar{x})^2$
15	-10	100
27	2	4
25	0	0
23	-2	4
35	10	100
		SS = 208

Applying (3-6), we obtain

$$s_B^2 = \frac{SS}{n-1} = \frac{208}{4} = 52$$

The variance for the gas mileages for car B is 52 square miles. Since the variance for car B is smaller than the variance for car A, car B got more consistent gas mileages. Notice that if we had used the range, we would have concluded that both cars obtained equally consistent gas mileages. ■

EXAMPLE 3-12   The data in Table 3-7 indicate the prices per pound (in U.S. dollars) for pork roast and cheddar cheese in 15 world capitals (*U.S. News and World Report*, Sept. 26, 1983). For which food, roast pork or cheddar cheese, are the world prices less variable (and more stable)?

Table 3-7 Price Data for Example 3-12

World capital	Pork roast (boneless)	Cheddar cheese
Bern	$6.61	$4.00
Bonn	2.38	2.74
Brasilia	1.27	1.08
Buenos Aires	1.36	2.03
Canberra	2.06	2.60
London	1.56	1.81
Madrid	2.33	3.15
Mexico City	1.08	2.29
Ottawa	1.99	3.98
Paris	2.47	2.37
Pretoria	1.95	1.76
Rome	2.46	2.96
Stockholm	5.35	2.54
Tokyo	4.19	2.38
Washington	3.29	2.69

Solution   As a consequence of formulas (3-4) and (3-6), we have

$$s_p^2 = 2.46$$
$$s_c^2 = .60$$

Thus, the world prices of cheddar cheese are more stable than the world prices of roast pork. ■

An Estimate of s

It is of interest to note that for a bell-shaped distribution, we have

$$s \simeq \frac{R}{4} \tag{3-7}$$

where R denotes the range. This relation can be used to check your computations for s and involves very little effort.

EXAMPLE 3-13   For the cheddar cheese data in Example 3-12, estimate s by using (3-7) and check the estimate by computing the value for s.

Solution   The range for the cheddar cheese prices is

$$R = U - L = 4.00 - 1.08 = 2.92$$

As a consequence of (3-7), we have

$$s_c \simeq \frac{R}{4} = \frac{2.92}{4} = .73$$

Since the standard deviation is the square root of the variance, we can use the result of Example 3-12 to obtain

$$s_c^2 = .60$$
$$s_c = \sqrt{.60} = .77$$

Since $R/4 = .73$ is in the same "ball park" as $s_c = .77$, we have little reason to suspect that an error has been committed.   ■

The relation (3-7) can be used as a practical check on the calculation of the standard deviation. Consider the next example.

EXAMPLE 3-14   Suppose that the largest measurement in a sample is 90 and the smallest measurement is 30. The standard deviation has been calculated to be 185. Does this answer seem reasonable? Explain.

Solution   No, the answer does not seem reasonable. The range is $90 - 30 = 60$, and using (3-7) we have

$$s \simeq \frac{R}{4} = \frac{60}{4} = 15$$

Thus, we suspect an error has been made in calculating s to be 185. The calculations should be rechecked.   ■

Variance and Standard Deviation for Data in Frequency Tables

Frequently, we will have occasion to find the variance or standard deviation for data displayed in a frequency table. Both these measures can be calculated once SS is known. To find SS for data that have measurements with repetitions, we first determine the frequency for each measurement. For example, to find SS for the data 2, 2, 2, 2, and 7, representing the number of walks given up by a baseball pitcher in the last five games, we need only find the deviation scores for the measurements 2 and 7. The squared deviation score for 2 can then be multiplied by its frequency, $f = 4$, to get the sum of the squared deviations for the four values of 2. This is then added to the squared deviation score for 7 to arrive at SS. Since the mean of the five data points is 3, we have

$$SS = \sum (x - \bar{x})^2$$
$$= 4(2 - 3)^2 + 1(7 - 3)^2$$
$$= 4 + 16 = 20$$

Based on the ideas of the previous paragraph, we have the following formulas for finding the sum of squares when data are organized in a frequency table:

$$\text{SS} = \sum f(x - \bar{x})^2 \qquad \text{SS} = \sum f(x - \mu)^2 \qquad (3\text{-}8)$$
$$\textit{Sample} \qquad\qquad \textit{Population}$$

EXAMPLE 3-15   The following measurements represent the number of days it took express mail shipped from the west coast to reach its destination on the east coast for the past ten mailings:

$$2 \quad 2 \quad 2 \quad 3 \quad 3 \quad 4 \quad 4 \quad 5 \quad 5 \quad 10$$

Use formulas (3-8) to determine SS.

Solution   We first construct Table 3-8, a frequency table to aid us with our calculations.

Table 3-8 Frequency Table for Example 3-15

x	f	$x - \bar{x}$	$(x - \bar{x})^2$	$f(x - \bar{x})^2$
2	3	-2	4	12
3	2	-1	1	2
4	2	0	0	0
5	2	1	1	2
10	1	6	36	36
				SS = 52

The sample mean is easily found to be $\bar{x} = 4$. The value of SS is the sum of the entries in the last column, SS = 52.

For illustrative purposes, we shall also determine the value of SS by using (3-3). The calculations are presented in Table 3-9. We see that SS = 52, as calculated using (3-8). Note that the first entry, 12, in the fifth column of Table 3-8 corresponds to the sum of the first three entries of 4 listed in the last column of Table 3-9, and so forth.

Table 3-9 Calculation of SS Using (3-3)

x	$x - \bar{x}$	$(x - \bar{x})^2$	
2	-2	4	
2	-2	4	$f = 3$ and $(3)(4) = 12$
2	-2	4	
3	-1	1	
3	-1	1	$f = 2$ and $(2)(1) = 2$
4	0	0	
4	0	0	$f = 2$ and $(2)(0) = 0$
5	1	1	
5	1	1	$f = 2$ and $(2)(1) = 2$
10	6	36	$f = 1$ and $(1)(36) = 36$
		SS = 52	

■

The following computational formula can be used to find the sum of squares for data displayed in a frequency table:

$$SS = \sum fx^2 - \frac{(\sum fx)^2}{\sum f} \qquad\qquad (3\text{-}9)$$

Formula (3-9) is usually more convenient to use than the formulas expressed by (3-8). Notice that only one subtraction is involved with (3-9) and that the sample mean does not need to be calculated first.

EXAMPLE 3-16   Find the sample variance for the following data representing the number of cigars smoked during a particular week by 15 cigar smokers:

x	10	15	17	20	22
f	1	3	5	2	4

Solution   The following table is used to organize the computations:

x	f	fx	fx^2
10	1	10	100
15	3	45	675
17	5	85	1445
20	2	40	800
22	4	88	1936
Sums 15		268	4956

As a consequence of (3-9), we have

$$SS = \sum fx^2 - \frac{(\sum fx)^2}{\sum f} = 4956 - \frac{(268)^2}{15} = 167.73$$

Hence, the sample variance is

$$s^2 = \frac{SS}{n-1} = \frac{167.73}{14} = 11.981 \qquad\blacksquare$$

Note that the entries in the fx^2 column can be found in either of the following two ways:

1. Multiply the corresponding entries in the x and fx columns; or
2. Square the entries in the x column and multiply by the corresponding values of f.

Chebyshev's Theorem

The sample standard deviation s indicates the dispersion of the data about the sample mean. If the data values are clustered closely about the mean, then s is small; if the values are considerably spread about the mean, then s is large. But how shall we determine what values of s are large and what values are small? A theorem named after the Russian mathematician Chebyshev (1821–1894) provides some useful insight as to how the magnitude of the standard deviation of any set of data relates to the concentration of the data about its mean. According to Chebyshev's theorem, the following statement is true for any set of quantitative data:

> The expression $1 - (1/k)^2$, where $k \geq 1$, represents the minimum proportion of the data that will lie within k standard deviations of the mean.

Note that the result of the calculation $1 - (1/k)^2$ is a fraction. Multiplying this fraction by 100 yields the minimum percentage of the data that lie within k standard deviations of the mean. If $k = 2$, then at least $1 - (\frac{1}{2})^2 = \frac{3}{4}$ or 75% of the data must fall within 2 standard deviations of the mean, as illustrated in Figure 3-9. Also, according to Chebyshev's theorem, for $k = 3$, at least $[1 - (\frac{1}{3})^2]100\% = 89\%$ of the data in any sample must fall within 3 standard deviations of its mean, as shown in Figure 3-10.

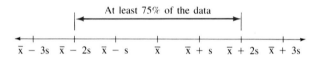

FIGURE 3-9 Illustration of Chebyshev's theorem for $k = 2$

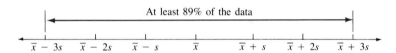

FIGURE 3-10 Illustration of Chebyshev's theorem for $k = 3$

EXAMPLE 3-17   Refer to the gasoline-cost data in Example 3-10.

a. Determine the interval specified by Chebyshev's theorem that will contain at least 75% of the data.

b. What percentage of the data values actually fall within 2 standard deviations of the mean?

Solution   By using a calculator, we easily determine the mean to be $\bar{x} = 52.89$ cents. Earlier we determined the sample variance to be $s^2 = 167.32$. Thus, the standard deviation is $s = \sqrt{167.32} = 12.94$ cents.

a. According to Chebyshev's theorem, at least $1 - \frac{1}{4} = \frac{3}{4} = 75\%$ of the data will lie within 2 standard deviations of the mean. For this data set,

$$\bar{x} - 2s = 52.89 - 2(12.94) = 27.01$$
$$\bar{x} + 2s = 52.89 + 2(12.94) = 78.77$$

Therefore, the interval (27.01, 78.77) will contain at least 75% of the data, as illustrated in the following diagram:

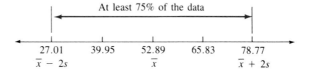

b. Seventeen of the 19 gasoline prices (89.47%) are found to fall between 27.01 and 78.77. This is consistent with our results in part (a). Chebyshev's theorem specifies only a lower bound for the percentage of data that will lie within 2 standard deviations of the mean.   ■

It is sometimes convenient to think of Chebyshev's theorem in different terms. The following statement is equivalent to Chebyshev's theorem:

At most $(1/k^2)100\%$ of the data in any data set will lie beyond k standard deviations of the mean.

For $k = 2$, we have the following diagram:

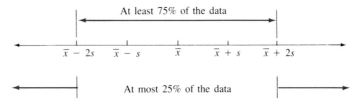

Chebyshev's theorem provides an explanation of how the standard deviation provides a measure of variation for a single sample or population. The validity of the theorem does not depend on the shape of the distribution. As such, it is a very useful and powerful result.

Summary of Notation Used

The following chart summarizes the notation frequently used in connection with samples and populations:

	Mean	Median	Variance	Standard deviation	Size
Sample	$\bar{x}$	$\tilde{x}$	s^2	s	n
Population	μ	$\tilde{\mu}$	σ^2	σ	N

Note that $\bar{x}$, $\tilde{x}$, s^2, s, and n are examples of statistics, while μ, $\tilde{\mu}$, σ^2, σ, and N are examples of parameters. Recall from Chapter 1 that statistics are values computed from a sample and parameters are values computed from a population. It is a popular convention in statistics to use Greek letters to denote most parameters. One exception to this is the notation for population size.

Problem Set 3.2

A

1. What is the sum of the deviation scores about the mean for any data set?

2. Is the value of the standard deviation always smaller than the corresponding value of the variance?

3. Why does the expression $\bar{x} - s^2$ not make sense?

4. What are the range, variance, and standard deviation of the sample 5, 2, 2, 1, 5, 3, 2, 3, and 4?

5. Find the range, variance, and standard deviation of the population 9, 6, 4, 6, 5, 8, 7, 6, 7, and 0.

6. Calculate $\bar{x}$, s^2, and s for each of the following situations:
  a. $\sum x^2 = 232$, $\sum x = 25$, $n = 15$
  b. $\sum x^2 = 515$, $\sum x = 101$, $n = 20$
  c. $\sum x^2 = 52$, $\sum x = 7$, $n = 9$

7. The following values have been found for a sample:

$$\sum x^2 = 428 \qquad \sum x = 75 \qquad n = 10$$

  Are they correct?

8. The following data represent the prices (in cents) for 1 pound of flour in 16 world capitals:

41	28	10	16	35	18	21	5
40	30	25	18	14	30	33	24

  Find the percentage of prices that are within 1 standard deviation of the mean. Now use Chebyshev's theorem for $k = 1$. Are the results consistent with the theorem?

9. Is it possible for the range and standard deviation of a population to be equal? If so, give an example.

10. Is it possible for the range and variance to be equal? If so, give an example.

11. If the standard deviation for a data set is 0, what must be true concerning the data?

12. The following values are the lap times (in minutes) on a 2.5-mile track for two cars, A and B:

A:	1.0	0.9	1.0	0.8	0.9	1.0	0.9	1.0
B:	1.3	1.3	1.0	0.9	1.1	0.9	1.4	1.3

  a. Find the average lap times for cars A and B.
  b. Find the variance of the lap times for cars A and B.
  c. Which car had a lower average lap time?
  d. Which car performed more consistently if consistency is measured by the variance?

13. Suppose one computed the variance of a sample of size 15 by dividing SS by 15 instead of 14, and obtained a result of 10. Find the correct value for s^2.

14. If a calculator has a built-in program for calculating the variance, how could it easily be determined which variance (s^2 or σ^2) it is computing?

15. The accompanying table indicates the annual salaries for a sample of 25 laborers. Find the sample standard deviation.

Annual salary	Frequency
$ 5,500	7
6,000	5
7,000	6
8,000	4
30,000	3

16. The table shows the distribution for the number of defective transistors found in 215 lots produced by an electronics manufacturer. Find the sample standard deviation.

Number of defective transistors	Number of lots
0	25
1	78
2	54
3	33
4	16
5	7
6	2

17. The accompanying grouped frequency table indicates the ages of new-car purchasers at a large automobile dealership. Find the approximate population standard deviation.

Age class	f
28–32	20
33–37	23
38–42	71
43–47	45
48–52	26

18. Verify Chebyshev's theorem for the data in Problem 4 using $k = 2$.

19. The number of patients admitted to Memorial Hospital during a weekday has an average of 32 and a standard deviation of 4. On a particular day, only 16 patients were admitted. Is this an unusual number of admittances for a weekday? Explain.

B

20. For the sample data illustrated by the accompanying line graph, find $\bar{x}$ and s for the number of children per family.

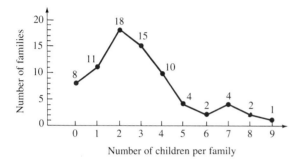

Number of children per family

21. In some situations, data are dichotomous, consisting of only two distinct values. For example, dichotomous data result if responses are recorded as male/female, true/false, up/down, on/off, and so forth. In such cases, it is customary to use 0 to represent one value and 1 to represent the other. If 1, 0, 0, 0, 1, 1, 1, 1, 1, and 0 represent a population of values, find μ and σ for the population of 0s and 1s. If p represents the proportion of 1s, show that $\mu = p$ and $\sigma = \sqrt{p(1 - p)}$.

22. For any finite collection of data, determine the value of c that makes $\sum(x - c)^2$ as small as possible.

23. Consider the following 3 data sets:

    A: 20   30   40   50   60
    B: −20   −10   0   10   20
    C: −2   −1   0   1   2

a. Find SS for each data set. Note that the values for set B were obtained by adding -40 to each measurement in set A and the values in set C were obtained by dividing each measurement in set B by 10.

b. What relationship exists between SS_A and SS_B? Between SS_A and SS_C?

c. What relationship exists between s_A^2 and s_B^2? Between s_A^2 and s_C^2?

d. What relationship exists between s_A and s_B? Between s_A and s_C?

24. If 3 is added to each measurement in a set of ten measurements having a standard deviation of 3, what is the standard deviation for the new data set?

25. Marlene has been following the price per pound of sirloin steak at two supermarkets for a period of 6 months. The information in the accompanying table was compiled. As a smart shopper, at which store is Marlene apt to pay the lower price for sirloin steak? Explain.

	$\bar{x}$	s
Store A	2.43	4.32
Store B	2.35	1.14

26. The average grade on a statistics examination was 75 and the standard deviation was 10. After returning the examination to the students, the professor determined that one question was scored incorrectly and that each grade should be increased by 5 points. Find the mean, variance, and standard deviation for the corrected grades.

27. Consider the population of measurements x:

1.233	1.236	1.230	1.236	1.234
1.237	1.233	1.235	1.238	1.238

Suppose each measurement is transformed using $y = 1000x - 1230$. Find the mean, variance, and standard deviation of the y measurements. In addition, show each of the following:

a. $\mu_y = 1000\mu_x - 1230$. As a result, $\mu_x = (.001)(\mu_y + 1230)$.

b. $\sigma_y^2 = (1000)^2\sigma_x^2$. As a result, $\sigma_x^2 = (.000001)\sigma_y^2$.

c. $\sigma_y = (1000)\sigma_x$. As a result, $\sigma_x = (.001)\sigma_y$.

28. If a constant C is added to each measurement in a data set, show that the variance of the new set of measurements is the same as the variance of the original set.

29. If each measurement in a data set is multiplied by a constant C, show that the sum of squares of the new set is equal to C^2 times the sum of squares of the original set.

30. If each measurement in a data set is multiplied by a constant C, is the standard deviation of the new set equal to C times the standard deviation of the original set?

31. Another measure of dispersion is the **mean absolute deviation**. It is defined by $\text{MAD} = \sum|x - \bar{x}|/n$. Compute the value of MAD for the data in Problem 4.

32. The **coefficient of variation** provides a measure of variability that is independent of the measuring unit. As such, it can be used to compare the variability of two groups of data involving different units of measure. For example, it can be used to compare the standard deviation of the distribution of annual incomes (in dollars) and the standard deviation of the years of service for all the employees of a certain company. The coefficient of variation (CV) expresses the standard deviation as a percentage of the mean and is defined by $\text{CV} = (s/\bar{x})(100)$. Suppose a financial analyst for a stock brokerage firm wants to compare the variation in the price-earnings ratios for a group of common stocks with the variation in their net returns on investment. For the price-earnings ratios, the mean is 9.8 and the standard deviation is 2.4. The mean net return on investments is 20%, and the standard deviation of the net returns on investments is 4.3%. Use the coefficient of variation to compare the relative variation for the price-earnings ratios and the net returns on the investments.

33. Suppose the board of directors of a large corporation wants to compare the dispersion of incomes for its top executives with the dispersion of incomes for its unskilled employees. For a sample of executives, the mean salary is $400,000 and the standard deviation is $50,000, while for a sample of unskilled employees, the mean is $11,000 and the standard deviation is $1200. In which group is the relative dispersion greatest?

34. Can the coefficient of variation be used with data sets involving negative numbers? Explain.

35. The degree of skewness of a distribution is commonly measured by **Pearson's coefficient of skewness**, denoted by CS. For a sample, it is defined by

$$\text{CS} = \frac{3(\bar{x} - \tilde{x})}{s}$$

For a skewed distribution, the sign of CS will correspond to the direction of skewness. A distribution that is symmetric will have $\text{CS} = 0$. The following data represent the starting salaries (in thousands of dollars) of a sample of 1986 college graduates from a large midwestern university:

29.2	27.8	29.0	20.3	16.9	28.7	19.6
24.8	17.4	24.4	20.8	17.8	16.2	17.8

Calculate the coefficient of skewness for the salary data.

36. Find a value for the constant C that minimizes $\sum|x - C|$ for the following sample of measurements: 2, 3, 7, 7, and 8.

37. Prove that $\sum(x - \bar{x})^2 = \sum x^2 - (\sum x)^2/n$.

38. If all the measurements in a population are within 1 standard deviation of the mean, characterize the population; i.e., determine what kinds of numbers comprise the population.

39. Consider the sample of measurements: 1.2, 2, 3, 4, and 4.9. Create another sample of measurements having:
    a. A mean 3 units higher.
    b. A variance 4 times as large.
    c. A mean 3 units higher and a variance 4 times as large.

40. For a population, can the standard deviation ever be larger than one-half the range? Explain.

41. Show that for a sample of two measurements, $s = R/\sqrt{2}$.

42. If s is the standard deviation of a sample, it can be shown that

$$\frac{R}{\sqrt{2(n-1)}} \leq s \leq \left(\frac{R}{2}\right)\sqrt{\frac{n}{n-1}}$$

where n is the sample size and R is the range. The following data represent the blood cholesterol levels for a sample of eight persons:

239	218	227	357	161	286	310	245

a. Find upper and lower bounds for s.
b. Estimate s by using the midpoint of the interval determined by the above result.
c. Calculate the value of s and compare the result with the estimated value found in part (b).

3.3 STANDARD SCORES

Consider the following problem: Bob scores 700 on the mathematics portion of the SAT and Jim scores 24 on the CPT test of mathematical ability. The mean and the standard deviation of the SAT are 500 and 100, respectively, and the mean and standard deviation of the CPT are 18 and 6, respectively. If both tests are assumed to measure the same kind of ability, which person ranks higher? To answer this question, we need some method of comparison that will allow us to compare scores from different distributions. It is clear that the deviation of each score from its mean is not a correct basis for comparison in this case, since Jim's deviation score is

$$x - \bar{x} = 24 - 18 = 6$$

and Bob's deviation score is

$$x - \bar{x} = 700 - 500 = 200$$

Neither deviation score takes into account the spread of the scores.

A measure that allows us to make comparisons from different distributions and takes into account the dispersion of the scores is the standard score. A **standard score** is defined as

$$\text{Standard score} = \frac{\text{Deviation score}}{\text{Standard deviation}}$$

and is usually denoted by z. This relationship can be expressed as

$$z = \frac{x - \mu}{\sigma} \qquad \text{or} \qquad z = \frac{x - \bar{x}}{s} \qquad\qquad (3\text{-}10)$$

$$\textit{Population} \qquad\qquad\qquad \textit{Sample}$$

depending on whether a population or sample is involved.

Since a standard score is defined as a deviation score divided by the standard deviation, it represents the number of standard deviations a score is from the mean. A standard score is sometimes called a **z score**. Jim's standard score or z score in the example described at the beginning of the section is

$$z = \frac{x - \mu}{\sigma}$$

$$= \frac{24 - 18}{6} = \frac{6}{6} = 1$$

and Bob's standard score is

$$z = \frac{x - \mu}{\sigma}$$

$$= \frac{700 - 500}{100} = 2$$

Jim's score of 24 is 1 standard deviation above the mean for the CPT test, and Bob's score of 700 is 2 standard deviations above the mean for the SAT. Since both z scores are positive and Bob's z score is higher than Jim's, Bob ranks higher than Jim in the ability measured by the test.

EXAMPLE 3-18   Suppose a set of scores has a mean of 10 and a standard deviation of 2.

a. Fill in the missing entries in the following chart:

x	4	6	8	10	12	14	16
z							

b. What does a z score of 0 indicate about a score?
c. What does a positive z score indicate about a score?
d. What does a negative z score indicate about a score?
e. Other than indicating whether a score is at, above, or below the mean, what additionally does a z score indicate?

Solution   **a.** By using formula (3-10), we obtain the following z-scores:

x	4	6	8	10	12	14	16
z	-3	-2	-1	0	1	2	3

b. A z score of 0 indicates the score is the mean.
c. A positive z score indicates the score is above the mean.
d. A negative z score indicates the score is below the mean.
e. A z score also indicates the number of standard deviations a score is from the mean.   ∎

EXAMPLE 3-19   If a distribution of numbers resulting from measuring weights of small children has a mean of 20 lb and a standard deviation of 2 lb, what is the unit associated with any z score?

Solution   If x denotes the weight of a child in pounds, then x pounds minus 20 pounds is $(x - 20)$ pounds. Dividing $(x - 20)$ pounds by 2 pounds yields a quotient of $(x - 20)/2$. Thus, a z score has no unit of measure; it represents a number.   ∎

EXAMPLE 3-20   For the data given in Example 3-12 concerning the prices of pork roast and cheddar cheese in 15 world capitals, use z scores to determine which grocery item has the higher relative price in Washington relative to the prices in the other world capitals.

Solution   It can be shown that $\bar{x}_p = \$2.69$ and $\bar{x}_c = \$2.56$. We showed earlier that $s_p = \$1.57$ and $s_c = \$.77$. Since pork roast costs $3.29 per pound in Washington, the z score for pork roast, z_p, is

$$z_p = \frac{x - \bar{x}}{s}$$

$$= \frac{3.29 - 2.69}{1.57} = .38$$

Cheddar cheese costs \$2.69 per pound in Washington. Its z score, z_c, is

$$z_c = \frac{x - \bar{x}}{s}$$

$$= \frac{2.69 - 2.56}{.77} = .17$$

Thus, the price of roast pork is relatively higher in Washington than the price of cheddar cheese.   ■

Suppose μ and σ are the mean and standard deviation, respectively, of a finite population. Each measurement x has a corresponding standard score z. The following important fact helps to characterize the collection of all standard scores for a population:

> The collection or population of all standard scores has a mean of 0 and a standard deviation of 1.

These facts are illustrated in the next example.

EXAMPLE 3-21   **a.** Find μ and σ for the population consisting of the values 1, 2, and 3.
b. Find the three standard scores.
c. Show that the mean of the standard scores is 0 and the standard deviation is 1.

Solution   **a.** The population mean is

$$\mu_x = \frac{1 + 2 + 3}{3} = 2$$

We use (3-5) to obtain the population variance:

$$\sigma_x^2 = \frac{SS}{N}$$

$$= \frac{(1 - 2)^2 + (2 - 2)^2 + (3 - 2)^2}{3}$$

$$= \frac{2}{3}$$

Thus, the standard deviation is

$$\sigma_x = \sqrt{\frac{2}{3}} = .816$$

b. We find the z scores using formula (3-10):

$$\text{For } x = 1, \quad z = \frac{1 - 2}{.816} = -1.225$$

$$\text{For } x = 2, \quad z = \frac{2 - 2}{.816} = 0$$

$$\text{For } x = 3, \quad z = \frac{3 - 2}{.816} = 1.225$$

c. The mean of the z scores is 0. To find SS for the z scores, we organize our computations in the following table and then use formula (3-4):

z	z^2
-1.225	1.50
0	0
1.225	1.50
0	3

Formula (3-4) yields:

$$\text{SS} = \sum z^2 - \frac{(\sum z)^2}{3} = 3 - 0 = 3$$

By using (3-5), we have

$$\sigma_z^2 = \frac{\text{SS}}{N} = \frac{3}{3} = 1$$

Hence,

$$\sigma_z = \sqrt{\text{Variance}} = \sqrt{1} = 1 \qquad \blacksquare$$

Converting z Scores to x Scores

Sometimes we want to convert z scores back to their original or **raw scores**. The following example illustrates the process.

EXAMPLE 3-22   If $\bar{x} = 10$ and $s = 2$, find the raw score x corresponding to a z score of $z = 16$.

Solution   We use the z score formula and solve for x.

$$z = \frac{x - \bar{x}}{s}$$

$$16 = \frac{x - 10}{2}$$

Multiplying both sides by 2, we have

$$32 = x - 10$$

And adding 10 to both sides, we get

$$x = 42 \qquad \blacksquare$$

If the z score formula is solved for x, we obtain the following formula, which can be used for finding the raw score x given a standard score z:

$$x = \sigma z + \mu \qquad\qquad (3\text{-}11)$$

EXAMPLE 3-23 If a population has a mean of 70 and a standard deviation of 5, find the raw score corresponding to a z score of 1.5.

Solution By using formula (3-11), we get

$$
\begin{aligned}
x &= \sigma z + \mu \\
&= (5)(1.5) + 70 \\
&= 7.5 + 70 = 77.5 \qquad \blacksquare
\end{aligned}
$$

Problem Set 3.3 _____

A

1. If $\mu = 47$ and $\sigma = 15$, fill in the missing values in the following table:

x	80	—	60	—	47	—
y	—	1.2	—	-2.37	—	3

2. Consider the following population of data: 4, 8, 12, 16, and 20. Find each of the following:
  a. μ
  b. σ
  c. The z score for each of the raw scores
  d. The mean and standard deviation for the z scores in part (c)

3. Sue scores 625 on exam A, in which $\mu = 600$ and $\sigma = 70$. Mary scores 525 on exam B, in which $\mu = 500$ and $\sigma = 25$. If Sue and Mary both apply for a job and all other factors for both candidates are equal, who should be offered the job, based on these exam scores? Explain your answer.

4. Dave and Rick are training for the Boston Marathon. Dave is training on a course in Cumberland, while Rick is training on a course in Frostburg. The mean time to complete the Cumberland course is 167.4 minutes, and the standard deviation is 25.9 minutes. The mean time to complete the Frostburg course is 143.1 minutes, and the standard deviation is 20.7 minutes. Dave says his course time on the Cumberland course is 91.5 minutes, and Rick says his course time on the Frostburg course is 86.2 minutes. Who do you think will do better in the Boston Marathon? Explain your answer.

5. The means and standard deviations of test scores for five classes are listed here. Suppose you obtain a score of 75. In which class would you have the highest relative standing?
  a. $\mu = 65$, $\sigma = 10$ **b.** $\mu = 70$, $\sigma = 5$
  c. $\mu = 55$, $\sigma = 15$ **d.** $\mu = 75$, $\sigma = 2$
  e. $\mu = 70$, $\sigma = 3$

6. Workers using machine A can produce daily quantities of product C with a mean of 75 and a standard deviation of 5, while workers using machine B can produce daily quantities of product C with a mean of 80 and a standard deviation of 8. Dick produced 83 units on machine A and John produced 92 units on machine B. Which worker produced the higher relative output? Explain.

B

7. Can a score of 5 have a standard score of 3 if it is a member of a population having a mean of 7? Explain.

8. If a score of 13 is a member of a population with a mean of 7 and has a standard score of 3, find the variance of the population.

9. If a score of 10 is a member of a population with a variance of 9 and has a standard score of 5, find the mean of the population.

10. A population has a mean equal to 7 and a variance equal to 1. Find the value of the score that has a standard score equal to twice its value.

CHAPTER SUMMARY _____

In this chapter we introduced the concepts of central tendency and variability. We learned four measures of central tendency: mean, median, mode, and midrange. These measures provide central values for data sets. We learned that the relative positions of the mean, median, and mode in a distribution determine the symmetry or skewness of the distribution. Next, we studied three measures of dispersion or variability: range, variance,

and standard deviation. These measures are used to describe the amount of spread in a data set. Chebyshev's theorem is important in understanding the concept of standard deviation. Finally, standard scores were introduced. These scores express the relative positions of measurements with respect to their mean. Standard scores are useful for making relative comparisons of data from two different populations or samples.

C   H   A   P   T   E   R   R   E   V   I   E   W

IMPORTANT TERMS _____

For each term, provide a definition in your own words. Then check your responses against those given in the chapter.

bimodal	measures of dispersion	position point	standard score
central tendency	median	positively skewed (histogram)	sum of squares
coefficient of variation	midrange		symmetric (histogram)
crude mode	mode	range	variability
deviation score	negatively skewed (histogram)	raw score	variance
geometric mean		standard deviation	z score
harmonic mean	Pearson's coefficient of skewness		
mean			
mean absolute deviation	percentile rank		

IMPORTANT SYMBOLS _____

$\bar{x}$, sample mean	μ, population mean	$\sum$, used to indicate addition	$\bar{x}_a$, approximate sample mean
$\tilde{x}$, sample median	$\tilde{\mu}$, population median	n, sample size	$\tilde{x}_a$, approximate sample median
R, range	σ^2, population variance	N, population size	
s^2, sample variance	σ, population standard deviation	SS, sum of squares	
s, sample standard deviation			

IMPORTANT FORMULAS AND FACTS

1. $\mu = \dfrac{\sum x}{N}$, mean of a population

2. $\bar{x} = \dfrac{\sum x}{n}$, mean of a sample

$\bar{x} = \dfrac{\sum fx}{\sum f}$, mean for grouped data

3. SS $= \sum(x - \mu)^2$, sum of squares for a population

SS $= \sum(x - \bar{x})^2$, sum of squares for a sample

4. SS $= \sum x^2 - \dfrac{(\sum x)^2}{n}$, computational formula (sample)

SS $= \sum x^2 - \dfrac{(\sum x)^2}{N}$, computational formula (population)

SS $= \sum f(x - \bar{x})^2$, sum of squares for data in a frequency table

SS $= \sum fx^2 - \dfrac{(\sum fx)^2}{\sum f}$, computational formula (frequency table)

5. $s^2 = \dfrac{\text{SS}}{n - 1}$, sample variance

6. $s = \sqrt{\text{Sample variance}}$, sample standard deviation

7. $\sigma^2 = \dfrac{\text{SS}}{N}$, population variance

8. $\sigma = \sqrt{\text{Population variance}}$, population standard deviation

9. $z = \dfrac{x - \mu}{\sigma}$, z score for a measurement from a population

$z = \dfrac{x - \bar{x}}{s}$, z score for a measurement from a sample

10. $x = \sigma z + \mu$, relationship between x and z

11. $s \simeq \dfrac{R}{4}$, approximation for s

12. For any finite collection of data, the sum of the deviation scores is 0.

13. A population of z scores has a mean of 0 and a standard deviation of 1.

14. Chebyshev's theorem: At least $[1 - (1/k^2)]100\%$ of any data set falls within k standard deviations of the mean.

REVIEW PROBLEMS

1. Calculate the mean, median, mode, midrange, range, variance, and standard deviation for each of the following populations:
a. 3, 7, 4, 6, 8, 2   **b.** 7, 8, 5, 2, 3
c. 9, 6, 0, 1, 4   **d.** 3, 3, 3

2. Calculate the mean, median, mode, range, variance, and standard deviation for each of the following samples:
a. 4, 7, 2, 2   **b.** 1, 8, 9, 4, 4
c. 0, 0, 1, 1, 10   **d.** 3, 3, 3
e. 8, 14, 15, 16, 22

3. Calculate the z score for x in each of the following situations:
a. $x = 22$, $\mu = 15$, $\sigma = 2$
b. $x = -10$, $\mu = 5$, $\sigma = 8$
c. $x = 0$, $\bar{x} = 12$, $s = 6$
d. $x = 12.5$, $\bar{x} = 22$, $s = .4$
e. $x = 17$, $\bar{x} = 15$, $s^2 = 4$

4. Calculate the mean, median, mode, variance, and standard deviation for the following frequency table of sample data:

x	f
0	1
1	3
2	2
3	4

5. Find the mean, median, mode, variance, and standard deviation for the sample data illustrated by the following line graph.

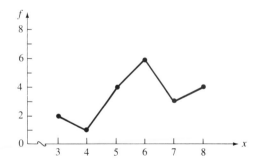

6. The following sample data were collected:

8	8	26	10	8
8	8	18	8	14
20	10	6	14	14

a. Find $\bar{x}$ and s.

b. If a mistake was made in collecting the data and the original measurement of 26 should have been 20, would s increase or decrease? Explain.

c. If a mistake was made in collecting the data and the original measurement of 26 should have been 8, would s increase or decrease? Explain.

7. A calculus class has 30 members. The following test scores are from the students who sit in the first row:

87	83	89	71	95

a. Is this collection of scores a sample or a population?

b. Calculate the mean and standard deviation for the data.

c. Find standard scores for the grades 71 and 95.

8. For each of the following data sets, specify an appropriate measure of central tendency and give its value. Justify your choice in each case.

a. Weight in pounds: 3, 2, 4, 13, 4, 4

b. Rank	Number
Professor	25
Associate professor	24
Assistant professor	13
Instructor	10

c. Party	Number
Democrat	200
Republican	300
Socialist	50
Independent	17

d. Grade	Number
A	2
B	3
C	1

e. Speed	Number
Fast	25
Slow	75

9. The following data represent the monthly charges (in U.S. dollars) for telephone service in 19 world cities: 7.28, 8.54, 15.28, 5.51, 3.17, 6.34, 3.80, 4.59, 5.12, 9.98, 7.04, 10.00, 11.96, 5.48, 2.30, 5.85, 9.39, 8.73, and 7.66 (*U.S. News and World Report*, June 25, 1984). Find the following:

a. $\bar{x}$

b. s

c. The z score for New York's monthly telephone service charge ($x = \$10.00$)

10. Fifty households were polled to determine the number of male inhabitants. The resulting data are listed here:

0	1	2	1	3	0	1	4	0	1
1	1	1	1	1	0	1	3	2	3
1	0	1	2	2	1	1	2	2	1
0	0	0	0	0	0	0	1	1	1
2	1	0	1	1	2	2	0	1	0

a. Find $\bar{x}$.

b. Find s.

c. How many measurements fall within 1 standard deviation of the mean?

11. The following are EPA average miles-per-gallon ratings for 15 1984 compact and subcompact automobiles: 30, 31, 34, 31, 35, 41, 27, 35, 20, 47, 27, 29, 34, 38, and 32. Find $\bar{x}$ and s.

12. The following data represent annual U.S. arms sales (in billions of dollars) to third-world nations from 1976 to 1983: 8.2, 9.8, 10.1, 9.2, 6.4, 6.8, 7.9, and 9.7. Find $\bar{x}$ and s.

CHAPTER ACHIEVEMENT TEST

(30 points) **1.** The final grades for a section of Math 209 are illustrated in the accompanying bar graph.

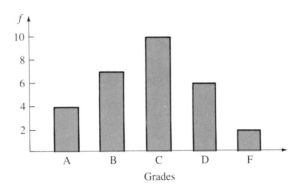

a. Which measure of central tendency should be used to describe the central grade?
b. Using your response for part (a), find the center grade(s).
c. How many students are represented in the graph?
d. What percentage of students received a grade of C?
e. What percentage of students received a grade of C or better?

(35 points) **2.** Consider the following sample: 3, 8, 7, 12, 10. Find each of the following:
a. Range   b. Mean   c. Median
d. Midrange   e. Variance   f. Standard deviation
g. Standard score for the measurement 10

(10 points) **3.** Consider the following frequency table for a population.

x	f
4	2
8	3
3	5

a. Find μ.   b. Find σ.

(15 points) **4.** What can be said about x in relation to the remainder of the data set if x:
a. Has a z score of 0?
b. Has a standard score of 2?
c. Has a z score of -1?

(5 points) **5.** In which of the following situations is the raw score x largest relative to its data set?
a. $x = 37$, $\bar{x} = 20$, $s = 10$
b. $x = 500$, $\bar{x} = 200$, $s = 250$
c. $x = 3.0$, $\bar{x} = 1.0$, $s = 0.7$

(5 points) **6.** If $\mu = 8$ and $\sigma^2 = 4$, find the raw score x corresponding to $z = -2$.

4

ANALYSES OF BIVARIATE DATA

Chapter Objectives

In this chapter you will learn:

- *how to determine the equation of a straight line*
- *what a scattergram is and how it is used*
- *what correlation is*
- *how to determine the correlation coefficient r*
- *the least squares method for determining the prediction equation*

- *how to determine the least squares equation which estimates how two variables are related*
- *how to use the regression equation for predictive purposes*
- *how the correlation coefficient and the slope of the regression line are related*
- *what the sum of squares for error is and how to calculate it*

Chapter Contents

Statistical analyses frequently involve quantitative data that are **bivariate** in nature; that is, for each unit in a sample, there corresponds a pair of measurements. The following are examples of bivariate data:

1. Salaries and ages of teachers in district A
2. Pulse rates and systolic blood pressures for Math 209 students
3. Heights and weights for a group of Cub Scouts
4. Daily rainfall and average daily temperatures for Frostburg for 10 days
5. 1985 spring and fall enrollments of 20 universities

This chapter will deal with graphs of bivariate data, measuring the strength of a linear relationship, and describing linear relationships between two variables. Throughout this chapter we will deal only with linear (**straight-line**) relationships.

We begin this chapter with a discussion of linear equations and their graphs.

4.1 LINEAR EQUATIONS AND THEIR GRAPHS

Coordinate Systems

A coordinate system in two dimensions can be constructed by drawing two perpendicular lines intersecting at a point called the **origin** (see Figure 4-1). The two lines are called **axes**; the horizontal line is called the **x-axis**, and the vertical line is called the **y-axis**. The four parts of the plane determined by the two axes are called **quadrants**, which are labeled I, II, III, and IV, as in Figure 4-1. Directed distances measured on the x-axis to the right of the origin are positive, while directed distances measured to the left of the origin are negative. Directed distances measured on the y-axis up from the origin are positive, and directed distances down from the origin are negative. Directed distances measured on

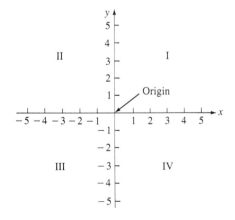

FIGURE 4-1 Coordinate system in two dimensions

the x-axis are called **abscissas,** and directed distances measured on the y-axis are called **ordinates.**

Every point in the coordinate plane has an abscissa and an ordinate. The abscissa at a point indicates the directed distance (left or right) of the point from the y-axis and the ordinate of a point measures the directed distance (up or down) of the point from the x-axis. We indicate these directed distances as an ordered pair (abscissa, ordinate). For example, refer to Figure 4-2. Point P is located in the first quadrant and is at a directed distance of $+2$ from the y-axis, while it is a directed distance of $+3$ from the x-axis; thus, the ordered pair (2, 3) denotes the point P. The point Q is at a directed distance of -1 from the y-axis and a directed distance of $+2$ from the x-axis; thus, the point $(-1, 2)$ uniquely identifies point Q. Similarly, the ordered pair $(-3, -2)$ in quadrant III uniquely identifies point R.

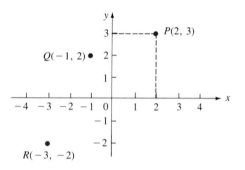

FIGURE 4-2 Illustration of ordered pairs

Graphs of Linear Equations

Coordinates can be used to draw graphs of straight lines. A linear equation $y = a + bx$ can be used to generate a set of ordered pairs. This set of ordered pairs can be plotted (located) on a coordinate system, and the resulting set of points is called the **graph** of $y = a + bx$.

For example, suppose $y = 1 + 2x$. Any pair (c, d) that satisfies $d = 1 + 2c$ is a solution of $y = 1 + 2x$ and can be plotted uniquely in a coordinate system. The point (1, 3) satisfies $y = 1 + 2x$ since $3 = 1 + (2)(1)$. Any number of pairs can be found by letting x take on arbitrary values and solving for the corresponding values for y. For example, let $x = -2, -1, 0, 1,$ and 2. The corresponding values of y are $-3, -1, 1, 3,$ and 5, since

$$-3 = 1 + 2(-2)$$
$$-1 = 1 + 2(-1)$$
$$1 = 1 + 2(0)$$
$$3 = 1 + 2(1)$$
$$5 = 1 + 2(2)$$

We generally express these pairs in tabulated form as follows:

x	y
-2	-3
-1	-1
0	1
1	3
2	5

The following ordered pairs are obtained: $(-2, -3)$, $(-1, -1)$, $(0, 1)$, $(1, 3)$, and $(2, 5)$. One could generate more pairs if desired. Note that the x value is always first in the ordered pair and the y value is always second. By plotting these ordered pairs, we note that they all lie on a straight line (see Figure 4-3). In fact, any pair (a, b) satisfying $b = 1 + 2a$ will lie on this line.

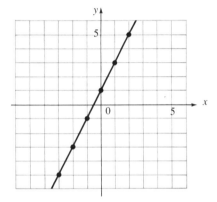

FIGURE 4-3 Graph of $y = 1 + 2x$

Any equation of the form $y = a + bx$ has a graph that is a straight line. This is the reason that the equation $y = a + bx$ is called a **linear equation**.

In order to draw the graph of a linear equation, we need to find only two ordered pairs of solutions to the equation. This is because two points determine a straight line. Often, the points $(0, y)$ and $(x, 0)$ are convenient to find. These points are called **intercepts**; $(0, y)$ is called the **y-intercept** and $(x, 0)$ is called the **x-intercept**.

EXAMPLE 4-1   Find the intercepts for the equation $y = 4 + 2x$ and draw its graph.

Solution   Since $y = 4 + 2x$ is a linear equation, its graph will be a straight line. The x-intercept is found by letting $y = 0$ and solving for x, and the y-intercept is found by letting $x = 0$ and solving for y:

x	y
?	0
0	?

In the equation $y = 4 + 2x$, if $y = 0$, $4 + 2x = 0$. Solving this equation for x, we get $x = -2$. Thus, $(-2, 0)$ is the x-intercept. If we let $x = 0$ in the equation $y = 4 + 2x$, we find that $y = 4$. Hence, $(0, 4)$ is the y-intercept. By plotting these two points on the coordinate system in Figure 4-4 and joining them by a straight line, we have the graph of the equation $y = 4 + 2x$.

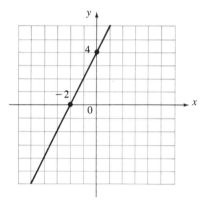

FIGURE 4-4 Graph of $y = 4 + 2x$ ■

Slope

All nonvertical straight lines have a measure of steepness. A line may be horizontal, have an uphill inclination, or have a downhill inclination. The property of steepness or inclination for nonvertical straight lines is called the **slope**. A vertical line has no slope. The slope of a horizontal line is 0, since it has no steepness or inclination. If the line has an uphill inclination, its slope is positive, while if the line has a downhill inclination, its slope is negative (see Figure 4-5).

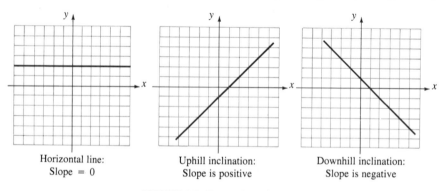

Horizontal line: Uphill inclination: Downhill inclination:
Slope = 0   Slope is positive   Slope is negative

FIGURE 4-5 Illustration of slope

The slope of a line is defined to be the change in y divided by the change in x, and is denoted by b:

$$b = \frac{\text{Change in } y}{\text{Change in } x}$$

Earlier we saw that the graph of $y = 1 + 2x$ is a straight line. What is the slope of this line? Toward answering this question, we choose any two points P and R on the line and compute the slope b. Suppose P is the point $(1, 3)$ and R is the point $(4, 9)$ (see Figure 4-6). We connect P and R by a line segment. The change in y (distance QR) is $9 - 3 = 6$, and the change in x (distance PQ) is $4 - 1 = 3$. Thus, the slope of the line is

$$b = \frac{\text{Change in } y}{\text{Change in } x} = \frac{6}{3} = 2$$

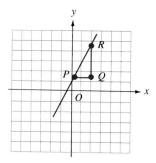

FIGURE 4-6 Slope of the graph of $y = 1 + 2x$

We are now in a position to give interpretations to a and b in the equation of a straight line, $y = a + bx$. From the previous discussion, we note that the line represented by $y = 1 + 2x$ has a slope of 2 ($b = 2$) and 2 appears as the coefficient of x. In general, the slope of the straight line represented by $y = a + bx$ is b. What does the a represent? If $x = 0$, then $y = a$. Thus, $(0, a)$ is the y-intercept. As a matter of convenience, we frequently refer to a as the y-intercept, instead of $(0, a)$.

If (x_1, y_1) and (x_2, y_2) are points on a straight line, the slope of the line joining these two points is given by

$$b = \frac{\text{Change in } y}{\text{Change in } x} = \frac{y_2 - y_1}{x_2 - x_1} = \frac{y_1 - y_2}{x_1 - x_2}$$

If $x_1 = x_2$, the slope is undefined and the line is vertical.

Consider the following examples.

EXAMPLE 4-2   Find the slope of the line through the points $(1, 2)$ and $(4, 6)$.

Solution   We use the formula for calculating the slope:

$$b = \frac{6 - 2}{4 - 1} = \frac{2 - 6}{1 - 4} = \frac{-4}{-3} = \frac{4}{3} \qquad \blacksquare$$

EXAMPLE 4-3   Find the equation of the line having slope 2 and y-intercept 8.

Solution   The slope is $b = 2$ and the y-intercept is $a = 8$. Since the general equation of the line is $y = a + bx$, the equation of the line we seek is

$$y = 8 + 2x \qquad \blacksquare$$

EXAMPLE 4-4   Find the equation of the line having slope 2 and passing through the point (3, 4).

Solution   Note that (3, 4) is not the *y*-intercept. But $y = a + 2x$ must be the equation. We need to find the value of *a*. Since the point (3, 4) is on the line, it must satisfy the equation $y = a + 2x$. Thus,

$$4 = a + 2(3)$$
$$4 - 6 = a$$
$$a = -2$$

Hence, the equation is $y = -2 + 2x$. Note that the slope is 2, and (3, 4) is on this line, since $4 = -2 + (2)(3)$.   ■

EXAMPLE 4-5   Graph the line passing through the point (2, 3) having slope $\frac{2}{3}$.

Solution   We will examine two different solutions.

First solution: Since $\frac{2}{3}$ = (Change in *y*)/(Change in *x*), the point (5, 5) must also be on the line. We therefore draw the line through points (2, 3) and (5, 5) (see Figure 4-7).

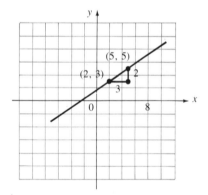

FIGURE 4-7 Graph of line through (2, 3) with slope $\frac{2}{3}$

Second solution: We will find the equation of the line and then plot two points to determine the line. Since $b = \frac{2}{3}$, $y = a + (\frac{2}{3})x$. Also, since the point (2, 3) is on the line, the coordinates of (2, 3) must satisfy the equation. Hence, $3 = a + (\frac{2}{3})(2)$. By solving for *a*, we find that $a = \frac{5}{3}$. The equation of the line is then $y = \frac{5}{3} + (\frac{2}{3})x$. To determine the graph, we find a point different from (2, 3), say (5, 5), and then draw the line containing these two points.   ■

The slope of a straight line has an interesting interpretation. If *x* increases by 1 unit, then *y* must change by *b* units, since b = (Change in *y*)/(Change in *x*). For example, if the equation $c = 20x + 3$ represents the cost in dollars of producing *x* items, the cost of producing an additional item will be $20.

Problem Set 4.1

1. Plot the graphs of the following equations:
  a. $y = 3 + x$ **b.** $y = 1 - 2x$
  c. $y = -2 - 3x$ **d.** $x = -3$
  e. $y = 2$ **f.** $x = 0$

2. Find the slopes of the lines represented by the following equations:
  a. $y = 1 - 2x$ **b.** $y = (\frac{1}{3})(x + 2)$
  c. $y = (3x - 1)/7$ **d.** $2x + 3y = 4$
  e. $x = 2y - 1$ **f.** $y = 2$

3. Graph the line and then find the equation if the line has:
  a. Slope $b = 4$ and goes through the point $(3, 2)$.
  b. Slope $b = -2$ and goes through the point $(-3, -4)$.
  c. Slope $b = -\frac{2}{3}$ and goes through the point $(1, 2)$.
  d. Slope $b = \frac{3}{2}$ and goes through the point $(-1, 2)$.

4. Find the equations of the lines that satisfy the given conditions.

  a. Points $(2, 1)$ and $(4, 2)$ are on the line.
  b. Points $(0, 0)$ and $(1, -2)$ are on the line.
  c. The line is parallel to the line $y = -2x + 1$ and passes through the point $(3, 4)$. [Hint: two lines are parallel if they have the same slope.]
  d. The line is parallel to the line $5x + 2y + 4 = 0$ and passes through the point $(0, 17)$.
  e. The line is parallel to the x-axis and passes through the point $(1, 2)$.
  f. The line is parallel to the y-axis and goes through the point $(-3, 4)$.

5. Find the slope and y-intercept for the lines represented by each of the following equations:
  a. $2x - 4y = 8$ **b.** $y = -2$
  c. $y + 3x + 4 = 0$ **d.** $y = x$
  e. $(x - 1)/(x + y) = 2$

6. If the cost equation is $c = 12x + 10$ and 1000 items are produced, what is the additional cost of producing one more item?

4.2 CORRELATION

One of the main objectives of statistics is to be able to estimate or predict the value of one variable from another variable. **Regression analysis** is a method used to study the relationship between two or more variables and to predict values for one of the variables. In many applications, a linear relationship exists between the variables that can be used for prediction purposes. **Correlation analysis** is a method used by statisticians to determine the strength of the linear relationship (or dependence) that exists between the variables. If the strength of the linear relationship is small, then it is usually not fruitful to use regression analysis to find the linear relationship to use for predictive purposes.

To determine if a linear relationship exists between two variables, a special graph called a scattergram is often used. A **scattergram** is a graph consisting of the points corresponding to all the ordered pairs from a finite sample of bivariate data. If all the points fall exactly on a straight line, then we say the two variables have **perfect linear correlation**. If the points lie close to a straight line, the two variables are said to have a **strong degree of linear correlation**. If the straight line has a positive slope, we say the two variables have **positive linear correlation**, and if the line has a negative slope, we say the variables have **negative linear correlation**. And if the straight line has a slope of 0, we say there is **no linear correlation** between the two variables.

The following application illustrates these ideas. The first world record for the 1-mile run was 4:56, recorded in 1864. Since that time the 1-mile run has been lowered to 3:47.3, and the year 1945 was the last year that the 1-mile record was over 4 minutes. Table 4-1 shows the progress in the world-record times for the 1-mile run from 1945 to 1981. A scattergram for the record data is shown in Figure 4-8.

**Table 4-1 World Records for the
1-Mile Run from 1945–1981**

Year	Country	Time
1945	Sweden	4:01.4
1954	United States	3:59.4
1954	Austria	3:58.0
1957	Great Britain	3:57.2
1958	Australia	3:54.5
1962	New Zealand	3:54.4
1964	New Zealand	3:54.1
1965	France	3:53.6
1966	United States	3:51.3
1967	United States	3:51.1
1975	Tanzania	3:50.0
1975	New Zealand	3:49.4
1981	Great Britain	3:47.3

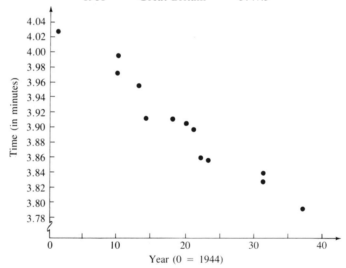

FIGURE 4-8 World-record times for the 1-mile run from
1945–1981

A look at the scattergram in Figure 4-8 suggests that a negative linear correlation exists for the data and that a linear approximation to the data would be reasonable. The correlation is negative since the points in the scattergram appear to lie closest to a straight line having a negative slope. Finding the linear approximation to the points in the scattergram involves regression analysis, which we will explore in Section 4.3. Much speculation has been made by sports enthusiasts about the year of the 3:40 mile. Some field and track experts, using regression analysis, have predicted the year 2000.

As another example, consider the SAT (math) scores and freshman grade point averages (GPAs) for each sophomore enrolled this semester at an eastern college. We might like to obtain answers to the following two questions:

1. Is there a linear relationship between SAT scores and GPAs?
2. If so, what is the relationship?

Answering the first question involves correlation and answering the second question involves regression. Consider the data in Table 4-2, which contains SAT math scores and freshman GPAs for a sample of ten sophomores enrolled at an eastern state college. The data can also be displayed in a scattergram such as that shown in Figure 4-9, where the SAT math scores are displayed on the horizontal axis and the GPAs are displayed on the vertical axis. Note that the correlation between GPAs and SATs appears to be positive, since the points seem to fall closest to a line having a positive slope.

Table 4-2 SAT Math Scores and GPAs for Ten Freshmen

Student number	SAT math score	GPA
1	450	2.5
2	600	3.0
3	550	2.0
4	400	3.0
5	350	2.5
6	650	2.5
7	300	1.5
8	400	2.0
9	700	3.5
10	250	1.0

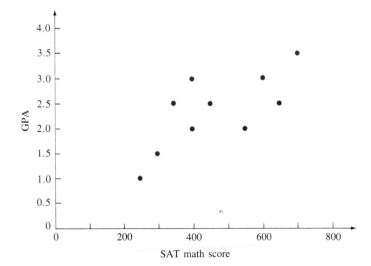

FIGURE 4-9 Scattergram for SATs and GPAs

The linear correlation or strength of linear relationship for the scattergrams displayed in Figures 4-8 and 4-9 can be measured by a special index r, which can be any value from -1 to 1, inclusive. The index r is called the **correlation coefficient**. For example, the scattergrams in Figure 4-10 exhibit perfect linear relationships between x and y. The scattergrams in Figure 4-11 exhibit no linear relationship, while the scattergrams in Figure 4-12 exhibit some linear relationship.

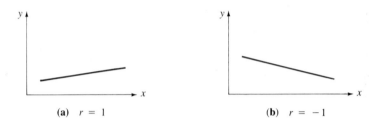

FIGURE 4-10 Scattergrams showing perfect linear relationship

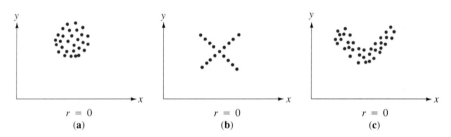

FIGURE 4-11 Scattergrams showing no linear relationship

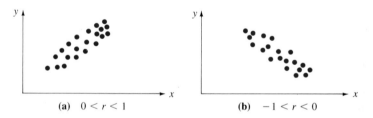

FIGURE 4-12 Scattergrams showing some linear relationship

If the value of r equals 1 or -1, then there is perfect linear correlation between the variables or a perfect linear relationship, while if $r = 0$, there is no linear correlation or relationship. This means that as x tends to increase there is no definite tendency for the values of y to increase or decrease. A value of $r = 0$ does not necessarily mean the lack of a relationship between x and y. A relationship that is nonlinear may exist [see Figure 4-11(c)]. If the points in a scattergram scatter from lower left to upper right, then there is a positive correlation (see Figure 4-9), and if the points scatter from the upper left to the lower right, then there is a negative correlation (see Figure 4-8). In other words, if the values of y tend to increase as the values of x increase, then we say the correlation is positive, whereas if the values of y tend to decrease as the values of x increase, we say the correlation is negative. If the points in a scattergram do not fall on a straight line, it is impossible to ascertain the magnitude of correlation r unless a formula is used.

Correlation Coefficient

By examining a scattergram for bivariate data, one can usually determine whether the correlation is positive or negative. If the points in the scattergram fall on a straight line which is not vertical or horizontal, then the correlation coefficient r must be -1 or $+1$. But if the points do not form a straight line, as is usually the case, a formula is needed to calculate the exact value of the correlation coefficient r. In order to develop such a formula, we need to reexamine the concept of sum of squares.

Recall from Chapter 3 that the sum of squares is found by using the following formula:

$$SS = \sum x^2 - \frac{(\sum x)^2}{n} \tag{4-1}$$

Since we are dealing with bivariate data (x, y), SS_x will denote the sum of squares for x, while SS_y will denote the sum of squares of y. Thus, we have

$$SS_y = \sum y^2 - \frac{(\sum y)^2}{n} \tag{4-2}$$

In order to obtain a dependency relationship between x and y, we need a formula similar to (4-1), but involving both x and y. The relationship between x and y is called the **sum of cross products** and is denoted by SS_{xy}. A formula for computing SS_{xy} is given by

$$SS_{xy} = \sum xy - \frac{(\sum x)(\sum y)}{n} \tag{4-3}$$

Formula (4-3) can be more easily remembered once the observation is made that the expression $\sum xy - (\sum x)(\sum y)/n$ can be obtained from the expression $\sum x^2 - (\sum x)^2/n$ by replacing one occurrence of x with a y, since

$$\sum x^2 - \frac{(\sum x)^2}{n} = \sum xx - \frac{(\sum x)(\sum x)}{n}$$

Replacing an occurrence of x by a y in the right-hand side of the above equation, we get

$$SS_{xy} = \sum xy - \frac{(\sum x)(\sum y)}{n}$$

The following example illustrates the calculations of SS_x, SS_y, and SS_{xy}.

EXAMPLE 4-6   Table 4-3 indicates the amounts (in billions of dollars) for U.S. agricultural imports and exports from 1974 to 1982 (*U.S. News and World Report*, Sept. 5, 1983). Find SS_x, SS_y, and SS_{xy} for the import/export data.

Table 4-3 U.S. Agricultural Imports (in Billions of Dollars) for Years 1974–1982

Year	Import costs	Export prices
1974	21.9	10.2
1975	21.9	9.3
1976	23.0	11.0
1977	23.6	13.4
1978	29.4	14.8
1979	34.7	16.7
1980	41.2	17.4
1981	43.3	16.8
1982	39.1	15.4

Solution   We first determine the column totals in the following table:

x	y	x^2	y^2	xy
21.9	10.2	479.61	104.04	223.38
21.9	9.3	479.61	86.49	203.67
23.0	11.0	529.00	121.00	253.00
23.6	13.4	556.96	179.56	316.24
29.4	14.8	864.36	219.04	435.12
34.7	16.7	1204.09	278.89	579.49
41.2	17.4	1697.44	302.76	716.88
43.3	16.8	1874.89	282.24	727.44
39.1	15.4	1528.81	237.16	602.14
278.1	125.0	9214.77	1811.18	4057.36

1. Find SS_x:

$$SS_x = \sum x^2 - \frac{(\sum x)^2}{n} = 9214.77 - \frac{(278.1)^2}{9} = 621.48$$

2. Find SS_y:

$$SS_y = \sum y^2 - \frac{(\sum y)^2}{n} = 1811.18 - \frac{(125.0)^2}{9} = 75.069$$

3. Find SS_{xy}:

$$SS_{xy} = \sum xy - \frac{(\sum x)(\sum y)}{n} = 4057.36 - \frac{(278.1)(125.0)}{9} = 194.86$$

The correlation coefficient r for bivariate data (x, y) is found by using the formula

$$r = \frac{SS_{xy}}{\sqrt{(SS_x)(SS_y)}} \qquad (4\text{-}4)$$

Formula (4-4) will always produce a value of r that falls between -1 and 1, inclusive. The correlation coefficient r for the bivariate data in Example 4-6 is

$$r = \frac{SS_{xy}}{\sqrt{(SS_x)(SS_y)}}$$

$$= \frac{194.86}{\sqrt{(621.48)(75.069)}}$$

$$= .902$$

This value suggests that as x increases, y increases. In fact, since the value of r is so high, it indicates a strong linear relationship between the dollar amounts of U.S. agricultural imports and exports.

Coding to Simplify Computations of r

Frequently the values of x and y make it extremely cumbersome to calculate SS_x, SS_y, and SS_{xy}. To simplify the computations as much as possible, coding is often used. **Coding** involves using linear transformations with the data. The transformations are of the following type:

$$U = ax + b$$
$$V = cy + d$$

where $a > 0$ and $c > 0$. Then the correlation coefficient between U and V is identically equal to the correlation coefficient between x and y. The following example illustrates the process.

EXAMPLE 4-7   Use coding to calculate the value of r for the bivariate data shown in the following table:

x	168	169	170	171
y	.6	.9	.2	.5

Solution   Let $U = x - 167$ ($a = 1$, $b = -167$), and let $V = 10y$ ($c = 10$, $d = 0$). To find the values for U, we substitute the values for x into the equation $U = x - 167$, and to find the values for V, we substitute the values for y into the equation $V = 10y$. Then the transformed data become the following:

U	1	2	3	4
V	6	9	2	5

According to (4-4), the value of r for the bivariate data (U, V) is $r = -.447$. Thus, the correlation coefficient for x and y is $r = -.447$.   ∎

EXAMPLE 4-8   For the 1-mile world-record data in Table 4-1, find the value of the correlation coefficient r.

Solution   The transformation $x = \text{year} - 1944$ will be used to code the data to simplify computations. The year 1945 will be coded as 1, the year 1946 as 2, and so on. In addition, the times will be expressed in minutes. In Table 4-4, x represents the coded year and y represents the time in minutes.

Table 4-4 Coded Data for Example 4-8

Year	Time	x	y	x^2	y^2	xy
1945	4:01.4	1	4.023	1	16.1845	4.023
1954	3:59.4	10	3.990	100	15.9201	39.900
1954	3:58.0	10	3.967	100	15.7371	39.670
1957	3:57.2	13	3.953	169	15.6262	51.389
1958	3:54.5	14	3.908	196	15.2725	54.712
1962	3:54.4	18	3.907	324	15.2646	70.326
1964	3:54.1	20	3.902	400	15.2256	78.040
1965	3:53.6	21	3.893	441	15.1554	81.753
1966	3:51.3	22	3.855	484	14.8610	84.810
1967	3:51.1	23	3.852	529	14.8379	88.596
1975	3:50.0	31	3.833	961	14.6919	118.823
1975	3:49.4	31	3.823	961	14.6153	118.513
1981	3:47.3	37	3.788	1369	14.3489	140.156
		251	50.694	6035	197.7410	970.711

The following steps are followed in calculating the value of r:

1. Calculate the value of SS_x:

$$SS_x = \sum x^2 - \frac{(\sum x)^2}{n} = 6035 - \frac{(251)^2}{13} = 1188.7692$$

2. Calculate the value of SS_y:

$$SS_y = \sum y^2 - \frac{(\sum y)^2}{n} = 197.7410 - \frac{(50.694)^2}{13} = .0577972$$

3. Calculate the value of SS_{xy}:

$$S_{xy} = \sum xy - \frac{(\sum x)(\sum y)}{n} = 970.711 - \frac{(251)(50.694)}{13} = -8.0731538$$

4. Calculate the value of r:

$$r = \frac{SS_{xy}}{\sqrt{(SS_x)(SS_y)}} = \frac{-8.0731538}{\sqrt{(1188.7692)(.0577972)}} = -.97$$

The computed value of r indicates a strong linear relationship between year and time. Note that the negative value of r means that as time in years increases, the world-record time for the 1-mile run decreases.   ∎

The correlation coefficient r should be interpreted only as a mathematical measure of the strength of the linear relationship between two variables. A high value of r should not be construed to mean necessarily that a cause-and-effect relationship exists between the variables, since both variables could have been influenced by other variables. Remember that a high value of r means that two variables tend simultaneously to vary in the same direction. The tendency is a mathematical phenomenon and does not necessarily imply a direct relationship between the variables. If the number of religious meetings and the number of violent crimes were recorded each month for a group of cities with widely varying populations, the data would probably indicate a high positive correlation. But it would be ridiculous to conclude that the number of violent crimes and the number of religious meetings are directly related. A third variable—namely, population—is causing crime and religious meetings to vary in the same direction. The correlation between crime and religious meetings is an example of **spurious correlation**, correlation caused by a third variable. As a result, we should be cautious about concluding a causation relationship from an observed correlation, since the correlation may be spurious.

Problem Set 4.2

A

1. For the bivariate data shown in the accompanying table, find each of the following:
   a. SS_x **b.** SS_y **c.** SS_{xy}

x	1	5	2	4	8	9
y	3	7	2	6	7	4

2. Would you expect positive correlation, negative correlation, or no correlation for each of the following bivariate data sets?
   a. Shoe sizes and hat sizes
   b. Average adult beer consumption for the 23 counties in Maryland last year and the number of births in the 23 counties in Maryland last year
   c. Average teacher salaries and average SAT mathematics scores for the public school systems in Pennsylvania
   d. Weights of cars and gas mileages
   e. Weights and heights of 6-year-old children
   f. Systolic blood pressures and heart rates for 30-year-old females
   g. Average rainfall in inches and average peach-tree yield in bushels for Clark County, Georgia over the past 10 years
   h. Diameters and areas of circles

3. Consider the following set of bivariate data:

x	0	1	4	-5	2	6
y	-2	-1	2	-7	0	4

   a. Draw a scattergram.
   b. Using the scattergram, determine if the correlation is positive or negative.
   c. Compute the value of r.

4. Consider the following set of bivariate data:

x	0	4	8	1	3	-1
y	2	6	-14	0	-4	4

   a. Draw a scattergram.
   b. Using the scattergram, determine if the correlation is positive or negative.
   c. Compute the value of r.

5. The grades of eight students in Math 101 (x) and English 101 (y) are as shown in the accompanying table.

x	77	81	94	50	72	63	88	95
y	82	47	85	66	65	72	89	95

   a. Draw a scattergram.
   b. Using the scattergram, determine if the correlation is positive or negative.
   c. Compute the value of r.

6. The data in the table represent SAT (math) scores (*x*) and GPAs (*y*) for a group of ten students:

Student	SAT (math), *x*	GPA, *y*
1	450	3.5
2	375	2.5
3	514	2.1
4	678	3.6
5	501	2.7
6	734	3.8
7	325	1.8
8	400	2.4
9	398	2.0
10	681	1.9

a. Draw a scattergram.
b. Using the scattergram, determine if the correlation is positive or negative.
c. Compute the value of *r*.

7. For the data in Example 4-7, verify that the correlation coefficient between *x* and *y* is equal to the correlation coefficient between *U* and *V*.

8. The accompanying data represent the engine sizes (in cubic inches) and estimated miles per gallon for seven 1984 subcompact automobiles. Determine the correlation coefficient *r*.

Car	Engine size	Miles/gallon
Chevette	98	31
Sentra	98	35
Colt	86	41
Isuzu I-Mark	111	27
Mercedes 190D	134	35
Firebird	173	20
VW Rabbit	97	47

9. The data in the accompanying table represent the 1984 price-earnings ratio (PE) and percentage yield for seven stocks. Calculate the correlation coefficient *r*.

PE ratio	2.4	2.4	3.4	2.9	4.0	3.8	2.7
Percentage yield	4.2	.7	10	4.6	6.2	6.3	8.4

10. Consider the following set of bivariate data:

x	1	2	3	4	5	6	7
y	12	7	4	3	4	7	12

a. Draw a scattergram.
b. Compute the value of the correlation coefficient *r*.
c. Discuss the results of parts (a) and (b).

B

11. For the data in Problem 5, show that $SS_{xy} = \sum(x - \bar{x})(y - \bar{y})$.

12. Prove that $SS_{xy} = \sum(x - \bar{x})(y - \bar{y})$.

13. Show that $SS_{xy} = \sum(x - \bar{x})y$.

14. Another measure of the tendency of two variables to increase or decrease together is called **covariance**. For a sample it is denoted by s_{xy} and defined as $s_{xy} = SS_{xy}/(n - 1)$. Find s_{xy} for the data in Problem 6.

15. Show that $r = s_{xy}/(s_x s_y)$, where s_x represents the sample standard deviation of *x* and s_y represents the sample standard deviation of *y*.

16. Let $U = ax + b$ and $V = cy + d$ with *a* and *c* greater than 0. Prove that the correlation coefficient between *U* and *V* is identically equal to the correlation coefficient between *x* and *y*.

4.3 REGRESSION AND PREDICTION

In Section 4.2, we learned how to determine the strength of the linear relationship between two variables by using scattergrams and the correlation coefficient *r*. If the strength of the linear relationship found by using the correlation coefficient *r* is determined to be high, then it may be desirable to describe the linear relationship in terms of an equation. Determining the linear relationship involves the study of regression. As we shall see later, a regression equation can be used for predictive purposes.

Suppose we are interested in studying the relationship between SAT math scores and freshman GPA scores. Further, suppose that after constructing a scattergram for the bivariate data from last year's freshman class and determining the correlation coefficient r, we decide to find the linear relationship, called the **regression equation**. By using the appropriate method (to be presented subsequently), suppose we determined the regression equation to be $\hat{y} = -1.33 + .007x$, where x represents the SAT score in mathematics and $\hat{y}$ (read "y-hat") represents the predicted GPA at the end of the freshman year. This equation can be used for predictive purposes, as the following example illustrates.

EXAMPLE 4-9   Suppose Mary is currently in high school and has applied for admission to college. She took the SAT test and received a score of 480 in mathematics. By using only the regression equation, $\hat{y} = -1.33 + .007x$, how successful would you predict her to be during her freshman year at college?

Solution   To determine her predicted GPA, we substitute $x = 480$ into the regression equation $\hat{y} = -1.33 + .007x$ and solve for $\hat{y}$:

$$\hat{y} = -1.33 + .007x = -1.33 + (.007)(480) = 2.03$$

We would predict Mary's GPA at the end of her freshman year to be 2.03.

■

Suppose the college wanted to adopt an admissions policy whereby no students would be admitted who would not have a predicted GPA of at least 1.25 their freshman year. What would be the college's SAT "cutoff" score, the SAT score above which a student would have a predicted GPA of at least 1.25? This can be found by solving the following equation for x when $\hat{y} = 1.25$:

$$\hat{y} = -1.33 + .007x$$

By substituting 1.25 for $\hat{y}$, we have

$$1.25 = -1.33 + .007x$$

And solving for x, we get

$$2.58 = .007x$$
$$x = 369$$

Thus, the SAT cutoff score would be 369 and any student with an SAT score below 369 would be denied admission because of predicted poor success. Of course, the situation of predicting success is not so simple and many other variables come into play that must be taken into account.

In most practical applications involving bivariate data, the points in a scattergram do not all fall on a straight line. The task then is to identify a straight line that comes closest to all the points in the scattergram, where "close" will be judged by using the squares of vertical distances of the points from a straight line. This line of best fit is represented by the equation $\hat{y} = a + bx$ and is called the **line of best fit** or the **regression line**. For this line, the sum of squares of the vertical distances is as small as possible. The procedure for determining the line of best fit is called the **method of least squares**. Of all the straight lines that can be drawn on a scattergram, this method will identify *the* line that produces the smallest sum of squared deviations of the points in the scattergram from any line contained in the scattergram.

For illustrative purposes, suppose our scattergram has only four points. If we let e_1 represent the distance of the first point from some line represented by $\hat{y} = a + bx$, then the vertical distance corresponds to the error in using the line to predict y using x_1 (see Figure 4-13). Thus, $e_1 = y - \hat{y}$ expresses the error when predicting y for $x = x_1$.

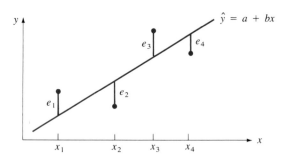

FIGURE 4-13 Vertical distances of points from regression line

If $\hat{y} = a + bx$ is the line of best fit, then by the principle of the method of least squares, $\sum e_i^2$ is a minimum. That is,

$$\sum e_i^2 = e_1^2 + e_2^2 + e_3^2 + e_4^2$$

is a minimum. In general, the sum $\sum e_i^2$ is called the **sum of squares for error** and is denoted by SSE. Thus, we have

$$\text{SSE} = \sum e_i^2 \tag{4-5}$$

It should be pointed out that since $e = y - \hat{y}$, points above the line will have positive errors, points on the line will have 0 errors, and points below the line will have negative errors. If for all the points, the errors are added, a sum of 0 will always result. This is the reason that the errors are squared before they are added.

Consider the following data:

x	3	2	4	1
y	2	3	2	5

The line represented by the equation $\hat{y} = -2 + 2x$ is drawn on the scattergram shown in Figure 4-14.

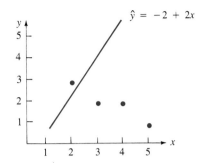

FIGURE 4-14 Scattergram showing line $\hat{y} = -2 + 2x$

The following table will be used to organize the computations:

x	y	$\hat{y}$	$y - \hat{y}$	$(y - \hat{y})^2$
3	2	4	-2	4
2	3	2	1	1
4	2	6	-4	16
1	5	0	5	25
			0	46

The first entry in the $\hat{y}$ column was found by substituting $x = 3$ into the equation $\hat{y} = -2 + 2x$ and solving for $\hat{y}$:

$$\hat{y} = -2 + 2x = -2 + (2)(3) = 4$$

Note that

$$\sum(y - \hat{y}) = -2 + 1 - 4 + 5 = 0$$

and that

$$\text{SSE} = \sum(y - \hat{y})^2 = 4 + 1 + 16 + 25 = 46$$

Thus, for the line represented by $\hat{y} = -2 + 2x$, SSE = 46. If no other line drawn on the scattergram produces a value of SSE smaller than 46, then the line represented by the equation $\hat{y} = -2 + 2x$ is *the* regression line or line of best fit. Clearly, a trial-and-error method for selecting the best line according to the least squares criterion would not be productive. Fortunately, the determination of a and b in the equation $\hat{y} = a + bx$ to minimize SSE can be accomplished by using partial derivatives from calculus, and the details will be omitted. The formulas for finding b and a are as follows:

$$b = \frac{\text{SS}_{xy}}{\text{SS}_x} \qquad (4\text{-}6)$$

$$a = \bar{y} - b\bar{x} \qquad (4\text{-}7)$$

EXAMPLE 4-10   For the data in the previous illustration, find the regression equation and compute SSE.

Solution   The computations are organized using the following table:

x	y	x^2	y^2	xy
3	2	9	4	6
2	3	4	9	6
4	2	16	4	8
1	5	1	25	5
Sums 10	12	30	42	25

From the table, we have

$$SS_x = \sum x^2 - \frac{(\sum x)^2}{n} = 30 - \frac{(10)^2}{4} = 5$$

and

$$SS_{xy} = \sum xy - \frac{(\sum x)(\sum y)}{n} = 25 - \frac{(10)(12)}{4} = -5$$

By (4-6),

$$b = \frac{SS_{xy}}{SS_x} = -\frac{5}{5} = -1$$

and by (4-7)

$$a = \bar{y} - b\bar{x} = \frac{12}{4} - (-1)\left(\frac{10}{4}\right) = 3 + 2.5 = 5.5$$

Thus, the regression equation is $\hat{y} = 5.5 - x$. Its graph appears on the scatter-gram in Figure 4-15.

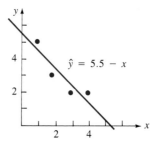

FIGURE 4-15 Scattergram for Example 4-10

To find SSE, we organize our calculations in the following table:

x	y	$\hat{y}$	$y - \hat{y}$	$(y - \hat{y})^2$
3	2	2.5	$-.5$	.25
2	3	3.5	$-.5$	.25
4	2	1.5	.5	.25
1	5	4.5	.5	.25

From the last column of the table, we have:

$$SSE = \sum (y - \hat{y})^2 = .25 + .25 + .25 + .25 = 1$$

For the data used in Example 4-10, of all the lines that can be drawn on the scattergram, the regression line produces the smallest sum of squares for error, which is 1. Any other line will produce a sum of squares for error greater than 1. The line whose equation is $\hat{y} = -2 + 2x$ produced a sum of squares for error of 46.

It is interesting to observe that the point $(\bar{x}, \bar{y})$ is always on the regression line. For the data in Example 4-10, $\bar{x} = 2.5$ and $\bar{y} = 3$, and the point $(2.5, 3)$ satisfies the equation $\hat{y} = 5.5 - x$, since $3 = 5.5 - 2.5$.

For a large set of bivariate data, it would be time-consuming to follow the above method for finding SSE. Instead, the following computational formula can be used to find SSE:

$$SSE = SS_y - bSS_{xy} \tag{4-8}$$

EXAMPLE 4-11   For the data in Example 4-10, find SSE by using formula (4-8).

Solution   By using the first table in Example 4-10, we first compute SS_y:

$$SS_y = \sum y^2 - \frac{(\sum y)^2}{n} = 42 - \frac{(12)^2}{4} = 6$$

By using (4-8), we obtain

$$SSE = SS_y - bSS_{xy} = 6 - (-1)(-5) = 6 - 5 = 1$$

Hence, SSE = 1, as was computed in the previous example.   ■

EXAMPLE 4-12   For the 1-mile world-record data in Example 4-8, find the regression equation and use it to determine the year for which the predicted time for the mile run will be 3:40.

Solution   From Example 4-8, we have

$$SS_x = 1188.7692$$
$$SS_{xy} = -8.0731538$$

Also, the mean of x is

$$\bar{x} = \frac{\sum x}{n} = \frac{251}{13} = 19.307692$$

and the mean of y is

$$\bar{y} = \frac{\sum y}{n} = \frac{50.694}{13} = 3.8995385$$

By applying (4-6), we obtain the slope of the regression line:

$$b = \frac{SS_{xy}}{SS_x} = \frac{-8.0731538}{1188.7692} = -.0067912$$

And by (4-7), the y-intercept is

$$a = \bar{y} - b\bar{x} = 3.8995385 - (-.0067912)(19.307692) = 4.0306609$$

Thus, the regression equation is

$$\hat{y} = a + bx = 4.0306609 - .0067912x$$

By letting $y = 3:40 = 3.667$ minutes and solving for x, we get $x = 53.5$. Since the transformation $x = \text{year} - 1944$ was used to code the years, the predicted year in which the 3:40 mile will occur is

$$\text{Year} = x + 1944$$
$$= 53.5 + 1944$$
$$= 1997.5$$
$$\simeq 1997$$

Thus, the predicted year for the 3:40 mile run is 1997. Note that the year 1997 was not arrived at by using prediction in the normal sense, where x is used to predict y. To do so, we would need to determine the regression equation using the least squares approach to predict y from x (see Problem, 20, page 123).

∎

The regression equation of Example 4-12 raises some interesting questions. What about the 3- or 2-minute mile? One finds it hard to believe that any human will ever run a 2-minute mile. Yet the regression equation will produce a value of $y = 2$ if $x = 299$. By decoding the value of x, we find that the regression equation estimates that a 2-minute mile will be accomplished in the year 2243. One should always be cautious about making predictions far removed from the values of the variable x contained in the sample data. In our example, the coded values for x represent selected years from 1945 to 1981. Only values of x equal to (or close to) these values should be used for predictive purposes.

The Relationship between r and b

The correlation coefficient r and the regression slope b both involve the quantities SS_{xy} and SS_x. As a result, it is possible to solve for one in terms of the other. Using some elementary algebra, it can be shown that the following relationship holds:

$$r = b\frac{s_x}{s_y} \qquad (4\text{-}9)$$

where s_x is the sample standard deviation of x and s_y is the sample standard deviation of y. Since s_x and s_y are greater than 0, the correlation coefficient r agrees in sign with the slope of the regression line. Thus, (4-9) offers another explanation of why the correlation is positive if the points in a scattergram cluster from the lower left to the upper right and negative if the points cluster from upper left to lower right.

By solving (4-9) for b, we have

$$b = \left(\frac{s_y}{s_x}\right)r$$

If $r = 1$, then $b = s_y/s_x$, as illustrated in the following diagram:

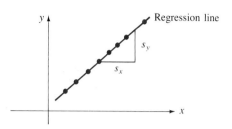

The next example illustrates the use of a computer package to perform a regression analysis which provides values for a, b, and r. The software used was developed for the TRS-80 Model III microcomputer by Radio Shack.

EXAMPLE 4-13   The maximum heart rate and age were recorded for ten individuals on an intensive exercise program. The data are as follows:

Age	10	20	20	25	30	30	30	40	45	50
Heart rate	210	200	195	195	190	180	185	180	170	165

The printout shown in Figure 4-16 contains the appropriate regression analysis. Note that different notation is used: m represents the slope of the regression line, n represents the number of data pairs, B represents the y-intercept, R represents the correlation coefficient, and S.D. represents population standard deviation. The concept of *degrees of freedom* will be discussed later in the text. Computer packages often provide more output than is needed and use different notations.

```
            CORRELATION AND LINEAR REGRESSION
VARIABLE X: AGE                 VARIABLE Y: HEART RATE
MEAN OF X = 30                  MEAN OF Y: 187
S.D. OF X = 11.61895            S.D. OF Y: 13.07670
               NUMBER OF PAIRS (N) = 10
          CORRELATION COEFFICIENT (R) = -.971
               DEGREES OF FREEDOM (DF) = 8
          SLOPE (M) OF REGRESSION LINE = -1.09259
          Y INTERCEPT (B) FOR THE LINE = 219.778
```

FIGURE 4-16 Computer printout for Example 4.13

Computer printouts from commercial software packages often do not contain certain desired information. But many times the required information can be calculated from the information provided by the printout. For example, from the printout in Figure 4-16, determine the regression equation and the sum of squares for error SSE. What peak heart rate would you predict for an age of 28?

Solution   The regression equation is $\hat{y} = 219.778 - 1.09259x$. For age 28, one would predict a peak heart rate of

$$\hat{y} = 219.778 - 1.09259(28)$$
$$= 189.185$$

To find SSE, we shall use (4-8). We first need to find SS_x, SS_y, and SS_{xy}. Since the population variance is defined by $\sigma^2 = SS/N$, to determine the value of SS given σ^2, we multiply σ^2 by N. That is, $SS = N\sigma^2$. Hence,

$$SS_x = N\sigma_x^2 = 10(11.61895)^2 = 1350$$
$$SS_y = N\sigma_y^2 = 10(13.07670)^2 = 1710$$

Since the slope of the regression equation is defined by

$$b = \frac{SS_{xy}}{SS_x}$$

we can solve this equation for SS_{xy} to get

$$SS_{xy} = bSS_x$$
$$= (-1.09259)(1350)$$
$$= -1474.9965$$

Thus, the sum of squares for error is

$$SSE = SS_y - bSS_{xy}$$
$$= 1710 - (-1.09259)(-1474.9965)$$
$$= 98.4336$$ ∎

Problem Set 4.3

A

1. For each of the following equations, find the slope and y-intercept of the line and sketch the graph.
a. $y = 2x - 3$ **b.** $y = x + 2$
c. $2x + 3y = 6$ **d.** $y = -2x$

2. For Problem 1 in Problem Set 4.2, find the regression equation and SSE. The data are reproduced here for convenience.

x	1	5	2	4	8	9
y	3	7	2	6	7	4

3. For Problem 3 in Problem Set 4.2, find the regression equation and SSE. The data are reproduced in the accompanying table.

x	0	1	4	-5	2	6
y	-2	-1	2	-7	0	4

4. The data for Problem 4 in Problem Set 4.2 are given in the table. Find the regression equation and SSE.

x	0	4	8	1	3	-1
y	1	6	-14	0	-4	4

5. The data for Problem 5 in Problem Set 4.2 are reproduced here. Find the regression equation and SSE.

x	77	81	94	50	72	63	88	95
y	82	47	85	66	65	72	89	95

6. The following statistics were obtained when nine pairs of bivariate data were analyzed:

$$\bar{x} = 7.27667 \qquad \bar{y} = 11.2722 \qquad r = .622$$
$$s_x = 2.60702 \qquad s_y = 5.24589 \qquad n = 9$$

Find the regression equation $\hat{y} = a + bx$ and SSE.

7. Consider the following bivariate data:

x	3	6	4	7
y	1	5	6	8

Let $x' = x - \bar{x}$ and $y' = y - \bar{y}$ (recall that x' and y' are the **deviation scores**).
a. Determine the regression equation $\hat{y}' = a + bx'$.
b. Determine $\hat{y}'$ if $x' = 5$.

8. The table shows the number of murders by a handgun in Maryland during each year from 1978 to 1982.

Year	1978	1979	1980	1981	1982
Number of murders	156	193	187	186	200

Find the regression equation and use it to predict the number of handgun murders in Maryland for 1983. [Hint: code the years using $x = $ year $- 1977$.]

9. The accompanying table lists the 400-meter freestyle Olympic swimming times (in seconds) for women since 1924. Find the regression equation and use it to predict the women's time for 1988. [Hint: code the years using $x = $ year $- 1923$.]

Year	Time	Year	Time
1924	362.2	1960	290.6
1928	342.8	1964	283.3
1932	328.5	1968	271.8
1936	326.4	1972	259.04
1948	317.8	1976	249.89
1952	312.1	1980	248.76
1956	294.6	1984	247.10

10. The table lists the 400-meter freestyle Olympic swimming times (in seconds) for men since 1924. Find the regression equation and use it to predict the men's time for 1988. [Hint: code the years using $x = $ year $- 1923$.]

Year	Time	Year	Time
1924	304.2	1960	258.3
1928	301.6	1964	252.2
1932	288.4	1968	249.0
1936	284.5	1972	240.27
1948	281.0	1976	231.93
1952	270.7	1980	231.31
1956	267.3	1984	231.23

B

11. Using the regression equations found in Problems 9 and 10, determine the year in which the women's predicted time will equal the men's predicted time for the 400-meter freestyle race. Discuss your results.

12. For the data in Problem 7, find the z scores z_x for x and the z scores z_y for y. Then determine the regression equation $z_y = a + bz_x$ and find r.

13. Refer to the accompanying computer printout.
a. Find the equation for the regression line.
b. Find SSE.

```
CORRELATION AND LINEAR REGRESSION
VARIABLE X: X          VARIABLE Y: Y
MEAN OF X = 20.13      MEAN OF Y = 10.31
S.D. OF X = 4.5927     S.D. OF Y = 8.33385
        NUMBER OF PAIRS (N) = 10
    CORRELATION COEFFICIENT (R) = .984
        DEGREES OF FREEDOM (DF) = 8
SLOPE (M) OF REGRESSION LINE = .333427
Y INTERCEPT (B) FOR THE LINE = 3.59811
```

14. Prove formula (4-9).

15. Show that $\text{SSE} = \text{SS}_y - b^2 \text{SS}_x$.

16. A measure of how the points in a scattergram are scattered about the regression line is the **standard error of estimate** s_e. It is defined by $s_e = \sqrt{\text{SSE}/(n-2)}$. Find s_e for the data in Problem 7.

17. Show that $\sum(y - \hat{y}) = 0$ where $\hat{y} = a + bx$.

18. Show that the point $(\bar{x}, \bar{y})$ is on the regression line.

19. Show that the regression equation can be written as $\hat{y} = \bar{y} + b(x - \bar{x})$.

20. For the 1-mile world record data of Example 4-8, find the regression equation to predict years given world record times. Use this equation to predict the year when the world record time for the mile run will be 3:40. Which year—1997 or the year obtained here—would you predict for a world record time of 3:40?

21. In 1985, Steve Cram of Great Britain set a new record time of 3:46.31 for the 1-mile run. Use this information and the 13 data pairs of Example 4-8 to:
a. Find a new regression equation.
b. Calculate the value of the correlation coefficient.
c. Determine the year of the 3:40 mile using the equation of part (a).

CHAPTER SUMMARY

The concepts of linear equations, linear regression, and linear correlation were introduced in this chapter. To determine if a linear relationship exists between two variables, a scattergram is often used. We saw that the strength of linear relationship can be measured by the correlation coefficient, r. The values of r can fall anywhere in the interval extending from -1 to 1, inclusive. If the points in the scattergram all fall on a straight line, the value of r is either -1 or 1, depending on whether the line has a positive or negative slope. A value of $r = 0$ indicates the lack of a linear relationship. We also learned that if the strength of linear relationship is determined to be strong, the equation of this linear relationship can be found by using the least squares method. The resulting equation is referred to as the regression equation. The sum of squares for error, SSE, is minimized when using this equation for prediction purposes.

C H A P T E R   R E V I E W

IMPORTANT TERMS

For each term, provide a definition in your own words. Then check your responses against those given in the chapter.

abscissa	least squares method	regression line	sum of squares for error
bivariate data	linear correlation	scattergram	x-axis
coding	linear equation	slope	y-axis
correlation analysis	ordinate	spurious correlation	x-intercept
correlation coefficient	origin	standard error of estimate	y-intercept
covariance	quadrant		
deviation score	regression analysis	straight line	
graph	regression equation	sum of cross products	

IMPORTANT SYMBOLS

r, correlation coefficient

$e = y - \hat{y}$, prediction error

SS_{xy}, sum of cross products

$\hat{y}$, predicted value of y

SSE, sum of squares for error

s_x, standard deviation of x

s_y, standard deviation of y

a, y-intercept

b, slope

IMPORTANT FACTS AND FORMULAS

1. $\text{SS}_{xy} = \sum xy - \dfrac{(\sum x)(\sum y)}{n}$, sum of cross products

2. $y = a + bx$, equation of a straight line

3. $r = \dfrac{\text{SS}_{xy}}{\sqrt{\text{SS}_x \text{SS}_y}}$, correlation coefficient

4. $b = \dfrac{y_1 - y_2}{x_1 - x_2}$, slope of a straight line

$= \dfrac{y_2 - y_1}{x_2 - x_1}$

5. $b = \dfrac{\text{SS}_{xy}}{\text{SS}_x}$, slope of regression equation

$a = \bar{y} - b\bar{x}$, y-intercept of regression equation

6. $\hat{y} = a + bx$, regression equation

7. $\text{SSE} = \sum(y - \hat{y})^2$, sum of squares for error

8. $\text{SSE} = \text{SS}_y - b\text{SS}_{xy}$, computational formula for SSE

9. $r = b\left(\dfrac{s_x}{s_y}\right)$, relationship between r and b

REVIEW PROBLEMS

1. In an attempt to determine the relationship between the amount spent on campaigning and the number of votes received during an election, the data in the accompanying table were collected.

Amount spent, x (in $1000)	3	4	2	5	1
Votes received, y (in 1000s)	14	12	5	20	4

a. Plot a scattergram.
b. Calculate the value of r.
c. Find the regression equation and SSE.
d. Predict the number of votes received if $3500 is spent on the campaign.
e. For each additional $1000 spent, how many additional votes can be expected?
f. Draw a graph of the regression line on the scattergram.

2. In order to study the relationship between the number of times students cut classes and their final course grades, a Mathematics 209 instructor obtained the data shown here.

Student	Number of classes cut	Grade
1	1	98
2	2	98
3	3	88
4	3	81
5	4	83
6	4	76
7	5	71
8	6	71
9	2	85
10	0	98

a. Plot a scattergram.
b. Calculate the value of r.
c. Find the regression equation and SSE.
d. Predict the final course grade if a student has cut 3 classes.
e. For each additional class cut, how much will the predicted final grade be affected?
f. Draw a graph of the regression line on the scattergram.

3. The number of students offered admission (x) and the number of actual first-time enrollments (y) for the past 7 years at a college are listed in the table.

Year	Students offered admission, x	First-time enrollments, y
1	3300	3000
2	4100	3500
3	5600	4200
4	5200	4800
5	5900	5000
6	5500	5100
7	5100	4700

a. Plot a scattergram.
b. Calculate r.
c. Find the regression equation and SSE.
d. Find $\hat{y}$ if $x = 5000$.
e. How many additional enrollments can be expected for every 1000 increase in offers made?
f. Draw a graph of the regression line on the scattergram.

4. A biometrician studied the effects of different doses (x) of a new drug on the pulse rate (y) of humans. The results for five individuals are as indicated in the table.

Dose, x	2.5	3	3.5	4	4.5
Drop in rate, y	8	11	9	16	19

a. Calculate r and interpret the result.
b. Find the regression equation.
c. Find SSE.
d. Find $\hat{y}$ if $x = 3.75$.
e. For each unit increase in dose, what is the predicted drop in pulse rate?
f. Draw a graph of the regression line on a scattergram.

5. The following information was determined in a regression analysis:

$$\hat{y} = 25.1875x - 878.8583$$
$$s_y = 278.5247 \qquad s_x = 9.3956$$
$$\bar{x} = 51.5 \qquad \bar{y} = 418.3 \qquad n = 10$$

a. Show that $(\bar{x}, \bar{y})$ lies on the regression line.
b. Find r.
c. Find SSE.
d. If $x = 45$, find $\hat{y}$.
e. For each unit increase in x, how much will $\hat{y}$ change?

6. The accompanying data represent the annual arms sales (in billions of dollars) by the United States to third-world nations.

Year	Sales
1976	8.2
1977	9.8
1978	10.1
1979	9.2
1980	6.4
1981	6.8
1982	7.9
1983	9.7

a. Find the equation for predicting sales.
b. By using the regression equation found in part (a), estimate the U.S. sales to third-world nations for the year 1984. [Hint: code the years using $x = \text{year} - 1975$.]

7. Nine goldfish were acclimated to a water temperature of $3°$ C and then subjected to gradually increasing water temperatures to determine if metabolic rate is related to temperature. Metabolic rate was determined by counting opercular beats per minute. The resulting data are listed in the accompanying table.

Temperature °C	Mean number of opercular beats/minute
5.0	33.0
7.5	44.8
10.0	54.0
12.5	52.5
15.0	70.2
17.5	99.8
20.0	110.5
22.5	117.0
25.0	129.1

a. Draw a scattergram.
b. Compute r.
c. Find the regression equation.

d. Determine SSE.
e. If the temperature was $0°$ C, how many opercular beats per minute would be expected?

The following data are for Problems 8 and 9.

Distance to nearest tree (feet)	Ht (feet)	DBH (inches)
1.0	39	5.7
19.5	72	9.2
4.0	69	9.3
5.0	67	9.5
10.5	73	9.5
9.0	78	9.7
8.0	79	9.8
15.0	81	10.4
12.5	65	10.7
18.0	60	11.7

8. A forester wants to determine the correlation between total height (ht) and diameter at breast height (DBH) of a sample of quaking aspen. For the tabled data, determine each of the following:
a. The value of r
b. The regression equation
c. SSE

9. A forester wants to determine the correlation between growth (measured as diameter at breast height) and distance to nearest tree for a sample of quaking aspen. Refer to the tabled data.
a. Draw a scattergram.
b. Compute the value of r.
c. Find the regression equation for predicting DBH.
d. Find SSE.

10. To determine if traffic flow, measured in number of vehicles per hour, and lead content contained in vegetation growing near highways are related, a study was undertaken at six different locations. The following data were obtained:

Number of vehicles	103	216	294	402	416	573
Lead content	4.6	7.4	26.1	37.2	24.8	38.7

Find the regression equation and use it to predict the lead content of vegetation experiencing a traffic flow of 300 vehicles per hour.

CHAPTER ACHIEVEMENT TEST

(20 points) **1.** The number of push-ups $\hat{y}$ a normal, healthy child should be able to perform, based on age x, is given by $\hat{y} = 1.4x - 0.9$, where $4 \le x \le 17$.
  a. A 10-year-old child should be expected to do how many push-ups?
  b. As age increases, will the number of push-ups increase or decrease?
  c. For each additional 1-year increase in age, a child is expected to do how many more push-ups?
  d. Does the y-intercept have a meaningful interpretation in this case? Why or why not?

(55 points) **2.** A study of the relationship between height (in inches) and weight (in pounds) of college males yielded the data given here:

Height, x	64	72	73	68	66	67
Weight, y	165	158	173	125	125	139

Determine the following:
  a. SS_x **b.** SS_y
  c. SS_{xy} **d.** b
  e. a **f.** The regression equation
  g. The correlation coefficient   **h.** SSE
  i. $\hat{y}$ when $x = 65$
  j. The error e for the male in the sample whose height is $x = 72$ inches

(10 points) **3.** There is a high positive correlation between teachers' salaries and annual consumption of beer in the United States.
  a. Does this mean that increased beer consumption has been caused by increasing teachers' salaries? Or, the more a teacher drinks, the more he or she will get paid? Explain.
  b. What additional factor(s) might cause teachers' salaries and beer consumption to increase together?

(5 points) **4.** Since $\hat{y} = bx + a$ and $a = \bar{y} - b\bar{x}$, we have $\hat{y} = bx + \bar{y} - b\bar{x}$. Thus, $\hat{y} - \bar{y} = b(x - \bar{x})$. With the regression equation written in this form, show that the point $(\bar{x}, \bar{y})$ is on the regression line.

(10 points) **5.** Find SSE for the following bivariate data (x, y):

x	1	3	5
y	2	4	0

5

INTRODUCTION TO ELEMENTARY PROBABILITY

Chapter Objectives

In this chapter you will learn:

- *what an experiment is*
- *what a sample space is*
- *what events are*
- *what a compound event is and how to form compound events*
- *what complementary events are*
- *how to use Venn diagrams to represent events*
- *what mutually exclusive events are*
- *what probability is*
- *the methods for assigning probabilities*

- *what mathematical odds are and how to calculate odds*
- *the fundamental theorem of counting*
- *how to use the fundamental theorem of counting to find probabilities*
- *some elementary rules of probability*
- *what independent events are*
- *what random variables are*

  how to find the mean and variance of a random variable

Chapter Contents

The purpose of this chapter is to develop the basic ideas that will be needed for an adequate understanding of inferential statistics. **Inferential statistics** is a body of knowledge that treats the methods for making probability statements about unknown population characteristics. For example, we could address the following questions:

1. What is the *chance* that Ajax Company will close down?
2. What is the *average* weight of 1-month-old babies?
3. Is brand A *better* than brand B?

Every day we are faced with decision-making and probability statements. Statements involving the words *chance, likelihood, odds, likely, expected, possible, uncertain*, and *probably* are all addressing the same issue: uncertainty. Every day we make or hear statements similar to the following:

1. What is the *likelihood* we will have a test today?
2. The *odds* are 1 in 2 million that you will be struck by lightning.
3. The job will *likely* be completed on time.
4. The *chance* for rain today is 50%.
5. If a coin is tossed, there is a 50–50 *possibility* that a head will occur.
6. What is the *probability* that the new proposed method will lead to better results?
7. I am *confident* that I can pass this course.

We already have a good intuitive feeling for probability. We will build on this basis to explore some of the not-so-obvious properties of probability theory. This will help us develop a better understanding of inferential statistics.

Probability provides the foundation for developing the science of inferential statistics. Using probability theory, we can deduce the likelihood of certain samples occurring with specified properties. Such information will enable us to draw inferences about the population.

We begin our discussion of probability with a discussion of experiments.

5.1 EXPERIMENTS AND EVENTS

An **experiment** is any planned process that results in observations being made or data being collected. A very simple example of an experiment is tossing a six-sided die and observing the number showing face up when the die comes to rest. For this experiment, there could be six possible outcomes recorded: 1, 2, 3, 4, 5, or 6. Since the number showing could be even or odd, the results of this experiment could also be recorded using just two outcomes: an even number or an odd number. As a result, note that what is called an **outcome** depends on how we view the experimental results. As another example of an experiment, suppose a finished product just off the assembly line is selected and we want to determine if it is defective or not. This experiment has two outcomes: the product is defective or it is not defective. In both of these situations, we can repeat the experiments many times. For our purposes, we will mainly be interested in experiments that can be repeated or be conceived as being repeatable. Experiments such as observing whether it will rain tomorrow or determining who will win the World Series in baseball next year are not repeatable; we will not consider such experiments in this section. Frequently we will be interested in the outcomes resulting from repeating an experiment a given number of times.

The outcomes of most experiments are uncertain and depend on chance. A collection of all possible outcomes for an experiment is called a **sample space** and is denoted by S. The simplest experiment involving uncertainty is one that has two outcomes. Observing the sex of the next baby born at Memorial Hospital is an experiment with two outcomes. A sample space for this experiment consists of the set $S = \{M, F\}$, where M represents a male, F represents a female, and the braces are used to indicate a collection or set. If we observed the births of the next two babies born at Memorial Hospital, then a sample space for the experiment could be the set $S_1 = \{MM, MF, FM, FF\}$, where, for example, FM indicates that the first baby born was a female and the second baby born was a male. Another sample space for this experiment might consist of the number of possible male births, $S_2 = \{0, 1, 2\}$. Notice that these two sample spaces for the experiment provide different amounts of information. If we know which outcome in S_1 occurs, then we know the outcome in S_2. But knowing which outcome occurs in S_2 does not necessarily help us to determine which outcome occurs in S_1. For the experiment of tossing a six-sided die, $S_1 = \{1, 2, 3, 4, 5, 6\}$ is a sample space for the experiment. The set $S_2 = \{$even number, odd number$\}$ is also a sample space. Sample space S_1 contains more information than sample space S_2. An experiment can have more than one sample space; that is, more than one sample space can be used to describe the outcomes of an experiment. It is usually desirable to choose a sample space that provides the most information concerning the experiment.

Consider the following example.

EXAMPLE 5-1   List a sample space for each of the following experiments:

    a. Toss a dime and a quarter, in that order.
    b. Toss a penny, a nickel, and a dime, in that order.
    c. Select a female college student at random and ask her how old she is in years.
    d. Toss a coin first, followed by a six-sided die.
    e. Toss a coin until a head occurs.

Solution   **a.** $S = \{HT, TH, TT, HH\}$. The outcome HT means the dime lands heads-up and the quarter lands tails-up.
    b. $S = \{HHH, HHT, HTH, HTT, THH, TTH, THT, TTT\}$. The outcome HTH means that the penny shows a head, the nickel shows a tail, and the dime shows a head.
    c. $S = \{10, 11, 12, 13, \ldots, 98, 99, 100\}$
    d. $S = \{H1, H2, H3, H4, H5, H6, T1, T2, T3, T4, T5, T6\}$
    e. $S = \{H, TH, TTH, TTTH, TTTTH, \ldots\}$. The outcome TTTH means a head was obtained on the fourth toss. ∎

Events

For a given experiment we may be interested in determining the likelihood of a collection of outcomes occurring instead of the likelihood of a single outcome happening. For example, when three coins are tossed once, we might be interested in the outcomes which indicate that at least two heads are obtained. This collection of outcomes, {HHT, HTH, THH, HHH}, is called an event. An **event** is any subcollection (or **subset**) of a sample space S. As another example, suppose the experiment is tossing a penny followed by tossing a dime. A sample space for this experiment could be $S = $ {HH, HT, TH, TT}. Some possible events are:

$$E_1 = \{HH\} \qquad E_4 = \{HH, TT\}$$
$$E_2 = \{HT\} \qquad E_5 = \{HT, TH\}$$
$$E_3 = \{TH\} \qquad E_6 = \{TT\}$$

Can you find any others? There are 16 possible events. The **empty set** $\varnothing$ and the sample space S are also events. The event E_6 can be described as getting a tail on the penny and a tail on the dime. The event E_4 can be described as getting two heads *or* two tails. An event is called a **simple event** if it contains only one outcome. The event $E_1 = $ {HH} is a simple event. Remember that an event is always a collection of outcomes from the collection of all outcomes identified as the sample space.

Venn diagrams can be used for graphically portraying sample spaces and relationships between events. A rectangle commonly denotes the sample space and events are represented by circles drawn inside the rectangle, as indicated in Figure 5-1.

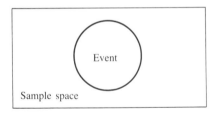

FIGURE 5-1 Venn diagram representing sample space and event

EXAMPLE 5-2   Suppose an experiment consists of examining three fuses. Each fuse can be defective (D) or nondefective (N). Eight possible outcomes are NDD, NDN, NND, NNN, DNN, DND, DDN, and DDD. List the outcomes that comprise each of the following events:

a. The first fuse is defective.
b. The first and last fuses are defective.
c. The fuses are all good.
d. At least one fuse is defective.
e. At most one fuse is defective.

Solution   **a.** $E_1 = \{\text{DNN, DND, DDN, DDD}\}$
b. $E_2 = \{\text{DDD, DND}\}$
c. $E_3 = \{\text{NNN}\}$
d. Note that *at least one* means one or more:

$$E_4 = \{\text{NDD, NDN, NND, DNN, DND, DDN, DDD}\}$$

e. Note that *at most one* means one or less:

$$E_5 = \{\text{NDN, NND, NNN, DNN}\}$$ ∎

EXAMPLE 5-3   Consider the experiment of tossing a red six-sided die and a black six-sided die. A sample space of 36 possible outcomes is as follows, where the first entry is the outcome on the red die and the second entry is the outcome on the black die:

(1, 1)	(1, 2)	(1, 3)	(1, 4)	(1, 5)	(1, 6)
(2, 1)	(2, 2)	(2, 3)	(2, 4)	(2, 5)	(2, 6)
(3, 1)	(3, 2)	(3, 3)	(3, 4)	(3, 5)	(3, 6)
(4, 1)	(4, 2)	(4, 3)	(4, 4)	(4, 5)	(4, 6)
(5, 1)	(5, 2)	(5, 3)	(5, 4)	(5, 5)	(5, 6)
(6, 1)	(6, 2)	(6, 3)	(6, 4)	(6, 5)	(6, 6)

The sample space S can be represented by the following diagram:

Give a description for the following events:

a. $\{(1, 1), (2, 1), (3, 1), (4, 1), (5, 1), (6, 1)\}$
b. $\{(1, 1), (2, 2), (3, 3), (4, 4), (5, 5), (6, 6)\}$
c. $\{(3, 4), (4, 3), (5, 2), (2, 5), (6, 1), (1, 6)\}$
d. $\{(5, 6), (6, 5)\}$
e. $\{(1, 1)\}$

Solution   **a.** The black die shows 1.
b. The two dice match.
c. The sum of the dice equals 7.
d. The sum of the dice equals 11.
e. Both dice show a 1 (a pair of 1s is called *snake eyes*). ∎

EXAMPLE 5-4   For the dice-tossing experiment in Example 5-3 list the outcomes for the following events:

   a. The sum is even.
   b. The sum is divisible by 5.
   c. The sum is a prime number. (A **prime number** is a number greater than 1 which is divisible only by 1 and itself.)
   d. The number on the black die is 2 greater than the number on the red die.
   e. The sum is not even.
   f. The sum is not exactly divisible by 5.

Solution   **a.** The following pairs have a sum that is even:

$(1, 1)$ $(1, 3)$ $(1, 5)$ $(2, 2)$ $(2, 4)$ $(2, 6)$
$(3, 1)$ $(3, 3)$ $(3, 5)$ $(4, 2)$ $(4, 4)$ $(4, 6)$
$(5, 1)$ $(5, 3)$ $(5, 5)$ $(6, 2)$ $(6, 4)$ $(6, 6)$

b. The following pairs have a sum that is divisible by 5:

$(1, 4)$ $(4, 1)$ $(3, 2)$ $(2, 3)$ $(5, 5)$ $(6, 4)$ $(4, 6)$

c. The following pairs have a sum that is a prime number:

$(1, 2)$ $(2, 1)$ $(1, 4)$ $(4, 1)$ $(1, 6)$ $(6, 1)$ $(2, 5)$
$(5, 2)$ $(3, 4)$ $(4, 3)$ $(5, 6)$ $(6, 5)$ $(2, 3)$ $(3, 2)$

d. The following pairs have a number on the black die that is 2 greater than the number on the red die:

$(1, 3)$ $(2, 4)$ $(3, 5)$ $(4, 6)$

e. All outcomes in S that are not listed in part (a) have a sum that is not even.
f. All outcomes in S that are not listed in part (b) have a sum that is not divisible by 5.   ∎

The Event (Not E)

If E is an event contained in a sample space S, then the event **not** E, denoted by $\bar{E}$, is the event containing all the outcomes in S that are not contained in E. In the Venn diagram shown in Figure 5-2, $\bar{E}$ is the shaded area inside of S and outside of E.

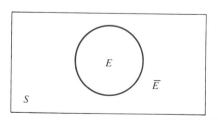

FIGURE 5-2 Venn diagram of the event "not E"

For example, consider the experiment of tossing a six-sided die. If E is the event of getting a 4 or a 6, then the event $\bar{E}$ contains the outcomes 1, 2, 3, and 5. That is, if $E = \{4, 6\}$, then $\bar{E} = \{1, 2, 3, 5\}$.

Suppose that E is an event for some experiment. Since an event E either does or does not occur, a sample space for the experiment is $S = \{E, \bar{E}\}$. As a result, any experiment has a sample space with just two outcomes, E and $\bar{E}$.

EXAMPLE 5-5   For each of the following experiments, list a sample space with only two outcomes.

    a. Toss a coin five times and observe the number of heads.
    b. Toss one die, followed by another.
    c. Spin the following spinner and observe where it lands:

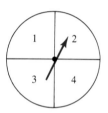

Solution   For each experiment, we list two of the many possible sample spaces containing two outcomes.

    a. $S_1 = \{$obtain 5 heads, do not obtain 5 heads$\}$
       $S_2 = \{$obtain 2 heads, do not obtain 2 heads$\}$
    b. $S_1 = \{$sum is even, sum is odd$\}$
       $S_2 = \{$get two 3s, do not get two 3s$\}$
    c. $S_1 = \{$lands on 1, does not land on 1$\}$
       $S_2 = \{$lands on 2, does not land on 2$\}$ ■

Compound Events

When two or more events are joined by the connectives *and* or *or*, the resulting event is called a **compound event**. If E and F are events, then the events (E or F) and (E and F) are examples of compound events.

    E or *F* is the event that *E* will occur *or* *F* will occur or they both occur.
    E and *F* is the event that both *E* *and* *F* will occur at the same time.

Venn diagrams can be used to illustrate compound events. In the Venn diagrams in Figure 5-3, the compound events (A or B) and (A and B) are represented by the shaded regions.

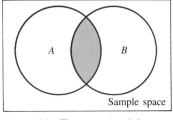

(a) The event A and B

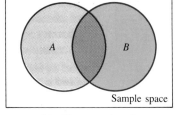

(b) The event A or B

FIGURE 5-3 Venn diagrams of compound events

EXAMPLE 5-6 Consider the dice-tossing experiment in Example 5-3. Let E be the event that the number on the red die is 4 and let F be the event that the sum of the numbers showing is 7. Find the outcomes making up the following events:

a. E
b. F
c. E or F
d. E and F

Solution See Example 5-3 for the listing of S.

a. $E = \{(4, 1), (4, 2), (4, 3), (4, 4), (4, 5), (4, 6)\}$
b. $F = \{(1, 6), (6, 1), (5, 2), (2, 5), (3, 4), (4, 3)\}$
c. E or $F = \{(4, 1), (4, 2), (4, 3), (4, 4), (4, 5), (4, 6), (1, 6), (6, 1), (5, 2), (2, 5), (3, 4)\}$
d. E and $F = \{(4, 3)\}$ ■

EXAMPLE 5-7 Suppose Harry goes to a quick-shop service station to buy a quart of transmission fluid and a quart of oil for his car. If there are three brands of transmission fluid available (X, Y, Z) and six brands of oil available (A, B, C, D, E, F), there are 18 different purchases that Harry can make. Suppose further that the three brands of transmission fluid are priced equally and that the six brands of oil are priced equally. Since Harry is not knowledgeable about oil for his car, he decides on his purchases by guessing. Let E be the event that Harry purchases brand X or brand Z transmission fluid and let F be the event that he purchases brand A oil or brand C oil. List each of the following:

a. A sample space for the experiment of choosing a quart of transmission fluid and a quart of oil
b. The event $\bar{E}$
c. The event $(E$ or $F)$
d. The event $(E$ and $F)$

Solution Figure 5-4 illustrates a sample space containing the events $\bar{E}$, $(E$ or $F)$, and $(E$ and $F)$. Note that S has 18 outcomes, event E has 12 outcomes, and event F has 6 outcomes.

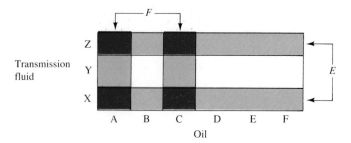

FIGURE 5-4 Sample space and events for Example 5-7

a. $S = \{$XA, XB, XC, XD, XE, XF, YA, YB, YC, YD, YE, YF, ZA, ZB, ZC, ZD, ZE, ZF$\}$
b. $\bar{E} = \{$YA, YB, YC, YD, YE, YF$\}$
c. $(E$ or $F) = \{$XA, XB, XC, XD, XE, XF, YA, YC, ZA, ZB, ZC, ZD, ZE, ZF$\}$
d. $(E$ and $F) = \{$XA, ZA, XC, ZC$\}$ ∎

Mutually Exclusive Events

If E and F are events that have no outcomes in common, then E and F are called **mutually exclusive events**. Note that as sets, if $(E$ and $F) = \varnothing$, then E and F are mutually exclusive events. Mutually exclusive events can be illustrated by a Venn diagram as in Figure 5-5.

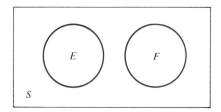

FIGURE 5-5 Mutually exclusive events

EXAMPLE 5-8   Suppose an experiment consists of tossing a penny, followed by tossing a quarter. Are the following pairs of events mutually exclusive?

a. $E =$ two heads; $F =$ two tails
b. $E = \{$HT, TT$\}$; $F = \{$HT, TH$\}$
c. $E = \varnothing$; $F = \{$TT$\}$
d. E; $\bar{E}$

Solution   **a.** Yes; they have no outcomes in common.
b. No; they have HT in common.
c. Yes; they have no outcomes in common.
d. Yes; they have no outcomes in common.   ∎

EXAMPLE 5-9   Suppose a town has three automobile dealerships—Ford, GM, and Chrysler. The GM dealer sells Pontiacs, Oldsmobiles, and Cadillacs; the Ford dealer sells Fords and Mercurys; and the Chrysler dealer sells Dodges, Plymouths, and Chryslers. The experiment consists of observing the order in which the next two cars are sold in town. If A is the event that the two cars are Ford products, B is the event that the two cars are GM products, and C is the event that the next two cars sold are a Pontiac and a Dodge, list the following events:

a. A

b. B

c. C

d. A and B

e. A or C

Are A and B mutually exclusive events?

Solution   Suppose F indicates that the next car sold is a Ford; M, that the next car sold is a Mercury; O, that the next car sold is an Oldsmobile; P, that the next car sold is a Pontiac; C, that the next car sold is a Cadillac; and D, that the next car sold is a Dodge.

a. $A = \{FM, MF\}$

b. $B = \{PO, OP, PC, CP, CO, OC\}$

c. $C = \{PD, DP\}$

d. $(A$ and $B) = \varnothing$

e. $(A$ or $C) = \{FM, MF, PD, DP\}$

Events A and B are mutually exclusive events, since $(A$ and $B) = \varnothing$.   ∎

Problem Set 5.1

A

1. Describe at least two different sample spaces for each of the following experiments:

   a. Choose a group of two students from a class of five students.

   b. Ten cards are numbered 1 through 10. Choose one card from the ten cards.

2. Let C be the event that tomorrow's weather is hot, and D the event that it rains tomorrow. Describe the following compound events:

   a. C or D **b.** C and D **c.** $\bar{C}$ and D
   d. $\bar{C}$ or $\bar{D}$ **e.** $\bar{C}$ or D **f.** $\bar{C}$ and $\bar{D}$
   g. $\overline{C \text{ or } D}$ **h.** $\overline{C \text{ and } D}$

3. Use Venn diagrams to illustrate each of the compound events in Problem 2.

4. For the sample space of 36 outcomes when two dice are tossed (see Example 5-3), find the number of outcomes for which:

   a. Both dice are even.

   b. Exactly one die is even.

   c. At most one die is even.

   d. Neither die is even.

   e. The sum is even.

   f. The quotient of the number on the red die divided by the number on the black die is a whole number.

5. A penny, a nickel, and a dime are tossed, in that order.

   a. List a sample space of eight outcomes.

   b. Count the number of outcomes for which the dime shows heads.

   c. Count the number of outcomes for which either the nickel or the dime shows heads.

   d. Count the number of outcomes for which either the nickel or the dime shows heads, but not both of them.

e. Count the number of outcomes for which the penny and the nickel do not agree.

f. Count the number of outcomes for which the penny agrees with the nickel, but not with the dime.

6. For the two-dice tossing experiment of Example 5-3, consider the sample space of 36 outcomes.

a. Illustrate three different pairs of mutually exclusive events.

b. Illustrate three different pairs of events that are not mutually exclusive.

7. For the accompanying Venn diagram containing the indicated events *A*, *B*, and *C*, list the following events by using the numbers contained in *S*.

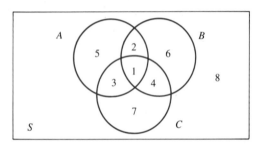

a. *A* and *B*
b. *A* or *B*
c. $\bar{A}$ and *B*
d. $\overline{A \text{ or } B}$ or *C*
e. $\bar{A}$ or *C*
f. *A* and *B* and *C*
g. (*A* or $\bar{B}$) and *C*
h. $\bar{A}$
i. (*A* and *B*) or (*C* and $\bar{A}$)
j. $\overline{A \text{ or } B}$

8. Which of the following pairs of events are mutually exclusive events?

a. *E* = Mrs. Smith gives birth to twins.
  F = A mother gives birth to a girl.
b. *E* = Henry fails the last statistics test.
  F = Henry passes the course.
c. *E* = Joe goes to the movies.
  F = Joe eats a candy bar.
d. *E* = Mr. Doe files the 1040EZ income-tax form.
  F = Mr. Doe files the 1040 long income-tax form.
e. *E* = Our team loses the last baseball game.
  F = Our term loses the baseball tournament.

9. An experiment consists of asking three shoppers at random if they buy brand A peanut butter. Let Y denote yes and N denote no. Let YYN denote the simple event that the first two people polled buy brand A and the third does not. Further, let *E* be the event that at least two people say yes and let *F* be the event that the first person polled says no. List the simple events making up the following events:

a. *E*
b. *F*
c. $\bar{E}$
d. *E* or *F*
e. *E* and *F*
f. $\bar{E}$ and *F*

B

10. If a sample space contains *n* outcomes, how many possible events are there?

11. If a coin is tossed *n* times, how many total ways can it fall?

5.2 THE CONCEPT OF PROBABILITY

Hardly a day goes by that some aspect of probability or chance is not encountered. The weather forecaster calls for an 80% chance of rain. The San Francisco Giants, according to a sports announcer, have a 50–50 chance of winning the World Series. The odds are good that you will pass Calculus 226 if you study hard.

The probability of an event is a number between 0 and 1, inclusive, that is assigned to the event. If *E* is an event, then *P(E)* denotes the probability of *E*; it is the likelihood of the occurrence of event *E*. If the probability is 0, then event *E* is certain not to occur, and if the probability is 1, the event is certain to occur. The closer *P(E)* is to 1, the more likely it is to occur. The closer *P(E)* is to 0, the more unlikely it is to occur, as shown in Figure 5-6.

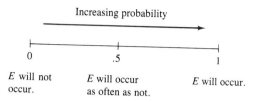

FIGURE 5-6 Interpretations of $P(E)$

Probability satisfies the following properties:

1. $P(E_i) \geq 0$
2. $P(E_i) \leq 1$
3. $\sum P(E_i) = 1$

where $\sum P(E_i)$ is the sum of the probabilities for all the outcomes (simple events) in a sample space.

The probability of an event A is defined to be the sum of the probabilities for the outcomes contained in A. The following example illustrates the assignment of probabilities to events once the probabilities of the outcomes in the sample space are known.

EXAMPLE 5-10   Suppose a six-sided die is tossed once and the probability of any side landing face up is $\frac{1}{6}$. If E is the event of getting an even number and F is the event of getting an odd number, find:

a. $P(E)$
b. $P(F)$
c. $P(E \text{ or } F)$
d. $P(E \text{ and } F)$

Solution   The sample space is $S = \{1, 2, 3, 4, 5, 6\}$, the event E is $\{2, 4, 6\}$, and the event F is $\{1, 3, 5\}$. We thus have:

a. $P(E) = P(2) + P(4) + P(6) = \frac{1}{6} + \frac{1}{6} + \frac{1}{6} = \frac{3}{6} = \frac{1}{2}$
b. $P(F) = P(1) + P(3) + P(5) = \frac{1}{6} + \frac{1}{6} + \frac{1}{6} = \frac{3}{6} = \frac{1}{2}$
c. $P(E \text{ or } F) = P(S) = 1$
d. $P(E \text{ and } F) = 0$, since $(E \text{ and } F) = \varnothing$ and $P(\varnothing) = 0$ ∎

Assigning Probabilities to Events

The three properties satisfied by probabilities do not tell us how to assign probabilities to the outcomes in a sample space. All they do is to rule out certain assignments that are not consistent with our intuitive notions. There are two general methods for assigning probabilities to events: the objective method and the subjective method. The **objective method** involves assigning probabilities to events based on counting or repeated experimentation. The **subjective method**, on the other hand, involves assigning probabilities to events based on intuition or personal belief. When using the subjective method to assign probabilities to events, two knowledgeable people may not agree on the assignments; this will not be the case for the objective method.

The Objective Method

Suppose a coin is tossed. The two outcomes are head (H) and tail (T). What number $P(H)$ should we assign to H and what number $P(T)$ should we assign to T? Suppose we assign .7 to H and .3 to T. Is this a valid assignment of probabilities? The answer is yes, based on the three given properties, since

1. Both numbers are greater than 0;
2. Both numbers, .7 and .3, are less than 1; and
3. The sum of .7 and .3 is 1.

But these assignments go against our intuition if we believe the coin is fair. The correct assignments, most of us would agree, are .5 for H and .5 for T. For if we tossed the coin a large number of times, N, and found the frequency, f, for the occurrence of a head, we would expect the **relative frequency** f/N for the occurrence of a head to be close to .5. Table 5-1 contains the number of heads obtained when a fair coin was tossed N times, as well as the relative frequency for the number of heads obtained in each case. The limiting value of the relative frequency f/N for obtaining a head when a fair coin is tossed N times will approach .5 as N becomes large, as illustrated in Figure 5-7.

Table 5-1 Frequency and Relative Frequencies of Heads in Repeated Tossing of a Coin

Number of tosses N	Number of heads f	Relative frequency f/N
51	29	.569
55	30	.545
64	33	.516
80	39	.488
105	51	.486
141	68	.482
190	91	.479
254	128	.504
335	173	.516
435	222	.510
556	276	.496
700	354	.506
869	435	.501
1065	540	.507
1290	650	.504
1546	771	.499
1835	906	.494
2159	1086	.503
2520	1268	.503

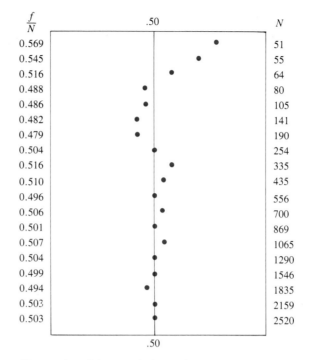

$$\frac{f}{N}$$

$\frac{f}{N}$		N
0.569		51
0.545		55
0.516		64
0.488		80
0.486		105
0.482		141
0.479		190
0.504		254
0.516		335
0.510		435
0.496		556
0.506		700
0.501		869
0.507		1065
0.504		1290
0.499		1546
0.494		1835
0.503		2159
0.503		2520

N = number of times a coin is tossed

FIGURE 5-7 Limiting value of *f/N* as *N* gets large

Since .7 is not equal to the limiting value of the relative frequency for a head occurring (which is .5), we would not recommend assigning .7 as the probability for the outcome *H*. With the value .5 assigned to *H*, what probability $P(T)$ should be assigned to *T*? As a consequence of the third probability property, we know that

$$P(H) + P(T) = 1$$

or

$$.5 + P(T) = 1$$

Hence, $P(T) = .5$ must be the probability assignment for getting a tail.

If an experiment is repeatable, we can assign probabilities to the outcomes in accordance with their limiting relative frequencies in a fashion similar to what we did for the coin above. The only problem with doing this is that the limiting values of relative frequencies are not always known. To use this method, we need to have much repetitive data available, and then only approximations to the relative frequency limits can be found.

Empirical Probability

When probability is based on experience and the exact limiting values of relative frequencies are unknown, approximations to these limiting values must be used. When this is done, the objective method of assigning probabilities is said to be **empirical**. According to the empirical method, if E is an event, $P(E)$ is approximately equal to f/N, where f is the number of favorable outcomes and N is the number of repetitions of the experiment. In this case, we have

$$P(E) \simeq \frac{f}{N}$$

For example, suppose we consider the experiment of tossing a thumbtack. It can land one of two ways:

How can we determine the probability that the thumbtack will land point up? We could ask someone. But who would know the answer? Probably nobody. We could toss the thumbtack ten times and record the number of times it landed point up. This relative frequency could serve as an estimate of the desired probability. But to get a better estimate, we could toss the tack 100 or 1000 times, or even more, and record the frequency of landing point up. We would thereby obtain a good approximation for the probability of the thumbtack landing point up.

As our experience changes, so does the relative frequency. For example, if we tossed a coin six times and obtained three heads, we would estimate the chance of getting a head to be $\frac{3}{6} = .5$. If the coin were tossed once more and it fell heads, then the estimated chance for getting a head would be $\frac{4}{7} = .5714$; or if it fell tails, then the estimated chance for getting a head would be $\frac{3}{7} = .4286$. This change in relative frequency reflects our changing knowledge. As was pointed out above, over the long run, the relative frequency should change very little, and the limiting value that the relative frequency is approaching is called the **probability**. Consider the following illustrative examples.

EXAMPLE 5-11   An insurance company wants to estimate the probability of a police car being involved in an accident in a certain city during a 1-month period. Last month, 7 out of 20 police cars were involved in accidents.

a. What would you estimate the desired probability to be?
b. What would you estimate to be the chance of a police car not being involved in an accident?

Solution   **a.** $\frac{7}{20} = .35$
b. $1 - .35 = .65$ ∎

EXAMPLE 5-12   The SAT math scores for students at a large university are displayed in the following grouped frequency table:

SAT	f
200–299	3,600
300–399	11,900
400–499	12,000
500–599	5,500
600–699	1,500
700–799	500

If a student is selected at random, what is the probability that the student's SAT math score

a. exceeds 399?
b. is at most 599?
c. is between 600–699, inclusive?
d. is not between 400–499, inclusive?
e. is less than or equal to 699?

Solution   We first form a grouped relative frequency table (Table 5-2). Recall that the relative frequency for a class is found by dividing the frequency f of the class by the total number of measurements in the data set, N.

Table 5-2 Grouped Relative Frequency Table for Example 5-12

SAT	f	Rel. f
200–299	3,600	.103
300–399	11,900	.340
400–499	12,000	.343
500–599	5,500	.157
600–699	1,500	.043
700–799	500	.014
	35,000	

a. $P(\text{SAT} > 399) = .343 + .157 + .043 + .014 = .557$
b. $P(\text{SAT} \leq 599) = .103 + .340 + .343 + .157 = .943$
c. $P(600 \leq \text{SAT} \leq 699) = .043$
d. $P(\text{SAT} < 400 \text{ or SAT} > 499) = 1 - .343 = .657$ (since the sum of the relative frequencies is 1)
e. $P(\text{SAT} \leq 699) = 1 - .014 = .986$ (since the sum of the relative frequencies is 1)

■

Probability Histograms

For data that have been displayed in a grouped frequency table, a special type of relative frequency histogram can be constructed and used to estimate the probability that a measurement falls in a particular class. If instead of letting the ith bar in a relative frequency histogram have a height of f_i/N, as was done in Section 2.3, we let the height h of the ith bar be

$$h = \frac{f_i}{wN}$$

where f_i is the frequency associated with the ith class, w is the class width, and N is the total number of measurements, then the area of each bar approximates the probability that a measurement will fall in that particular class. In addition, the total area of the histogram is then equal to 1, the sum of the probabilities. For this reason, such a histogram is referred to as a **probability histogram**. Note that the area of a relative frequency histogram, as constructed in Section 2.3, has an area equal to w, the class width.

EXAMPLE 5-13   For the grouped data in Example 5-12, construct a probability histogram.

Solution   Table 5-3 contains the class boundaries, the heights of the bars, and the areas of the bars. The heights h were found by dividing the relative frequencies by $w = 100$, since $h = f_i/(wN)$.

The graph of the probability histogram is shown in Figure 5-8. The shaded area represents the probability $P(299.5 \leq x \leq 499.5) = .683$.

Table 5-3 Areas of Bars for Probability Histogram of Example 5-13

Boundaries	h	Area of bar
199.5–299.5	.00103	.103
299.5–399.5	.00340	.340
399.5–499.5	.00343	.343
499.5–599.5	.00157	.157
599.5–699.5	.00043	.043
699.5–799.5	.00014	.014
		1.000

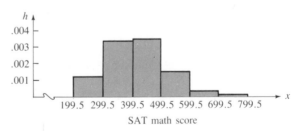

FIGURE 5-8 Probability histogram for Example 5-13

EXAMPLE 5-14   In a small city, each person was classified according to religion and political party affiliation. The results are summarized in the following table:

Religion	Political party		
	Democrat	Republican	Independent
Protestant	10,000	8,000	2,000
Jewish	5,500	6,000	500
Catholic	8,500	9,500	1,500

If a person is chosen at random from this city, what is the probability that the person is a:

a. Republican?
b. Catholic?
c. Protestant and a Republican?
d. Catholic and an Independent?

Solution   We first find the total for each row and column, as indicated in Table 5-4.

Table 5-4 Row and Column Totals for Example 5-14

Religion	Political party			Total
	Democrat	Republican	Independent	
Protestant	10,000	8,000	2,000	20,000
Jewish	5,500	6,000	500	12,000
Catholic	8,500	9,500	1,500	19,500
Total	24,000	23,500	4,000	51,500

a. There are 23,500 Republicans out of a total of 51,500 persons. Therefore,

$$P(R) = \frac{23,500}{51,500} = .456$$

b. There are 19,500 Catholics out of a total of 51,500 persons. Therefore,

$$P(C) = \frac{19,500}{51,500} = .379$$

c. There are 8,000 people out of a total of 51,500 that are Protestant and Republican. Therefore,

$$P(P \text{ and } R) = \frac{8,000}{51,500} = .155$$

d. There are 1,500 people out of a total of 51,500 that are Catholic and Independent. Therefore,

$$P(C \text{ and } I) = \frac{1,500}{51,500} = .029$$

■

Classical Probability

If an experiment has a finite number, n, of outcomes that we believe are equally likely to occur, we can assign each outcome in the sample space S a probability value of $1/n$. This is a consequence of probability property 3, which states that the sum of the probabilities for a sample space must be 1. Then if E is an event containing f outcomes, the probability of E occurring is simply f/n. Thus, we have the following basic fact:

If S is a sample space of **equally likely outcomes**, then

$$P(E) = \frac{f}{n} \tag{5-1}$$

where f is the number of outcomes in E and n is the number of outcomes in S. It is common to call the outcomes contained in E **favorable outcomes**.

The objective method for assigning probabilities using sample spaces of equally likely outcomes is referred to as the **classical probability method**.

Suppose a penny and a dime are both tossed once. They can fall in four ways. A sample space for this experiment is $S = \{HH, TH, HT, TT\}$. The letter written first in a pair stands for the outcome on the penny and the letter written second stands for the outcome on the dime. The possible outcomes may be visualized as follows:

		Dime	
		H	T
Penny	H	HH	HT
	T	TH	TT

Possible outcomes

The four outcomes are equally likely to occur if both coins are fair and they are tossed so that the outcome on one is not influenced by and does not influence the outcome of the other. As a result, we assign a probability value of $\frac{1}{4}$ to each of the four outcomes. If E is the event $\{TH, HT\}$, then we have the following two methods for assigning a probability value to E:

1. Add the probabilities for the outcomes contained in E. Thus,

$$P(E) = \frac{1}{4} + \frac{1}{4} = \frac{1}{2}$$

2. Use formula (5-1). The number of favorable outcomes is $f = 2$, the number of outcomes in E, and the number of equally likely outcomes in S is $n = 4$. Hence,

$$P(E) = \frac{f}{n} = \frac{2}{4} = \frac{1}{2}$$

The following examples illustrate the classical probability method.

EXAMPLE 5-15 Consider the previous experiment of tossing a penny and a dime once. When two coins are tossed, 0, 1, or 2 heads can result. As a result, a sample space is $S = \{0H, 1H, 2H\}$. Since there are three outcomes in S, we may be tempted to conclude that $P(1H) = \frac{1}{3}$. But by the above argument, $P(1H) = P\{HT, TH\} = \frac{1}{2}$. Resolve this apparent contradiction.

Solution The three outcomes in S are not equally likely to occur. The outcome 1H is twice as likely to occur as 0H or 2H. Thus, formula (5-1) is not valid for computing $P(1H)$, and we have $P(1H) = \frac{1}{2}$. ∎

EXAMPLE 5-16 For the two-dice-tossing experiment described in Example 5-3, find the probability that:

a. Both show an even number.
b. A sum of 7 shows.
c. A sum of 7 or 11 shows.
d. Both show a prime number (a prime number is a positive number different from 1 that has no divisors other than 1 and itself).
e. The red die shows a 2.
f. A sum of 13 shows.

Solution A sample space is as follows:

(1, 1)	(1, 2)	(1, 3)	(1, 4)	(1, 5)	(1, 6)
(2, 1)	(2, 2)	(2, 3)	(2, 4)	(2, 5)	(2, 6)
(3, 1)	(3, 2)	(3, 3)	(3, 4)	(3, 5)	(3, 6)
(4, 1)	(4, 2)	(4, 3)	(4, 4)	(4, 5)	(4, 6)
(5, 1)	(5, 2)	(5, 3)	(5, 4)	(5, 5)	(5, 6)
(6, 1)	(6, 2)	(6, 3)	(6, 4)	(6, 5)	(6, 6)

There are 36 equally likely outcomes in S.

a. Let E_1 represent the event that both dice show an even number. A favorable outcome is one in which both dice show an even number. There are 9 favorable outcomes. Hence,

$$P(E_1) = \frac{f}{n} = \frac{9}{36} = \frac{1}{4}$$

b. Let E_2 represent the event that a sum of 7 shows. The set of favorable outcomes is $E_2 = \{(3, 4), (4, 3), (5, 2), (2, 5), (6, 1), (1, 6)\}$. Therefore,

$$P(E_2) = \frac{f}{n} = \frac{6}{36} = \frac{1}{6}$$

c. Let E_3 represent the event that a sum of 7 or 11 shows. The set of favorable outcomes is $E_3 = \{(1, 6), (2, 5), (3, 4), (4, 3), (5, 2), (6, 1), (5, 6), (6, 5)\}$. Hence,

$$P(E_3) = \frac{8}{36} = \frac{2}{9}$$

d. Let E_4 represent the event that both dice show a prime number. Since $E_4 = \{(2, 2), (2, 3), (2, 5), (3, 2), (3, 5), (5, 2), (5, 3), (5, 5)\}$, we have

$$P(E_4) = \frac{f}{n} = \frac{8}{36} = \frac{2}{9}$$

e. Let E_5 represent the event that the red (first) die shows a 2. The set of favorable outcomes is $E_5 = \{(2, 1), (2, 2), (2, 3), (2, 4), (2, 5), (2, 6)\}$. Thus,

$$P(E_5) = \frac{f}{n} = \frac{6}{36} = \frac{1}{6}$$

f. Let E_6 represent the event that a sum of 13 shows. Since a sum of 13 is impossible, $E_6 = \emptyset$. Thus,

$$P(E_6) = 0 \qquad\blacksquare$$

EXAMPLE 5-17   What is the probability of correctly guessing all the answers to a true/false test containing four questions?

Solution   Let T denote true and F denote false. Since only one of these outcomes represents a favorable outcome, $f = 1$. A sample space of equally likely outcomes contains the following 16 outcomes:

TTTT	TFFT	FFTF	FTFT
TTFT	TFFF	FFFT	FTTF
TTTF	TFTT	FFTT	FTFF
TTFF	TFTF	FFFF	FTTT

By using (5-1), we have

$$P(E) = \frac{f}{n} = \frac{1}{16} \qquad\blacksquare$$

The Subjective Method

In many situations we have limited or no information concerning the outcomes of an experiment. We would have no frequency data available if the situation had never occurred before. A doctor treating a patient with a rare disease must make a prognosis based on his experience with the patient and the patient's overall medical record. A candidate running for public office for the first time will estimate his chance for election based on his intuition and hearsay. When probabilities are assigned to events based on intuition and personal beliefs, the assignment method is called **subjective**. Of course, when using the subjective method for assigning probabilities to events, the assignments will depend on the individual making the assignments and may vary widely from individual to individual.

Mathematical Odds

In gambling situations, such as horse races and sporting contests, the chance of an event occurring is often arrived at by using subjective probability. And these probabilities are often stated in terms of odds. If E is an event, then the **odds in**

favor of *E*, written as Odds(*E*), is defined by

$$\text{Odds}(E) = \frac{P(E)}{P(\bar{E})}$$

The **odds against** *E*, written as Odds(*Ē*), is defined by

$$\text{Odds}(\bar{E}) = \frac{P(\bar{E})}{P(E)} = \frac{1}{\text{Odds}(E)}$$

For example, if *E* is the event that a favorite horse wins the race and $P(E) = \frac{1}{3}$, then the odds in favor of winning, Odds(*E*), is

$$\text{Odds}(E) = \frac{P(E)}{P(\bar{E})} = \frac{(\frac{1}{3})}{(\frac{2}{3})} = \frac{1}{2}$$

This result is sometimes written as 1:2.

Odds can be converted to probabilities. As a general rule, if Odds(*E*) = *a*:*b*, then

$$P(E) = \frac{a}{a + b}$$

For example, if the odds in favor of the Cincinnati Reds winning the baseball game are 2:3, then the probability that the Reds will win, *P*(*E*), is

$$P(E) = \frac{a}{a + b} = \frac{2}{2 + 3} = \frac{2}{5} = .4$$

EXAMPLE 5-18   If the odds are 3:2 against a favorite horse winning a race, what is the probability of

a. the horse winning the race?
b. the horse losing the race?

Solution   Let *E* be the event the horse will win the race.

a. $P(E) = \dfrac{2}{2 + 3} = \dfrac{2}{5}$

b. $P(\bar{E}) = \dfrac{3}{2 + 3} = \dfrac{3}{5}$ ■

EXAMPLE 5-19   In 1967 approximately 15 million men were eligible for the military draft during any month. If 20,000 men were drafted each month, what were the odds against John Jones being drafted in April 1967?

Solution   Let *E* be the event that John was drafted. Then

$$P(E) = \frac{20,000}{15,000,000} = \frac{1}{750} \quad \text{and} \quad P(\bar{E}) = \frac{749}{750}$$

Hence, the odds against John being drafted were

$$\text{Odds}(\bar{E}) = \frac{P(\bar{E})}{P(E)} = \frac{\frac{749}{1}} = 749{:}1$$ ■

Problem Set 5.2 _____

A

1. Match each of the following probabilities with one of the statements that follow:

   0   .01   .3   .99   1

   a. The event is impossible. It can never occur.
   b. The event is certain.
   c. The event is very unlikely, but it will occur once in a while in a long sequence of trails.
   d. The event will occur more often than not.

2. Which of the following numbers cannot be the probability of some event?
   a. .74   **b.** −1   **c.** 1.02
   d. .5   **e.** 0   **f.** 1
   g. $\frac{2}{3}$ **h.** .999 . . .   **i.** .67

3. A bag contains three red marbles, two white marbles, and five blue marbles. One marble is selected at random. What is the probability that the marble is
   a. red?   **b.** white?   **c.** blue?

4. For Problem 3, list two sample spaces for the experiment, one containing equally likely outcomes, and one not containing equally likely outcomes. In each case, list the probabilities associated with the outcomes.

5. The odds against being dealt three of a kind in a five-card poker hand are about 49:1. What is the probability of being dealt three of a kind?

6. An American roulette wheel contains compartments numbered 1 through 36 plus 0 and 00. Of the 38 compartments, 0 and 00 are colored green, 18 of the others red, and 18 are black, A ball is spun in the direction opposite to the wheel's motion, and bets are made on the number where the ball comes to rest. Suppose the wheel is fair.
   a. What is the probability of a black outcome?
   b. What are the odds against a red outcome?
   c. What is the probability of a red outcome?
   d. What are the odds in favor of a red outcome?
   e. What is the probability of a red or black outcome?
   f. What is the probability of a 0?
   g. What are the odds in favor of a green outcome?
   h. What are the odds against a green outcome?

7. To estimate the number of books required per course for a small college, the Student Council took a sample of 100 courses and obtained the following results:

Number of books required (x)	Number of courses (f)
0	5
1	48
2	21
3	12
4	9
5	3
6	2

Use empirical probabilities to find:
   a. The probability value for each x
   b. $\sum P(x)$
   c. $P(x$ is at least 5)
   d. $P(x$ is 2 or 4)

8. If a meteorologist forecasts rain with a probability of 70%, what are the odds in favor of rain? Against rain?

9. Each teacher at a college was classified according to sex and academic rank. The results are displayed in the following table:

		Rank		
Sex	Instr.	Asst. Prof.	Assoc. Prof.	Prof.
Male	300	400	700	300
Female	350	450	300	200

If a teacher is selected at random from this college, what is the probability that the teacher is a
   a. male?   **b.** female?
   c. professor?   **d.** male professor?
   e. female assistant professor?

10. For each of the following situations, express the situation by using a probability value in decimal form.

   a. There is a 30–70 chance for getting funding for our project.
   b. The odds against striking oil are 100 to 1.

c. There is a 75% chance that surgery will be needed to correct the problem.

d. The odds that the Giants will win the game tomorrow are 5 to 3.

B

11. Fingerprints are classified into eight generic types and possess many characteristics. One of these is the ridge count, which is useful in criminal investigation work. The accompanying grouped frequency table displays the ridge counts for 800 males.

Ridge count	f
0–19	10
20–39	12
40–59	24
60–79	40
80–99	73
100–119	100
120–139	90
140–159	112
160–179	124
180–199	95
200–219	67
220–239	36
240–259	10
260–279	4
280–299	3
	800

a. Suppose fingerprints are found at the scene of a crime and it is determined that the ridge count is at least 220. If a male suspect has a ridge count of 241, what would you conclude? Why?

b. What is the probability of a male having a ridge count of at least 200?

c. What is the probability that a male has a ridge count of at least 120 and no more than 199?

12. Roll a pair of dice 100 times and record the sum of the numbers showing for each roll. What is the relative frequency of a sum of 8? (By mathematical logic, in the long run this relative frequency will approach the probability of a sum of 8, which is about .14, if the dice are fair.)

13. Two dice are tossed and the larger number showing face up is recorded. If {1, 2, 3, 4, 5, 6} is a sample space for the experiment, find the probability for each of the six outcomes. Are the outcomes equally likely? Why or why not?

14. If a deck of 52 playing cards is well shuffled, what would you estimate to be the probability that the top three cards contain a king or a queen, or that a king and queen are next to each other somewhere among the remaining 49 cards? Test your guess by simulating the experiment 20 times.

15. Can an experiment have two different sample spaces of equally likely outcomes? Explain.

16. To decide whether to bunt in a baseball game, a manager studies the accompanying data obtained from several hundred baseball games.[1]

Base(s) occupied	Number of outs	Proportion of cases no runs scored in inning	Average number of runs in inning	Number of cases observed
First	0	.604	.813	1728
Second	1	.610	.671	657
Second	0	.381	1.194	294

a. If a player is on first base with no outs, in how many cases did runs score?

b. If a player is on first base with no outs, find the probability that at least one run scored.

c. If a player is on first base with no outs and a sacrifice bunt succeeds in the normal way of advancing the player on first base, is this a better situation than we had (in the sense of average number of runs scored)?

d. If a player is on second base with no outs, find the probability that at least one run scored in the inning.

e. If a player is on second base with one out, in how many cases did no runs score in the inning?

17. The following data represent the average annual salaries (in thousands of dollars) for workers in the 50 states covered by unemployment insurance (*U.S.*

[1] Source: G. R. Linsey, "An Investigation of Strategies in Baseball," *Operations Research*, II, 1963, pp. 477–501.

News and World Report, Sept. 24, 1984):

28.7	18.0	17.1	16.0	15.2
19.7	17.9	16.9	15.9	15.0
19.7	17.8	16.8	15.9	14.8
19.3	17.3	16.8	15.5	14.7
17.8	17.3	16.7	15.5	14.6
18.8	17.2	16.5	15.5	14.6
18.7	17.2	16.5	15.5	14.3
18.1	17.2	16.4	15.4	14.1
18.1	17.1	16.1	15.2	13.9
18.1	17.1	16.1	15.2	13.2

Construct a probability histogram with eight bars.

18. When two people seat themselves at a table with a square top and with one chair at each side, it is not unusual for them to sit in adjacent, rather than opposite, seats. If two people randomly choose seats at such a table, determine the probability that they occupy adjacent seats at the table.

5.3 COUNTING

Suppose three six-sided dice are tossed and we are asked to determine the number of ways they can fall. A sample space, as we shall see later, may contain 216 equally likely outcomes. Such a problem represents a sophisticated counting problem.

As another example, consider the experiment of going to your new-car dealer to buy a new car. Once there, you find you can select from 4 models, each with 15 power options, with a choice of 5 exterior colors and 8 interior colors. How many different choices (outcomes) are possible for you to make? The answer is 2400. In this case, the choices may not all be equally likely unless you choose randomly.

Fundamental Theorem of Counting

Solutions to the above two examples may be obtained by applying the **fundamental theorem of counting** (FTC). It is stated as follows:

> If an event can occur in any one of *m* ways, and if after it has occurred, a second event can occur in any one of *n* ways, then the two events can occur together, in the order stated, in *mn* different ways. (This rule can be extended to any number of events.)

Since the three dice in the first example above are in no way related when tossed, and since each can fall in six different ways, the total number of ways they can fall, one following another, is $6 \times 6 \times 6 = 216$ by the FTC.

For the second example above, there are four events that can occur, one following another. They are as follows:

E_1 = Choose one of 4 models.
E_2 = Choose one of 15 power options.
E_3 = Choose one of 5 exterior colors.
E_4 = Choose one of 8 interior colors.

The first event can occur in 4 ways, the second event in 15 ways, the third event in 5 ways, and the fourth event in 8 ways. Therefore, as a consequence of the FTC, the four events, one following another, can occur in $4 \times 15 \times 5 \times 8 = 2400$ different ways.

Let's look at a simpler example.

EXAMPLE 5-20   Suppose there are three roads from town A to town B and two roads from town B to town C. How many different ways can a person travel from A to C by way of B?

Solution   Consider the following diagram, called a **tree diagram**:

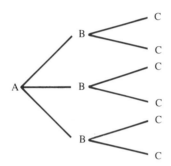

There are two decisions to be made:

1. At A, which road to B?
2. At B, which road to C?

The first decision can be made in 3 ways, and after this decision is made, the second decision can be made in 2 ways. According to the FTC, the total number of ways to go from A to C is $3 \times 2 = 6$. This can also be seen by counting the branches on the tree diagram going from A to C.   ■

EXAMPLE 5-21   There are five different books a teacher wants to arrange on his desk from left to right. How many different arrangements are possible?

Solution   This problem can be viewed two different ways: as an arrangement problem or as a selection problem.

> ***An arrangement problem.***   The five events are
>
> E_1 = Arrange the first book in one of 5 spaces.
> E_2 = After the first book is arranged, arrange the second book in one of the 4 remaining spaces.
> E_3 = After the first two books are arranged, arrange the third book in one of the 3 remaining spaces.
> E_4 = After the first three books are arranged, arrange the fourth book in any of the 2 remaining spaces.
> E_5 = Arrange the last book in only 1 way.

E_1 can be done in 5 ways, E_2 in 4 ways, E_3 in 3 ways, E_4 in 2 ways, and E_5 in 1 way. Thus, by the fundamental theorem of counting, the total number of arrangements is $5 \times 4 \times 3 \times 2 \times 1 = 120$.

A selection problem.   The five events are

E_1 = Select a book for the first space.

This can be done in 5 ways.

E_2 = After the first space is filled, select the second book from the remaining four books.

This can be done in 4 ways.

E_3 = After the first two spaces are filled, select the third book from the remaining three books.

This can be done in 3 ways.

E_4 = After the first three spaces are filled, select the fourth book from the remaining two books.

This can be done in 2 ways.

E_5 = Place the fifth book in the fifth space.

This can be done in 1 way.

By the FTC, the total number of selections is $5 \times 4 \times 3 \times 2 \times 1 = 120$.   ■

EXAMPLE 5-22   In how many ways can three items be chosen from a group of six items and arranged on a shelf from left to right?

Solution   There are three events:

E_1 = Choose an item from six items for the first space.

This can be done in 6 ways.

E_2 = Choose an item from the remaining five items for the second space.

This can be done in 5 ways.

E_3 = Choose an item from the remaining four items for the third space.

This can be done in 4 ways.

By using the FTC, we see that the total number of ways is $6 \times 5 \times 4 = 120$.

■

By examining the previous examples, we note that there are two questions to be asked when solving any counting problem:

1. How many events are there?
2. In how many ways can each occur?

After determining the answers to questions 1 and 2, we use the FTC to solve the problem.

EXAMPLE 5-23   There are five candidates for president of a club and two candidates for vice-president. Use a tree diagram to determine the number of ways the two offices can be filled.

Solution   There are two events of interest: choosing a president and choosing a vice-president. The president can be chosen in 5 ways and the vice-president can be chosen in 2 ways. As a consequence of the FTC, the two offices can be filled in $(5)(2) = 10$ different ways. A tree diagram can be constructed to illustrate the ten possibilities. Let A_1, A_2, A_3, A_4, and A_5 represent the candidates for president, and let B_1 and B_2 represent the candidates for vice-president. Then we have the tree diagram shown in Figure 5-9.

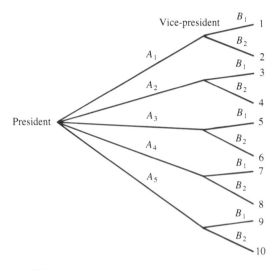

FIGURE 5-9 Tree diagram for Example 5-23   ■

Problem Set 5.3

A

1. How many different homes can be built if a builder offers a choice of five basic plans, three roof styles, and two exterior finishes?

2. A business school gives courses in typing, shorthand, transcription, business English, technical writing, and accounting. In how many ways can a student choose three courses to take during three class periods?

3. In how many ways can seven of ten monkeys be arranged in a row for a genetics experiment?

4. In an experiment of social interaction, six people will sit in six seats in a row. In how many ways can this be done?

5. In some electronic equipment, six wires enter a box that has six terminals. In how many ways can the wires be connected to the terminals, one wire to each terminal?

6. In how many ways can a judge award first, second, and third places in a contest with 15 entries?

7. A clothing store stocks socks made of either cotton or wool, each in five colors and seven sizes. How many items must be stocked in order to have available a complete assortment?

8. A man tries to choose the winner of each of eight football games. Excluding ties, how many different predictions are possible? Including ties?

9. Answer Problem 4, if the six people sit around a circular table and the relative positions of the people make a difference and not the seats they sit in.

10. If you are taking a true/false test with ten questions, in how many different ways can you fill in your answer sheet if you know none of the answers and guess at each one?

B

11. The Upper Crust Pizza Shop advertises ten different pizza toppings. How many different pizzas can be ordered?

12. For Problem 2, in how many ways can a schedule of three courses be selected?

5.4 FINDING PROBABILITIES USING THE FUNDAMENTAL THEOREM OF COUNTING

The fundamental theorem of counting can be used to solve some of the more difficult probability problems. The following examples illustrate the use of the FTC to solve probability problems.

EXAMPLE 5-24 If four coins are tossed, what is the probability of getting four heads?

Solution We first determine the number of ways the four coins can land when tossed, one following another. Let E_1, E_2, E_3, and E_4 represent tossing the four coins, respectively. Each can land in exactly 2 ways. By the FTC, the four coins can fall in $2^4 = 16$ different ways. Of these, there is only one favorable outcome—namely, HHHH. Hence,

$$P(\text{HHHH}) = \frac{1}{16} = .0625$$ ∎

EXAMPLE 5-25 If four coins are tossed, what is the probability that the first two coins are heads?

Solution The total number of outcomes is 16, from Example 5-24. A favorable outcome is an outcome in which the first two coins are heads. The first coin must land heads up, the second coin must land heads up, and the third and fourth coins can land either heads up or tails up. By the fundamental theorem of counting, the total number of favorable outcomes is $1 \times 1 \times 2 \times 2 = 4$. Therefore, the probability that the first two coins are heads when four coins are tossed is $\frac{4}{16} = \frac{1}{4}$.   ∎

EXAMPLE 5-26 From a set of 100 cards numbered 1 to 100, one is selected at random. What is the probability that the number on it

a. is exactly divisible by 5?
b. ends in a 1 or a 2?

Solution   a. Let E represent the event that the number on a card is divisible by 5. A number is divisible by 5 if it ends in a 0 or a 5. The number of 1-digit numbers divisible by 5 is 1. The number of 2-digit numbers divisible by 5 is $9 \times 2 = 18$. The number of 3-digit numbers divisible by 5 is 1. Therefore, the total number of cards with numbers divisible by 5 is $f = 1 + 18 + 1 = 20$. Since there are 100 cards, we have

$$P(E) = \frac{f}{n} = \frac{20}{100} = \frac{1}{5}$$

b. Let F represent the event that the number on the card ends in a 1 or a 2. A number ending in a 1 or 2 can be a 1-digit number or a 2-digit number. There are $f_1 = 2$ 1-digit numbered cards ending in a 1 or 2. For the 2-digit numbered cards, the units digit can be one of two values and the tens digit can be one of nine values. By the FTC, the number of 2-digit numbered cards ending in a 1 or 2 is $f_2 = (9)(2) = 18$. Hence,

$$P(F) = \frac{f_1 + f_2}{n} = \frac{2 + 18}{100} = \frac{20}{100} = \frac{1}{5}$$ ■

EXAMPLE 5-27   There are ten different books on a bookshelf; seven are mathematics books and three are science books. If the books are randomly placed on the bookshelf, what is the probability that the science books are all together?

Solution   Let E be the event that the science books are arranged together. By the FTC, the ten books can be arranged in $(10)(9)(8)(7)(6)(5)(4)(3)(2)(1) = 3{,}628{,}800$ different ways. To find the number of ways of arranging the ten books so that the science books are together, we identify the following events:

$E_1 =$ Arranging the mathematics books
$E_2 =$ Arranging the science books
$E_3 =$ Inserting the set of science books among the mathematics books

The number of ways that E_1 can occur is $f_1 = (7)(6)(5)(4)(3)(2)(1) = 5040$. The number of ways that E_2 can occur is $f_2 = (3)(2)(1) = 6$. And the number of ways that E_3 can occur is $f_3 = 8$ (putting the science books before the mathematics books, after the first mathematics book, after the second mathematics book, after the third mathematics book, and so forth). By the FTC, the number of ways of arranging the ten books so that the science books are together is

$$f = (f_1)(f_2)(f_3) = (5040)(6)(8) = 241{,}920$$

Thus, the probability that the science books are arranged together is

$$P(E) = \frac{f}{n} = \frac{241{,}920}{3{,}628{,}800} \simeq .06667$$ ■

Problem Set 5.4 _____

A

1. The tickets in a box are numbered 1 to 20 inclusive, and two tickets are drawn, one following the other without replacing the first ticket drawn. What is the probability that
a. both are even?

b. the first is even and the other odd?
c. both are even or both are odd?

2. Do Problem 1, if the first ticket is replaced before the second ticket is drawn.

3. Four nuts and four bolts are mixed together. If two parts are chosen at random, what is the probability that
   a. both are nuts?
   b. the first is a nut and the second is a bolt?
   c. one will be a nut and one a bolt?
   d. either both will be nuts or both will be bolts?

4. The letters of the word *Chance* are written on cards, one letter to a card. The cards are shuffled and land face up one after another. What is the probability that they spell *Chance*? (Note that the first letter is a capital.)

5. What is the probability that a student will spell the word *math*, by randomly arranging the letters *m*, *a*, *t*, and *h*?

6. What is the probability that a four-letter "word" chosen from the letters in *figure* starts with a consonant?

7. If ten pool balls, each having a unique number from 1 to 10, are lined up along the side of a pool table, what is the probability that the balls numbered 5 and 6 occur together?

8. a. What is the probability that a 3-digit number formed from the digits 1, 2, 3, 4, and 5 is even?

b. What is the probability that it is odd? (Assume in both cases that digits can be repeated.)

9. Solve Problem 8 under the assumption that the digits cannot be repeated.

B

10. If four men and four women are to be arranged in a line-up, what is the probability that a random arrangement of the eight individuals has
   a. the men and women alternating?
   b. the men all next to each other?

11. From a set of cards numbered from 1 to 10,000, one is selected at random. What is the probability that the number on it
   a. is exactly divisible by 5?
   b. ends in an even number?

12. In an eight-cylinder engine, the even-numbered cylinders are on the left side and the odd-numbered cylinders are on the right side. A good firing order is an arrangement in which the two sides of the engine alternate when fired, starting with cylinder 1. (For example, 1, 4, 5, 8, 3, 2, 7, and 6 is a good firing order.) If the engine is wired by a novice who knows nothing about what he is doing, determine the probability that a good firing order will be chosen.

5.5 SOME PROBABILITY RULES

Probability of (E or F)

If *E* and *F* are two events, we would like to develop a formula for $P(E \text{ or } F)$, the probability of (*E* or *F*) occurring. Let's examine Figure 5-10, a Venn diagram for (*E* or *F*). Let the area of circle *E* represent $P(E)$, and let the area of circle *F* represent $P(F)$. The event (*E* or *F*) is illustrated by the entire shaded region within the two circles and $P(E \text{ or } F)$ is the area of this region. Also, $P(E \text{ and } F)$ is the area common to both regions *E* and *F* and represents the area of the common region. Thus,

$$P(E \text{ or } F) = \text{Area } E + \text{Area } F - \text{Area common to both } E \text{ and } F$$
$$= P(E) + P(F) - P(E \text{ and } F)$$

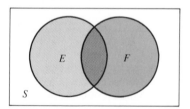

FIGURE 5-10 Venn diagram for (*E* or *F*)

Hence, the probability of (E or F) can be found by using the following rule:

$$P(E \text{ or } F) = P(E) + P(F) - P(E \text{ and } F) \tag{5-2}$$

Rule (5-2) is referred to as the **sum rule**. The following example illustrates its use.

EXAMPLE 5-28 If a single card is drawn from an ordinary deck of playing cards, find the probability that it will be red or a face card (jack, queen, or king).

Solution Let E represent drawing a red card and F represent drawing a face card. Then $P(E) = \frac{26}{52}$, $P(F) = \frac{12}{52}$, and $P(E \text{ and } F) = \frac{6}{52}$. Thus, using the sum rule, we have

$$P(E \text{ or } F) = P(E) + P(F) - P(E \text{ and } F)$$
$$= \frac{26}{52} + \frac{12}{52} - \frac{6}{52} = \frac{32}{52} = \frac{8}{13} \qquad \blacksquare$$

If E and F are mutually exclusive events, then (E and F) = $\varnothing$, the empty set. In this case $P(E \text{ and } F) = 0$ and the sum rule (5-2) becomes $P(E \text{ or } F) = P(E) + P(F)$. Thus, we have the following rule:

If E and F are mutually exclusive events, then

$$P(E \text{ or } F) = P(E) + P(F) \tag{5-3}$$

Probability of $\bar{E}$

Since E and $\bar{E}$ are mutually exclusive events, we can apply (5-3) to obtain

$$P(E \text{ or } \bar{E}) = P(E) + P(\bar{E})$$

In addition, since $P(E \text{ or } \bar{E}) = P(S) = 1$, we have

$$P(E) + P(\bar{E}) = 1$$

or

$$P(\bar{E}) = 1 - P(E) \tag{5-4}$$

Consider the following example.

EXAMPLE 5-29 The probability that Bob will finish his term paper is $\frac{3}{7}$. Find the probability that he will not finish his term paper.

Solution Let E be the event that Bob will finish his term paper. Then $\bar{E}$ is the event that he will not finish. Since $P(E) = \frac{3}{7}$, we have

$$P(\bar{E}) = 1 - P(E) = 1 - \frac{3}{7} = \frac{4}{7} \qquad \blacksquare$$

Conditional Probability

We would now like to determine a rule for computing $P(E \text{ and } F)$, the probability of (E and F). In order to do so, we need the notion of conditional probability. The probability of event E occurring when we know that event F has already occurred is called **conditional probability** and is written as $P(E|F)$. $P(E|F)$ means the probability that the event E will occur given the condition that the event F has occurred, or simply the probability of E given F. The following examples illustrate conditional probability by using selection-type problems. For these examples, **selection with replacement** means that the first object is returned before the second object is drawn. Similarly, **selection without replacement** means that the first object is not returned before the second object is drawn.

EXAMPLE 5-30   Two balls are drawn without replacement from a bag containing three white and two black balls. Find the probability that

a. the second ball is black given that the first ball is white.
b. the second ball is black given that the first ball is black.

Solution   **a.** If the first ball selected is white, there are four balls left in the bag, of which two are black. Therefore,

$$P(B|W) = \frac{2}{4} = \frac{1}{2}$$

See the following diagram:

$$4 \begin{cases} \cancel{W} \\ W \\ W \\ \left. \begin{matrix} B \\ B \end{matrix} \right\} 2 \end{cases}$$

b. If the first ball selected is black, there are four balls left in the bag, of which one is black. Therefore,

$$P(B|B) = \frac{1}{4}$$

See the following diagram:

$$4 \begin{cases} W \\ W \\ W \\ B \} \ 1 \\ \cancel{B} \end{cases}$$

■

EXAMPLE 5-31   A study was undertaken at a certain college to determine what relationship, if any, exists between mathematics ability and interest in mathematics. The ability and interest for 150 students were determined, with the results summarized in the following table:

		Interest	
Ability	**Low**	**Average**	**High**
Low	40	8	12
Average	15	17	18
High	5	10	25

If one of the participants in the study is chosen at random, what is the probability

a. of selecting a person who has low interest in mathematics?

b. of selecting a person with average ability?

c. that the person has high ability in mathematics given that the person selected has high interest in mathematics?

d. that the person has high interest in mathematics given that the person selected has average ability in mathematics?

Solution   We first find the row and column totals, as shown in Table 5-5.

Table 5-5 Row and Column Totals for Example 5-31

		Interest		
Ability	**Low**	**Average**	**High**	**Total**
Low	40	8	12	60
Average	15	17	18	50
High	5	10	25	40
Total	60	35	55	150

a. Since there are 60 participants with low interest out of a total of 150, the probability is $\frac{60}{150} = \frac{2}{5}$.

b. Since there are 50 participants with average ability out of a total of 150, the probability is $\frac{50}{150} = \frac{1}{3}$.

c. Of the 55 participants with high interest, 25 have high ability. Therefore, the probability is $\frac{25}{55} = \frac{5}{11}$.

d. Since 50 participants have average ability and of these, 18 have high interest, the probability is $\frac{18}{50} = \frac{9}{25}$.   ■

Note that parts (c) and (d) of Example 5-31 involve conditional probability. For part (c), we know the person chosen has high interest in mathematics. There are 55 college students who have high interest in mathematics; of these, 25 have high ability. Let HA denote high ability, HI denote high interest, and AA denote average ability. We therefore have

$$P(HA\,|\,HI) = \frac{25}{55} = \frac{5}{11}$$

This probability can be written as

$$P(HA|HI) = \frac{P(HA \text{ and } HI)}{P(HI)}$$

$$= \frac{\left(\frac{25}{150}\right)}{\left(\frac{55}{150}\right)}$$

$$= \frac{25}{55} = \frac{5}{11}$$

Also, with part (d) of Example 5-31, we find

$$P(HI|AA) = \frac{P(HI \text{ and } AA)}{P(AA)}$$

$$= \frac{\left(\frac{18}{150}\right)}{\left(\frac{50}{150}\right)}$$

$$= \frac{18}{50} = \frac{9}{25}$$

These results suggest the following formula for computing $P(E|F)$:

$$P(E|F) = \frac{P(E \text{ and } F)}{P(F)} \tag{5-5}$$

Notice that condition F in $E|F$ has the effect of reducing the size of the sample space. In Example 5-31, part (c), the size of the original sample space is 150, and after the condition HI is stipulated, the size of the reduced sample space (only those with high interest) is 55.

Conditional probability $P(E|F)$ can be illustrated using a Venn diagram in which probabilities are interpreted as areas (see Figure 5-11). If we are given the condition F, we are restricted to the outcomes in F. Hence, the elements in F comprise the reduced sample space. Then we are interested in the outcomes of E that are also contained in F. These are just the outcomes in (E and F). The probability of E given F is then obtained by dividing the area of (E and F) by the area of F. Thus,

$$P(E|F) = \frac{\text{Area } (E \text{ and } F)}{\text{Area } F}$$

$$= \frac{P(E \text{ and } F)}{P(F)} \tag{5-6}$$

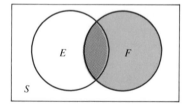

FIGURE 5-11 Venn diagram illustrating conditional probability

Probability of (E *and* F)

The probability of (E *and* F) is given by the formula

$$P(E \text{ and } F) = P(F)P(E|F) \tag{5-7}$$

Formula (5-7) was obtained by multiplying both sides of formula (5-6) by $P(F)$. Rule (5-7) is often referred to as the **product rule**. Note that (E *and* F) is logically equivalent to (F *and* E). As a consequence, we have

$$P(E \text{ and } F) = P(F)P(E|F)$$
$$= P(E)P(F|E)$$

A crucial point to remember when using the product rule for calculating $P(E \text{ and } F)$ is the location of the symbol for the event occupying the "$*$" position in the following expression:

$$P(E \text{ and } F) = P(*)P(\quad|*) \tag{5-8}$$

For example, if "$*$" is replaced by E, then (5-8) would read as

$$P(E \text{ and } F) = P(E)P(F|E)$$

If "$*$" is replaced by F, then expression (5-8) would read as

$$P(E \text{ and } F) = P(F)P(E|F)$$

The following example illustrates the product rule (5-7).

EXAMPLE 5-32   If two balls are drawn without replacement from a bag containing three red, two black, and one white ball, what is the probability of getting two red balls?

Solution   Let R_1 denote the event of getting a red ball on the first draw and R_2 denote the event of getting a red ball on the second draw. By using (5-7), we have

$$P(R_1 \text{ and } R_2) = P(R_1)P(R_2|R_1) = \left(\frac{3}{6}\right)\left(\frac{2}{5}\right) = \frac{1}{5} \qquad\blacksquare$$

EXAMPLE 5-33   Find the probability of drawing two red balls in Example 5-32 if the balls are drawn with replacement.

Solution   By using (5-7), we have

$$P(R_1 \text{ and } R_2) = P(R_1)P(R_2|R_1) = \left(\frac{3}{6}\right)\left(\frac{3}{6}\right) = \frac{1}{4}$$

Note that in this case we have

$$P(R_2|R_1) = P(R_2) \qquad\blacksquare$$

EXAMPLE 5-34   Refer to Example 5-31.

   a. Find the probability of selecting a person with high ability and high interest in mathematics.

b. Find the probability of selecting a person with low interest in mathematics and high ability in mathematics.

Solution   **a.** By using the product rule, we have

$$P(\text{HA and HI}) = P(\text{HA})P(\text{HI}|\text{HA})$$

$$= \left(\frac{40}{150}\right)\left(\frac{25}{40}\right) = \frac{25}{150} = \frac{1}{6}$$

b. By the product rule, we have

$$P(\text{LI and HA}) = P(\text{HA})P(\text{LI}|\text{HA})$$

$$= \left(\frac{40}{150}\right)\left(\frac{5}{40}\right) = \frac{5}{150} = \frac{1}{30} \qquad \blacksquare$$

The following two observations should be kept in mind when finding probabilities associated with compound events:

1. To find $P(E \text{ or } F)$, we associate "or" with the addition rule.
2. To find $P(E \text{ and } F)$, we associate "and" with the product rule.

Problem Set 5.5

A

1. Let $P(E) = .4$, $P(F|E) = .3$, and $P(E|F) = .4$.
  a. Find:
  (i) $P(E \text{ and } F)$ (ii) $P(F)$ (iii) $P(E \text{ or } F)$
  b. Are E and F mutually exclusive events?

2. If a six-sided die is tossed, find the probability of rolling
  a. a 2, given that the number rolled was odd.
  b. a 4, given that the number rolled was even.
  c. an even number, given that the number rolled was 6.

3. If a ball is drawn from a box containing three red, two white, and five black balls, find the probability that
  a. the ball is red.
  b. the ball is not red.
  c. the ball is red or black.
  d. the ball is not red and not white.

4. Two balls are drawn without replacement from a bag containing two white balls and two black balls. Find the probability that
  a. the second ball is white given that the first ball is white.
  b. the second ball is white given that the first ball is black.

5. If two cards are drawn without replacement from a 52-card deck, find the probability that
  a. the second card is a heart, given that the first is a heart.
  b. they are both hearts.
  c. the second is black, given that the first is a spade.
  d. the second is a face card, given that the first is a jack.
  e. the first card is a jack and the second card is a face card.

6. Two marbles are drawn without replacement from a jar with four black and six white marbles. Find the probability that
  a. both are white.
  b. both are black.
  c. the second is white, given that the first is black.
  d. they are different colors.

7. Suppose a hospital survey indicates that 35% of patients admitted have high blood pressure, 53% have heart trouble, and 22% have both high blood pressure and heart trouble. What is the probability that a patient selected at random has
  a. either high blood pressure or heart trouble or both?

b. high blood pressure given that he has heart trouble?

c. heart trouble given that he has high blood pressure?

8. The probability that a person smokes cigarettes is .45, and the probability that a person drinks beer is .58. If the probability that a person smokes given that he drinks is .21:

a. Find the probability that a person drinks and smokes.

b. Find the probability that a person drinks given that he smokes.

9. The accompanying table displays relative frequencies for red/green color blindness for males and females, where M represents males, $\bar{M}$ represents females, C represents color-blind, and $\bar{C}$ represents not color-blind.

	M	$\bar{M}$
C	.042	.007
$\bar{C}$	.485	.466

If a person is chosen at random, use the table to find the following probabilities:

a. $P(M)$ **b.** $P(M|C)$

c. $P(\bar{M}|C)$ **d.** $P(C)$

e. $P(\bar{M}|\bar{C})$ **f.** $P(M \text{ and } C)$

g. $P(C|M)$

10. A bank has observed that most customers at the tellers' windows either cash a check or make a deposit. The table indicates the transactions for teller A in 1 day.

	Check cashed	No check cashed
Deposit	50	20
No deposit	30	10

Let C represent cashing a check and D represent making a deposit. Suppose a customer is chosen at random from teller A's customers. Express each of the following in words and find its value:

a. $P(C)$ **b.** $P(D)$

c. $P(C \text{ and } D)$ **d.** $P(C \text{ or } D)$

e. $P(C|D)$ **f.** $P(\bar{C}|D)$

g. $P(\bar{D}|D)$

11. In an attitudinal survey on strict gun-control legislation administered to 800 U.S. adult males, the following results were obtained:

	Stand	
	In favor	Against
Shot a gun	75	200
Never shot a gun	425	100

If one of the 800 males is chosen at random, use relative frequencies to approximate probabilities and determine each of the following:

a. P(in favor)

b. P(shot a gun and against)

c. P(against | never shot a gun)

d. P(shot a gun and in favor)

e. P(shot a gun)

B

12. Let $P(E) = .2$ and $P(F) = .3$. Answer each of the following questions. Where appropriate, give an example.

a. Can $P(E \text{ or } F) = .5$?

b. Can $P(E \text{ or } F) = .7$?

c. Can $P(E \text{ or } F) = .4$?

d. Can $P(E \text{ and } F) = .2$?

e. Can $P(E \text{ and } F) = .3$?

f. Can $P(E \text{ and } F) = .1$?

g. Can $P(E \text{ and } F) = .4$?

13. During a class lecture, a history teacher stated that the probability of Israel and Syria both sending representatives to a peace conference is .8. Later during the same lecture, he stated that the probability of Syria sending a representative is .5. Do you believe these two statements? Explain.

14. If $P(E) = .2$, $P(F) = .3$, and $P(F|E) = .5$, rank the following events according to increasing probability: E, F, $\bar{E}$, $(E \text{ and } F)$, $(E \text{ or } F)$, $(E \text{ or } \bar{E})$, and $(F \text{ and } \bar{F})$.

15. Give convincing arguments for the following two facts, and illustrate each using an example:

a. If $P(E) < P(F)$, then $P(E \text{ or } F)$ is at least as large as $P(E)$.

b. If $P(E) < P(F)$, then $P(E \text{ and } F)$ is at most as large as $P(E)$.

16. Determine a formula for $P(E \text{ and } \bar{F})$ not involving conditional probability. [*Hint:* Draw a Venn diagram.]

5.6 INDEPENDENT EVENTS ─────────────────────────────────

If E and F are events such that the occurrence of F in no way influences the occurrence of E, then E and F are called **independent events**. Stated differently, E and F are independent events if

$$P(E|F) = P(E) \tag{5-9}$$

For independent events E and F, we also have, as a consequence of (5-7),

$$P(E \text{ and } F) = P(F)P(E|F)$$
$$= P(F)P(E)$$

Thus, for independent events E and F, we have the rule

$$P(E \text{ and } F) = P(E)P(F) \tag{5-10}$$

Consider the following examples.

EXAMPLE 5-35   Which of the following pairs of events are independent?

    a. E = getting a head on a toss of a penny
       F = getting a head on a toss of a dime
    b. E = Mary's first child being a boy
       F = Mary's second child being a girl
    c. E = it will rain in Frostburg today
       F = John fails his Math 101 exam today
    d. Two balls are drawn without replacement from a bag containing white balls and black balls.
       E = drawing a white ball on the first draw
       F = drawing a black ball on the second draw
    e. The same events as in part (d), but the balls are drawn with replacement

    Solution   **a.** The two events are independent.
       b. The two events are independent.
       c. The two events are independent.
       d. The two events are not independent; they are dependent.
       e. The two events are independent.   ■

EXAMPLE 5-36   If two coins are tossed, find $P(HT)$.

    Solution   The probability $P(HT)$ is the probability of getting a head on the first coin and a tail on the second coin. Since getting a head on one coin and getting a tail on another are independent events, by using (5-10) we have,

$$P(HT) = P(H)P(T) = \left(\frac{1}{2}\right)\left(\frac{1}{2}\right) = \frac{1}{4} \qquad ■$$

EXAMPLE 5-37   Refer to the problem illustrated in Example 5-31. If a student is selected and E is the event that the student has low ability in mathematics and F is the event that the student has high interest in mathematics, are E and F independent events?

Solution   No. The probability of E given F is given by

$$P(E|F) = \frac{12}{55}$$

and

$$P(E) = \frac{60}{150} = \frac{2}{5}$$

As a result of (5-9), E and F are not independent events, since $P(E|F) \neq P(E)$, i.e, $\frac{12}{55} \neq \frac{2}{5}$. ∎

Care must be taken not to confuse the concepts of mutually exclusive events and independent events. There is no general relationship between the two types of events. Events can be mutually exclusive and not independent or they can be independent and not mutually exclusive. Or events can be dependent and not mutually exclusive. (See Problems 10–13.)

Problem Set 5.6

A

1. If two cards are drawn with replacement from a 52-card deck, find the probability that
  a. the second card is a heart, given that the first is a heart.
  b. both cards are hearts.
  c. the second is black, given that the first is a spade.
  d. the second is a face card, given that the first is a jack.
  e. the first card is a club or the second card is an ace.

2. Two marbles are drawn with replacement from a jar with four black and six white marbles. Find the probability that
  a. both are white.
  b. both are black.
  c. the second is white, given that the first is black.
  d. the first is black and the second is white.
  e. one is black and the other is white.

3. Suppose $P(E) = .5$, $P(F) = .6$, and $P(E \text{ and } F) = .1$.
  a. Are E and F independent events? Why?
  b. Are E and F mutually exclusive events? Why?

4. The accompanying table displays relative frequencies for red/green color blindness, where C represents the event a person is color-blind and M represents the event a person is male. Are events C and M independent events?

	M	$\bar{M}$
C	.042	.007
$\bar{C}$	.485	.466

5. If E and F are events such that $P(E) = .4$, $P(F) = .3$, and $P(E|F) = .4$, are E and F independent events? Explain.

6. If three coins are tossed once, find:
  a. $P(\text{HHH})$
  b. $P(\text{THT})$ [*Note:* "THT" means getting a tail on the first coin, a head on the second coin, and a tail on the third coin.]
  c. $P(\text{exactly one head})$
  d. $P(\text{at least one head})$
  e. $P(\text{at most two heads})$

7. Freshmen at a small college are classified according to GPA in high school and SAT score in mathematics. The results are summarized in the accompanying table. A student is selected at random. If E is the event that the student has an average high school GPA and F is the event that the student has a low SAT mathematics score, are E and F independent events? Explain.

SAT score	High school GPA		
	Low	Average	High
Low	50	30	50
Average	20	30	20
High	30	40	30

8. If events E and F are mutually exclusive events with $P(E) = .2$ and $P(F) = .3$, find:
  a. $P(E \text{ or } F)$ **b.** $P(E|F)$

9. Workers at a small industrial plant are classified according to religion and sex. The results are displayed in the table. If E is the event a person is male and F is the event a person is Jewish, are E and F independent events?

	Sex	
Religion	Male	Female
Protestant	30	20
Catholic	45	30
Jewish	45	30
Other	7	8

B

10. Suppose a coin is tossed once. Let $E = \emptyset$ and $F = \{H\}$.
    a. Are E and F independent events? Explain.
    b. Are E and F mutually exclusive events? Explain.

11. Two coins are tossed. Let E represent the event of getting a head on the first coin and F represent getting two heads or two tails. Show that:
    a. E and F are independent events.
    b. E and F are not mutually exclusive events.

12. Suppose a coin is tossed twice. Let $E = \{HH\}$ and $F = \{TT\}$.
    a. Are E and F mutually exclusive events? Explain.
    b. Are E and F independent events? Explain.

13. Suppose two six-sided dice are tossed once. Let $E = \{(1, 2), (2, 1)\}$ and $F = \{(1, 2), (3, 4)\}$.
    a. Are E and F independent events? Explain.
    b. Are E and F mutually exclusive events? Explain.

14. If E and F are mutually exclusive events, what is the value of $P(E|F)$?

15. Suppose E and F are events with $P(E) > 0$ and $P(F) > 0$. If E and F are independent events, can E and F be mutually exclusive events? Explain.

16. A given manufacturing process produces 5% defective parts. Twenty percent of all parts are produced by machine A. There is a 10% probability that a part is defective given that it was produced by machine A.
    a. What is the probability that a tested part is defective and was produced by machine A?
    b. What is the probability that a part came from machine A given that it is defective?

5.7 RANDOM VARIABLES

It will be convenient in later work to relate the outcomes in an experiment to real numbers. When outcomes of an experiment can be associated with real numbers, they are easier to analyze. Unfortunately, not all experiments result in outcomes that are real numbers. For example, suppose you work in the quality control department for a manufacturer of microcomputers. You have been assigned the task of inspecting four microcomputers chosen at random from the last batch produced and classifying them as defective or nondefective. This experiment results in the following sample space of qualitative data:

DNDN	DDND	NDND	NNDN
DNND	DDDN	NDDN	NNND
DNDD	DDNN	NDDD	NNDD
DNNN	DDDD	NDNN	NNNN

These outcomes are not real numbers, but if each outcome is associated with the number of defective microcomputers, we can associate a unique real number with each outcome. For example, the outcome DNDN can be assigned the number 2, the outcome DNND can be assigned the number 2, the outcome NDNN can be assigned the number 1, and the outcome NNNN can be assigned the number 0. Such a pairing of the outcomes in a sample space of an experiment with unique real numbers is called a *random variable*.

Random Variable

A **random variable** is a rule (or function) that assigns unique real numbers to each outcome in a sample space of an experiment. The random variable in the quality control at the beginning of this section is the *number* of defective microcomputers in a batch of four. It has the five possible values 0, 1, 2, 3, and 4.

The following are other examples of random variables:

Experiment	Random variable
1. Toss five coins.	Number of heads obtained
2. Observe customers at a bank during 1 hour.	Number of customers
3. Buy 12 computers.	Number of defective computers
4. Weigh a person.	Weight in pounds
5. Hit a golf ball.	Distance it travels
6. Audit ten tax returns.	Number of returns containing errors
7. Audit ten tax returns.	Number of mistakes on line 33
8. Observe an employee work for an 8-hour period.	Time spent by the employee in unproductive work
9. Poll ten people concerning candidate A's chances for election.	Proportion in favor of A
10. Treat 50 people with a headache using pill A.	The number cured

The values of a random variable are typically equated with the random variable. This usually causes no problems and is a matter of convenience. For example 1 above, if we let X denote the number of heads obtained when five coins are tossed, the possible values of the random variable are 0, 1, 2, 3, 4, and 5; we denote this by $X = \{0, 1, 2, 3, 4, 5\}$. Strictly speaking, this is incorrect, since X is the rule (function) and $\{0, 1, 2, 3, 4, 5\}$ represents its range values. Throughout this section, capital letters such as X, Y, and Z will denote random variables, and small letters such as x, y, and z will denote their values (range values).

If the values of a random variable can be counted, the random variable is called **discrete**; if the values of the random variable cannot be counted, the random variable is called **continuous**. Examples 4, 5, and 8 in the previous list represent continuous random variables, while the remaining random variables are discrete. The number of units sold, the number of defective computers, and any other random variable concerned with counting are discrete. Any random variable dealing with measurement, such as weight, time, temperature, and distance, is continuous. Example 8 above involves a continuous random variable since the number of hours of unproductive work may be any value in the interval 0–8.

Probability Distributions

Probabilities can be associated with the values of a discrete random variable. If X is a discrete random variable and we are interested in determining the probability of the outcomes associated with k, a value of the random variable X, we write $P(X = k)$. When there is no chance for confusion we write "$P(X = k)$" as "$P(k)$." For example, suppose the random variable X represents the number of heads obtained when two coins are tossed. The possible values for X are 0, 1, 2. Then,

$$P(X = 0) = P(0) = P(\text{TT}) = \frac{1}{4}$$

$$P(X = 1) = P(1) = P(\text{HT, TH}) = \frac{1}{2}$$

$$P(X = 2) = P(2) = P(\text{HH}) = \frac{1}{4}$$

A **probability distribution table** for a discrete random variable is a two-column table, with one column representing the values of the random variable and the other representing the associated probabilities. For the two-coin-toss example, a probability distribution table is shown in Table 5-6. Notice that the probability column must always sum to 1.

Table 5-6 Probability Distribution for Two-Coin-Toss Example

x	$P(x)$
0	$\frac{1}{4}$
1	$\frac{1}{2}$
2	$\frac{1}{4}$

Probability Functions

Sometimes it is convenient to express the relationship between the values of a random variable and its associated probabilities in terms of a rule. Such a rule is called a **probability function** or **probability distribution function**. Consider the following example.

EXAMPLE 5-38   Suppose five balls of the same size are numbered 1 to 5 and placed in a bag. An experiment involves selecting a ball at random. Let the random variable X denote the value of the ball selected.

a. Construct a probability table for X.
b. Find the probability function for X.

Solution   The five balls are equally likely to be selected.

a. The following is the probability table for the discrete random variable X:

x	$P(x)$
1	$\frac{1}{5}$
2	$\frac{1}{5}$
3	$\frac{1}{5}$
4	$\frac{1}{5}$
5	$\frac{1}{5}$

b. The function that relates the values of X with its associated probabilities is $g(x)$ $= \frac{1}{5}$; $x = 1, 2, 3, 4, 5$. Thus, the probability function for the random variable X is g.   ■

Probability Graphs

Graphs can be constructed for probability functions. For a discrete random variable, a graph of the probability function can be constructed using vertical line segments. The values of the random variable are displayed on the horizontal axis and the probabilities are displayed on the vertical axis. At each value of the random variable, a vertical line segment is constructed with a height equal to the probability of the random variable. The sum of the lengths of the vertical line segments must equal 1. For example, suppose the following is a probability table for a discrete random variable X:

x	$P(x)$
1	.2
2	.3
3	.4
4	.1

The graph of the corresponding probability function is shown in Figure 5-12. Note that the sum of the lengths of the vertical line segments is 1.

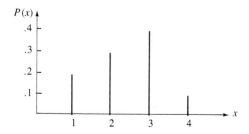

FIGURE 5-12 Probability function for discrete random variable

For continuous random variables, we use areas instead of vertical line segments to represent probabilities. For example, suppose the values of the random variable X are all real numbers between 0 and 1 and the probability function of X is f, defined by

$$f(x) = \begin{cases} 2x & \text{if } 0 \le x \le 1 \\ 0 & \text{elsewhere} \end{cases}$$

A graph for the probability function f is shown in Figure 5-13. The area of the region bounded by the line $f(x) = 2x$, the x-axis, and the lines $x = 0$ and $x = 1$ is 1. The region is a right triangle whose area is found by using the formula

$$A = \left(\frac{1}{2}\right)(\text{Base})(\text{Height})$$

$$= \left(\frac{1}{2}\right)(1)(2) = 1$$

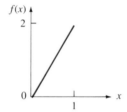

FIGURE 5-13 Probability function for continuous random variable

Since probabilities are associated with areas for continuous random variables, probabilities can be found only for intervals, such as $a < x < b$. The probability of any particular value x for a continuous random variable must necessarily equal 0, since a vertical line segment drawn at the value x and extending upward to the probability function must have an area of 0. As a result, if X is a continuous random variable, then

$$P(a < X < b) = P(a \le X \le b)$$

For the continuous random variable of Figure 5-13, suppose we want to find the probability that the random value X is between .25 and .5, i.e., $P(.25 \le X \le .5)$. This probability is equal to the area under the line whose equation is $f(x) = 2x$ bounded by the x-axis and the lines $x = .25$ and $x = .5$, as illustrated in Figure 5-14. The figure formed is a trapezoid. Recall that the area of a trapezoid is given by $A = h(b_1 + b_2)/2$, where h is the altitude and b_1 and b_2 are the bases. This formula yields the area (or probability) for the shaded region:

$$A = \frac{(.25)(.5 + 1)}{2} = \frac{(.25)(1.5)}{2} = .1875$$

A thorough study of continuous random variables involves calculus and is beyond the scope of this text. For the remainder of this section, we will deal only with discrete random variables and their probabilities.

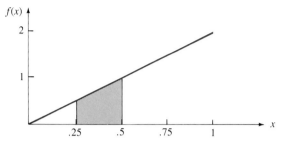

FIGURE 5-14 $P(.25 \le X \le .5)$

EXAMPLE 5-39   Consider the following probability function for a discrete random variable X:

$$f(x) = \frac{x}{10} \quad \text{for } x = 1, 2, 3, 4$$

    a. What are the values of the random variable X?
    b. What probability is associated with each value of X?
    c. Construct a probability table for X.
    d. Construct a probability graph for X.

Solution   **a.** The values of X are 1, 2, 3, and 4.
    b. The probability associated with $x = 1$ is

$$P(X = 1) = f(1) = \frac{1}{10} = .1$$

    Similarly, the probabilities for $x = 2$, 3, and 4 are found to be .2, .3, and .4, respectively.
    c. A probability table is as follows:

x	$P(x) = f(x)$
1	.1
2	.2
3	.3
4	.4
	$\overline{1.0}$

d.

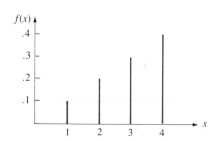

Mean of a Discrete Random Variable

Recall that one formula for calculating the population mean μ is

$$\mu = \frac{\sum(fx)}{N}$$

where f is the frequency of a distinct measurement x and N is the size of the population. This can be rewritten as

$$\mu = \sum\left[x\left(\frac{f}{N}\right)\right]$$

If we let the relative frequency f/N represent the probability of x occurring, $P(x)$, then we have the following formula for the **mean of a random variable** X:

$$\mu = \sum[xP(x)] \tag{5-11}$$

where μ represents the mean of the random variable X.

A large industrial plant ran a campaign encouraging car-pooling among its employees. The data in Table 5-7 were recorded for all plant employees to monitor the effects of the campaign.

Table 5-7 Car-Pooling Data

Number of occupants per car (x)	f	xf	Relative frequency (f/N)
1	425	425	.450
2	235	470	.249
3	205	615	.217
4	52	208	.055
5	22	110	.023
6	6	36	.006
	945	1864	1.000

The population mean μ as defined in Chapter 3 is identical to (5-11). This is demonstrated as follows:

$$\mu = \frac{\sum(xf)}{N}$$

$$= \frac{425 + 470 + 615 + 208 + 110 + 36}{945}$$

$$= \frac{425}{945} + \frac{470}{945} + \frac{615}{945} + \frac{208}{945} + \frac{110}{945} + \frac{36}{945}$$

$$= (1)\left(\frac{425}{945}\right) + (2)\left(\frac{235}{945}\right) + (3)\left(\frac{205}{945}\right) + (4)\left(\frac{52}{945}\right) + (5)\left(\frac{22}{945}\right) + (6)\left(\frac{6}{945}\right)$$

$$= 1.97$$

This can be rewritten as

$$\mu = (1)(.450) + (2)(.249) + (3)(.217) + (4)(.055) + (5)(.023) + (6)(.006)$$
$$= (1)P(1) + (2)P(2) + (3)P(3) + (4)P(4) + (5)P(5) + (6)P(6)$$

But this is just an instance of formula (5-11).

The mean of a random variable X is also called the **expected value of X** and is denoted by $E(X)$. As a result, we have the following results for a discrete random variable:

$$\mu = E(X) = \sum [xP(x)]$$

Analogous formulas also exist for the case of continuous random variables. The symbol "X" can be used to describe a population, where X represents one randomly selected value from the population. So, $P(5 < X < 10)$ can be interpreted as the proportion of values in the population between 5 and 10.

EXAMPLE 5-40   A large hotel has found that the number of air-conditioning units that must be replaced each summer has the following probability table:

Number replaced	Probability
0	.35
1	.30
2	.20
3	.10
4	.05

Find the expected value of the number of air conditioners that must be replaced each summer and interpret your answer.

Solution   Let the random variable X denote the number of air conditioners replaced. We will compute the product of x and $P(x)$ for each value of X. The results are organized in Table 5-8. From the table, we see that $E(X) = \sum xP(x) = 1.20$. Hence, if the number of air-conditioning units that must be replaced is recorded for a large number of years, the average number of replacements would be 1.20 units per year.

Table 5-8 Calculation of $E(X)$ for Example 5-40

x	$P(x)$	$xP(x)$
0	.35	0
1	.30	.30
2	.20	.40
3	.10	.30
4	.05	.20
		1.20

Variance of a Discrete Random Variable

Recall from Chapter 3 that the population variance can be defined as

$$\sigma^2 = \frac{\sum f(x - \mu)^2}{N}$$

where f is the frequency associated with the measurement x and N is the size of the population. This can be rewritten as

$$\sigma^2 = \sum \left[\frac{(x - \mu)^2 f}{N} \right]$$

Again, if we let the relative frequency f/N represent the probability of x occurring, $P(x)$, then we have the following formula for the **variance of a random variable X**:

$$\sigma^2 = \sum [(x - \mu)^2 P(x)] \tag{5-12}$$

where σ^2 is the variance of the random variable X.

For the car-pooling data presented in Table 5-7, the population mean is $\mu = 1.97$ and the population variance is

$$\sigma^2 = \frac{\sum [(x - \mu)^2 f]}{N}$$

$$= [(1 - 1.97)^2(425) + (2 - 1.97)^2(235) + (3 - 1.97)^2(205)$$
$$+ (4 - 1.97)^2(52) + (5 - 1.97)^2(22) + (6 - 1.97)^2(6)]/945$$

$$= 1.197$$

These computations can be rewritten as

$$\sigma^2 = (1 - 1.97)^2 \left(\frac{425}{945} \right) + (2 - 1.97)^2 \left(\frac{235}{945} \right) + (3 - 1.97)^2 \left(\frac{205}{945} \right)$$

$$+ (4 - 1.97)^2 \left(\frac{52}{945} \right) + (5 - 1.97)^2 \left(\frac{22}{945} \right) + (6 - 1.97)^2 \left(\frac{6}{945} \right)$$

$$= (1 - 1.97)^2 P(1) + (2 - 1.97)^2 P(2) + (3 - 1.97)^2 P(3) + (4 - 1.97)^2 P(4)$$
$$+ (5 - 1.97)^2 P(5) + (6 - 1.97)^2 P(6)$$

But this is just an instance of formula (5-12).

The **standard deviation** of a random variable is defined to be the positive square root of the variance and is denoted by σ.

Consider the following example.

EXAMPLE 5-41   For the random variable in Example 5-40, find the variance σ^2 and standard deviation σ.

Solution   In Example 5-40, μ was found to be 1.2. We organize our computations for σ^2 in the following table:

x	$P(x)$	$x - \mu$	$(x - \mu)^2$	$(x - \mu)^2 P(x)$
0	.35	-1.2	1.44	.504
1	.30	$-.2$	.04	.012
2	.20	.8	.64	.128
3	.10	1.8	3.24	.324
4	.05	2.8	7.84	.392
				1.360

Hence, $\sigma^2 = 1.360$ and $\sigma = \sqrt{1.360} = 1.17$.   ∎

The following formula is computationally much easier to use than (5-12) to find the variance σ^2 of a random variable and is mathematically equivalent to (5-12):

$$\sigma^2 = \sum [x^2 P(x)] - \mu^2 \tag{5-13}$$

We will use this formula to find σ^2 for the random variable illustrated in Example 5-41. The calculations are organized in the following table:

x	x^2	$P(x)$	$x^2 P(x)$
0	0	.35	0
1	1	.30	.30
2	4	.20	.80
3	9	.10	.90
4	16	.05	.80
			2.80

The mean μ was previously found to be $\mu = 1.2$. Thus, the variance of the random variable X is

$$\sigma^2 = \sum x^2 P(x) - \mu^2 = 2.80 - (1.2)^2 = 1.36$$

Note that this agrees with the result obtained in Example 5-41.

Problem Set 5.7

A

1. The probability table for the number of telephone calls, X, received by Mr. Jones in a day is as follows:

x	$P(x)$
0	.40
1	.23
2	.17
3	.09
4	.11

Construct a probability graph and find each of the following:
a. $P(X = 1)$ **b.** $P(1 < X \le 3)$
c. $P(X \ge 1)$ **d.** $E(X)$
e. σ^2

2. The number X of fish dinners sold in 1 hour at a local restaurant is described by the following probability table:

x	0	1	2	3	4	5	6
$P(x)$	.14	.16	.30	.14	.13	.08	.05

Construct a probability graph and find:

a. $P(X \leq 1)$ **b.** $P(X > 1)$

c. $P(2 \leq X \leq 4)$ **d.** $P(X \leq 5)$

e. μ **f.** σ

3. Let X represent the number of heads obtained when three coins are tossed. Find μ and σ for X.

4. Consider the probability function f defined by $f(x) = (5 - x)/10$ for $x = 1, 2, 3, 4$. Graph the probability function and find the expected value and standard deviation of X.

5. Based on past history, the accompanying frequency table reports the number of automobiles sold per day, X, for a car dealer. Construct a probability table for the random variable X.

X	Number of days
0	44
1	87
2	128
3	234
4	297
5	155
6	30
7	25

6. Refer to Problem 5. Find

a. $P(X \leq 5)$

b. the expected number of cars sold. Interpret your result.

c. the standard deviation

7. A raffle offers a first prize of $1000, two second prizes of $500, and 20 prizes of $20 each. If 10,000 tickets are sold at $.50 each, find the expected winnings if one ticket is bought. Also find σ^2.

8. A builder is considering a job that promises a profit of $25,000 with a probability of .8 or a loss (due to poor weather, strikes, and so forth) of $10,000 with a probability of .2.

a. What is the builder's expected profit?

b. What is the standard deviation of the profit?

9. Suppose X is a discrete random variable with probability function f defined by

$$f(x) = \begin{cases} \dfrac{x}{55} & \text{for } x = 1, 2, 3, \ldots, 10 \\ 0 & \text{elsewhere} \end{cases}$$

a. Show that $\sum P(x) = 1$.   **b.** Find $P(X = 4)$.

c. Find $P(X \leq 3)$.   **d.** Find $P(X \leq 9)$.

e. Find $P(2 \leq X \leq 4)$.   **f.** Find $P(2 < X < 4)$.

g. Find μ.   **h.** Find σ.

10. Find the expected number of male children in a family with five children. Assume that male and female births are equally likely to occur.

11. You have to be at school in 20 minutes and there are two routes that you can take to get there. The mean times to reach school are 12 and 16 minutes, respectively. Is the 12-minute route the better route to take? Explain.

B

12. Mr. Baker wants to insure his house for $80,000. The insurance company estimates a total loss may occur with a probability of .004, a 50% loss with a probability of .02, and a 25% loss with a probability of .08. If the insurance company will pay no benefits for any other partial loss, what premium will Mr. Baker be required to pay each year if the insurance company wants to make an average profit of $500 per year on all policies of this type?

13. If two dice are tossed and the sum is recorded, let the random variable X denote this sum. Find μ and σ for X.

14. If X is a continuous random variable, with $P(X \leq 3) = .45$, and $P(X \geq 4) = .40$, find:

a. $P(X < 3)$

b. $P(4 < X)$

c. $P(3 \leq X \leq 4)$

15. If two cards are drawn with replacement from a standard deck of 52 cards, find the expected number of diamond cards that can result.

16. Prove that $\sigma^2 = \sum x^2 P(x) - \mu^2$.

17. The following problem is known as the St. Petersburg Paradox. Peter agrees to toss a coin until a head shows face up. If a head shows on the first toss, he agrees to pay you $1. If not, he agrees to give you $2 if he gets a head on the second toss, $4 if a head shows on the third toss, $8 if a head shows on the fourth toss, and so on. The number of dollars he will give you doubles with each additional throw. In order to break even, how much should Peter charge to play the game? Explain.

CHAPTER SUMMARY _____

The concepts and principles of elementary probability were introduced in this chapter. The fundamental theorem of counting was introduced as an aid in calculating more difficult probabilities. We saw that random variables can be used to provide numerical descriptions of experimental outcomes. We learned how to associate probability distributions with random variables and how to calculate means, variances, and standard deviations for discrete random variables.

C   H   A   P   T   E   R   R   E   V   I   E   W

IMPORTANT TERMS _____

classical probability method
compound event (E and F)
compound event (E or F)
conditional probability
continuous random variable
discrete random variable
empirical probability
empty set
equally likely outcomes
event

event (not E)
expected value of a random variable
experiment
favorable outcomes
fundamental theorem of counting
independent events
inferential statistics
mean of a random variable
mutually exclusive events
objective method

odds
outcome
prime number
probability distribution table
probability function
probability histogram
product rule
random variable
relative frequency
sample space
selection with replacement

selection without replacement
simple events
standard deviation
subjective method
subset
sum rule
tree diagram
variance of a random variable
Venn diagram

IMPORTANT SYMBOLS _____

$\bar{E}$, the event (not E)
$a:b$, the odds are a to b
$P(E)$, the probability of event E

$P(E|F)$, conditional probability
μ, mean of a random variable

σ^2, variance of a random variable
σ, standard deviation of a random variable

$E(X)$, expected value of X
(E or F), event (E or F)
(E and F), event (E and F)

IMPORTANT FACTS AND FORMULAS _____

1. *Fundamental theorem of counting*: If one event can be performed in m different ways, and after it has been done, another event can be performed in n different ways, then the two events can be performed, one following the other, in a total of mn different ways.

2. *Sum rule*: $P(E \text{ or } F) = P(E) + P(F) - P(E \text{ and } F)$

3. If E and F are mutually exclusive events, $P(E \text{ or } F) = P(E) + P(F)$.

4. $P(E|F) = \dfrac{P(E \text{ and } F)}{P(F)}$, provided $P(F) \neq 0$.

5. *Product rule:* $P(E \text{ and } F) = P(F)P(E|F)$

6. If E and F are independent events, then $P(E \text{ and } F) = P(E)P(F)$

7. $\mu = E(X) = \sum[xP(x)]$, mean of a random variable

8. $\sigma^2 = \sum[(x - \mu)^2 P(x)]$, variance of a random variable

9. $\sigma^2 = \sum x^2 P(x) - \mu^2$, computational formula for variance of a random variable

REVIEW PROBLEMS

1. The records of a particular hospital indicate that 18% of all its patients are admitted for surgery, 30% are admitted for obstetrics, and 5% are admitted for both surgery and obstetrics.
 a. What is the probability that a randomly selected patient will be admitted for surgery, obstetrics, or both?
 b. What is the probability that a randomly selected patient is not admitted for surgery?
 c. What is the probability that a randomly selected patient will be a surgery patient and will not receive obstetrical treatment?

2. Suppose E and F are two events with $P(E) = \frac{1}{2}$ and $P(F) = \frac{1}{3}$.
 a. If E and F are mutually exclusive events, find $P(E \text{ or } F)$.
 b. If E and F are independent events, find $P(E|F)$, $P(F|E)$, $P(E \text{ or } F)$, and $P(E \text{ and } F)$.
 c. Can $P(E \text{ or } F) = \frac{11}{12}$? Explain.
 d. Can $P(E \text{ or } F) = \frac{1}{6}$? Explain.
 e. Can $P(E \text{ and } F) = .6$? Explain.

3. Seventy-five students were polled and asked to name their favorite drink. Their responses are as follows:

Beer	13	Wine	2
Soft drink	40	Water	7
Iced tea	4	Hot tea	1
Coffee	7	Whiskey	1

Suppose one of the surveyed students is chosen at random.
 a. Find the probability that the student indicated beer or wine.
 b. Find the probability that the student did not indicate coffee.

4. Employees at a particular plant were classified according to sex and political party affiliation. The results follow:

	Political affiliation		
Sex	**Democrat**	**Republican**	**Independent**
Male	40	50	5
Female	18	8	4

If an employee is chosen at random, find the probability that the employee is a
 a. male.
 b. Republican.
 c. female Democrat.
 d. Republican given that she is a female.
 e. male given that he is a Republican.

5. If a six-sided die is tossed, let the random variable X denote the number showing face up.
 a. Construct a probability table for X.
 b. Construct a probability graph for X.
 c. Find $E(X)$ and interpret your result.
 d. Find σ.

6. A person tosses three coins. If he gets three heads or three tails, he receives $10. If he does not receive three heads or three tails, he pays out $10. What are his expected winnings?

7. Mr. Jones can make $5000 with probability .4 or lose $2000 with probability .6 if he invests in Ajax Company stock.
 a. What is his expected gain?
 b. What is the standard deviation of his gain?

8. If three dice are tossed, let the random variable X denote the number of 2s obtained.
 a. Construct a probability table for X.
 b. Construct a probability graph for X.
 c. Find $E(X)$.
 d. Find σ^2.

9. A coin is biased (not fair) in such a way that a head is twice as likely to occur as a tail when the coin is tossed. If the coin is tossed twice, let the

random variable X denote the number of heads obtained.
  a. Construct a probability table for X.
  b. Find μ and σ.
  c. Suppose the random variable Y denotes the number of tails obtained. Find μ and σ for Y.

10. The Ace Bank stock is currently selling for $10 a share. An investor plans to buy shares and to hold the stock for 1 year. Let X denote the price of the stock (in dollars) after 1 year. The probability table for X is shown here.

x	$P(x)$
10	.35
11	.25
12	.20
13	.15
14	.05

  a. Find the expected price of the stock after 1 year.
  b. What is the expected gain per share of stock over the 1-year period?
  c. What percentage return on the investment is reflected by the expected gain per share?
  d. Find the variance in the price of the stock over the 1-year period.
  e. Find the variance in the gain per share of the stock over the 1-year period.
  f. Construct a graph for the probability function.

CHAPTER ACHIEVEMENT TEST

(6 points) **1.** A restaurant offers six sandwiches, ten types of drinks, four varieties of soups, and three desserts on its menu. How many different lunches can be ordered if one type of sandwich, one drink, one soup, and one dessert are desired?

(32 points) **2.** A study is made of religious affiliation and political party. The following results were tabulated:

Political party	Religion		
	Protestant (*P*)	Catholic (*C*)	Jewish (*J*)
Democrat (*D*)	10	15	25
Republican (*R*)	20	30	40
Independent (*I*)	5	15	5

A person is randomly selected from the group surveyed.
  a. Find $P(R)$. b. Find $P(J)$.
  c. Find $P(R \text{ and } J)$. d. Find $P(R \text{ or } J)$.
  e. Find $P(R|J)$. f. Find $P(C|D)$.
  g. Are events J and R independent?
  h. Are J and R mutually exclusive events? Why?

(20 points) **3.** Two males and three females apply for two teaching positions. Since all the applicants are equally qualified, two are randomly selected to fill the positions.
  a. List a sample space for this experiment.
  b. How many ways can two males be selected?
  c. How many ways can two females be selected?
  d. Find the probability that two females are hired.
  e. Find the probability that one female and one male are hired.

(20 points) **4.** The following table reports the annual deaths of male cancer patients over age 65 from the five major types of cancer:

Location of cancer	Number of deaths
Colon	8,000
Lung	12,500
Pancreas	3,000
Prostate	10,800
Stomach	3,200

If a deceased male cancer patient is selected, what is the probability that he
a. died from lung cancer?
b. died from one of the two leading causes?
c. died from cancer of the colon or from cancer of the pancreas?
d. did not die from cancer of the stomach?

(12 points) **5.** You pay $.50 to spin a roulette wheel with 38 equally-spaced positions. If the wheel stops on the position you choose, you win $5.00; otherwise, you lose. If you play one game, find your expected winnings.

(10 points) **6.** If two cards are drawn with replacement from a standard deck of playing cards, let the random variable X denote the number of hearts. Find μ and σ for X.

6

BINOMIAL DISTRIBUTIONS

Chapter Objectives

In this chapter you will learn:

- *what a binomial experiment is*
- *the language associated with binomial experiments*
- *how to calculate binomial coefficients*
- *how to calculate binomial probabilities*
- *how to use the binomial tables for calculating binomial probabilities*

- *how to calculate the mean of a binomial random variable*
- *how to calculate the variance and standard deviation of a binomial random variable*
- *how to construct probability distribution graphs for binomial random variables*

Chapter Contents

Binomial distributions form an important class of discrete distributions in statistics. They are used to describe a wide variety of processes in many areas.

6.1 BINOMIAL EXPERIMENTS

Consider the experiment of tossing a fair coin three times and observing the number of heads that result. This experiment possesses the following characteristics:

1. It consists of three identical trials of tossing one coin.
2. Each trial results in exactly one of two outcomes—heads or tails.
3. The trials are independent; the outcome of one trial does not affect the outcome of any other.
4. The probability of getting a head remains constant from trial to trial, the probability being .5.

There are eight different, equally likely possibilities that can result from tossing a fair coin three times. They are listed as follows:

HHT   THT   HTH   TTH   HHH   THH   HTT   TTT

If the discrete random variable X denotes the number of heads obtained, then the following probability distribution table can be used to organize the results:

x	$P(x)$
0	$\frac{1}{8}$
1	$\frac{3}{8}$
2	$\frac{3}{8}$
3	$\frac{1}{8}$

This probability distribution has the graph shown in Figure 6-1. Notice that the graph is symmetric about $x = 1.5$ heads.

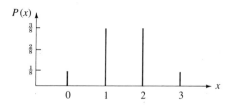

FIGURE 6-1 Probability distribution for $X =$ number of heads in three tosses of a fair coin

The experiment just described is an example of a class of experiments called **binomial experiments**. In general, a binomial experiment is an experiment that has the following four properties:

1. The experiment consists of *n* identical trials.
2. Each trial results in exactly one of two outcomes, called *success* or *failure*. A success is denoted by *S* and a failure is denoted by *F*.
3. The *n* trials are independent.
4. The probability of success, *p*, remains constant from trial to trial.

The probability distribution for the number of successes is called a **binomial distribution**. Each of the following situations can be modeled using a binomial experiment:

- Administering a cold medication with a success rate of .90 to ten people with colds and observing the number of cures
- Tossing a thumbtack eight times and observing the number of times it lands point up
- Guessing on a ten-question true/false test and observing the number of correct answers
- Tossing a die six times and observing the number of 3s that result
- Observing a baseball player with a .400 batting average and counting the number of base hits he gets in the next six times at bat

The following symbols are used to describe binomial experiments:

$$S = \text{Success}$$
$$F = \text{Failure}$$
$$p = P(\text{success})$$
$$1 - p = P(\text{failure})$$
$$n = \text{Number of trials}$$
$$x = \text{Number of successes}$$

Note that for a binomial experiment with *n* trials, *x* can be any one of the $(n + 1)$ values $0, 1, 2, 3, \ldots, n$. Consider the following examples.

EXAMPLE 6-1   A cross-fertilization of related species of white-flowered and blue-flowered plants produces offspring of which 20% are white-flowered plants. Six blue-flowered plants were paired and crossed with six white-flowered plants, and it was found that there were two white-flowered plants among the six offspring. Is this a binomial experiment? If so, identify a trial, a success, and the values for *p*, *n*, and *x*.

Solution   This problem is an application of Mendel's theory of inherited characteristics. A trial consists of crossing a blue-flowered plant with a white-flowered plant. A *success* is obtaining a white-flowered offspring and $p = P(S) = .20$. According to Mendelian theory, the trials are independent. As a result, the experiment is binomial with $n = 6$ and $x = 2$.   ■

EXAMPLE 6-2   Mary tossed a six-sided die ten times to determine the number of 1s resulting. She obtained three 1s. Is this a binomial experiment? If so, what constitutes a trial, a success, and a failure? What are the values of *n*, *p*, and *x*?

Solution   This is a binomial experiment in which a trial consists of tossing a die. A success is getting a 1, and a failure is getting an outcome other than 1. The probability of success is $p = \frac{1}{6}$, while the probability of failure is $\frac{5}{6}$. There are $n = 10$ trials, and *x*, the number of successes, is 3. If instead of recording the number of 1s that resulted, Mary had recorded the number of times each number on the die appeared face up, the experiment would not have been binomial, since each trial would have resulted in one of six outcomes.   ∎

Success is the term used to describe the outcome of interest for a binomial experiment. It does not necessarily correspond to a "good" event, as the following example illustrates.

EXAMPLE 6-3   A certain type of medication will not cause a skin reaction in 90% of the people who use it. We are interested in the number of people out of the next five who use it and have a skin reaction. Identify a trial, a success, the values for *p* and *n*, and the possible values of *x*.

Solution   A trial is treating a person with the medication, and a success is having a skin reaction. Hence, $p = 1 - .90 = .10$, $n = 5$, and *x* can be any one of the six values 0, 1, 2, 3, 4, or 5.   ∎

The assumption of independence of trials for a binomial experiment implies that the probability of success, *p*, remains constant from trial to trial. However, the converse is not true. The probability of success can remain constant from trial to trial, but the trials need not be independent.

In many situations the independence assumption is not satisfied, particularly in situations where sampling is done without replacement. For example, suppose a town has five licensed restaurants, of which two currently have at least one serious health-code violation. There are two inspectors, each of whom will inspect one restaurant during the coming week. The names of the restaurants are written on different slips of paper and thoroughly mixed. Each inspector randomly draws one of the slips without replacement of the first before the second is drawn. A trial is inspecting a restaurant, and the number of trials is 2. A success *S* is interpreted as observing no health-code violations. Let S_1 denote a success on the first trial, S_2 denote a success on the second trial, and F_1 denote a failure on the first trial. Then the probability of getting a success on the first trial is

$$P(S_1) = \frac{3}{5} = .6$$

There are two possibilities for getting a success on the second trial:

$$S_2 = (S_1 \text{ and } S_2) \quad \text{or} \quad (F_1 \text{ and } S_2)$$

By using the sum rule, we have

$$P(S_2) = P(S_1 \text{ and } S_2) + P(F_1 \text{ and } S_2)$$

And by using the product rule, we obtain

$$P(S_2) = \left(\frac{3}{5}\right)\left(\frac{2}{4}\right) + \left(\frac{2}{5}\right)\left(\frac{3}{4}\right)$$

$$= \frac{3}{10} + \frac{3}{10} = .6$$

However, the trials are not independent, since

$$P(S_2 | S_1) = \frac{P(S_1 \text{ and } S_2)}{P(S_1)}$$

$$= \frac{\left(\frac{3}{10}\right)}{\left(\frac{6}{10}\right)} = .5$$

and therefore

$$P(S_2 | S_1) \neq P(S_2)$$

Binomial Coefficients

Suppose it has been determined over a long period of time that 80% of the people making plane reservations follow through with their flight plans; 20% of the potential customers do not show up at departure time. If four people make plane reservations and we are interested in the number of people who show up at flight time, we have an example of a binomial experiment with $n = 4$ and $p = .80$. The number of people who show up at flight time is a random variable with values $x = 0, 1, 2, 3,$ and 4. Let's label the outcome "shows up" as a success (S) and the outcome "no-show" as a failure (F). If x denotes the number of successes, the following 16 possibilities can result:

$x = 0$	$x = 1$	$x = 2$	$x = 3$	$x = 4$
FFFF	SFFF	SSFF	SSSF	SSSS
	FSFF	SFSF	SSFS	
	FFSF	SFFS	SFSS	
	FFFS	FSSF	FSSS	
		FSFS		
		FFSS		

The number of possibilities that can result for each value of x is as follows:

Value of x (number of S's)	0	1	2	3	4
Number of outcomes	1	4	6	4	1

For a given value of x, the number of possible outcomes containing x successes is called a **binomial coefficient** and is denoted by $\binom{n}{x}$, where n is the number of trials and x is the number of successes. For our illustration, we have $\binom{4}{0} = 1$, $\binom{4}{1} = 4$, $\binom{4}{2} = 6$, $\binom{4}{3} = 4$, and $\binom{4}{4} = 1$.

A binomial coefficient $\binom{n}{x}$ counts the number of outcomes for a binomial experiment containing n trials that can result in x successes. In other words, a binomial coefficient indicates the number of combinations or unordered selections of n trials that can result in x successes.

Toward the goal of developing a formula for calculating binomial coefficients, we need the use of the factorial symbol $n!$ (read "**n factorial**"), which is defined by

$$n! = (n)(n - 1)(n - 2)(n - 3) \cdots (2)(1)$$

Thus, for example,

$$4! = (4)(3)(2)(1) = 24$$
$$3! = (3)(2)(1) = 6$$
$$2! = (2)(1) = 2$$
$$1! = 1$$

We define $0!$ to equal 1.

The formula for the binomial coefficient $\binom{n}{x}$ is then given by

$$\binom{n}{x} = \frac{n!}{x!(n - x)!} \tag{6-1}$$

where n is the number of trials, x is the number of successes, and $(n - x)$ is the number of failures. Note that the number of successes plus the number of failures equals the number of trials:

$$x + (n - x) = n$$

EXAMPLE 6-4   Evaluate the following binomial coefficients:

a. $\binom{6}{4}$

b. $\binom{6}{2}$

c. $\binom{5}{3}$

d. $\binom{5}{2}$

Solution   By using (6-1), we have

a. $\binom{n}{x} = \frac{n!}{x!(n - x)!}$

$\binom{6}{4} = \frac{6!}{4!(6 - 4)!} = \frac{(6)(5)(4!)}{4!2!} = \frac{(6)(5)}{2} = 15$

b. $\binom{6}{2} = \dfrac{6!}{2!4!} = 15$

c. $\binom{5}{3} = \dfrac{5!}{3!2!} = \dfrac{(5)(4)3!}{3!2!} = \dfrac{(5)(4)}{2} = 10$

d. $\binom{5}{2} = \dfrac{5!}{2!3!} = 10$, by part (c).   ■

Binomial coefficients have a general use in mathematics. It can be shown that the binomial coefficient, $\binom{n}{x}$, represents the number of combinations of n objects taken x at a time. By using this fact, it is easy to demonstrate that $\binom{n}{x}$ is also equal to the number of combinations of n objects taken $n - x$ at a time (see Problem 9, Problem Set 6.1). That is,

$$\binom{n}{x} = \binom{n}{n - x}$$

By using this fact, we can often simplify computations involving binomial coefficients. Consider the following examples:

a. $\binom{10}{8} = \binom{10}{2} = \dfrac{10!}{2!8!} = \dfrac{(10)(9)8!}{2!8!} = 45$

To get 45, we multiply the largest two factors of 10!, 10, and 9, and divide this produced by $2! = (2)(1) = 2$. Note that both the numerator and denominator of $\binom{10}{2}$ have the same number of factors—namely, 2.

b. $\binom{10}{7} = \binom{10}{3} = \dfrac{10!}{3!7!} = \dfrac{(10)(9)(8)7!}{3!7!} = 120$

To get 120, we multiply the largest three factors of 10!, 10, 9, and 8, and divide this product by $3! = (3)(2)(1) = 6$. Note again that both the numerator and denominator of $\binom{10}{3}$ have three factors.

Pascal's Triangle

Binomial coefficients can also be obtained from a triangular array of numbers called **Pascal's triangle**. The following array is a partial triangle containing seven rows:

	\multicolumn{7}{c}{*x* **successes**}
n **trials**	**0**	**1**	**2**	**3**	**4**	**5**	**6**
0	1						
1	1	1					
2	1	2	1				
3	1	3	3	1			
4	1	4	6	4	1		
5	1	5	10	10	5	1	
6	1	6	15	20	15	6	1

It is easy to construct the triangle. The first column ($x = 0$) has all 1s. In order to obtain a number in any row (not in the first column), we just add the number directly above the one we want and the one above and to the left. For example, to find $\binom{6}{2}$, we go to the row corresponding to $n = 5$ ($6 - 1 = 5$) and add 5 and 10. Thus, $\binom{6}{2} = 15$. To find $\binom{6}{5}$ we add 5 and 1. The process of adding numbers in the previous row may be continued to construct as many rows as desired. The entries in the row corresponding to $n = 7$ are

$$1 \quad 7 \quad 21 \quad 35 \quad 35 \quad 21 \quad 7 \quad 1$$

EXAMPLE 6-5   Trees in a particular forest are infested with a certain parasite. If 15 trees are selected at random for study, how many outcomes can result in

a. 3 infested trees?
b. no infested trees?
c. 15 infested trees?
d. at most 2 infested trees?

Solution   By using Pascal's triangle or formula (6-1), we have

a. $\binom{15}{3} = \dfrac{(15)(14)(13)}{(3)(2)(1)} = 455$

b. $\binom{15}{0} = 1$

c. $\binom{15}{15} = 1$

d. $\binom{15}{0} + \binom{15}{1} + \binom{15}{2} = 1 + 15 + 105 = 121$ ■

Problem Set 6.1

A

1. Interpret the following binomial coefficients in terms of a binomial experiment. Do not calculate the coefficients.

a. $\binom{8}{4}$ **b.** $\binom{6}{0}$

c. $\binom{10}{1}$ **d.** $\binom{7}{7}$

For Problems 2–7, determine if the experiment is binomial. If the experiment is binomial, indicate the values for n, p, and x, and identify a trial and a success.

2. Forty percent of all campers at a certain summer camp contract poison ivy. Eight students attend and we are interested in the number who contract poison ivy.

3. A survey of the residents in a certain town indicated that 30% of the residents favor building a community center and 70% are opposed. Ten residents are randomly surveyed and asked if they favor the proposed new community center.

4. A six-sided die is tossed five times and the sum of the faces showing is to be determined.

5. Out of the next ten babies born at Memorial Hospital, the number of males is to be determined. (Assume that male and female births are equally likely.)

6. At a certain college, 40% of entering freshmen eventually graduate. Of 30 freshmen who enroll next semester, the number who eventually graduate is to be determined.

7. Four candidates are running for governor. A survey is conducted to determine voters' support for the four candidates.

8. Evaluate each of the following:

   a. 5!   **b.** 0!   **c.** $\binom{4}{0}$ **d.** $\binom{5}{5}$

   e. $\binom{11}{4}$ **f.** $\binom{18}{7}$ **g.** $\binom{12}{5}$ **h.** $\binom{12}{7}$

   i. $\binom{25}{23}$ **j.** $\binom{4}{2}(.2)^2(.8)^2$

B

9. Show that $\binom{n}{x} = \binom{n}{n-x}$. How does this fact relate to Pascal's triangle? Explain.

10. For a given value of n, find the sum of all the binomial coefficients.

11. Show that $\binom{n+1}{x} = \binom{n}{x} + \binom{n}{x-1}$. How does this fact relate to Pascal's triangle? Explain.

6.2   CALCULATING BINOMIAL PROBABILITIES

In this section, we will be concerned with calculating the probabilities associated with the outcomes in a binomial experiment.

Binomial Probability Formula

Let's return to the plane-reservation illustration in Section 6.1, a binomial experiment with $n = 4$ and $p = .80$. Recall the following 16 possibilities:

$x = 0$	$x = 1$	$x = 2$	$x = 3$	$x = 4$
FFFF	SFFF	SSFF	SSSF	SSSS
	FSFF	SFSF	SSFS	
	FFSF	SFFS	SFSS	
	FFFS	FSSF	FSSS	
		FSFS		
		FFSS		

Note that each outcome associated with a given value of x has the same probability, since the trials are independent. For example, if $x = 2$,

$$P(SSFF) = P(S)P(S)P(F)P(F) = (.80)^2(.20)^2$$

and

$$P(SFSF) = P(S)P(F)P(S)P(F) = (.80)^2(.20)^2$$

The other four possibilities having $x = 2$ successes also have probabilities equal to $(.80)^2(.20)^2$. The probabilities associated with the five possible values of x are summarized in Table 6-1.

Table 6-1 Probabilities of Individual Outcomes for Binomial Experiment with $n = 4$ and $p = .80$

x	Probability of each outcome
0	$(.80)^0(.20)^4$
1	$(.80)^1(.20)^3$
2	$(.80)^2(.20)^2$
3	$(.80)^3(.20)^1$
4	$(.80)^4(.20)^0$

Four observations concerning the entries in the table should be made:

1. The exponent of $p = P(S)$ is equal to the number of successes x.
2. The exponent of $1 - p = P(F)$ is equal to the number of failures $(n - x)$.
3. The two bases, p and $(1 - p)$, must sum to 1.
4. The sum of the exponents, x and $(n - x)$, must be n.

Since for each value of x there are $\binom{4}{x}$ possible outcomes containing x successes, we have the probabilities shown in Table 6-2 for the number of successes, x. (Recall from algebra that $p^0 = 1$ for all nonzero values of p.) Observe from the table that the probabilities fit the following pattern:

$$\binom{n}{x} p^x (1 - p)^{n-x}$$

Table 6-2 Probabilities for Binomial Experiment with $n = 4$ and $p = .80$

x	$P(x)$
0	$(1)(.80)^0(.20)^4 = .002$
1	$(4)(.80)^1(.20)^3 = .026$
2	$(6)(.80)^2(.20)^2 = .154$
3	$(4)(.80)^3(.20)^1 = .410$
4	$(1)(.80)^4(.20)^0 = .410$

A general formula for calculating $P(x)$, the probability of obtaining x successes in a binomial experiment having n trials with $P(S) = p$, can now be stated:

$$P(x) = \binom{n}{x} p^x (1 - p)^{n-x} \tag{6-2}$$

\# ways to get x successes in n trials   \# successes   \# failures

$P(S)$ $P(F)$

This formula is referred to as the **binomial probability formula**.

As an example to illustrate the use of (6-2), let's find the probability of $x = 2$ successes when $n = 5$ and $p = .3$. By using (6-2) we have

$$P(x) = \binom{n}{x} p^x (1 - p)^{n-x}$$

$$P(2) = \binom{5}{2} (.3)^2 (.7)^{5-2}$$

$$= (10)(.09)(.343) = .3087 \simeq .309$$

Note that we will round all binomial probabilities to three decimal places in this chapter.

The following procedure will prove to be useful for determining probabilities associated with a binomial distribution problem:

1. Determine what constitutes a trial and a success.
2. Determine the probability of success, p.
3. Determine the number of trials, n.
4. Determine the value(s) of x, the number of successes.
5. Use the binomial probability formula to find $P(x)$:

$$P(x) = \binom{n}{x} p^x (1 - p)^{n-x}$$

Consider the following examples.

EXAMPLE 6-6   A recent survey showed that 60% of college students smoke. What is the probability that of five students surveyed, three of them smoke?

Solution   By using the above five-step procedure, we have:

1. A trial consists of determining whether a college student smokes. A success is finding a student who smokes.
2. $p = .60$
3. $n = 5$
4. $x = 3$
5. $P(x) = \binom{n}{x} p^x (1 - p)^{n-x}$

$$P(3) = \binom{5}{3} (.60)^3 (.40)^{5-3}$$

$$= (10)(.216)(.16) = .3456 \simeq .346$$ ■

EXAMPLE 6-7   If

$$P(12) = \frac{18!}{12!6!} (.3)^6 (.7)^{12}$$

represents a binomial probability, find:

a. the number of successes.
b. $P(S)$.
c. the number of failures.
d. $P(F)$.

Solution   **a.** $P(12)$ indicates that $x = 12$.
b. $P(S) = p = .7$, the base that has the exponent of 12.
c. $(n - x) = 6$, the exponent of the base .3.
d. $P(F) = 1 - .7 = .3$, the other base.   ■

Binomial Probability Tables

Table 1 of Appendix B gives values for $P(x)$ for selected values of p and values of n through 25. To use the table, locate the proper section for the value of n. Then under the column labeled by p and across the row labeled by x, find the value of $P(x)$. Consider the following example.

EXAMPLE 6-8   If a baseball player with a batting average of .600 comes to bat five times in a game, what is the probability he will get three hits?

Solution   The following values are identified: $n = 5$, $p = .6$, and $x = 3$. By using the binomial probability tables, we find

Thus, $P(3) = .346$.

By using the binomial probability formula, we get

$$P(3) = \binom{5}{3}(.6)^3(.4)^2$$

$$= (10)(.6)^3(.4)^2 = .3456 \simeq .346$$

which agrees with the answer obtained from the binomial probability tables.   ■

Note that for given values of n and p, the sum of the probabilities for all values of x in the table may not equal 1.000 exactly. For example, for $n = 4$ and $p = .8$, the sum of the table values is $.002 + .026 + .154 + .410 + .410 =$

1.002. This is because the entries in the table have been rounded to three decimal places. An entry listed in the table as .000 means that the entry is 0 when rounded to three decimal places.

Problem Set 6.2

A

For each of the following problems, you should verify that the problem involves a binomial experiment and identify the values of p, n, and x before solving the problem. Your answers to the probability problems may differ from those given in the text, depending on whether you use the binomial probability formula or the binomial probability tables. The differences are due to rounding errors in the tables.

1. It was found that 40% of the campers at a certain summer camp contracted poison ivy. If eight students attend camp this summer, find the probability that
   a. all will contract poison ivy.
   b. two will contract poison ivy.
   c. at most three will contract poison ivy.
   d. at least seven will contract poison ivy.

2. A survey of the residents in a certain town showed that 30% favored brand X toothpaste. If ten of the residents are randomly surveyed at a grocery store, what is the probability that
   a. no one favors brand X?
   b. four favor brand X?
   c. at least eight favor brand X?
   d. at most two favor brand X?
   e. all favor brand X?

3. Dick has been observed to make 65% of his free-throw shots during basketball games. What is the probability that Dick will make
   a. three of the next six shots?
   b. five of the next ten shots?
   c. all of the next four shots?

4. In a certain city, 40% of the registered voters are Democrats. If nine voters are randomly selected, find the probability that
   a. two of them are Democrats.
   b. at least one of them is a Democrat.
   c. at least eight are Democrats.
   d. at most three are Democrats.

5. Assume that males and females are equally likely to be born (technically this is not correct, since approximately 100 female births occur for every 106 male

births). Find the probability that of six babies born,
   a. there are three boys.
   b. there are at most three boys.
   c. there are four girls.
   d. none are boys.

6. At a certain college, 35% of the entering freshmen graduate. Of the next five freshmen who enroll, what is the probability that
   a. they all graduate?
   b. four graduate?
   c. three do not graduate?
   d. at least four graduate?
   e. none graduate?

7. A certain type of seed has a germination rate of 83%. If 12 seeds are planted, find the probability that
   a. they all germinate.
   b. 10 germinate.
   c. 11 germinate.
   d. at most 2 germinate.
   e. at least 10 germinate.

8. If five dice are tossed, what is the probability that
   a. three of them show a 1?
   b. all of them show a 1?
   c. at least three of them show a 1?
   d. four of them do not show a 1?

9. A test consists of ten multiple-choice questions with five possible answers. If a person guesses at each question, what is the probability that
   a. all questions are answered correctly?
   b. at most three questions are answered correctly?
   c. five questions are answered correctly?
   d. seven questions are answered incorrectly?

10. If 20 fair coins are tossed, find the probability that
   a. 11 heads are obtained.
   b. 15 heads are obtained.
   c. at least 16 heads are obtained.
   d. either no heads or 20 heads are obtained.

B

11. A **Poisson distribution** often results when counting the number of times a particular event occurs in

a given time period. For example, the number of phone calls arriving at a switchboard between 1 and 2 P.M., the number of customers between 11 A.M. and 1 P.M. at a McDonald's restaurant, and the number of alpha-particle emissions of carbon 14 in .5 second all follow Poisson distributions. If X denotes the number of events occurring in a given time period and $\lambda = E(X)$, then the probability formula for X is given by

$$f(x) = \frac{e^{-\lambda}\lambda^x}{x!}, \quad x = 0, 1, 2, \ldots$$

where $e \simeq 2.718$. If $\lambda = 4$, find:
a. $P(X = 1)$ **b.** $P(X = 3)$
c. $P(X \leq 3)$ **d.** $P(X \geq 4)$

12. Telephone calls enter a switchboard at an average

rate of one every 3 minutes. If the number of calls follows a Poisson distribution, what is the probability of two or more calls arriving in a 6-minute period?

13. If the number of trials in a binomial experiment is greater than or equal to 9, a Poisson distribution can be used to approximate the binomial distribution, where $\lambda = np$. That is, if X is a binomial random variable with parameters n and p, then

$$\binom{n}{x}p^x(1 - p)^{n-x} \simeq \frac{(np)^x e^{-np}}{x!}$$

If a manufacturer of integrated circuits knows that 2% of its integrated circuits are defective, approximate the probability that a batch of 150 integrated circuits contains at most one defective.

6.3 *FINDING PARAMETERS FOR BINOMIAL DISTRIBUTIONS*

Recall that the probability distribution for the number of successes x resulting from a binomial experiment is referred to as a **binomial distribution**. A binomial distribution can be thought of as an infinite population consisting of all possible values of x resulting from repeating a binomial experiment indefinitely. In this section we want to find the mean and variance of a binomial distribution and examine the shape of its probability graph.

A binomial probability distribution can be displayed as a table or as a collection of $(n + 1)$ ordered pairs $(x, P(x))$ whose first component is the number of successes and whose second component is the probability associated with the given number of successes. This display can then be used to construct a graph for the distribution.

As an example, consider the binomial experiment of tossing a fair coin five times. We first construct a probability distribution table for the number of heads obtained (Table 6-3). We note that since $p = .5$, the probability of obtaining any particular arrangement of the five coins is $(\frac{1}{2})(\frac{1}{2})(\frac{1}{2})(\frac{1}{2})(\frac{1}{2}) = \frac{1}{32}$. Thus, $P(x) = \binom{5}{x}(\frac{1}{32})$ for $x = 0, 1, 2, 3, 4$, and 5. This information can be used to construct the probability distribution graph for the number of successes (see Figure 6-2). Note that the probability distribution is symmetric about $x = 2.5$, which appears to be the "center" of the distribution.

A binomial distribution accurately describes, to a reasonable degree, the repetitions of binomial experiments involving repeated trials, each with two outcomes. A binomial distribution describes all possible outcomes and corresponding probabilities for a binomial experiment. It is very similar to a relative frequency distribution that describes what has occurred. But it differs from a

Table 6-3 Probability Distribution Table for Number of Heads Obtained in Five Tosses of Fair Coin

x	$\binom{5}{x}$	$P(x)$
0	1	$\frac{1}{32}$
1	5	$\frac{5}{32}$
2	10	$\frac{10}{32}$
3	10	$\frac{10}{32}$
4	5	$\frac{5}{32}$
5	1	$\frac{1}{32}$

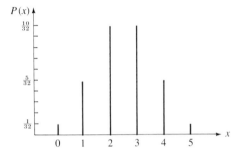

FIGURE 6-2 Probability distribution graph for number of heads in five tosses of a fair coin

relative frequency distribution since it projects into the future and describes— using probability—what theoretically should occur.

If a binomial experiment is repeated a fixed number of times, the resulting distribution of relative frequencies for each of the $(n + 1)$ possible values for the number of successes x is called an **empirical binomial distribution**. As the number of repetitions for the binomial experiment increases, the empirical distributions approach the theoretical binomial distribution. The two distributions are illustrated in Figure 6-3 for the binomial experiment consisting of tossing a fair coin six times and recording the number of heads obtained. For the empirical distribution, the experiment was repeated 100 times. Note that the graph of the empirical distribution resembles the graph of the theoretical distribution.

Many people confuse binomial experiments with binomial distributions; there is a difference. A binomial experiment involving n trials results in exactly one of the $(n + 1)$ possible outcomes for the associated binomial random variable. A binomial distribution, on the other hand, describes the probabilities associated with the $(n + 1)$ values (outcomes) for the random variable X denoting the number of successes that can result.

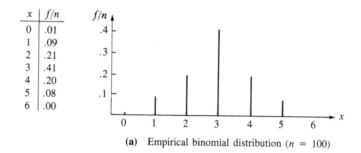

(a)   Empirical binomial distribution ($n = 100$)

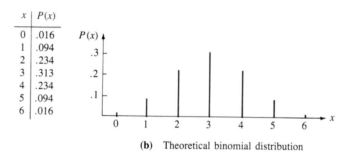

(b)   Theoretical binomial distribution

FIGURE 6-3 Comparison of empirical and theoretical binomial distributions

Mean of a Binomial Distribution

We now want to find the average of a binomial distribution; that is, we want to find the mean for a binomial random variable. For a binomial distribution, the mean μ is the expected value $E(x)$ for the number of successes x. Recall from Section 5.7 that

$$E(X) = \sum x P(x)$$

Thus, the mean of a binomial distribution is given by

$$\mu = E(X) = \sum x P(x) \tag{6-3}$$

For the experiment of tossing a fair coin five times, the mean of the random variable X denoting the number of heads is

$$\mu = \sum x P(x)$$
$$= 0P(0) + 1P(1) + 2P(2) + 3P(3) + 4P(4) + 5P(5)$$
$$= 0 + 1\left(\frac{5}{32}\right) + 2\left(\frac{10}{32}\right) + 3\left(\frac{10}{32}\right) + 4\left(\frac{5}{32}\right) + 5\left(\frac{1}{32}\right)$$
$$= \frac{80}{32} = 2.5$$

How should we interpret this answer of 2.5 heads? Surely, it does not mean that if we toss a fair coin five times, we will obtain 2.5 heads. Instead, if a large number of people each tossed a coin five times and we recorded the number of heads obtained for each person, the average of the number of heads obtained would be close to 2.5.

Formula (6-3) for calculating the mean μ for a binomial distribution could require considerable computation. It is intuitively reasonable that a machine that produces defective parts 1% of the time when batches of 500 are produced would, over a long period of time, produce an average number of defective parts per batch equal to $(.01)(500) = 50$. Similarly, it is reasonable that in a large number of repeated experiments of tossing a fair coin 300 times, we can expect to obtain $(.5)(300) = 150$ heads. In both of these situations, the expected value (or average) was obtained by multiplying the number of trials n by the probability of success p. As a consequence, to find the mean for a binomial distribution, we have the following formula, which is mathematically equivalent to (6-3) and is simpler to use:

$$\mu = np \tag{6-4}$$

For our coin-tossing example above,

$$\mu = np = 5(.5) = 2.5$$

Note that this answer agrees with the answer found by using formula (6-3).

EXAMPLE 6-9   The probability that a patient recovers from lung surgery is .95. If 25 people have this surgery, find the mean number of recoveries and interpret the result.

Solution   By using formula (6-4), we have

$$\mu = np = (25)(.95) = 23.75$$

We interpret this as follows: if a large number of hospitals each performed lung surgery on 25 patients and recorded the number who recovered, the average number of recoveries for all the hospitals involved would be close to 23.75.

∎

Variance of a Binomial Distribution

Recall from Section 5.7 that the variance of a random variable X is given by

$$\sigma^2 = \sum[(x - \mu)^2 P(x)] \tag{6-5}$$

The following example illustrates finding the variance for a binomial distribution.

EXAMPLE 6-10   Dave and Rick are gambling by rolling a die. If the die shows a 2, 3, or 4, Dave wins; otherwise, Rick wins. If the die is tossed three times, find the mean and variance for the distribution of the number of times Dave wins.

Solution   The experiment is binomial with $n = 3$ and $p = \frac{1}{2}$. By using (6-4), we have

$$\mu = np = (3)\left(\frac{1}{2}\right) = 1.5$$

To find the variance, we need the following probability distribution table for X, the number of times Dave wins:

x	$P(x)$
0	$\frac{1}{8}$
1	$\frac{3}{8}$
2	$\frac{3}{8}$
3	$\frac{1}{8}$

The following table will be used to facilitate the computations involved in calculating σ^2:

x	$P(x)$	$x - \mu$	$(x - \mu)^2$	$(x - \mu)^2 P(x)$
0	$\frac{1}{8}$	-1.5	2.25	.28125
1	$\frac{3}{8}$	$-.5$	.25	.09375
2	$\frac{3}{8}$	.5	.25	.09375
3	$\frac{1}{8}$	1.5	2.25	.28125
				.75

Hence, $\sigma^2 = .75$.   ∎

For a binomial distribution, formula (6-5) is mathematically equivalent to the following formula, which is simpler to use:

$$\sigma^2 = np(1 - p) \tag{6-6}$$

This is the preferred formula for computing the variance σ^2 of a binomial distribution, because it involves fewer computations. Note that by using (6-6) to calculate the variance σ^2 for the binomial random variable in Example 6-10, we get

$$\sigma^2 = np(1 - p) = (3)(.5)(.5) = .75$$

which agrees with the result obtained by using (6-5).

The **standard deviation for a binomial distribution** is given by

$$\sigma = \sqrt{np(1 - p)} \tag{6-7}$$

Consider another example.

EXAMPLE 6-11   Refer to Example 6-10. Suppose Dave wins if the die shows a 2 or 4. Find the mean and standard deviation for the distribution of the number of times Dave wins.

Solution   For this binomial experiment, $p = \frac{1}{3}$. By using (6-4), we get

$$\mu = np = 3\left(\frac{1}{3}\right) = 1$$

Applying (6-6), we obtain

$$\sigma^2 = np(1 - p) = 3\left(\frac{1}{3}\right)\left(\frac{2}{3}\right) = \frac{2}{3}$$

Hence, by using (6-7), we find the standard deviation to be

$$\sigma = \sqrt{\frac{2}{3}} \simeq .82$$

∎

Let's consider one more example.

EXAMPLE 6-12   A student takes a multiple-choice test with 50 questions, each with five possible choices. If the student guesses at each question, find the mean and standard deviation of the distribution of the number of questions answered correctly. Also, find the mean and standard deviation for the distribution of the number of questions missed by the student.

Solution   For this binomial experiment, $n = 50$ and $p = .20$. According to formula (6-4), the mean is

$$\mu = np = (50)(.2) = 10$$

To determine the standard deviation, we use (6-7):

$$\sigma = \sqrt{np(1 - p)} = \sqrt{(50)(.2)(.8)} = 2.83$$

The number of questions missed forms a binomial distribution with $n = 50$ and $p = .80$. Thus,

$$\mu = np = (50)(.80) = 40$$

and

$$\sigma = \sqrt{np(1 - p)} = \sqrt{(50)(.80)(.20)} = 2.83$$

Note that the standard deviations are equal and the sum of the two means equals n.

∎

Shapes of Binomial Distribution Graphs

The values of n and p determine the shape of a binomial distribution. Figures 6-4(a)–(g) indicate how the graphs of binomial distributions change for various values of p when $n = 6$. Seven different values for p are used.

Notice that in Figure 6-4(g) the graph is symmetric about its mean, $\mu = 3$. In general, if $p = \frac{1}{2}$, then we have

$$\mu = np = \frac{n}{2}$$

and the graph of the binomial distribution is symmetric about its mean $\mu = n/2$. And if $p \neq .5$, a binomial graph is not symmetric about its mean. In Figures

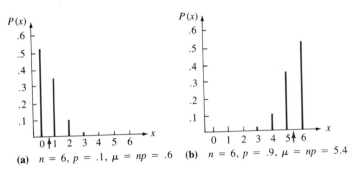

(a) $n = 6, p = .1, \mu = np = .6$ (b) $n = 6, p = .9, \mu = np = 5.4$

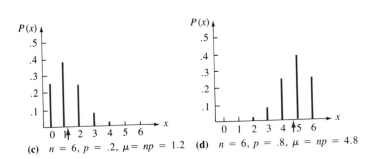

(c) $n = 6, p = .2, \mu = np = 1.2$ (d) $n = 6, p = .8, \mu = np = 4.8$

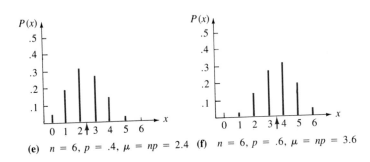

(e) $n = 6, p = .4, \mu = np = 2.4$ (f) $n = 6, p = .6, \mu = np = 3.6$

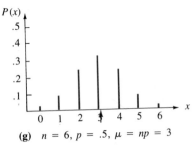

(g) $n = 6, p = .5, \mu = np = 3$

FIGURE 6-4 Binomial distribution graphs for $n = 6$ and
different values of p

6-4(a), (c), (e), and (g), notice that, as the value of p approaches the value .5 from the left side, the mean μ (indicated by the arrow) approaches the value $n/2 = 3$ from the left side. By examining the graphs in Figures 6-4(b), (d), (f), and (g), we also see that as the value of p approaches .5 from the right, the value of μ also approaches the value $n/2 = 3$ from the right side.

The graphs in Figures 6-5(a)–(f) illustrate the effect on a binomial distribution of varying the values of n for the fixed value of $p = .1$. Probability graphs are given for $n = 2$, 5, 10, 20, 25, and 30. Notice that as the sample size n increases, the mean (indicated by the arrow) increases and the distributions appear to become more symmetric about their means.

The next example illustrates the probability graph for an additional binomial distribution.

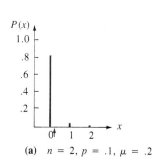
(a) $n = 2, p = .1, \mu = .2$

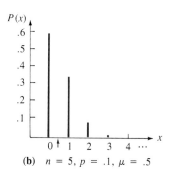
(b) $n = 5, p = .1, \mu = .5$

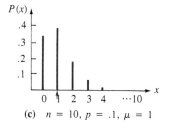
(c) $n = 10, p = .1, \mu = 1$

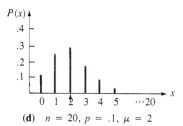
(d) $n = 20, p = .1, \mu = 2$

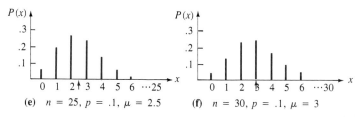
(e) $n = 25, p = .1, \mu = 2.5$ (f) $n = 30, p = .1, \mu = 3$

FIGURE 6-5 Binomial distribution graphs for $p = .1$ and different values of n

EXAMPLE 6-13   Construct a graph for the binomial distribution having $n = 10$ and $p = .6$ and find μ.

Solution   The following probabilities were found using the binomial probability table (Table 1 of Appendix B):

x	$P(x)$
0	.000
1	.002
2	.011
3	.042
4	.111
5	.201
6	.251
7	.215
8	.121
9	.040
10	.006

The corresponding graph is shown in Figure 6-6. The mean μ is

$$\mu = np = (10)(.6) = 6$$

While the graph is not symmetric about $\mu = 6$, the graph reveals that the distribution is not badly skewed. This is not surprising, since the value $p = .6$ is close to the value $p = .5$.

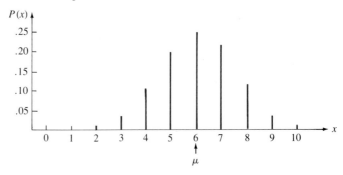

FIGURE 6-6 Binomial distribution with $n = 10$ and $p = .6$ ■

Problem Set 6.3

A

1. Assume that the conditions for a binomial experiment hold in the following situations. For each, find μ, σ^2, and σ.
   a. $n = 10$, $p = .6$ b. $n = 40$, $p = .3$
   c. $n = 15$, $p = .9$ d. $n = 500$, $p = .1$
   e. $n = 20$, $p = .85$ f. $n = 5$, $p = .2$
   g. $n = 6$, $p = .45$ h. $n = 4$, $p = .60$

2. If 2% of all radios manufactured by Ajax Company are defective, find the mean and standard deviation of the number of defective radios in a group of 50.

3. Find the mean and variance of the number of girls in families with six children. Assume that for every 106 males born, there are 100 females born.

4. What are the mean and standard deviation of the number of heads that will occur when 1000 fair coins are tossed?

5. The probability that Joe shoots a bull's-eye with his rifle is 68%. If he shoots in rounds of ten shots, find the mean and standard deviation of the number of perfect hits per round.

6. Over a long period of time it has been determined that 90% of all students enrolled in Math 209 pass the course. If groups of 30 students take the course, determine the mean and standard deviation of the number of students per course who pass.

7. Construct graphs similar to those of Figures 6-5–6-6 for the following probability distributions. Assume that the binomial conditions hold. Locate μ on each graph.
   a. $n = 5$, $p = .2$
   b. $n = 6$, $p = .45$
   c. $n = 4$, $p = .60$

8. Construct a graph of the probability distribution for each of the following binomial distributions:

   a. $n = 20$, $p = .5$ b. $n = 20$, $p = .4$
   c. $n = 20$, $p = .6$ d. $n = 20$, $p = .3$
   e. $n = 20$, $p = .8$ f. $n = 20$, $p = .2$

B

9. Suppose a binomial distribution results from n trials, where the probability of success is p. Then consider the binomial distribution with n trials in which the probability of success is $1 - p$.
   a. How do the means of the two binomial distributions compare?
   b. How do their standard deviations compare?

10. For a fixed value of n, are there values of p for which $\sigma^2 = 0$? Explain.

11. For a fixed value of n, for what value of p is σ^2 largest? Explain.

12. If $b(x; n, p)$ denotes the probability of x successes for a binomial experiment with n trials and $P(S) = p$, show that

$$b(x; n, 1 - p) = b(n - x; n, p)$$

CHAPTER SUMMARY

In this chapter we studied binomial experiments and their properties. We learned how to associate the outcomes of a binomial experiment with a discrete binomial random variable. We also learned how to calculate probabilities associated with a binomial random variable. Finally, we learned how to find the mean, variance, and standard deviation of binomial probability distributions.

C H A P T E R   R E V I E W

IMPORTANT TERMS

For each of the following terms, provide a definition in your own words. Then check your responses against the definitions given in the chapter.

binomial coefficient
binomial distribution
binomial experiment
binomial probability
   formula

empirical binomial
   distribution
mean of a binomial
   distribution
n factorial

Pascal's triangle
Poisson distribution
standard deviation for a
   binomial distribution

variance of a binomial
   distribution

IMPORTANT SYMBOLS

n, the number of trials

p, the probability of success

$1 - p$, the probability of failure

x, the number of successes

$E(x)$, the mean number of successes

S, success

F, failure

$n!$, n factorial

$\binom{n}{x}$, binomial coefficient

IMPORTANT FACTS AND FORMULAS

1. $n! = n(n - 1)(n - 2)(n - 3) \cdots 1$

2. $\binom{n}{x} = \dfrac{n!}{x!(n - x)!}$

3. $P(x) = \binom{n}{x} p^x (1 - p)^{n-x}$, binomial probability formula

4. $\mu = E(x) = np$, mean of a binomial random variable

5. $\sigma^2 = np(1 - p)$, variance of a binomial random variable

6. $\sigma = \sqrt{np(1 - p)}$, standard deviation of a binomial random variable

7. $\binom{n}{x} = \binom{n}{n - x}$

REVIEW PROBLEMS

1. Thirty-five percent of all automobiles sold in the United States are foreign-made. Five new automobiles are randomly selected.
   a. Construct a binomial probability distribution table for the number of automobiles in the sample that are foreign-made.
   b. Construct a graph for the binomial distribution.
   c. Find the mean number of automobiles in a sample of five that are foreign-made.
   d. Find the standard deviation of the number of automobiles in a sample of five that are foreign-made.
   e. Find the probability that at most four are foreign-made.

2. The probability is $\frac{1}{2}$ that a newborn baby will be a boy. If eight babies are born at a local hospital, find
   a. the probability that three are boys.
   b. the probability that none are boys.
   c. the probability that two are girls.
   d. the expected number of boys.
   e. the variance of the number of boys born.

3. If a baseball player with a batting average of .250 comes to bat ten times in a series, what is the probability that he will get
   a. three hits?
   b. at least one hit?
   c. at most two hits?

4. If it is assumed that a golfer will hit a drive into a sand trap 17% of the time, what is the probability that the player will hit the ball into a sand trap exactly four out of the first nine holes?

5. Refer to Problem 4. Find the expected number of times that the ball will be hit into a sand trap.

6. A particular type of birth-control pill is 90% effective. If 500 people used the pill, how many unplanned births would you expect?

7. It is estimated that 85% of all household plants are overwatered. For ten household plants, find the mean and standard deviation of the number of plants that are overwatered and interpret your results.

8. If there are 5 red balls and 15 black balls in a bag and three are selected one at a time with replacement, what is the probability that all three are red?

9. If an unprepared student takes a multiple-choice test, each question having five choices, and guesses at each of the 25 questions, what is the probability that she receives a score of 28%?

10. On the average, a store has had gross sales of over $500 a day on 7 days in 10 over the past several months. If we assume this trend to continue, what is the probability that the store will have gross sales over $5000 at least 4 times in the next 7 days?

CHAPTER ACHIEVEMENT TEST

(28 points)   **1.** It is estimated that 5% of the type A transistors manufactured by Ajax Company are defective. If a lot of 20 transistors is examined, find the probability that
   a. none are defective.
   b. at most 1 is defective.
   c. 19 are defective.
   d. at least 5 are defective.

(28 points)   **2.** A multiple-choice test contains ten questions, each with five choices. If an unprepared student guesses at each question, find the probability that he
   a. gets five correct.
   b. gets at most two correct.
   c. gets at most nine correct.
   d. gets two or three correct.

(16 points)   **3.** For Problem 2, find and interpret your answers for
   a. the expected number of correct answers.
   b. the variance of the number of correct answers.

(28 points)   **4.** Each face of a four-sided pyramid is numbered 1 to 4. Suppose this pyramid is tossed five times and each side is equally likely to land face down.
   a. Construct a probability table for the number of 1s obtained.
   b. Construct a graph depicting the probabilities for the number of 1s obtained.
   c. Find μ and σ for the number of 1s obtained.

7

NORMAL DISTRIBUTIONS

Chapter Objectives

In this chapter you will learn:

- *what a normal distribution is*
- *what the standard normal distribution is*
- *how to associate areas under the standard normal curve with probabilities*
- *how to calculate probabilities associated with the standard normal distribution*

- *how to calculate probabilities associated with any normal distribution*
- *practical applications for normal distributions*
- *how to use normal distributions to approximate binomial distributions*

Chapter Contents

An experiment involving measurements that can assume any value in a continuum can be identified with a continuous random variable. The values of the random variable consist of the real-number measurements falling in the continuum. The infinite set of measurements that comprise the possible outcomes of an experiment involving some form of measurement is called a **continuous distribution**. Recall from Section 5.7 that probabilities associated with a continuous distribution are identified as areas under a curve. The total area under the curve above the x-axis must equal 1.

As an example of a continuous distribution, consider the time it takes an individual to react to a given stimulus and suppose the maximum range in reaction times is 2 seconds. If reaction times were measured to the nearest millisecond, there would be 2000 possible outcomes; there are only 2000 outcomes because of the limitations of the measuring device. Theoretically, the reaction times could be any of the infinite number of values from the shortest reaction time to the longest reaction time. The theoretical reaction times comprise a continuous distribution.

One of the most important classes of continuous distributions is the **normal distribution**. These distributions were studied by the French mathematician Abraham de Moivre (1667–1754) and later by the German mathematician Karl Gauss (1777–1855). The normal distributions grew out of studies concerned with errors in experiments.

Many real and natural occurrences have frequency distributions that are approximately normal. For example, the frequency distribution of the nitrogen content of leaves on a tree tends to be normal. Physical measurements are often normally distributed. Systolic and diastolic blood pressures, heart rates, blood cholesterol levels, heights of adult men, weights of 2-year-old girls, scores on IQ tests, and SAT scores are all examples of distributions of data that tend to be normal.

7.1 NORMAL DISTRIBUTIONS

A normal distribution is mound-shaped or takes on the appearance of a bell, as illustrated in Figure 7-1. The equation of the bell-shaped curve is given by

$$y = \frac{e^{-(x-\mu)^2/(2\sigma^2)}}{\sqrt{2\pi}\sigma}$$

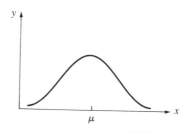

FIGURE 7-1 Normal distribution

where μ represents the mean of the population, σ is the standard deviation of the population, $e \simeq 2.718$, and x is any real number.

The two parameters μ and σ completely specify the position and shape, respectively, for a normal distribution. A small value of σ means that the normal curve is a narrow, peaked bell, whereas a large value of σ means that the normal curve is a wide, flat bell (see Figure 7-2). Figure 7-3 illustrates three normal distributions having a mean of 20 but different standard deviations. And Figure 7-4 illustrates three normal distributions with different means but the same standard deviation.

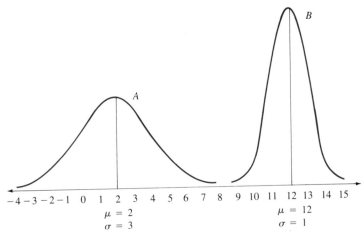

FIGURE 7-2 Effect of μ and σ on position and shape of normal distribution

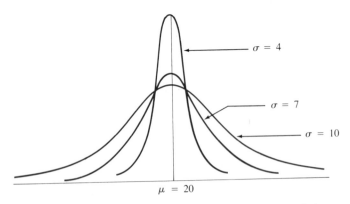

FIGURE 7-3 Normal distributions with same mean but different standard deviations

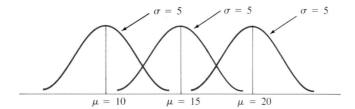

FIGURE 7-4 Normal distributions with different means but
same standard deviation

In many practical applications, we have collections of measurements whose graphs approximate a bell-shaped curve. For example, consider the following collection of 100 systolic blood-pressure measurements of a group of individuals aged 20–24 years:

98	141	127	112	107	111	116	104	87	130
131	124	126	129	65	126	139	136	137	102
128	152	115	124	96	114	116	119	108	113
126	116	107	134	82	141	85	126	131	141
128	85	145	150	97	125	109	119	159	114
117	131	122	104	145	119	132	138	109	102
115	109	143	128	136	118	125	124	109	133
131	127	100	110	109	112	125	166	93	92
113	118	116	136	96	131	141	153	85	119
92	112	140	121	108	92	98	159	91	125

The following stem-and-leaf diagram for this data set suggests that the blood-pressure measurements are normally distributed:

```
 6 | 5
 7 |
 8 | 2 5 5 5 7
 9 | 1 2 2 2 3 6 6 7 8 8
10 | 0 2 2 4 4 7 7 8 8 9 9 9 9 9
11 | 0 1 2 2 2 3 3 4 4 5 5 6 6 6 6 7 8 8 9 9 9 9
12 | 1 2 4 4 4 5 5 5 5 6 6 6 6 7 7 8 8 8 9
13 | 0 1 1 1 1 1 2 3 4 6 6 6 7 8 9
14 | 0 1 1 1 1 3 5 5
15 | 0 2 3 9 9
16 | 6
```

A grouped frequency table for the data is as follows:

Class limits	f
65–75	1
76–86	4
87–97	9
98–108	11
109–119	27
120–130	20
131–141	19
142–152	5
153–163	3
164–174	1

A frequency histogram for the data is shown in Figure 7-5. Although the frequency histogram is not symmetric, it is approximately bell-shaped and suggests that the blood-pressure measurements may be normally distributed.

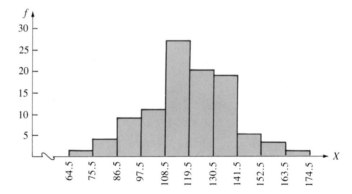

FIGURE 7-5 Histogram for blood-pressure measurements

Properties of Normal Distributions

Some of the more important properties of normal distributions are listed here:

1. A normal distribution is mound- or bell-shaped.
2. The area under a normal curve always equals 1.
3. The mean is located at the center of the distribution and a curve is symmetric about its mean.
4. The mean, median, and mode are all equal.
5. A curve for a normal distribution extends indefinitely to the left and to the right of the mean and approaches the horizontal axis.
6. A curve for a normal distribution never touches the horizontal axis.
7. The shape and position of a normal distribution depend on the parameters μ and σ; as a result, there are an infinite number of normal distributions.

EXAMPLE 7-1   Explain whether each of the following curves could represent a normal distribution.

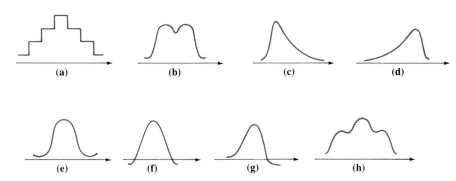

      (a)   (b)   (c)   (d)

      (e)   (f)   (g)   (h)

Solution   **a.** No; the curve is not mound-shaped.
   b. No; the curve is not mound-shaped.
   c. No; the curve is not symmetric.
   d. No; the curve is not symmetric.
   e. No; the curve does not get close to the horizontal axis.
   f. No; the curve crosses the horizontal axis.
   g. No; the curve crosses the horizontal axis.
   h. No; the curve is not mound-shaped.   ∎

EXAMPLE 7-2   Find the area of the shaded regions under the normal curves shown here.

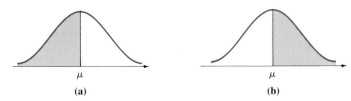

          μ μ
          (a)   (b)

Solution   In both cases the shaded area is .5, since the total area under a normal curve is 1 and a curve is symmetric about its central measurement μ.   ∎

Empirical Rule

The following **Empirical Rule** applies to any normal distribution:

a. Approximately 68% of the measurements fall within 1 standard deviation of the mean—that is, within $\mu \pm \sigma$.
b. Approximately 95% of the measurements fall within 2 standard deviations of the mean—that is, within $\mu \pm 2\sigma$.
c. Approximately 99.7% of the measurement fall within 3 standard deviations of the mean—that is, within $\mu \pm 3\sigma$.

Figure 7-6 illustrates the Empirical Rule.

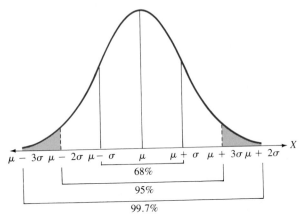

FIGURE 7-6 Graphical illustration of Empirical Rule

EXAMPLE 7-3   Suppose $\sigma = 2$ and $\mu = 30$ for the normal distribution pictured here:

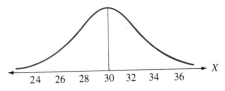

Find the percentage of the measurements that fall

a. between 28 and 32.
b. between 30 and 32.
c. between 30 and 34.
d. between 32 and 34.
e. below 24.
f. above 32.

Solution   **a.** Sixty-eight percent of the distribution falls between 28 and 32, since $28 = \mu - \sigma$ and $30 = \mu + \sigma$.

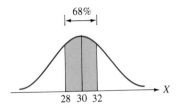

b. Since the curve is symmetric about $\mu = 30$ and approximately 68% of the scores fall between 28 and 32, approximately $(\frac{1}{2}) \times 68\% = 34\%$ of the measurements fall between 30 and 32.

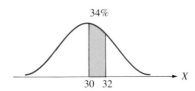

34%

30 32

c. Approximately 95% of the scores fall between 26 and 34. Since the curve is symmetric about $\mu = 30$, $(\frac{1}{2})(.95) = .475 = 47.5\%$ of the scores fall between 30 and 34.

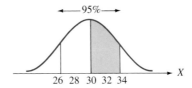

←——95%——→

26 28 30 32 34

d. This problem requires us to find the shaded areas illustrated in the following drawings:

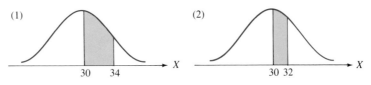

(1)   (2)

30   34   30 32

As a result of part (c), the percentage illustrated in drawing (1) is 47.5%. As a result of part (b), the percentage illustrated in drawing (2) is 34%. Therefore, we subtract 34% from 47.5% to get 13.5%, the percentage of measurements that fall between 32 and 34.

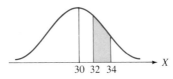

30 32 34

e. Since 99.7% of the scores fall between 24 and 36, .3% remain for the upper and lower tails of the distribution. Since the curve is symmetric, $(\frac{1}{2})(.3\%) = .15\%$ of the scores fall below 24.

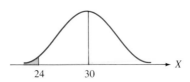

24   30

f. By part (b), 34% of the scores fall between 30 and 32. Since 50% fall above 30, it follows that $.50 - .34 = .16 = 16\%$ of the scores fall above 32.

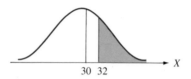

Recall from Section 3.3 that a z score for a population is defined by

$$z = \frac{x - \mu}{\sigma}$$

and indicates the number of standard deviations a score is from the mean. Thus, a score equal to $\mu + \sigma$ has a z score equal to

$$z = \frac{x - \mu}{\sigma}$$

$$= \frac{(\mu + \sigma) - \mu}{\sigma}$$

$$= \frac{\sigma}{\sigma}$$

$$= 1$$

By the same reasoning, $\mu + 2\sigma$ has a z score equal to

$$z = \frac{x - \mu}{\sigma}$$

$$= \frac{(\mu + 2\sigma) - \mu}{\sigma}$$

$$= 2$$

and $\mu + 3\sigma$ has a z score equal to

$$z = \frac{x - \mu}{\sigma}$$

$$= \frac{(\mu + 3\sigma) - \mu}{\sigma}$$

$$= 3$$

As a result, the Empirical Rule can be restated as follows:

    a. In any normal distribution, approximately 68% of the z scores fall between -1 and $+1$.

    b. In any normal distribution, approximately 95% of the z scores fall between -2 and $+2$.

c. In any normal distribution, approximately 99.7% of the z scores fall between -3 and $+3$.

Estimating σ and s

In Section 3.3 it was stated that $R/4$ provides a good estimate for s. In part, this is due to the fact that nearly all the scores in a normal distribution fall within $\mu \pm 3\sigma$. As a result, the range R of a normal distribution is approximately equal to 6σ, as illustrated in Figure 7-7. Thus, we have

$$\sigma \simeq \frac{R}{6}$$

Hence, $R/6$ provides an estimate for σ and also for s.

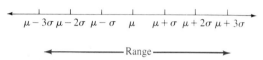

FIGURE 7-7 Range $\simeq 6\sigma$ for a normal distribution

In addition, since 95% of the measurements in a normal distribution fall within $\mu \pm 2\sigma$, $R/4$ also provides an estimate for s based on the same line of reasoning. Since $R/4$ always provides a larger—and thus more conservative—estimate for s than $R/6$, the estimate $R/4$ is typically used. In most cases when a distribution is symmetric, $R/4$ provides a reasonable estimate for s. But for badly **skewed distributions**—distributions in which measurements occur with greater frequency either above or below the mean—$R/4$ provides a poor estimate for s most of the time.

Probability and Area

Probabilities related to a normal distribution are identified with areas. For example, the probability that a measurement x from a normal distribution is between 20 and 40 can be identified with the shaded area in the following figure:

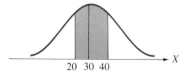

Similarly, the shaded area under the following normal curve can be interpreted as the probability that a measurement x is between 6 and 8 and is written as $P(6 < x < 8)$:

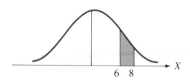

EXAMPLE 7-4   Write a corresponding probability statement for each of the following areas under a normal curve:

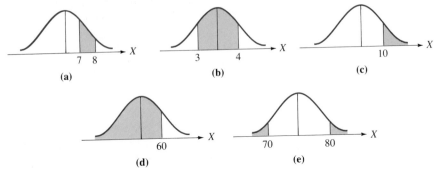

Solution   The following are the probability statements and their interpretations:

Using inequalities	Interpretation
a. $P(7 < x < 8)$	the probability that x is between 7 and 8
b. $P(3 < x < 4)$	the probability that x is between 3 and 4
c. $P(x > 10)$	the probability that x is greater than 10
d. $P(x < 60)$	the probability that x is less than 60
e. $P(x < 70 \text{ or } x > 80)$	the probability that x is less than 70 or greater than 80

Standard Normal Distribution

The normal distribution with a mean of 0 and a standard deviation of 1 is called the **standard normal distribution**. The symbol Z is used to denote the standard normal distribution.

Finding Probabilities

We want to be able to determine the probability of the random variable Z having a value in a specified interval. The Empirical Rule can be used to find certain probabilities or areas associated with the standard normal distribution. However, most practical applications will not involve intervals of exactly 1, 2, or 3 standard deviations from the mean. For example, the Empirical Rule cannot be used to find $P(0 < z < 1.23)$. In such a case, the probability must be found using a **standard normal probability table** (z table; see the front endpaper). This table gives the area under the standard normal curve to the right of 0 and below a specified z value, as illustrated in Figure 7-8.

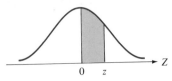

FIGURE 7-8 Area under standard normal curve given in the z table

To find an area (or probability) in the z table, we locate the units and tenths digits of z down the left-hand column and the hundredths digit across the top; the number where the row and column intersect is the desired area. For example, consider the shaded area in the following figure:

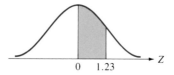

To find the area between 0 and $z = 1.23$ under a standard normal curve, we locate the 1.2 row at the left of the z table 1 and the .03 column along the top; the entry in the table body at the intersection of the 1.2 row and the .03 column is .3907, as shown here:

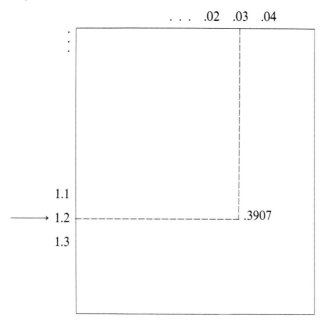

Thus, .3907 is the area of the shaded region, which is also the probability that z is between 0 and 1.23.

The following example illustrates the use of the z table to find probabilities associated with the standard normal distribution.

EXAMPLE 7-5   Find the following probabilities:

a. $P(z$ is between -1.34 and $1.68)$
b. $P(z$ is between -1.34 and $1.34)$
c. $P(z$ is between 1.34 and $2.38)$
d. $P(z$ is greater than $2.12)$
e. $P(z$ is between -2.14 and $-1.17)$
f. $P(z$ is less than $-1.25)$

Solution   **a.** An important first step to finding any normal probability value is to draw a diagram illustrating the area corresponding to the probability to be found:

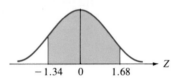

The steps are as follows: (1) Find the area from 0 to 1.68. (2) Find the area from −1.34 to 0. (3) Add these areas.

The area from 0 to 1.68 is .4535, and the area from −1.34 to 0 is .4099. Adding these areas, we obtain .8634.

b. The desired area is shaded in the figure:

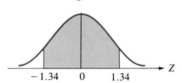

The steps are as follows: (1) Find the area from 0 to 1.34. (2) Double this area (because of symmetry).

The area from 0 to 1.34 is .4099. Thus, the probability (area) of interest is 2(.4099) = .8198.

c.

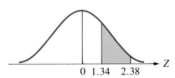

The steps are as follows: (1) Find the area from 0 to 2.38. (2) Find the area from 0 to 1.34. (3) Subtract the second area from the first.

Thus, the area between 1.34 and 2.38 is .4913 − .4099 = .0814.

d.

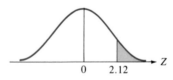

The steps are as follows: (1) Find the area from 0 to 2.12. (2) Subtract this area from .5.

Thus, $P(z > 2.12) = .5 − .4830 = .017$.

e.

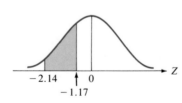

The steps are as follows: (1) Find the area between -2.14 and 0. (2) Find the area between -1.17 and 0. (3) Subtract the second area from the first. Thus, $P(-2.14 < z < -1.17) = .4838 - .3790 = .1048$.

f.

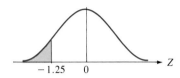

The steps are as follows: (1) Find the area between -1.25 and 0. (2) Subtract this area from .5.
   Thus, $P(z < -1.25) = .5 - .3944 = .1056$.   ■

Occasionally we need to calculate probabilities associated with z values that fall in the extreme regions of the standard normal distribution. For example, we might need to determine $P(z < -5)$ or $P(z > 6)$. In both these cases the probabilities are small, but not equal to 0, since the curve extends indefinitely far in both directions. For very small probabilities (probabilities written as .0000 to four decimal places), we shall use the symbol 0^+. In addition, we shall indicate cumulative probabilities associated with large z values, such as $P(z < 7)$, by using the symbol 1^-.

Verifying the Empirical Rule

Probabilities listed in the standard normal table (z table; see front endpaper) can be used to verify the Empirical Rule. Note that:

   a. $P(-1 < z < 1) = 2P(0 < z < 1) = 2(.3413) = .6826 \simeq .68$
   b. $P(-2 < z < 2) = 2P(0 < z < 2) = 2(.4772) = .9544 \simeq .95$
   c. $P(-3 < z < 3) = 2P(0 < z < 3) = 2(.4987) = .9974 \simeq .997$

Computational Patterns for Finding Probabilities

The following computational patterns serve to illustrate the steps needed to compute probabilities associated with the standard normal curve. The values of a and b are assumed to be greater than 0.

   1. To find the probability that z is greater than some nonzero value:
      a. $P(z > -a)$

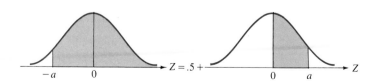

b. $P(z > a)$

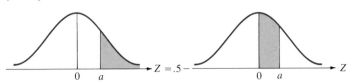

2. To find the probability that z is between two values:
   a. $P(-a < z < b)$

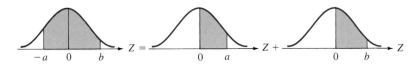

   b. $P(a < z < b)$

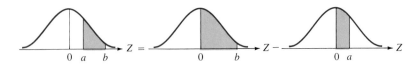

   c. $P(-a < z < -b)$

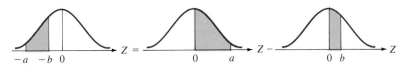

3. To find the probability that z is less than some value:
   a. $P(z < -a)$

   b. $P(z < a)$

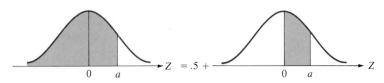

Finding z Scores, Given Areas

Frequently, we must find the z value, given an area under the standard normal curve. This is the reverse process of finding probabilities or areas corresponding to z values. Consider the following examples.

EXAMPLE 7-6   Find a standard normal value a so that

$$P(z \text{ is between } -a \text{ and } a) = .95$$

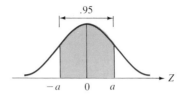

Solution   The shaded area in the accompanying figure is .95. Since the z table lists areas under the normal curve from $z = 0$ to $z = a$, we divide .95 by 2 to get an area of .475 between 0 and a. We now locate the area .475 in the body of the z table:

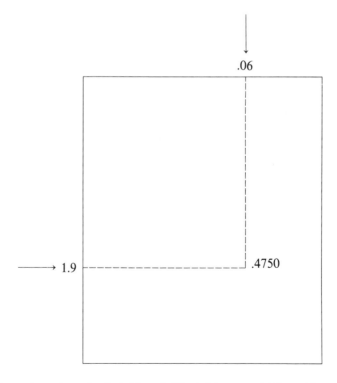

Thus, the value of a is 1.96 and $P(-1.96 < z < 1.96) = .95$.   ∎

If the area we seek is equally close to two values in the body of the table, we choose the larger z value, as illustrated in the next example.

EXAMPLE 7-7   Find a standard normal value a so that $P(z > a) = .05$.

Solution $P(0 < z < a) = .45$, since $P(0 < z < a) + P(z > a) = .5$. From the z table, we find

z value	Area
1.64	.4495
	.45
1.65	.4505

Since .45 is midway between .4495 and .4505, and 1.65 is the larger z value, we choose $a = 1.65$. Hence, $P(z > 1.65) \simeq .05$.   ∎

EXAMPLE 7-8   Find a standard normal value a so that $P(z < a) = .75$.

Solution   Since $P(z < 0) = .5$, $P(0 < z < a) = .25$. Checking the z table, we find $a = .67$, since .25 is closer to .2486 than to .2517:

	Area	z value
distance = .0014 ⟶	.2486	.67
	.2500	
distance = .0017 ⟶	.2517	.68

Thus, $P(z < .67) \simeq .75$.   ∎

EXAMPLE 7-9   Find a standard normal value a so that $P(-1.14 < z < a) = .25$.

Solution   By symmetry, $P(-1.14 < z < 0) = P(0 < z < 1.14) = .3729$. Hence, a must be less than 0, and we have the following drawing:

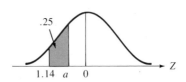

Thus, $P(0 < z < -a) = .3729 - .25 = .1229$. Reading the z table, we see that the area closest to .1229 is .1217:

Area	z value
.1217	.31
.1229	
.1255	.32

Hence, $-a = .31$ and $a = -.31$.   ∎

Problem Set 7.1

A

1. Suppose a given normal distribution has a mean of 20 and a standard deviation of 4. By using the Empirical Rule, find the following probabilities:
   a. $P(x > 24)$ **b.** $P(16 < x < 24)$
   c. $P(12 < x < 24)$ **d.** $P(x > 28)$
   e. $P(8 < x < 12)$ **f.** $P(24 < x < 32)$
   g. $P(x < 12)$ **h.** $P(20 < x < 28)$

2. Find the following probabilities using the z table on the front endpaper.
   a. $P(z$ is greater than 0)
   b. $P(z$ is between -2.5 and 1.5)
   c. $P(z$ is between -2.7 and -1.3)
   d. $P(z$ is between 2 and 3)
   e. $P(z$ is greater than 2.8)
   f. $P(z$ is greater than 3.7)
   g. $P(z$ is between -4 and 4)

3. Find the following probabilities using the z table on the front endpaper.
   a. $P(-1.23 < z < 0)$ **b.** $P(0 < z < 2.34)$
   c. $P(-1.78 < z < 2.38)$ **d.** $P(-1.12 < z < -1.01)$
   e. $P(1.23 < z < 2.75)$ **f.** $P(z < 2.34)$
   g. $P(z > 2.97)$ **h.** $P(z < 4.38)$

4. For each problem, find the standard normal value a so that:
   a. $P(z$ is between $-a$ and $a) = .90$
   b. $P(z$ is greater than $a) = .025$
   c. $P(z > a) = .10$
   d. $P(-a < z < a) = .99$
   e. $P(z > a) = .85$
   f. $P(z$ is between $-a$ and 2) = .94$
   g. $P(z < a$ or $z > 1.1) = .90$

B

5. A handy approximation for areas under the standard normal curve from 0 to z ($z > 0$) is given by the following rule:

z value	Approximate area
$0 \le z \le 2.2$	$z(4.4 - z)/10$
$2.2 < z < 2.6$	.49
$z \ge 2.6$	.50

The maximum absolute error for the approximation is .0052 (see *The American Statistician*, Vol. 39, No. 1, Feb. 1985, p. 80). Use this approximation to find the probabilities in Problem 3.

7.2 APPLICATIONS OF NORMAL DISTRIBUTIONS

The standard normal table cannot be directly used to calculate probabilities associated with a normal distribution other than the standard normal. For example, suppose we are interested in finding the following area for the normal distribution with a mean of 50 and a standard deviation of 5:

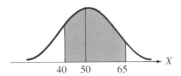

We cannot directly use the standard normal table (z table; see the front endpaper), since the random variable of interest is not the standard normal random variable. In order to determine probabilities associated with a nonstandard normal distribution, we need a rule that will enable us to transform any normal distribution to the standard normal distribution, as indicated in Figure 7-9. Then we could determine probabilities by using the z table.

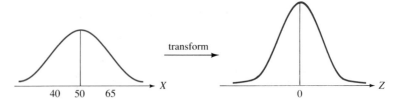

FIGURE 7-9 Transformation to standard normal distribution

The transformation rule that we use is the standard-score formula:

$$z = \frac{x - \mu}{\sigma}$$

It is used to transform any normal x score to a standard normal z score. Thus, for the above illustration, we have

$$z = \frac{40 - 50}{5} = -2$$

and

$$z = \frac{65 - 50}{5} = 3$$

Hence, we have the following diagrams where the two shaded areas are equal:

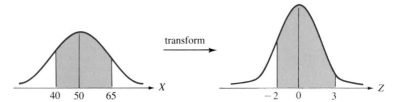

That is,

$$P(40 < x < 65) = P(-2 < z < 3)$$

The probability that z is between -2 and 3 is

$$P(-2 < z < 3) = .4772 + .4987 = .9759$$

Thus,

$$P(40 < x < 65) = .9759$$

It is always a good idea to draw normal-curve illustrations when attempting to calculate probabilities associated with a normal distribution. The illustrations need not be drawn accurately, since their primary use is to identify the appropriate regions to calculate areas. The normal-curve illustrations used throughout the text may not be technically accurate; they are intended for conceptual use only.

The z score formula allows us to think of the standard normal probability table as a list of probabilities associated with intervals beginning with the mean and extending to z standard deviations to the right of the mean for any normal curve.

Consider the following applications of normal distributions.

EXAMPLE 7-10   A company manufactures light bulbs with a mean life of 500 hours and a standard deviation of 100 hours. If it is assumed that the useful lifetimes of the light bulbs are normally distributed (that is, the lifetimes form a normal distribution), find the number of bulbs out of 10,000 that can be expected to last between 650 and 780 hours.

Solution

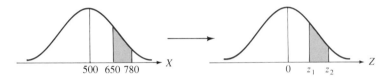

We first compute the z scores for $x = 650$ and $x = 780$:

$$z_1 = \frac{650 - 500}{100} = 1.5$$

$$z_2 = \frac{780 - 500}{100} = 2.8$$

Thus, the area between z_1 and z_2 is $.4974 - .4332 = .0642$. Therefore, we can expect $(10{,}000)(.0642) = 642$ light bulbs to last between 650 and 780 hours.

■

In practical applications of statistics we deal primarily with finite distributions of raw data measurements. These distributions can be only approximations to normal distributions, since normal distributions are continuous and contain infinite numbers of measurements. It is common practice in statistical literature to refer to such distributions as "normal" even when the context makes it clear that such distributions can be only approximately normal. Such is the case in the statement of the problem in Example 7-10.

EXAMPLE 7-11   If systolic blood-pressure measurements x for the 20–24-year age group are normally distributed with a mean of 120 and a standard deviation of 20, find:

a. $P(x > 135)$
b. $P(x < 146)$
c. $P(105 < x < 110)$

Solution   **a.** The z score for $x = 135$ is

$$z = \frac{x - \mu}{\sigma} = \frac{135 - 120}{20} = .75$$

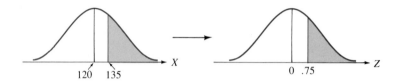

By using the z table (see the front endpaper), we find

$$P(x > 135) = P(z > .75)$$
$$= .5 - P(0 < z < .75)$$
$$= .5 - .2734 = .2266$$

b. The z score for $x = 146$ is

$$z = \frac{x - \mu}{\sigma} = \frac{146 - 120}{20} = 1.3$$

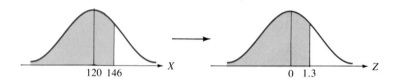

From the z table, we find

$$P(x < 146) = P(z < 1.3)$$
$$= .5 + P(0 < z < 1.3)$$
$$= .5 + .4032 = .9032$$

c. The z score for $x = 105$ is

$$z = \frac{105 - 120}{20} = -.75$$

and the z score for $x = 110$ is

$$z = \frac{110 - 120}{20} = -.5$$

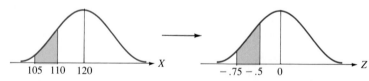

By using the z table, we find

$$P(105 < x < 110) = P(-.75 < z < -.5)$$
$$= P(0 < z < .75) - P(0 < z < .5)$$
$$= .2734 - .1915 = .0819$$

Measures of Relative Standing

Percentiles

Percentiles are numbers that divide an interval of measurements ranging from the smallest measurement to the largest measurement into 100 parts so that each part contains 1% of the measurements. There are 99 percentile points, denoted by $P_1, P_2, \ldots, P_{99}$. The 75th percentile, P_{75}, is that measurement below which 75% of the measurements fall. A given measurement x has a **percentile rank** of n if x equals (or is approximately equal to) the nth percentile—that is, if $x = P_n$.

EXAMPLE 7-12   Math scores on the Scholastic Aptitude Test (SAT) are normally distributed with a mean of 500 and a standard deviation of 100. Find P_{80}, the 80th percentile.

Solution P_{80} is that number (or score) below which 80% of the scores fall. Then the area under the normal curve between $x = 500$ and P_{80} is .30 (see Figure 7-10). By locating the z score in the z table that corresponds to an area of .30, we will be able to use the z score formula to solve for P_{80}, since

$$z = \frac{P_{80} - 500}{100}$$

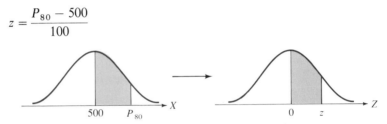

FIGURE 7-10 Determination of P_{80} for SAT math scores

The approximate value of z that produces an area of .30 is found in the z table to be .84. Hence, we have

$$.84 = \frac{P_{80} - 500}{100}$$

Multiplying both sides of this equation by 100 and then adding 500 to both sides, we get

$$P_{80} = 584$$ ∎

EXAMPLE 7-13   John took the SAT math test and received a score of 572. Find his percentile rank.

Solution   Consider Figure 7-11. To find the percentile rank of John's score, we find the shaded area in the figure and round it to the nearest percentage. We will find this area by first finding the z score for $x = 572$:

$$z = \frac{x - \mu}{\sigma} = \frac{572 - 500}{100} = .72$$

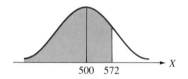

FIGURE 7-11 Determination of percentile rank for SAT math
score of 572

The shaded area in Figure 7-11 is

$$P(x < 572) = P(z < .72)$$
$$= .5 + P(0 < z < .72)$$
$$= .5 + .2642 = .7642$$

Hence, $P_{76} \simeq 572$ and the percentile rank of 572 is 76. ∎

Quartiles

Quartiles are numbers that divide an interval of measurements extending from
the smallest measurement to the largest measurement into four parts so that
each part contains 25% of the measurements. There are three quartile points,
denoted by Q_1, Q_2, and Q_3. The second quartile is the 50th percentile or the
median. The first quartile Q_1 is that number such that 25% of the measurements
fall below Q_1. The first quartile Q_1 is also equal to P_{25}, the 25th percentile.

Deciles

Deciles are numbers that divide an interval of measurements extending from the
smallest measurement to the largest measurement into ten parts so that each
part contains 10% of the measurements. There are nine decile points, denoted
by D_1, D_2, D_3, ..., and D_9.

EXAMPLE 7-14 For the SAT math scores with $\mu = 500$ and $\sigma = 100$, find:

a. D_3, the third decile.
b. Q_1, the first quartile.

Solution **a.**

We find the z score for D_3:

$$z = \frac{D_3 - 500}{100}$$

By using the z table (see the front endpaper), we find $P(0 < z < .52) \simeq 20$.

Thus, the z score for the third decile is

$$-.52 = \frac{D_3 - 500}{100}$$

Multiplying both sides by 100 and adding 500 to both sides, we get

$$D_3 = 448$$

b.

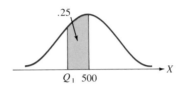

We find the z score for Q_1:

$$z = \frac{Q_1 - 500}{100}$$

From the z table, we find

$$P(0 < z < .67) = .25$$

Thus,

$$-.67 = \frac{Q_1 - 500}{100}$$

Solving for Q_1, we have

$$Q_1 = 433$$ ∎

Checking the Plausibility That a Sample Is from a Normal Distribution

To determine whether a normal distribution can serve as a reasonable model for the population that produced the sample, graphical procedures are helpful in detecting serious departures from normality. Histograms can be checked for lack of symmetry, and the Empirical Rule can be used to check the proportions of the data that fall within the intervals determined by $\bar{x} \pm s$, $\bar{x} \pm 2s$, and $\bar{x} \pm 3s$.

A **z score plot** corresponding to specific percentile points of the standard normal distribution can be an effective procedure to check the plausibility of a normal model when the sample size is at least 20. If the sample size is n, n values of z can be found that divide the standard normal distribution into $(n + 1)$ intervals, each interval having the same probability. The measurements in the sample are ordered from smallest to largest and paired with the n percentile points. The pairs are then plotted to obtain a z score plot. For purposes of discussion, let's suppose the sample is of size 4. We would first find the four percentile points, P_{20}, P_{40}, P_{60}, and P_{80}, that divide the standard normal distribution into five intervals, each having a probability of .20. The smallest sample value is paired with P_{20}, the next largest sample value is paired with P_{40}, the next with P_{60}, and the largest sample value with P_{80}. These four pairs are then plotted in a graph. A straight-line pattern supports the plausibility that the sample was drawn from a normal population. A departure from normality is indicated by a graph displaying a curved appearance. The following example illustrates the ideas only; in practical applications the sample size should be at least 20.

EXAMPLE 7-15   Construct a z score plot of the following sample of data to determine if the sample could have been produced by a normal population.

Y: 48   55   57   60

Solution   **1.** The four percentile points that divide the distribution into five intervals, each corresponding to an area of 20, are found with the aid of the z table on the front endpaper (see the accompanying figure).

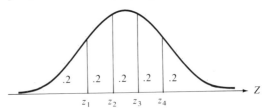

2. These z scores are then paired with the four ordered observations:

Sample observations, Y	Percentile points, z scores
48	$-.84$
55	$-.26$
57	$.26$
60	$.84$

3. The z score plot is shown in Figure 7-12.

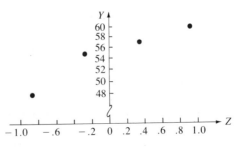

FIGURE 7-12 z score plot for Example 7-15

Since the four data points appear to follow a linear trend, we have no reason to believe that the sample wasn't produced by a normal population. If the distribution (X) that produced the sample has mean μ and standard deviation σ, each percentile point z can be converted to an x score by using the transformation $x = \mu + \sigma z$. We would expect each sample value to be close to its corresponding x value if the population is normal. That is, a plot of the sample values versus the normal x scores should produce a straight-line pattern $(Y = \mu + \sigma z)$, where the slope of the line is σ and the vertical intercept is μ.

Sample value (Y)	x value
48	$\mu + \sigma z_1$
55	$\mu + \sigma z_2$
57	$\mu + \sigma z_3$
60	$\mu + \sigma z_4$

■

Problem Set 7.2

A

1. A machine produces bolts whose diameters are normally distributed with an average diameter of .25 inch and a standard deviation of .02 inch. What is the probability that a bolt will be produced with a diameter
   a. greater than .3 inch?
   b. between .2 and .3 inch?
   c. less than .19 inch?
   d. greater than .1 inch?

2. For the SAT math scores, which are normally distributed with a mean of 500 and a standard deviation of 100, find:
   a. P_{30} b. P_{50} c. P_{75} d. Q_1 e. D_7 f. D_3

3. If x scores are normally distributed with a mean of 70 and a standard deviation of 5, find the probability that
   a. x is between 60 and 80.
   b. x is less than 65.
   c. x is less than 81.
   d. x is greater than 67.5.
   e. x is less than 67 or greater than 77.

4. Refer to Problem 1. If 100,000 bolts were manufactured, find the expected number of bolts with a diameter
   a. greater than .3 inch.
   b. between .2 and .3 inch.
   c. less than .19 inch.
   d. greater than .1 inch.

5. The Apex Taxi Company has found that taxi fares

are normally distributed with $\mu = \$4.30$ and $\sigma = \$1.25$. If a driver takes a taxi call at random, what is the probability the fare will be less than \$3.05?

6. The mean GPA at a certain university is 2.2 and the standard deviation is .5. At a special honors convocation, students with the top 3% of the GPAs are to be given special recognition. If the GPAs are normally distributed, what is the minimum GPA a student needs in order to receive recognition?

7. The heights of adult males are normally distributed with a mean of 70 inches and a standard deviation of 2.6 inches. How high should a doorway be constructed so that 90% of the men can pass through it without having to bend?

8. The number of hours a college student studies per week is normally distributed with a mean of 25 hours and a standard deviation of 10 hours.
   a. What percentage of students will study less than 30 hours?
   b. What percentage will study less than 20 hours?
   c. What percentage will study more than 30 hours?
   d. Out of a class of 100 students, approximately how many will study between 12 and 38 hours per week?

9. A coffee dispenser is set to fill cups with an average of 8 ounces of coffee. If the amounts dispensed are normally distributed with a standard deviation of .4 ounce, what percentage of 9-ounce cups will overflow?

10. If x is normally distributed with $\mu = 30$ and $\sigma = 5$, find the value a so that:

a. $P(-a < x < a) = .95$
b. $P(x > a) = .05$
c. $P(x < a) = .005$

B

11. The useful life of a particular brand of CRT used in color televisions is normally distributed with a mean life of 7.7 years and a standard deviation of 1.9 years.

a. What is the probability that a CRT will last more than 8 years?
b. If the CRTs are guaranteed for 2 years, what percentage of CRTs will have to be replaced?
c. If the manufacturer will replace only 2% of the CRTs, what guarantee period (to the nearest month) should be stated on the warranty card?

12. A manufacturer of electric motors wants to determine the length of time to guarantee its motors so no more than 4% of them will have to be replaced. If the lives of the motors are normally distributed with a mean life of 10 years and a standard deviation of .75 year, find the guarantee period to the nearest month.

13. If the random variable x is normally distributed with mean μ and variance σ^2, find the percentile rank of

a. $\mu + \sigma$ **b.** $\mu - \sigma$ **c.** $\mu + 2\sigma$ **d.** $\mu - 2\sigma$

14. Construct a z score plot to determine if a normal distribution could serve as a model for the following sample. That is, determine if the following sample could have been drawn from a normal population:

36	58	42	37	24
52	49	56	34	32
44	53	51	47	37
30	31	47	50	43
31	53	44	27	

7.3 USING NORMAL DISTRIBUTIONS TO APPROXIMATE BINOMIAL DISTRIBUTIONS

There are many binomial experiments for which the calculation of binomial probabilities involves a number of long and tedious calculations. For example, consider the following binomial experiment:

Over a long period of time, it has been determined that 70% of the lawyers who take the bar examination pass the examination. Of 500 lawyers who take the examination next, find the probability that at least 370 pass.

In order to solve this problem, we need to find $P(370) + P(371) + P(372) + \cdots + P(499) + P(500)$. This would involve using the binomial probability formula 132 times. Let's *attempt* to calculate $P(370)$ first. By using (6-2), we have

$$P(x) = \binom{n}{x} p^x (1 - p)^{n-x}$$

$$P(370) = \binom{500}{370} (.7)^{370} (.3)^{130}$$

We now *try* to compute the binomial coefficient $\binom{500}{370}$:

$$\binom{500}{370} = \frac{500!}{(370)!(130)!}$$

A hand-held calculator does not help us with this calculation. And do not be surprised if the college's computer also fails you. We will use a normal distribution to help us approximate the probability that at least 370 lawyers will pass the bar examination.

Bar Graphs for Binomial Distributions

Recall that we used vertical line segment graphs in Chapter 6 to represent binomial probability distributions. Another type of probability graph, called a **bar graph** or **histogram**, employs vertical bars instead of vertical line segments. At each value of x on the horizontal axis, two points are marked—at $(x - .5)$ and $(x + .5)$. The distance between these points, which is 1, becomes the width of one side of a rectangular bar and the length of the other side becomes $P(x)$. The area of the xth bar is then $P(x)$, as indicated in Figure 7-13.

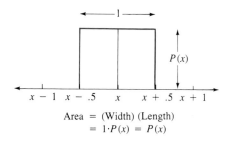

Area $=$ (Width) (Length)
$= 1 \cdot P(x) = P(x)$

FIGURE 7-13 Probability bar graph

Note that the area under a probability bar graph is 1, since $\sum P(x) = 1$. Although areas are generally associated with probabilities of continuous distributions, we should be careful not to classify a binomial distribution as continuous. The use of a bar graph for a binomial distribution is for convenience only, especially with approximating binomial probabilities.

For a binomial distribution that is symmetric (or approximately symmetric) about its mean, the bar graph closely resembles a normal distribution. As a result, a normal distribution can be used to approximate a binomial distribution (see Figure 7-14).

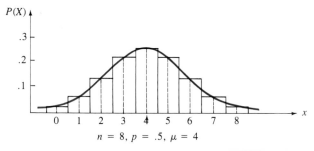

$n = 8, p = .5, \mu = 4$

FIGURE 7-14 Approximating a binomial distribution with a normal distribution

Recall that in any normal distribution with mean μ and standard deviation σ, 99.7% of the distribution (or, for practical purposes, almost all of the distribution) lies between $\mu - 3\sigma$ and $\mu + 3\sigma$. Suppose a binomial distribution has mean $\mu = np$ and standard deviation $\sqrt{np(1-p)}$. If it is to be approximated by a normal distribution, the normal distribution should have a mean equal to np and a standard deviation equal to $\sqrt{np(1-p)}$. Since the values for the binomial distribution are greater than or equal to 0 and less than or equal to n, we must have $0 \le \mu - 3\sigma$ and $\mu + 3\sigma \le n$, where μ and σ represent the mean and standard deviation, respectively, for the binomial. Thus, the following two inequalities must be satisfied:

1. $0 \le np - 3\sqrt{np(1-p)}$
2. $np + 3\sqrt{np(1-p)} \le n$

If $p \le .5$, it can be shown that the inequalities are equivalent to $np \ge 4.5$ and if $p \ge .5$, they are equivalent to $n(1-p) \ge 4.5$. Thus, we state the following rule:

> A binomial distribution can be approximated by a normal distribution if $np \ge 5$ and $n(1-p) \ge 5$.

Note that for large values of n, both inequalities should hold, especially for small values of p or $(1-p)$. For example, if $p = .01$ and $n = 600$, then $np = 600(.01) = 6$ and $n(1-p) = (600)(.99) = 5.94$.

To use a normal distribution to approximate a binomial distribution, we follow these steps:

1. Check to see if $np \ge 5$ and $n(1-p) \ge 5$ hold.
2. If so, calculate μ and σ for the binomial distribution.
3. The normal distribution with mean μ and standard deviation σ is used (the parameters were found in step 2) as the approximating distribution.
4. Construct a bar graph for the binomial distribution and determine the limits for finding areas under the normal curve.
5. Find z scores for these limits and use the z table to find the corresponding area(s). This number is the approximation to the binomial probability sought.

The following examples illustrate the process.

EXAMPLE 7-16   Find $P(x \ge 370)$ for the binomial experiment discussed at the beginning of this section, where $n = 500$ and $p = .70$.

Solution   Since $np = (500)(.7) = 350 \ge 5$ and $n(1-p) = (500)(.3) = 150 \ge 5$, a normal approximation is possible. By using (6-4), we have

$$\mu = np = 350$$
$$\sigma = \sqrt{np(1-p)} = \sqrt{(350)(.3)} = 10.25$$

Thus, the normal distribution with $\mu = 350$ and $\sigma = 10.25$ will be used for approximation purposes.

A bar graph for the binomial is shown here, along with the normal distribution with $\mu = 350$ and $\sigma = 10.25$:

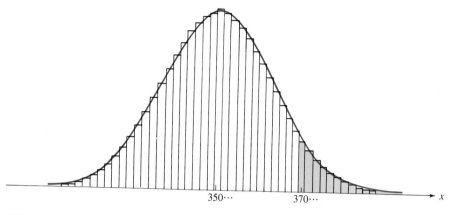

We want to find the sum of the areas of the bars that are shaded, and as an approximation to the sum of these areas, we will find the area under the normal curve to the right of 369.5. (To see this, note that the bar labeled 370 has a width of 1 and extends from 369.5 to 370.5, a distance of 1, as indicated in the following figure.)

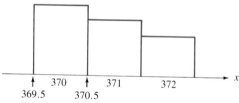

Hence, we find the shaded area under the following normal curve:

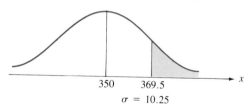

To find the shaded area, we find the z score for 369.5:

$$z = \frac{x - \mu}{\sigma} = \frac{369.5 - 350}{10.25} = 1.90$$

From the z table, we find

$$P(z \geq 1.90) = .5 - .4713 = .0287$$

Thus,

$$P(x \geq 370) \simeq .0287$$

∎

EXAMPLE 7-17   If it is known that 60% of cattle inoculated with a serum are protected from a certain disease, find the probability that 8 or 9 of 15 cows inoculated will not contract the disease by using:

a. the binomial probability formula.
b. a normal distribution to approximate the binomial.
c. the binomial probability tables.

Then compare the results.

Solution   We have $n = 15$, $p = .6$, and $x = 8, 9$.

a. By using the binomial probability formula (6-2), we have

$$P(8) = \binom{15}{8}(.6)^8(.4)^7$$
$$= (6435)(.6)^8(.4)^7 = .177$$

and

$$P(9) = \binom{15}{9}(.6)^9(.4)^6$$
$$= (5005)(.6)^9(.4)^6 = .207$$

Hence, the probability that 8 or 9 of the 15 inoculated cows will not contract the disease is

$$P(8) + P(9) = .177 + .207 = .384$$

b. We first note that $np = (15)(.6) = 9$ and $n(1 - p) = (15)(.4) = 6$, both of which are greater than 5. By using (6-4), we have

$$\mu = np = 9$$

and by using (6-7), we have

$$\sigma = \sqrt{np(1 - p)} = \sqrt{(9)(.4)} = 1.897$$

Next, we draw part of the binomial distribution to determine the limits for the area under the normal curve with $\mu = 9$ and $\sigma = 1.897$:

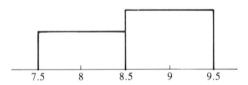

From the figure we see that we want the area under the normal curve between 7.5 and 9.5:

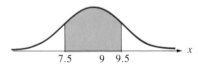

Next, we find the z scores for 7.5 and 9.5. The z score for $x = 7.5$ is

$$z = \frac{7.5 - 9}{1.897} = -.79$$

And the z score for $x = 9.5$ is

$$z = \frac{9.5 - 9}{1.897} = .26$$

From the z table, we find that the area under the standard normal curve between $z = -.79$ and $z = .26$ is

$$P(-.79 < z < .26) = .2852 + .1026 = .3878$$

The error of this value, compared with .3840, is .0038, which represents a $.0038/.3840 = .99\%$ error; the approximation is good, even for an n as small as 15.

c. By using the binomial probability tables, we find $P(8) = .177$ and $P(9) = .207$. Thus,

$$P(8, 9) = P(8) + P(9) = .177 + .207 = .384$$

which agrees exactly with the result in part (a).   ■

Problem Set 7.3

A

1. The probability that an adverse reaction will result from a flu shot is .02. If 1000 individuals are randomly selected to receive the flu shot, find the approximate probability that:
   a. at most 20 people will have an adverse reaction.
   b. at least 24 people will have an adverse reaction.
   c. between 21 and 29 (inclusive) will have an adverse reaction.
   d. exactly 210 will have an adverse reaction.

2. Suppose the probability that a certain thumbtack will land point up is .7. If 1000 thumbtacks are tossed, find the approximate probability that:
   a. 715 or more will land point up.
   b. between 685 and 720 (inclusive) will land point up.
   c. exactly 675 will land point up.
   d. at most 730 will land point up.

3. If a fair coin is tossed 1000 times, find the approximate probability that:
   a. at most 530 tosses will land heads up.
   b. between 485 and 520 (inclusive) tosses will land heads up.
   c. exactly 525 will land heads up.
   d. the number of tosses that land heads up is greater than 490.

4. In a certain city, 60% of the families own their homes. If 500 families are selected at random from the city, what is the approximate probability that:
   a. at least 315 families own their homes?
   b. between 290 and 310 families (inclusive) own their homes?
   c. exactly 305 families own their homes?
   d. at most 320 families own their homes?

5. For a binomial experiment with $p = .5$ and $n = 10$, find $P(4 \leq x \leq 6)$ by using:
   a. the binomial probability formula.
   b. a normal approximation.
   c. the binomial probability tables.
   Compare your results.

6. In a quality control program for a production process, if a sample of 100 items produced yields less than 10 defective items, the production is considered satisfactory; otherwise it is considered unsatisfactory. What is the approximate probability that in a production of 100 items the production is considered:
   a. satisfactory when in fact the process produces 15% defective items?
   b. unsatisfactory when in fact the process produces 15% defective items?

7. To determine whether a new drug is effective, it will be given to 500 patients. If 440 or more achieve positive results, the drug will be classified as effective. Otherwise, the drug will be classified as ineffective. Find the approximate probability that the drug will be classified ineffective if, in fact, the drug is 90% effective.

8. The probability that a certain brand of thumbtack will land point up when dropped is .60. If 500 tacks are dropped, what is the approximate probability that 325 or less land point up?

9. If 1000 coins are tossed, the most likely outcome is 500 heads. Use the normal approximation to find the probability of getting 500 heads.

B

10. The equation for the standard normal curve is

$$f(z) = \left(\frac{1}{\sqrt{2\pi}} \right) e^{-z^2/2}$$

The result in Problem 9 could be approximated by multiplying $f(0)$ by $1/\sigma$, since the width of the 500th bar in x units is 1 and in z units is $1/\sigma$. Perform this calculation and compare your answer to the answer obtained in Problem 9.

11. Prove that if $0 \le np - 3\sqrt{np(1-p)}$ and $np + 3\sqrt{np(1-p)} \le n$, then $np \ge 4.5$ or $n(1-p) \ge 4.5$.

CHAPTER SUMMARY

In this chapter we studied normal probability distributions. We learned how to associate areas under a normal curve with normal probabilities. We learned how to find probabilities associated with the standard normal curve. We also learned how to use the standard normal distribution to compute probabilities associated with any normal distribution. Finally, we learned how to use the normal distributions to approximate binomial distributions. These approximations can be used whenever np and $n(1-p)$ are greater than or equal to 5.

C H A P T E R   R E V I E W

IMPORTANT TERMS

For each of the following terms, provide a definition in your own words. Then check your responses against the definitions given in the chapter.

bar graph	histogram	quartiles	standard normal table
continuous distribution	normal distribution	skewed distribution	z score plot
deciles	percentile rank	standard normal	
Empirical Rule	percentiles	probability distribution	

IMPORTANT SYMBOLS

D_n, the nth decile P_n, the nth percentile Q_n, the nth quartile

IMPORTANT FACTS AND FORMULAS

1. $z = \dfrac{x - \mu}{\sigma}$ (transformation rule)

2. *Empirical Rule* In any normal distribution, approximately:
   68% of the measurements fall within $\mu \pm \sigma$;
   95% of the measurements fall within $\mu \pm 2\sigma$;
   99.7% of the measurements fall within $\mu \pm 3\sigma$.

3. The area under a normal curve is 1.

4. A normal distribution with $\mu = np$ and $\sigma = \sqrt{np(1-p)}$ can be used to approximate a binomial distribution with parameters n and p if both $np \ge 5$ and $n(1-p) \ge 5$.

REVIEW PROBLEMS

1. Find the following probabilities:
   a. $P(-1.2 < z < 2.2)$
   b. $P(z > 1.24)$
   c. $P(z > 1.78)$
   d. $P(z < -2.14)$
   e. $P(z < 1.58)$
   f. $P(z > 1.58)$
   g. $P(z < -2.14 \text{ or } z > 2.18)$
   h. $P(1.14 < z < 2.76)$
   i. $P(-2.11 < z < -1.17)$

2. For each of the following, find the value C so that:
   a. $P(-C < z < C) = .70$
   b. $P(z > C) = .60$
   c. $P(z > C) = .40$
   d. $P(z < C) = .80$
   e. $P(z < C) = .30$

3. Let x be normally distributed with $\mu = 40$ and $\sigma = 5$. Find the following probabilities:
   a. $P(x > 47)$
   b. $P(x > 32)$
   c. $P(x < 36)$
   d. $P(x < 51)$
   e. $P(36 < x < 48)$
   f. $P(42 < x < 61)$

4. Let x be normally distributed with $\mu = 50$ and $\sigma = 10$. For each of the following, find the value of C so that:
   a. $P(-C < x < C) = .68$
   b. $P(-C < x < C) = .95$
   c. $P(x > C) = .70$
   d. $P(x > C) = .30$
   e. $P(x < C) = .30$
   f. $P(x < C) = .60$

5. If the test scores of 400 students are normally distributed with a mean of 100 and a standard deviation of 10, approximately how many students scored between 90 and 110?

6. A statistics professor announces that 10% of the grades he gives are As. The results of the final examination indicate that the mean score is 73 and the standard deviation is 6. What minimum score must a student obtain to receive an A? (Assume that the grades are normally distributed.)

7. The daily output of an assembly line is normally distributed with a mean of 165 units and a standard deviation of 5 units. Find the probability that the number of units produced per day is:
   a. less than 162 units.
   b. greater than 173 units.
   c. between 152 units and 174 units.
   d. between 171 units and 177 units.

8. For Problem 7, find the number of units x so that the production output is at least x units on 75% of the days.

9. Suppose the typical speeds on Maryland highways are normally distributed with a mean of 59 mph and a standard deviation of 4 mph. If the state police are instructed to ticket the fastest 10% of the motorists, what is the fastest you could travel on Maryland highways and not receive a ticket?

10. A student with an IQ of 135 claims her IQ score is in the top 3% of her class. Is her claim true if the scores of her classmates are normally distributed with a mean of 120 and a standard deviation of 7?

11. A particular type of birth-control pill is 90% effective. If 50 people use the pill, what is the approximate probability that the pill will result in one birth, assuming that male and female births are equally likely? What is the probability that this birth is a male?

12. It is estimated that 85% of all household plants are overwatered. In a group of 500 plants, what is the approximate probability that:
    a. 433 are overwatered?
    b. between 420 and 435 (inclusive) are overwatered?
    c. at least 430 are overwatered?
    d. at most 440 are overwatered?

13. If it is estimated that 80% of all students who take the Math 209 final exam pass, what is the approximate probability that at least half of a group of 30 students who take the final exam pass?

CHAPTER ACHIEVEMENT TEST

(20 points) **1.** Find the following probabilities:
   a. $P(z < .51)$ b. $P(-1.2 < z < 2.3)$ c. $P(-2.4 < z < -.78)$
   d. $P(z > -1.14)$ e. $P(z < -.24)$

(20 points)   **2.** For each of the following, find the value of C so that:
  a. $P(z > C) = .30$ **b.** $P(z > C) = .75$
  c. $P(z < C) = .40$ **d.** $P(-C < z < C) = .80$

(20 points)   **3.** Suppose x is normally distributed with $\mu = 20$ and $\sigma = 3$. Find:
  a. $P(x > 27)$ **b.** $P(x > 16)$
  c. $P(13 < x < 15)$ **d.** $P(25 < x < 28)$

(20 points)   **4.** For Problem 3, find the value of C for each of the following:
  a. $P(x > C) = .90$ **b.** $P(x > C) = .40$
  c. $P(-C < x < C) = .60$ **d.** $P(x < C) = .78$

(10 points)   **5.** If x is normally distributed with $\mu = 40$ and $\sigma = 16$, find:
  a. P_{75} **b.** P_{40}

(5 points)   **6.** The useful lifetimes of a certain brand of light bulb are normally distributed with $\mu = 250$ hours and $\sigma = 50$ hours. If the manufacturer guarantees that its lights will last at least 125 hours, what percentage of the bulbs does it expect to replace under the guarantee?

(5 points)   **7.** A manufacturer of computer chips claims that only 1% of its chips are defective. If we assume the claim to be true, what is the probability that no more than 25 chips are defective in the next batch of 2000 chips?

8

SAMPLING THEORY

Chapter Objectives

In this chapter you will learn:

- *what bias is*

- *what sampling error is*

- *how to use the random number table*

- *various methods for drawing a random sample*

- *what sampling distributions are*

- *what the sampling distribution of the mean is*

- *what the mean, variance, and standard deviation for the sampling distribution of the mean are*

- *how to find probabilities associated with the sampling distribution of the mean when the population is normal*

- *what t distributions are*

- *the central limit theorem*

- *applications of the central limit theorem*

- *what the sampling distribution of the sample proportion is*

- *what the mean, variance, and standard deviation of the sampling distribution of the sample proportion are*

- *applications using the sampling distribution of the sample proportion*

Chapter Contents

243

A major concern of inferential statistics is to estimate unknown population characteristics (parameters) by examining information gathered from a sub-collection (sample) of the population. The focus of concern is on the sample. A sample should be representative of the population if it is to be used to study the population. We will follow certain selection procedures to ensure that our samples accurately reflect the sampled population. Only when representative samples are used can probabilistic statements be made about the population being sampled. For example, suppose we are interested in the attitudes of college students toward studying. The relevant population in this case is the collection of responses from all college students regarding studying. A sample is a subcollection from the population that we will use to gauge the attitudes of all college students. If we use the responses of only social fraternity members, then we will most likely obtain biased or atypical results. Or, if we use only the responses from members of honorary fraternities, we would probably obtain results biased in favor of studying. By using a biased sample, we would obtain a distorted view of the attitudes of the entire student body.

When it is desirable to study the characteristics of large populations, samples are used for many reasons. A complete enumeration of the population, called a **census**, may not be economically possible; or, there may not be enough time to examine the whole population. In some situations, a census may not be possible; for example, a census of the sea-life population in the Atlantic Ocean is not possible.

The following examples illustrate the uses of sampling in different fields:

1. *Politics* Samples of voters' opinions are used by candidates to gauge public opinion and support in elections.
2. *Education* Samples of students' test scores are used to determine the effectiveness of a teaching technique or program.
3. *Industry* Samples of products coming off an assembly line are used for quality control purposes.
4. *Medicine* Samples of blood-sugar measurements for diabetic patients are used to test the efficacy of a new technique or drug.
5. *Agriculture* Samples of corn yields per plot are used to test the effects of a new fertilizer on crop yields.
6. *Government* Samples of voters' opinions are used to determine public opinions on matters related to national welfare and security.

8.1 TYPES OF ERRORS AND RANDOM SAMPLES

Two general types of errors may occur when sample values (or statistics) are used to estimate population values (or parameters): sampling error and non-sampling error. **Sampling error** refers to the natural variation inherent among samples taken from the same population. This type of error will exist when the sample is not a perfect copy of the population. Even if great care is taken to ensure that two samples of the same size are representative of the same population, one would not expect both samples to be identical in every detail. For example, we would not expect the two sample means to be identical. Sampling error is an important concept that will help us better understand the nature of inferential statistics. We will further examine the concept of sampling error later in this chapter.

Errors that arise with sampling but which cannot be classified as sampling errors are called **nonsampling errors**. Statistical bias is one type of nonsampling error. **Statistical bias** refers to a systematic tendency inherent in a sampling method which gives estimates of a parameter that are, on the average, either smaller (negative bias) or larger (positive bias) than the actual parameter. Errors arising from data collection fall into this category. For example, if we wanted to obtain information concerning attitudes toward abortion and we obtained a sample consisting predominantly of men, then we would encounter a form of statistical bias called **sampling bias**. As a classic example of sampling bias, consider the survey of the attitudes of several million people taken by *Literary Digest*, a popular journal of the period, to forecast the 1936 presidential winner. Republican Alfred Landon was running against Democrat Franklin Roosevelt. Names of people to be included in the survey were obtained by the *Digest* from telephone directories and magazine subscription lists. A majority of those surveyed indicated their preference for Landon, and the journal predicted he would win by a landslide. As we know, Landon lost. Many people who were most likely to vote for Roosevelt did not have telephones or subscribe to magazines. The sample thus contained strong bias in favor of Landon.

Errors that result from data accumulation or processing are also classified as nonsampling errors. For example, such errors could result when the instruments used in making measurements are out of adjustment or not properly calibrated, data are misplaced or lost in storage, or answers obtained from people during a survey are not truthful. The latter case might arise with questions dealing with age, to which some people will lie out of vanity.

Sampling bias can be removed (or minimized) by using randomization. **Randomization** refers to any process for selecting a sample from the population that involves impartial or unbiased selection. A sample chosen by randomization procedures is called a **random sample**. The most popular types of sampling techniques involving randomization are simple random sampling, stratified sampling, cluster sampling, and systematic sampling. Each of these will now be explained.

Random Sample

If a random sample is chosen in such a way that all samples of the same size are equally likely to be chosen, then the sample is called a **simple random sample**. Suppose it is desired to select a random sample of 5 students from a statistics class of 20 students. The binomial coefficient $\binom{20}{5}$ gives the total number of ways of selecting an unordered sample of 5 students from a class of 20 students:

$$\binom{20}{5} = \frac{20!}{(5!)(15!)}$$
$$= \frac{(20)(19)(18)(17)(16)}{(5)(4)(3)(2)(1)}$$
$$= 15{,}504$$

If we listed the 15,504 possibilities on separate pieces of paper (a tremendous task), placed them into a container, and thoroughly mixed them, then we could choose a random sample of 5 by selecting one slip with five names. A simpler procedure to select a random sample would be to write each of the 20 names on separate pieces of paper, place them into a container, thoroughly mix them, and then draw 5 slips at once.

Another method of obtaining a random sample of 5 students from a class of 20 students is based on a table of random numbers. Table 8-1 is a brief table of random numbers. It was created by using a computer with a built-in random number generator. Numbers are grouped into sets of five digits for ease of reading. For example, the shaded digit 5 in the table is in the third row and tenth column.

Table 8-1 Random Number Table

50242	94818	05825	87975	77496
11021	62231	88043	88062	71692
59323	52225	32913	02586	31934
00576	29013	84384	00445	98538
53228	13826	36564	74019	32067
26406	32693	68126	99353	23898
64122	44928	96161	39435	31767
90911	26356	91554	43233	39124
13835	62884	31326	28818	05167
36471	28587	30432	74161	07649
27201	65409	40210	72802	20198
95370	73018	94933	15544	26640
60880	56768	47370	38451	63468
15130	08631	55031	48008	04123
01239	72330	94403	71123	37141
02928	17216	79729	27076	58220
10220	70101	10902	80993	19104
88707	83518	68517	40570	40291
61461	91259	97508	56729	71972
51245	79283	10366	38290	30673

A **random number table** can be constructed without using a computer. If the ten digits from 0 to 9 are written on slips of paper, placed into a container, and thoroughly mixed, then the first slip chosen determines the first number in the table. The slip of paper is returned to the container and a second slip is drawn after the slips are again mixed thoroughly. The second slip determines the second number in the random number table. This process is followed until a table with the desired number of digits is obtained.

We will illustrate the use of the random number table by choosing a random sample of 5 students from the following list of 20 students enrolled in Statistics 209:

01	Mike Able	11	Sarah Kemp
02	Mary Baker	12	James Lum
03	Joe Cable	13	Robert Moon
04	Ed Doe	14	Karen Nie
05	Jake Eldon	15	Beth Oboe
06	Sue Fum	16	Dave Poe
07	Pete Gum	17	Rick Quest
08	Harry Hoe	18	Bart Rat
09	Ora Ida	19	Stella Star
10	Helen Jewel	20	Maud Tuck

The names of the 20 students are listed in some order (we have used alphabetical order for convenience) and each student is assigned an identification number. In our example, we need only two-digit identification numbers. A starting place in Table 8-1 is selected at random, perhaps by closing your eyes and pointing to some number in the table with the tip of your pencil. Random numbers in groups of two are then read either horizontally or vertically starting at the random starting place. If a two-digit number greater than 20 or a number that has been previously obtained is chosen, it is omitted and the next pair is selected. Suppose the digit 2 in the sixth row and first column is selected as the starting point. If two-digit numbers are read horizontally starting at this point, then the following names constitute a random sample of 5 students:

12	James Lum
17	Rick Quest
13	Robert Moon
07	Pete Gum
01	Mike Able

The above procedure for selecting a random sample of size 5 from a class of 20 students could be made more efficient in terms of the time required to identify the five students and, as a result, require less effort. Instead of assigning each of the 20 students a single two-digit number, we could assign each of them five two-digit numbers as follows:

Student	Two-digit number assignments
Able	00, 01, 02, 03, 04
Baker	05, 06, 07, 08, 09
Cable	10, 11, 12, 13, 14
Doe	15, 16, 17, 18, 19
Eldon	20, 21, 22, 23, 24
Fum	25, 26, 27, 28, 29
⋮	
Star	90, 91, 92, 93, 94
Tuck	95, 96, 97, 98, 99

If we use this procedure to identify the students for our random sample and start with digit 2 in the sixth row and first column of the random number table, we will select the following five students:

Random digits	Student
26	Sue Fum
40	Ora Ida
63	Robert Moon
93	Stella Star
68	Karen Nie

Note that we had to draw six two-digit numerals to choose our sample since the random number 26 was chosen twice.

It is impossible to determine by inspection whether a sample is random or not. The following example illustrates this point.

EXAMPLE 8-1   Suppose it is desired to select four months of the year to study certain biological phenomena and that March, June, September, and December are chosen. Do these four months represent a random sample of months?

Solution   It is impossible to tell from the information given whether the sample is random. To determine whether a sample is random, we must know what selection process was used. These months might have been chosen because they are spread through the year; if so, the sample is not random. However, if these months were selected with the aid of a random number table or other randomization procedure, then they do represent a random sample.   ■

There are many situations in which simple random sampling is impractical, impossible, or undesirable. Although it would be desirable to use simple random samples for national opinion polls on product information or presidential elections, such would be very costly and time-consuming.

Stratified sampling involves separating the population into nonoverlapping groups, called **strata**, and then selecting a simple random sample from each stratum. The information from the collection of simple random samples would then constitute the overall sample. For example, suppose we want to sample the opinions of faculty at a large university on an important issue. It might be difficult to sample all university faculty, so let's suppose we select a random sample from each college (or academic department). The strata would be the colleges (or academic departments) within the university.

Stratified sampling is commonly used for national opinion polls because opinions tend to vary more among different localities than within localities. For this application, the strata might be counties, states, or regions. The criteria for forming the strata should ensure that the observations within each stratum are as much alike as possible. The observations within a given stratum should have less variation than the observations among strata.

Cluster sampling involves selecting a simple random sample of heterogeneous subunits, called **clusters**, from the population. Each element in the population belongs to exactly one cluster, and the elements within each cluster are usually heterogeneous or unalike. For example, suppose a cable television service company is considering opening a branch office in a certain large city. The company plans to conduct a survey to determine the percentage of households that would use its service. Since it is not practical to survey every household, the company decides to choose a city block at random and then to survey every household in the block. The city block is a cluster.

With cluster sampling, clusters are formed to represent, as closely as possible, the entire population. A simple random sample of the clusters is then used to study the population. Studies of social institutions, such as churches, hospitals, schools, and prisons, generally involve cluster sampling. In such studies, a single cluster, or several clusters, are selected randomly for study. The entire population can be effectively studied by studying its miniature copies or clusters. If a cluster is too large to be studied effectively, elements within the cluster(s) can be randomly selected.

Systematic sampling is a sampling technique that involves an initial random selection of observations followed by a selection of observations obtained by using some rule or system. For example, to obtain a sample of telephone subscribers within a large city, a random sample of page numbers within the phone directory might first be obtained. Then by selecting the 20th name on each page, we would obtain a systematic sample. Or we could select one name at random from the first page of the directory and then select every 100th name thereafter. An alternate approach would involve first selecting a number at random from those numbered 1 to 100; assume the number 40 is selected. Then select names from the directory that correspond to the numbers 40, 140, 240, 340, and so forth.

Sampling Error

Any measurement involves some error. If the sample mean is used to measure (estimate) the population mean μ, then the sample mean, as a measurement, involves some error. As an example, suppose a random sample of size 25 is drawn from a population with a mean of $\mu = 15$. If the mean of the sample is $\bar{x} = 12$, then the observed difference $\bar{x} - \mu = -3$ is referred to as **sampling error**. A sample mean $\bar{x}$ can be thought of as the sum of two quantities, the population mean μ and sampling error. If e denotes the sampling error, then we have

$$\bar{x} = \mu + e$$

For the sake of illustration, suppose random samples of size 2 are selected from a population consisting of three values, 2, 4, and 6. In order to simulate a "large" population so that sampling can be carried out a large number of times, we will assume that sampling is done with replacement—that is, the sampled number is replaced before the next number is drawn. In addition, we will select ordered samples. In an ordered sample, the order in which observations are selected is important. Thus, the ordered sample $\{2, 4\}$ is different from the ordered sample $\{4, 2\}$. In the ordered sample $\{4, 2\}$, 4 was drawn first, and then 2. Table 8-2 contains a list of all possible ordered samples of size 2 that can be selected with replacement from the population of values 2, 4, and 6. In addition, the table contains sample means and corresponding sampling errors. Note that the population mean is equal to $\mu = (2 + 4 + 6)/3 = 4$.

Table 8-2 Ordered Samples of Size 2 from Population of Values 2, 4, 6

Ordered samples	$\bar{x}$	Sampling error ($\bar{x} - \mu$)
$\{2, 2\}$	2	$2 - 4 = -2$
$\{2, 4\}$	3	$3 - 4 = -1$
$\{2, 6\}$	4	$4 - 4 = 0$
$\{4, 2\}$	3	$3 - 4 = -1$
$\{4, 4\}$	4	$4 - 4 = 0$
$\{4, 6\}$	5	$5 - 4 = 1$
$\{6, 2\}$	4	$4 - 4 = 0$
$\{6, 4\}$	5	$5 - 4 = 1$
$\{6, 6\}$	6	$6 - 4 = 2$

Note the following interesting relationships contained in Table 8-2:

1. The mean of the collection of all sample means is 4, the mean of the population from which the samples were drawn. If $\mu_{\bar{x}}$ denotes the mean of all the sample means, then we have

$$\mu_{\bar{x}} = \frac{2 + 3 + 4 + 3 + 4 + 5 + 4 + 5 + 6}{9}$$

$$= \frac{36}{9} = 4$$

2. The sum of the sampling errors is 0:

$$e_1 + e_2 + e_3 + \cdots + e_9 = (-2) + (-1) + 0 + (-1) + 0 + 1 + 0 + 1 + 2$$
$$= 0$$

Thus, if $\bar{x}$ is used to measure (estimate) the population mean μ, the average of all the sampling errors is 0.

When a statistic, such as $\bar{x}$, is used to measure or estimate a parameter and the average of all the sampling errors is 0, the statistic is said to be **unbiased**. In addition, we say a statistic has **zero bias** when used to measure or estimate a parameter if the average of all the sampling errors is 0. Thus, in the above example, $\bar{x}$ has zero bias and is an unbiased statistic. Strictly speaking, when the sample mean is used for estimation purposes, it is referred to as an **estimator** and a value of the sample mean is referred to as an **estimate**. Since the value of the sample mean varies from sample to sample, it is an example of a random variable. All estimators are random variables. We will discuss estimators in more detail in Chapter 9.

If two different random samples of size 25 are drawn from the same population and the mean of each sample is calculated, one would not expect both sample means to be identical. Rather, each mean would reflect a different sampling error. If $\bar{x}_1$ denotes the mean of the first sample, $\bar{x}_2$ denotes the mean of the second sample, e_1 denotes the sampling error associated with $\bar{x}_1$, and e_2 denotes the sampling error associated with $\bar{x}_2$, then we have

$$\bar{x}_1 = \mu + e_1$$
$$\bar{x}_2 = \mu + e_2$$
$$\bar{x}_1 - \bar{x}_2 = e_1 - e_2$$

The difference between sample means reflects only the difference between the sampling errors. If repeated pairs of random samples were obtained and the differences in sample means were obtained for each pair, we would expect to obtain an average difference in sample means (sampling errors) of 0.

Problem Set 8.1

A

1. By using the first digit in the fourth row of Table 8-1 as the starting point and moving horizontally to the right, select a random sample of size 10 from the list of Statistics 209 students given in this section.

2. A new-car dealer wants to select a random sample of opinions concerning a new model from 15 of the 97 customers on the mailing list who have purchased the new model during the past year. Explain how the sample could be selected with the help of a random number table.

3. Would the unordered sample {2, 4, 6, 8, 10} constitute a random sample from the population of whole numbers from 1 to 10, inclusive? Explain.

4. Consider the first 10 rows in Table 8-1 and record the frequency that each digit occurs. How many times would you expect each digit to occur?

5. For Problem 4, do you think the variation between the observed frequency and expected frequency for each digit indicates a variation due to sampling error? Complete the accompanying table and find the average of the sampling errors.

Digit	Frequency	Expected frequency	Sampling error
0			
1			
2			
3			
4			
5			
6			
7			
8			
9			

6. A random sample of 5 teachers is to be selected from a population of 200 teachers to participate in a special workshop.

   a. Label the teachers from 001 to 200. Which teachers will be selected for the workshop if Table 8-1 is used and the starting point is the first digit in the fourth row and second column and the digits are read horizontally moving to the right?

   b. A more efficient selection process involves labeling the teachers as in part (a) and assigning the numbers 001, 201, 401, 601, 801 to the first teacher, numbers 002, 202, 402, 602, 802 to the second teacher, numbers 003, 203, 403, 603, 803 to the third teacher, ..., and numbers 200, 400, 600, 800, 000 to the last teacher. By using this scheme and starting at the same point as in part (a), select a random sample of 5 teachers.

7. Start at the first digit in the sixth row and move horizontally to the right in Table 8-1 to select a random sample of 12 tosses of a six-sided die. Construct a table similar to the one in Problem 5 and find the average of the sampling errors.

8. Consider all possible ordered samples of size 2 drawn with replacement from the population of values 0, 2, 4, 6. Construct a table similar to Table 8-2 and determine if $\bar{x}$ is an unbiased estimator of μ.

9. Do the following procedures produce random samples? Why or why not?

   a. To obtain a random sample of students in a class, select all students in the class who wear glasses.

   b. To obtain a random sample of coeds, select every third person entering the gym through the front door.

   c. To obtain a random sample of college students attending a large university, select students by using a random number table and the last four digits of their Social Security numbers.

   d. To select a random sample of college athletes, select every second person coming out of the gym through the front door.

   e. To select a random sample of people who have phones, select every other person on a randomly selected page of the phone book.

B

10. Starting at the first digit in the sixth row of Table 8-1 and reading horizontally, select two random

samples of ten distinct two-digit numbers, one sample at a time, and compute the means for both samples. If a sample mean is used to estimate the population mean, μ, what is the difference between the sampling errors for these two sample means? If this experiment is repeated a large number of times, what kind of bias would you expect for the differences between sample means? Explain.

11. Indicate an appropriate sampling technique (simple random sample, stratified sample, cluster sample, or systematic sample) for obtaining the following samples. For each, give reasons for your choice.

   a. A sample of students in a college to study student opinions on nuclear power as an alternative power source

   b. A sample of students to study student attitudes on campus parking

   c. A sample of corn plants to study yields and quality

   d. A sample of rabbits in western Maryland to study rabbit weights

   e. A sample of fish in Piney Dam to study species of fish

   f. A sample of school teachers in Allegheny County to study teachers' opinions on the issue of merit pay

   g. A sample of school teachers in Maryland to study teachers' opinions on the issue of merit pay

   h. A sample of car owners to study their preferences for engine oil

   i. A sample of retired persons to study use of leisure time by retired persons

12. Suppose two upper-level mathematics classes, A and B, take the same 10-point quiz and the results are as follows:

Class A	3	9	6	5	4	8	4	7	6	5
Class B	6	5	4	2	3	7	7	3	5	6

If in fact both classes know the material equally well, estimate the probability that $\bar{x}_A - \bar{x}_B \geq .9$. [*Hint:* Write each of the above 20 scores on a 3×5 card, shuffle the 20 cards well, and deal two hands of ten cards. Then calculate the mean for each hand, A and B, and determine if $\bar{x}_A - \bar{x}_B \geq .9$. After dealing 20 pairs of ten-card hands, determine the proportion of ten-card hands resulting in $\bar{x}_A - \bar{x}_B \geq .9$. This number estimates $P(\bar{x}_A - \bar{x}_B \geq .9)$ if in fact both classes did equally well.]

13. The device shown in the accompanying figure is called a **hextat**. A ball is dropped into the device. At each peg it hits, it goes either left or right. After hitting five pegs, it lands in one of the slots A through F. Simulate dropping 16 balls into the hextat as follows: use the random number table starting with the digit 1 in the second row and first column and moving horizontally to the right to choose 80 digits, which determine the final positions of the 16 balls. Let an even digit represent left and an odd digit represent right. The first five digits will determine the position of the first ball; the next five digits, the position of the second ball; and so on. For example, the first five random digits drawn are 1, 1, 0, 2, and 1. This means that the first ball moves right, right, left, left, and right, as shown in the figure. [If a large number of balls are used and the final position of each ball measures its horizontal distance from the starting point, then the horizontal distances have a mean of 0 and form a normal distribution. The hextat is used to depict, among other things, Brownian movement of plant spores or small particles suspended in a liquid.]

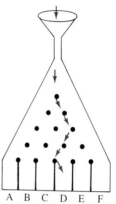

8.2 SAMPLING DISTRIBUTIONS

Random samples drawn from a population are by their basic nature unpredictable. One would not expect two random samples of the same size taken from the same population to possess the same sample mean or to be exactly alike. Any statistic (such as the sample mean) computed from the measurements in a random sample can be expected to vary in value from random sample to random sample. Accordingly, we shall want to study the distribution of all possible values of a statistic. Such distributions will be very important in the study of inferential statistics, since inferences about populations will be made using sample statistics. By studying the distributions associated with sample statistics, we will be able to judge the reliability of a sample statistic as an instrument for making an inference about an unknown population parameter.

Since a statistic, such as $\bar{x}$, varies from random sample to random sample, it can be considered a **random variable** with a corresponding frequency distribution. Frequency distributions for sample statistics are referred to as **sampling distributions**. As an example, suppose random samples of size 20 are selected from a large population. For each sample, the sample mean $\bar{x}$ is computed. The collection of all these sample means is called the **sampling distribution of the sample mean**. This concept is illustrated as in Figure 8-1. For convenience, we shall denote the population by X and the sampling distribution of the mean by $\bar{X}$.

As another example, suppose samples of size 20 are randomly selected from a large population and the sample standard deviation of each sample is computed. The collection of all these sample standard deviations is called the

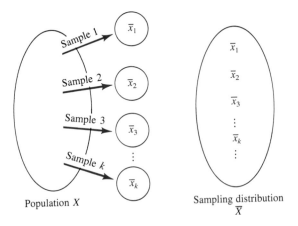

FIGURE 8-1 Sampling distribution of the sample mean

sampling distribution of the sample standard deviation. This can be illustrated as in Figure 8-2.

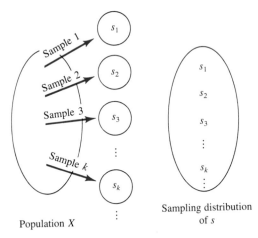

FIGURE 8-2 Sampling distribution of the sample standard deviation

In general, the **sampling distribution of a statistic** is the distribution of all possible values of the statistic computed from samples of the same size. Consider the following example.

EXAMPLE 8-2   Suppose ordered samples of size 2 are selected, with replacement, from the population of values 0, 2, 4, 6. Find:

a. μ, the population mean.
b. σ, the population standard deviation.

c. $\mu_{\bar{x}}$, the mean of the sampling distribution of the mean.
d. $\sigma_{\bar{x}}$, the standard deviation of the sampling distribution of the mean.

In addition, draw frequency graphs for the population distribution of X and the sampling distribution of $\bar{X}$.

Solution   **a.** The population mean is

$$\mu = \frac{0 + 2 + 4 + 6}{4} = 3$$

b. The population standard deviation is

$$\sigma = \sqrt{\frac{(0 - 3)^2 + (2 - 3)^2 + (4 - 3)^2 + (6 - 3)^2}{4}}$$

$$= \sqrt{\frac{9 + 1 + 1 + 9}{4}} = \sqrt{5}$$

c. We next list the elements of the sampling distribution of the mean and the corresponding frequency distribution:

Sample	$\bar{x}$		$\bar{x}$	f
{0, 0}	0		0	1
{0, 2}	1		1	2
{0, 4}	2		2	3
{0, 6}	3		3	4
{2, 0}	1		4	3
{2, 2}	2		5	2
{2, 4}	3		6	1
{2, 6}	4			
{4, 0}	2		$N = 16$	
{4, 2}	3		sample means	
{4, 4}	4			
{4, 6}	5			
{6, 0}	3			
{6, 2}	4			
{6, 4}	5			
{6, 6}	6			

The mean of the sampling distribution of the sample mean is

$$\mu_{\bar{x}} = \frac{\sum(f\bar{x})}{\sum f}$$

$$= \frac{(0)(1) + (1)(2) + (2)(3) + (3)(4) + (4)(3) + (5)(2) + (6)(1)}{16}$$

$$= \frac{48}{16} = 3$$

d. To find the standard deviation of the distribution of the sample mean, we organize the calculations in the following table:

$\bar{x}$	f	$\bar{x} - \mu_{\bar{x}}$	$(\bar{x} - \mu_{\bar{x}})^2$	$(\bar{x} - \mu_{\bar{x}})^2 f$
0	1	$0 - 3 = -3$	9	9
1	2	$1 - 3 = -2$	4	8
2	3	$2 - 3 = -1$	1	3
3	4	$3 - 3 = 0$	0	0
4	3	$4 - 3 = 1$	1	3
5	2	$5 - 3 = 2$	4	8
6	1	$6 - 3 = 3$	9	9
				$\overline{40}$

The standard deviation of the sampling distribution of the sample mean is

$$\sigma_{\bar{x}} = \sqrt{\frac{\sum (\bar{x} - \mu_{\bar{x}})^2 f}{N}}$$

$$= \sqrt{\frac{40}{16}}$$

$$= \sqrt{2.5} \simeq 1.58$$

The frequency graphs for the population and the sampling distribution of the sample mean are shown in Figure 8-3.

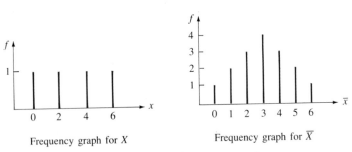

Frequency graph for X Frequency graph for $\overline{X}$

FIGURE 8-3 Population distribution and sampling distribution of sample mean for Example 8-2   ■

As with any random variable, the distribution of the sample mean has a mean (or expected value), a variance, and a standard deviation. It can be shown that the sampling distribution of the sample mean has a mean equal to the population mean. That is,

$$\mu_{\bar{x}} = E(\bar{x}) = \mu$$

Notice that for Example 8-2, $\mu_{\bar{x}} = 3 = \mu$.

EXAMPLE 8-3   For the sampling distribution of the mean in Example 8-2, find:

a. the sampling error for each mean.
b. the mean of the sampling errors.
c. the standard deviation of the sampling errors.

Solution   a. The samples, sample means, and sampling errors are shown in the following table:

Sample	Sample mean, $\bar{x}$	Sampling error, $e = \bar{x} - \mu$
{0, 0}	0	$0 - 3 = -3$
{0, 2}	1	$1 - 3 = -2$
{0, 4}	2	$2 - 3 = -1$
{0, 6}	3	$3 - 3 = 0$
{2, 0}	1	$1 - 3 = -2$
{2, 2}	2	$2 - 3 = -1$
{2, 4}	3	$3 - 3 = 0$
{2, 6}	4	$4 - 3 = 1$
{4, 0}	2	$2 - 3 = -1$
{4, 2}	3	$3 - 3 = 0$
{4, 4}	4	$4 - 3 = 1$
{4, 6}	5	$5 - 3 = 2$
{6, 0}	3	$3 - 3 = 0$
{6, 2}	4	$4 - 3 = 1$
{6, 4}	5	$5 - 3 = 2$
{6, 6}	6	$6 - 3 = 3$

b. The mean of the sampling errors, μ_e, is

$$\mu_e = \frac{(-3) + (-2) + (-1) + 0 \cdots + 1 + 3}{16} = 0$$

c. The standard deviation of the sampling errors is simply found after condensing the information obtained thus far into the following frequency table:

e	f	$e - \mu_e$	$(e - \mu_e)^2 f$
-3	1	$-3 - 0 = -3$	9
-2	2	$-2 - 0 = -2$	8
-1	3	$-1 - 0 = -1$	3
0	4	$0 - 0 = 0$	0
1	3	$1 - 0 = 1$	3
2	2	$2 - 0 = 2$	8
3	1	$3 - 0 = 3$	9
	$N = 16$		40

The standard deviation of the distribution of sampling errors, σ_e, is thus

$$\sigma_e = \sqrt{\frac{\sum (e - \mu_e)^2 f}{N}} = \sqrt{\frac{40}{16}} = 1.58 \qquad \blacksquare$$

The standard deviation of the sampling distribution of a statistic is referred to as the **standard error of the statistic**. For Example 8-2, the standard error of the mean, denoted by $\sigma_{\bar{x}}$, is 1.58. Notice that for Example 8-3, the standard deviation of the 16 sampling errors, σ_e, is also equal to 1.58. Thus, it is reasonable to call $\sigma_{\bar{x}}$ the **standard error of the mean**. It can be shown that if samples of size n are selected with replacement from a population, then the standard error of the mean is equal to the standard deviation of the distribution of sampling errors when the sample mean $\bar{x}$ is used to estimate μ. Thus, in general, we have

$$\sigma_e = \sigma_{\bar{x}}$$

where $e = \bar{x} - \mu$.

As an illustration, suppose a fair coin is tossed ten times and the number of heads obtained, X, is recorded. If this experiment is repeated a large number of times, the number of heads obtained each time will vary from 0 to 10. Thus, X, the number of heads obtained when ten fair coins are tossed, is a random variable with a sampling distribution. The sampling distribution is binomial with a mean equal to

$$\mu_x = np = 10(.5) = 5$$

and a standard deviation equal to

$$\sigma_{\bar{x}} = \sqrt{np(1 - p)} = \sqrt{10(.5)(.5)} \simeq 1.58$$

As another illustration, suppose samples of size 1 are drawn from a normal population with mean $\mu = 10$ and standard deviation $\sigma = 2$ and for each sample, a z score $Z = (x - \mu)/\sigma$ is recorded. Since the values for Z will vary with the sample, Z can be considered a random variable with a sampling distribution. The sampling distribution of Z is the standard normal distribution. As was seen in Chapter 7, the mean of Z is

$$\mu_Z = 0$$

and the standard deviation of Z is

$$\sigma_Z = 1$$

EXAMPLE 8-4   Suppose ordered samples of size 2 are selected with replacement from the population of values 0, 2, 4, 6, and 8. For each sample, the range R is recorded. For the sampling distribution of R

a. construct a frequency table for R and a frequency graph for R.
b. find the mean of the sampling distribution of R.
c. find the standard error of R.

Solution **a.** We first list the ordered samples along with their corresponding ranges in the following table:

Sample	Range R	Sample	Range R
{0, 0}	0	{4, 6}	2
{0, 2}	2	{4, 8}	4
{0, 4}	4	{6, 0}	6
{0, 6}	6	{6, 2}	4
{0, 8}	8	{6, 4}	2
{2, 0}	2	{6, 6}	0
{2, 2}	0	{6, 8}	2
{2, 4}	2	{8, 0}	8
{2, 6}	4	{8, 2}	6
{2, 8}	6	{8, 4}	4
{4, 0}	4	{8, 6}	2
{4, 2}	2	{8, 8}	0
{4, 4}	0		

The frequency table and frequency graph are shown here:

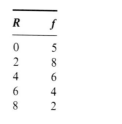
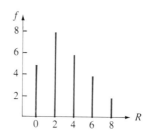

R	f
0	5
2	8
4	6
6	4
8	2

b. The mean of the sampling distribution of the range is

$$\mu_R = \frac{\Sigma(fR)}{\Sigma f}$$

$$= \frac{(5)(0) + (8)(2) + (6)(4) + (4)(6) + (2)(8)}{25}$$

$$= \frac{80}{25} = 3.2$$

c. The standard deviation of the sampling distribution of R (the standard error of R) is found next:

$$\sigma_R = \sqrt{\frac{\Sigma(R - \mu_R)^2 f}{\Sigma f}}$$

$$= \sqrt{\frac{(0 - 3.2)^2(5) + (2 - 3.2)^2(8) + (4 - 3.2)^2(6) + (6 - 3.2)^2(4) + (8 - 3.2)^2(2)}{25}}$$

$$= 2.4$$

Hence, the standard error of R is $\sigma_R = 2.4$.

Note that the sampling distribution of a statistic must necessarily involve samples of the same size. If the samples involved are of different sizes, then it makes no sense to talk about the shape of the sampling distribution.

The Sampling Distribution of the Mean

The sampling distribution of the sample mean has parameters that are related to the parameters of the population sampled. As we previously pointed out, the mean of the sampling distribution of the sample mean is equal to the mean of the population from which the samples were selected. This can be expressed by the following formula:

$$\mu_{\bar{x}} = \mu_x = \mu \tag{8-1}$$

This fact was illustrated in Example 8-2.

As another illustration, suppose a large hardware store carries three different brands of paint, priced at $8, $10, and $14 per gallon. Based on past experience, the probability distribution table for the price paid by a randomly selected customer is as follows:

x	8	10	14
$P(x)$	.5	.3	.2

The price a randomly selected customer pays for paint is a discrete random variable with values 8, 10, and 14. Suppose two randomly chosen customers make independent paint purchases. Let x_1 denote the price paid by the first customer and let x_2 denote the price paid by the second customer. Table 8-3 lists the possible purchases with their corresponding probabilities and average purchase price $\bar{x}$.

Table 8-3 Probabilities of Average Purchase Price for Paint Example

x_1	x_2	$P(x_1, x_2)$	$\bar{x}$
8	8	$(.5)(.5) = .25$	8
8	10	$(.5)(.3) = .15$	9
8	14	$(.5)(.2) = .10$	11
10	8	$(.3)(.5) = .15$	9
10	10	$(.3)(.3) = .09$	10
10	14	$(.3)(.2) = .06$	12
14	8	$(.2)(.5) = .10$	11
14	10	$(.2)(.3) = .06$	12
14	14	$(.2)(.2) = .04$	14

Note that if E is the event that the first customer purchases paint priced at $8, and F is the event that the second customer purchases paint priced at $10, then since the paint purchases are independent, we have $P(E \text{ and } F) = P(E)P(F)$.

In addition, since $P(E) = P(x_1 = 8)$ and $P(F) = P(x_2 = 10)$, to calculate the value of $P(8, 10) = P(x_1 = 8$ and $x_2 = 10)$, we need only multiply $P(8)$ by $P(10)$. Thus, $P(8, 10) = P(8)P(10) = (5.)(.3) = .15$. The remaining probabilities in Table 8-3 were found in a similar fashion.

A probability table for the sampling distribution of $\bar{x}$ for the paint example is shown below:

$\bar{x}$	8	9	10	11	12	14
$P(\bar{x})$	.25	.30	.09	.20	.12	.04

The probability $P(9) = P(\bar{x} = 9)$ was found by adding .15 and .15 obtained from Table 8-3 to get .30. The other probabilities were found similarly. A probability histogram for the sampling distribution of mean purchases is shown in Figure 8-4.

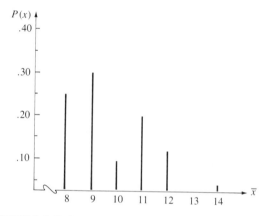

FIGURE 8-4 Probability histogram for sampling distribution of $\bar{x}$ in paint example

The mean of the sampling distribution of the sample mean is

$$\mu_{\bar{x}} = \sum \bar{x} P(\bar{x})$$
$$= (8)(.25) + (9)(.30) + (10)(.09) + (11)(.20) + (12)(.12) + (14)(.04)$$
$$= 9.8$$

This result agrees with the value for μ_x, since

$$\mu_x = \sum x P(x)$$
$$= (8)(.5) + (10)(.3) + (14)(.2) = 9.8$$

The standard error of the mean is found by using formula (5-11).

$$\sigma_{\bar{x}} = \sqrt{\sum [\bar{x}^2 P(\bar{x})] - \mu_{\bar{x}}^2}$$
$$= \sqrt{98.62 - 9.8^2} = \sqrt{2.58} = 1.61$$

Method of Sampling

The standard deviation of the sampling distribution of the mean, $\sigma_{\bar{x}}$, is influenced by the method of sampling. If sampling is done with replacement from a small population, then each measurement in the sample is independent of any other value, and the sampling can be done indefinitely, as would be the case for a very large population. In large populations where sampling is done without replacement, the statistical dependence of any one value on another is so slight that it is usually ignored. But if sampling is done without replacement from a small population, the sample values are not statistically independent, and this fact must be taken into consideration when computing $\sigma_{\bar{x}}$.

From a practical point of view, what really matters when sampling is done without replacement and $\sigma_{\bar{x}}$ is calculated is the size of the samples relative to the size of the population. When sampling is from large populations, statistical independence of the sample values is usually assumed as an approximation to reality. For example, suppose the sample $\{5, 68\}$ were drawn without replacement from the population of positive integers less than or equal to 1000. Then, $P(68) = .001$ and $P(68|5) = \frac{1}{999} \simeq .001001$. For most practical purposes, $P(\text{choosing 68 given 5 has been selected})$ can be considered to be equal to $P(68)$ and statistical independence of the sample values can be assumed.

When we sample without replacement from small populations, the statistical dependence between values must be taken into consideration. For example, suppose a sample of size 2 is drawn without replacement from a population having the three values 2, 6, and 8. The probability of choosing 2 first is $\frac{1}{3}$, but the probability for the next value chosen cannot be $\frac{1}{3}$. It must be $\frac{1}{2}$. Here $P(\text{choosing 3 given that 2 has been chosen}) = \frac{1}{2}$, instead of $\frac{1}{3}$. Hence, the sampled values cannot be considered independent of one another.

The following formula can be used to calculate $\sigma_{\bar{x}}$ when sampling is done from a large population (or a small population with replacement):

$$\sigma_{\bar{x}} = \frac{\sigma}{\sqrt{n}} \tag{8-2}$$

where σ is the standard deviation of the population sampled and n is the sample size.

To find the standard error of the mean $\sigma_{\bar{x}}$ for the paint illustration introduced at the beginning of this section, we first find the standard deviation of the population, σ:

$$\sigma = \sqrt{\sum[x^2 P(x)] - \mu^2}$$
$$= \sqrt{101.2 - 9.8^2} = \sqrt{5.16} = 2.27$$

Then, by using (8-2), we have

$$\sigma_{\bar{x}} = \frac{\sigma}{\sqrt{n}} = \frac{2.27}{\sqrt{2}} = 1.61$$

Note that this value agrees with the result found earlier.

As another illustration consider Example 8-5.

EXAMPLE 8-5   For Example 8-2, use formula (8-2) to find $\sigma_{\bar{x}}$, the standard error of the mean (the standard deviation of the sampling distribution of the sample mean).

Solution   The standard error of the mean is

$$\sigma_{\bar{x}} = \frac{\sigma}{\sqrt{n}} = \frac{\sqrt{5}}{\sqrt{2}} = \sqrt{2.5} \simeq 1.58$$

Note that this result agrees with the answer found in Example 8-2 by finding the standard deviation of the 16 measurements in the sampling distribution of the mean.   ∎

The standard error of the mean is a very important concept in statistics. Let's examine how the size of the samples used to form the sampling distribution of the mean relates to the standard error of the mean. Suppose the standard deviation of a large sampled population is $\sigma = 12$ and that samples of size $n = 4$ are selected. The standard error of the mean is

$$\sigma_{\bar{x}} = \frac{\sigma}{\sqrt{n}} = \frac{12}{\sqrt{4}} = 6$$

If instead of using samples of size $n = 4$, we increased the sample size to 16, the standard error of the mean would become

$$\sigma_{\bar{x}} = \frac{\sigma}{\sqrt{n}} = \frac{12}{\sqrt{16}} = 3$$

We see from this demonstration that as the sample size n increases, the standard error of the mean $\sigma_{\bar{x}}$ decreases. This means that as the values of n increase, the sample means cluster more closely about their mean μ. We shall rely heavily on this fact in Chapter 9 when we use a sample mean to estimate an unknown population mean. If the size of the random sample is large, we shall feel confident that our sample mean is as close to the population mean as any other sample mean in the same sampling distribution of the mean, since the sample means cluster closely about μ. Of course, the larger the sample, the closer our sample mean should be to the population mean.

When sampling is done from a small population without replacement, the following formula can be used to find $\sigma_{\bar{x}}$:

$$\sigma_{\bar{x}} = \left(\frac{\sigma}{\sqrt{n}} \right) \sqrt{\frac{N - n}{N - 1}} \tag{8-3}$$

where σ is the standard deviation of the sampled population, n is the sample size, and N is the size of the population.

As a rule of thumb, if sampling is done without replacement and the size of the population is at least 20 times as large as the sample size (i.e., if $N \geq 20n$), then either formula (8-2) or (8-3) can be used. Otherwise, formula (8-3) is used. Formula (8-3) is equivalent to formula (8-2) when N is infinitely large because the values of $\sqrt{(N-n)/(N-1)}$ approach 1 as the values of N get large. The factor $\sqrt{(N-n)/(N-1)}$ is sometimes referred to as a **correction factor** for the population being finite.

EXAMPLE 8-6 Suppose the lengths of service (in years) of three college mathematics teachers are as follows:

Math teacher	Length of service
Smith	6
Jones	4
Doe	2

Further, suppose we select random samples of size 2 without replacement and compute the mean length of service for each. The collection of sample means will constitute the sampling distribution of the mean. Find the standard error of the mean $\sigma_{\bar{x}}$:

a. without using formula (8-3).
b. using formula (8-3).

Solution All possible samples of size 2 are listed in the table with their sample means.

Sample names	Lengths of service	Sample mean, $\bar{x}$
Jones, Doe	4, 2	3
Doe, Jones	2, 4	3
Smith, Doe	6, 2	4
Doe, Smith	2, 6	4
Jones, Smith	4, 6	5
Smith, Jones	6, 4	5

The population mean is

$$\mu = \frac{2+4+6}{3} = 4$$

a. The following table organizes the computations for finding the standard error of the sample means:

Sample	$\bar{x}$	$\bar{x} - \mu_{\bar{x}}$	$(\bar{x} - \mu_{\bar{x}})^2$
4, 2	3	$3-4$	1
2, 4	3	$3-4$	1
6, 2	4	$4-4$	0
2, 6	4	$4-4$	0
4, 6	5	$5-4$	1
6, 4	5	$5-4$	1

The standard error of the mean is

$$\sigma_{\bar{x}} = \sqrt{\frac{\sum(\bar{x} - \mu_{\bar{x}})^2}{N}} = \sqrt{\frac{4}{6}} \simeq .82$$

b. We first find the population standard deviation. The calculations are organized in the following table:

x	$x - \mu$	$(x - \mu)^2$
2	$2 - 4 = -2$	4
4	$4 - 4 = 0$	0
6	$6 - 4 = 2$	4
		$\overline{8}$

The standard deviation of the population is

$$\sigma = \sqrt{\frac{\sum(x - \mu)^2}{N}} = \sqrt{\frac{8}{3}} \simeq 1.63$$

By using formula (8-3), with $n = 2$, $N = 3$, and $\sigma = 1.63$, we have

$$\sigma_{\bar{x}} = \left(\frac{\sigma}{\sqrt{n}}\right)\sqrt{\frac{N - n}{N - 1}}$$

$$= \left(\frac{1.63}{\sqrt{2}}\right)\sqrt{\frac{3 - 2}{3 - 1}} \simeq .82$$

Note that this agrees with the result found in part (a). ∎

EXAMPLE 8-7   If the breaking point of 4000 fuses has a mean of 5 amperes and a standard deviation of 1.5 amperes, and samples of 150 fuses are selected from this population, find $\mu_{\bar{x}}$ and $\sigma_{\bar{x}}$.

Solution   From (8-1), we have $\mu_{\bar{x}} = 5$ amperes.

Since $4000 > 20(150)$, we can use either formula (8-2) or formula (8-3) to find $\sigma_{\bar{x}}$. By using (8-2), we have

$$\sigma_{\bar{x}} = \frac{\sigma}{\sqrt{n}} = \frac{1.5}{\sqrt{150}} \simeq .122$$

By formula (8-3), we have

$$\sigma_{\bar{x}} = \left(\frac{\sigma}{\sqrt{n}}\right)\sqrt{\frac{N - n}{N - 1}}$$

$$= \left(\frac{1.5}{\sqrt{150}}\right)\sqrt{\frac{4000 - 150}{4000 - 1}} \simeq .120$$

Note that the answers agree to two decimal places. When the condition $N \geq 20n$ is satisfied, using formula (8-3) instead of formula (8-2) is usually not worth the additional effort. ∎

The flowchart in Figure 8-5 summarizes the decisions to be made when the value of the standard error of the mean is to be calculated.

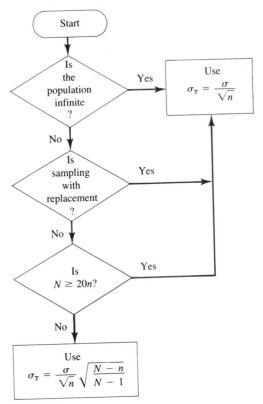

FIGURE 8-5 Flowchart for calculating standard error of the mean

As a final illustration, we note that a six-sided die can be used to simulate sampling from the population of values 1, 2, 3, 4, 5, and 6. By rolling the die five times, we can generate a sample of size 5. And by repeatedly generating samples and recording the sample means, we can generate an empirical sampling distribution. The following 32 samples were generated by using a die:

Sample	$\bar{x}$	Sample	$\bar{x}$	Sample	$\bar{x}$	Sample	$\bar{x}$
1, 2, 3, 5, 4	3.0	5, 1, 1, 3, 2	2.4	5, 2, 4, 3, 5	3.8	5, 3, 1, 6, 5	4.0
3, 3, 6, 5, 2	3.8	2, 2, 4, 4, 1	2.6	2, 6, 3, 6, 2	3.8	4, 5, 3, 3, 4	3.8
2, 1, 3, 5, 4	3.0	5, 3, 5, 4, 1	3.6	1, 4, 2, 2, 1	2.0	2, 2, 5, 6, 1	3.2
3, 3, 4, 6, 3	3.8	2, 4, 5, 6, 1	3.6	1, 6, 6, 5, 1	3.8	1, 5, 5, 3, 8	4.4
3, 5, 4, 6, 2	4.0	4, 4, 4, 2, 5	3.8	4, 2, 2, 6, 4	3.6	3, 4, 6, 4, 6	4.6
1, 2, 5, 2, 5	3.0	2, 3, 6, 1, 2	2.8	5, 3, 3, 1, 5	3.4	1, 3, 1, 1, 1	1.4
6, 5, 6, 6, 5	5.6	4, 6, 3, 4, 5	4.4	1, 4, 4, 3, 1	2.6	6, 2, 6, 5, 1	4.0
1, 6, 3, 3, 3	3.2	2, 5, 3, 1, 2	2.6	4, 5, 5, 1, 4	3.8	3, 1, 1, 2, 3	2.0

A frequency table for the 32 sample means is as follows:

$\bar{x}$	f	$\bar{x}$	f
1.4	1	3.4	1
2.0	2	3.6	3
2.4	1	3.8	8
2.6	3	4.0	3
2.8	1	4.4	2
3.0	3	4.6	1
3.2	2	5.6	1

The empirical distribution can be thought of as a sample from the sampling distribution of the mean. As a sample, it has a mean of 3.42 and a standard deviation of .85. The standard deviation of the population of values 1, 2, 3, 4, 5, and 6 is $\sigma = 1.71$. As a result, the standard error of the mean is

$$\sigma_{\bar{x}} = \frac{\sigma}{\sqrt{n}} = \frac{1.71}{\sqrt{5}} = .76$$

Note that the mean and standard deviation for the empirical distribution can be used to estimate $\mu_{\bar{x}}$ and $\sigma_{\bar{x}}$, respectively:

Empirical distribution of $\bar{x}$	Sampling distribution of $\bar{x}$
$\bar{x} = 3.42$	$\mu_{\bar{x}} = 3.5$
$s_{\bar{x}} = .85$	$\sigma_{\bar{x}} = .76$

Problem Set 8.2

A

1. Characterize or identify the elements in the following sampling distributions
   a. the sampling distribution of the median
   b. the sampling distribution of the variance
   c. the sampling distribution of the range
   d. the sampling distribution of the midrange

2. Under what conditions will the sampling distribution of the mode exist?

3. For each of the following situations, in which sampling is done without replacement, specify which formula(s) can be used for determining $\sigma_{\bar{x}}$:
   a. $N = 5000, n = 500$
   b. $N = 3500, n = 100$
   c. $N = 1000, n = 40$

4. What do the following symbols denote?
   a. $\mu_{\bar{x}}$ and μ
   b. $\sigma_{\bar{x}}$ and σ

5. A certain population has a standard deviation of 15. Random samples of size n are taken and the means of the samples are computed. What happens to the standard error of the mean as n is increased from 400 to 900? What happens to $\sigma_{\bar{x}}$ as n increases?

6. If $\sigma = 25$, find the standard error of the mean if:
   a. $N = 75, n = 15$
   b. $N = 500, n = 45$
   c. $N = 1000, n = 100$

7. If ordered samples of size 2 are selected with replacement from the population whose values are 0, 2, 4, and 8, find $\mu_{\bar{x}}$ and $\sigma_{\bar{x}}$ without using formulas (8-2) or (8-3). [*Hint:* There are 16 ordered samples.]

8. If the sampling in Problem 7 is done without replacement, find $\mu_{\bar{x}}$ and $\sigma_{\bar{x}}$ without using formulas (8-2) or (8-3).

9. Repeat Problem 7, but use formula (8-2) or (8-3) to find $\sigma_{\bar{x}}$.

10. Repeat Problem 8, but use formula (8-2) or (8-3) to find $\sigma_{\bar{x}}$.

11. Costs for a tune-up at a large service station are $30 for a 4-cylinder car, $36 for a 6-cylinder car, and $42 for an 8-cylinder car. From past records, it is known that 10% of its tune-ups are done on 4-cylinder cars, 40% on 6-cylinder cars, and 50% on 8-cylinder cars. Suppose two cars that need a tune-up are randomly chosen.
   a. Construct a probability distribution for the average service charge.
   b. Find $\mu_{\bar{x}}$.
   c. Find $\sigma_{\bar{x}}$.

B

12. If ordered samples of size 2 are drawn with replacement from the population of values 0, 2, 4, and 6 and the range is computed for each sample, find:
   a. a frequency table for values of the range.
   b. the mean of the sampling distribution of the range.
   c. the standard error of the range.
   d. a graph for the frequency distribution of the range.

13. Use a die to simulate sampling from the population of values 1, 2, 3, 4, 5, and 6. Let the size of each sample be 10, select 25 samples, and find the mean for each sample. Then find the mean and standard deviation for the empirical sampling distribution of the mean. Compare these values with the theoretical values. Why do they differ?

14. Show that if $20n \leq N$, then
$$.97 \leq \sqrt{(N - n)/(N - 1)} < 1.$$

15. Suppose a finite population consists of the values 0, 3, and 9.
   a. Find the parameters μ and σ^2 for the population.
   b. Suppose ordered samples of size 2 are selected with replacement from the population. Construct the sampling distribution of $\bar{x}$ by listing the nine samples and finding the mean for each.
   c. For the sampling distribution of $\bar{x}$, find $\mu_{\bar{x}}$ and $\sigma_{\bar{x}}$.
   d. Show that $\bar{x}$ is an unbiased estimate of μ.
   e. For each sample in part (b), find s^2.
   f. For the sampling distribution of s^2, show that s^2 is an unbiased estimator for σ^2.
   g. Show that s is a biased estimator of σ.
   h. Find the standard error of the sample variance.

16. Refer to Problem 11(a).
   a. Construct a probability distribution table for the total service charge for the two cars.
   b. Find the mean of the sampling distribution of the total service charge.
   c. Find the standard error of the total service charge.

17. Prove that $\sigma_e = \sigma_{\bar{x}}$, where $e = \bar{x} - \mu$.

8.3 SAMPLING FROM NORMAL POPULATIONS

Many sampling distributions in statistics are based on the assumption that sampling is done from a normal population. It is reasonable to expect that if the sampled population is normal, then the sampling distribution of the sample mean is also normal. This is the substance of the following important result:

> If samples of size n are drawn from a normal population, then the sampling distribution of the mean is normal.

In Section 8.2 we learned that if sampling is done from a large population with mean μ and standard deviation σ, then the mean of the sampling distribution of the mean is equal to

$$\mu_{\bar{x}} = \mu$$

and the standard deviation of the distribution of sample means (standard error of the mean) is given by

$$\sigma_{\bar{x}} = \frac{\sigma}{\sqrt{n}}$$

where n is the size of the samples.

As an illustration, suppose the heights of American males are normally distributed with a mean of 70 inches and a standard deviation of 2.6 inches. If samples of size 20 are selected from the population of heights, the sampling distribution of the mean is normally distributed with a mean equal to $\mu = 70$ inches and a standard deviation equal to $\sigma_{\bar{x}} = \sigma/\sqrt{n} = 2.6/\sqrt{20} = .58$ inch. These two distributions are illustrated in Figure 8-6. Note that the sampling distribution of the mean has a smaller standard deviation than the population distribution.

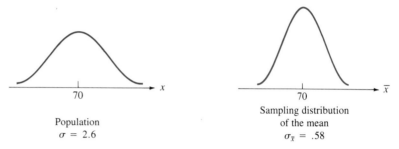

FIGURE 8-6 Population distribution and sampling distribution of mean for heights of American males

Recall from Section 7.2 that the z-score formula

$$z = \frac{x - \mu}{\sigma}$$

can be used to transform any normal random variable X with mean μ and standard deviation σ into a standard normal variable Z with mean 0 and standard deviation 1. As a result, the sampling distribution of z is the standard normal.

If a sample of size n is taken from a normal population and the sample mean $\bar{x}$ is calculated, then the z score for $\bar{x}$ is given by

$$z = \frac{\bar{x} - \mu_{\bar{x}}}{\sigma_{\bar{x}}}$$

Since $\mu_{\bar{x}} = \mu$ and $\sigma_{\bar{x}} = \sigma/\sqrt{n}$, we have

$$z = \frac{\bar{x} - \mu}{\sigma/\sqrt{n}}$$

Hence, the sampling distribution of the z statistic $(\bar{x} - \mu)/(\sigma/\sqrt{n})$ is the standard normal. Note that the z statistic can also be written as

$$z = \frac{\sqrt{n}(\bar{x} - \mu)}{\sigma}$$

Consider the following applications.

EXAMPLE 8-8   Suppose a random sample of size 9 is selected from a normal population with mean $\mu = 25$ and standard deviation $\sigma = 6$. What is the probability that the sample mean $\bar{x}$ is greater than 28?

Solution   The sampling distribution of the mean is normal with $\mu_{\bar{x}} = \mu = 25$ and $\sigma_{\bar{x}} = \sigma/\sqrt{n} = 6/\sqrt{9} = 2$. To determine the probability $P(\bar{x} > 28)$, we need to find the area under the normal curve to the right of $\bar{x} = 28$, as indicated in Figure 8-7.

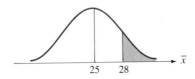

FIGURE 8-7 $P(\bar{x} > 28)$ for Example 8-8

The z value for $\bar{x} = 28$ is

$$z = \frac{\bar{x} - \mu}{(\sigma/\sqrt{n})}$$

$$= \frac{(28 - 25)}{2} = 1.5$$

Thus, we have

$$P(\bar{x} > 28) = P(z > 1.5)$$
$$= .5 - .4332 = .0668$$

Hence, the probability that a random sample of size 9 has a mean greater than 28 is $P(\bar{x} > 28) = .0668$.   ∎

EXAMPLE 8-9   The times required for workers to complete a certain task are normally distributed with a mean of 30 minutes and a standard deviation of 9 minutes. If a random sample of 25 workers is drawn from the work force, find the probability that the mean completion time for the sample of workers is between 28 and 33 minutes.

Solution   The population distribution of task completion times and the sampling distribution of the mean are shown in Figure 8-8. Since $\mu = 30$ and $\sigma = 9$, $\mu_{\bar{x}} = 30$

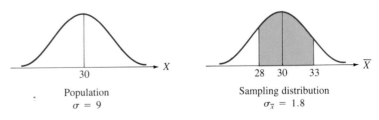

FIGURE 8-8 Population distribution and sampling distribution for Example 8-9

The standard error of the mean is

$$\sigma_{\bar{x}} = \frac{\sigma}{\sqrt{n}} = \frac{9}{\sqrt{25}} = 1.8$$

As a result, the z value for $\bar{x} = 28$ is

$$z = \frac{28 - 30}{1.8} = -1.11$$

and the z value for $\bar{x} = 33$ is

$$z = \frac{33 - 30}{1.8} = 1.67$$

Thus,

$$P(28 < \bar{x} < 33) = P(-1.11 < z < 1.67)$$
$$= .3665 + .4525 = .8190 \qquad \blacksquare$$

It is important to recognize when to use the sampling distribution of the mean. Consider the following two problems:

a. If the weights of newborn children are normally distributed with a mean of 115 ounces and a standard deviation of 12 ounces, find the probability that a randomly selected newborn child weighs more than 125 ounces.
b. If the weights of newborn children are normally distributed with a mean of 115 ounces and a standard deviation of 12 ounces, find the probability that the average weight of a random sample of 16 newborn children exceeds 125 ounces.

Which problem involves the sampling distribution of the mean? If you chose (b) you are correct. Problem (b) involves locating a *sample* mean based on a sample of size 16 and problem (a) involves locating a *single* measurement. In (b) we are interested in the location of $\bar{x}$ within the sampling distribution of the mean. Both problems involve finding normal probabilities.

The solution to problem (a) is as follows (see Figure 8-9). The z value for $x = 125$ is

$$z = \frac{125 - 115}{12} = .83$$

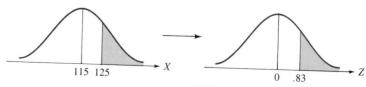

FIGURE 8-9 Normal distribution and transformed normal distribution for problem (a)

The probability that a newborn child weighs more than 125 ounces is

$$P(x > 125) = P(z > .83) = .5 - .2967 = .2033$$

The solution to problem (b) refers to Figure 8-10. The z value for a mean of 125 ounces is

$$z = \frac{\bar{x} - \mu_{\bar{x}}}{\sigma_{\bar{x}}}$$

$$= \frac{125 - 115}{3} = 3.33$$

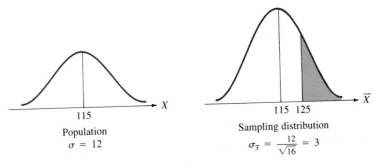

FIGURE 8-10 Population distribution and sampling distribution for problem (b)

The probability that a sample mean is greater than 125 ounces is

$$P(\bar{x} > 125) = P(z > 3.33)$$
$$= .5 - .4996 = .0004$$

Thus, it is rather unlikely that the mean weight of a sample of 16 newborn children will exceed 125 ounces.

t Distributions

When sampling is from a normal population, the sampling distribution of the mean is normal. However, if the standard deviation of the population is unknown, we cannot standardize the sample mean to a standard normal. In most practical situations, the population standard deviation σ is unknown, and the sample standard deviation s is used to estimate σ. Consequently, the following statistic does not have the standard normal sampling distribution:

$$\frac{\bar{x} - \mu}{(s/\sqrt{n})}$$

This statistic is denoted by t and is called the **t statistic**. Thus, the t statistic is given by the formula

$$t = \frac{\bar{x} - \mu}{(s/\sqrt{n})}$$

In 1908, W. Gosset, a staff member of an Irish brewery, published a research paper concerning the equation for the probability distribution of t. Since employees of the brewery were not permitted to publish research findings, Gosset published his results under the pen name "Student." Since that time, the sampling distribution of the t statistic has been referred to as the **Student-t distribution**, or simply the **t distribution**. The actual equation of the t distribution is very complicated and will be omitted here. Instead, the t table (see the back endpaper) contains a collection of t values and their associated probabilities.

The sampling distribution of t is similar to the standard normal. Both are bell-shaped, both have means equal to 0, and both are symmetric about their means. The sampling distribution of t is more variable than the standard normal distribution. For the statistic z, $\bar{x}$ is the only quantity that varies from sample to sample, whereas for t, both $\bar{x}$ and s vary from sample to sample. The exact shape of a t distribution is completely specified by a single value (parameter) referred to as the number of **degrees of freedom** (df). The sample size n is related to df by the following relationship:

df $= n - 1$

Actually, df is the divisor in the formula for computing the sample variance s^2:

$$s^2 = \frac{SS}{n - 1} \leftarrow df$$

The variance of a t distribution depends on the sample size and is always greater than 1, the variance of the standard normal distribution. When df > 2, the variance of a t distribution is given by $\sigma_t^2 = df/(df - 2)$.

In summary, the sampling distributions of t have the following properties:

1. They have a mean $\mu = 0$.
2. They are symmetric about $\mu = 0$.
3. They are more variable than the standard normal.
4. They are bell-shaped.
5. Their exact shape depends on df $= n - 1$.
6. Their variances depend on df and $\sigma^2 = df/(df - 2)$, provided df > 2.
7. As n gets large, the sampling distributions of t approach the standard normal distribution Z.

Since the sampling distributions for t are more variable than the standard normal distribution, the t distributions have thicker tail areas than the standard normal, as Figure 8-11 illustrates.

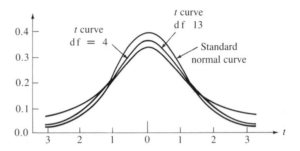

FIGURE 8-11 *t* distributions compared to standard normal

EXAMPLE 8-10   Suppose a random sample of size 11 is selected from a normal population with a mean of 14. If the sample mean is 18 and the sample standard deviation is 14.3, calculate the value of the *t* statistic.

Solution   The value of *t* is

$$t = \frac{\bar{x} - \mu}{(s/\sqrt{n})}$$

$$= \frac{18 - 14}{(14.3/\sqrt{11})} \simeq .93 \qquad \blacksquare$$

To locate a *t* value in the *t* table (see the back endpaper) for one of the five given probabilities listed at the top of the table, we locate the row labeled by df at the left of the table and the column identified by the probability at the top of the table; the *t* value is determined by where the row and column intersect. For example, to locate the *t* value for a probability (or area) of $\alpha = .025$ and a sample size of 13, we locate the row labeled df $= n - 1 = 13 - 1 = 12$ and the column labeled $\alpha = .025$; the *t* value is listed at the intersection of this row and column, as shown here:

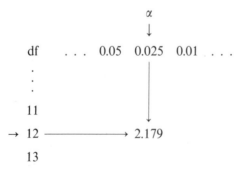

Thus, $t = 2.179$.

EXAMPLE 8-11   Suppose a random sample of size 16 is selected from a normal population with $\mu = 20$. If the standard deviation of the sample is $s = 4$, find $P(\bar{x} > 21.753)$.

Solution   We find the t value for 21.753:

$$t = \frac{\bar{x} - \mu}{(s/\sqrt{n})}$$

$$= \frac{21.753 - 20}{(4/\sqrt{16})} = 1.753$$

Thus,

$$P(\bar{x} > 21.753) = P(t > 1.753)$$

Since df $= n - 1 = 16 - 1 = 15$, we read across the row labeled 15 in the table until we locate 1.753 under the column heading 0.05. (See Table 8-4 for a portion of the t table.)

Table 8-4 Portion of the *t* Table (See the Back Endpaper)

df	0.10	0.05	0.025	0.01	0.005
⋮					
12	1.356	1.782	2.179	2.681	3.055
13	1.350	1.771	2.160	2.650	3.012
14	1.345	1.761	2.145	2.624	2.977
→ 15	1.341	(1.753)	2.131	2.602	2.947
16	1.337	1.746	2.120	2.583	2.921

with α as the column heading.

Thus,

$$P(\bar{x} > 21.753) = .05$$ ∎

Unlike the standard normal table (z table; see the front endpaper), the t table contains only five probabilities (or areas). This may appear at first to be a big disadvantage, but for our applications, these five probabilities will be sufficient. Our future applications involving a t distribution will require t values corresponding to df $= n - 1$ for one of the five probability values in the table.

Problem Set 8.3

A

1. If a sample of size 37 is taken from a normal population that has a mean of 50 and a standard deviation of 30, find:
   a. $P(\bar{x} > 60)$ **b.** $P(45 < \bar{x} < 58)$
   c. $P(\bar{x} < 47)$ **d.** $P(\bar{x} > 45)$
   e. $P(\bar{x} < 62)$

2. A sample of size 100 is taken from a normal population that has a mean of 25 and a standard deviation of 15. Find the probability that the sample mean is between 26 and 27.

3. If $n = 6$, find $P(t > 4.032)$.

4. If $n = 21$, find $P(t < 2.845)$.

5. If $n = 16$, find $P(-2.131 < t < 2.131)$.

6. If $P(t < -1.796) = .05$, find n.

7. If $n = 13$, find $P(1.782 < t < 3.055)$.

8. If $P(-1.96 < t < 1.96) = .95$, find n.

9. If $\bar{x} = 15$, $\mu = 20$, $s = 7$, and df $= 8$, calculate the value of the t statistic.

10. A sample of size 26 is taken from a normal population that has a mean of 30. If the sample standard deviation is 10, find the probability that the sample mean is less than 34.04.

11. If it is assumed that SAT math scores are normally distributed with a mean of 500 and a standard deviation of 100, find the probability that the sample mean obtained from a sample of 49 scores is
a. greater than 520.   **b.** between 475 and 520.
c. less than 525.   **d.** less than 470.

12. The average life of a certain type of electric heater is 10 years and the standard deviation is 1.5 years. If the lives of the electric heaters are known to be normally distributed, find:
a. the probability that the mean life of a random sample of 16 heaters is less than 10.5 years.
b. the value of $\bar{x}$ so that 20% of the means computed from random samples of size 25 fall below $\bar{x}$.

c. P_{60}, the 60th percentile of the sampling distribution of the mean with $n = 9$.

13. Suppose the hourly wages of all workers in a certain auto plant are normally distributed with a mean of $12.50 and a standard deviation of $.95. If a random sample of 100 workers is selected from this plant, find the probability that the sample mean hourly wage is
a. less than $12.60.
b. between $12.45 and $12.65.
c. more than $12.30.

14. Given a random sample of size 12 from a normal distribution, find a value for c such that:
a. $P(1.363 < t < c) = .09$
b. $P(c < t < 3.106) = .045$

15. If a random sample of size 80 is obtained from a normal distribution with $\mu = 32$ and $\sigma = 10$, find a value for c so that:
a. $P(\bar{x} > c) = .30$
b. $P(-c < \bar{x} < c) = .80$
c. $P(\bar{x} < c) = .40$
d. $P(\bar{x} < c) = .60$
e. $P(33 < \bar{x} < c) = .05$

B

16. Is one likely to obtain the random sample 27, 25, 26, 24, 23, and 22 from a normal population that has a mean of 22? Explain.

8.4 SAMPLING FROM NONNORMAL POPULATIONS

In actual practice we sample without replacement from populations that are usually large, but whose exact shape and parameters are usually unknown. As a result, one might suspect that the sampling distribution of the mean would not follow any basic shape. But if the sample size n is large ($n \geq 30$), the sampling distribution of the mean is bell-shaped. This remarkable fact is the substance of the well-known **central limit theorem**, which is stated as follows:

> The sampling distribution of the mean is approximately normal for nonnormal sampled populations. The larger the sample size, the closer the sampling distribution is to being normal.

For most purposes, the normal approximation is considered good provided $n \geq 30$. That the shape of the sampling distribution of the mean is approximately normal, even in cases where the original population is bimodal (see Figure 8-12), is indeed remarkable. Recall from Section 8.3 that if the sampled population is normal, the sampling distribution of the mean is normal, regard-less of the sample size.

FIGURE 8-12 Bimodal population and sampling distribution
of mean

The following three situations illustrate the central limit theorem for three nonnormal sampled populations. For each of the nonnormal populations, three sampling distributions of the mean are illustrated, one for $n = 2$, one for $n = 5$, and one for $n = 25$. For each population, a computer simulation was used to generate an empirical sampling distribution of the mean involving 100 samples for each of the three sample sizes. Notice the shape of the sampling distribu-tion of the mean in each situation as the sample size n increases. For $n = 2$, the sampling distributions of the mean take on appearances different from the sampled populations. For $n = 5$, the sampling distributions of the mean take on bell-shaped appearances. And for $n = 25$, the sampling distributions of the mean appear to be approximately normal. For each of the three situations, notice also that the variance of the sampling distribution of the mean decreases as n increases.

For convenience, the histograms are drawn horizontally using asterisks to form the bars. Each asterisk represents one measurement. The mean and stan-dard error of the theoretical sampling distribution of the mean and the empirical sampling distribution of the mean are given for each sample size for purposes of comparison.

Situation A   The histogram for the population has a rectangular shape, as shown here:

Class boundaries	Measurements
0 to 10	**********
10 to 20	**********
20 to 30	**********
30 to 40	**********
40 to 50	**********
50 to 60	**********
60 to 70	**********
70 to 80	**********
80 to 90	**********
90 to 100	**********

The empirical sampling distribution of the mean based on 100 samples of size $n = 2$ is as follows:

Class boundaries	Sample means
0 to 10	*
10 to 20	******
20 to 30	************
30 to 40	******************
40 to 50	*******************
50 to 60	*********************
60 to 70	**************
70 to 80	*****
80 to 90	**
90 to 100	

Empirical distribution	Theoretical distribution
Mean = 45.6998	Mean = 49.8716
Standard error = 17.1722	Standard error = 19.9839

The empirical distribution of 100 sample means for a sample size of $n = 5$ is as follows:

Class boundaries	Sample means
0 to 10	
10 to 20	
20 to 30	***
30 to 40	**************
40 to 50	*********************************
50 to 60	*******************************
60 to 70	***************
70 to 80	***
80 to 90	*
90 to 100	

Empirical distribution	Theoretical distribution
Mean = 49.9644	Mean = 49.8716
Standard error = 11.4430	Standard error = 12.6389

The empirical distribution for 100 sample means for a sample size of $n = 25$ is as follows:

Class boundaries	Sample means
35 to 39	****
39 to 43	****
43 to 47	************************
47 to 51	***************************
51 to 55	**********************
55 to 59	***********
59 to 63	*****
63 to 67	
67 to 71	*

Empirical distribution	Theoretical distribution
Mean = 49.8392 Standard error = 5.6512	Mean = 49.8716 Standard error = 5.6523

Situation B   The histogram for the population is U-shaped:

Class boundaries	Measurements
0 to 10	************************
10 to 20	***************
20 to 30	*****
30 to 40	***
40 to 50	**
50 to 60	**
60 to 70	***
70 to 80	*****
80 to 90	***************
90 to 100	*************************

The empirical distribution of sample means for 100 samples of size $n = 2$ is as follows:

Class boundaries	Sample means
0 to 10	*************
10 to 20	*************
20 to 30	***
30 to 40	*********
40 to 50	**********************
50 to 60	************
60 to 70	**
70 to 80	****
80 to 90	***************
90 to 100	******

Empirical distribution	Theoretical distribution
Mean = 45.9345 Standard error = 28.1235	Mean = 49.4953 Standard error = 26.7562

The empirical distribution of 100 sample means for a sample size of $n = 5$ is given here:

Class boundaries	Sample Means
0 to 10	*
10 to 20	*
20 to 30	**********
30 to 40	******************
40 to 50	*************************
50 to 60	************************
60 to 70	************
70 to 80	******
80 to 90	**
90 to 100	*

Empirical distribution	Theoretical distribution
Mean = 48.8509 Standard error = 15.6455	Mean = 49.4953 Standard error = 16.9221

The empirical distribution for 100 sample means for a sample size of $n = 25$ is as follows:

Class boundaries	Sample means
26 to 31	*
31 to 36	*******
36 to 41	*******
41 to 46	*********************
46 to 51	******************************
51 to 56	*****************
56 to 61	*****************
61 to 66	****
66 to 71	*

Empirical distribution	Theoretical distribution
Mean = 49.3831 Standard error = 7.9903	Mean = 49.4953 Standard error = 7.5678

Situation C   The histogram for the population is J-shaped:

Class boundaries	Population measurements
0 to 10	***
10 to 20	*******************
20 to 30	**********
30 to 40	*****
40 to 50	****
50 to 60	***
60 to 70	**
70 to 80	**
80 to 90	**
90 to 100	**

The empirical distribution for 100 sample means based on a sample size of $n = 2$ is as follows:

Class boundaries	Sample means
0 to 10	***************************************
10 to 20	*************************
20 to 30	*****************
30 to 40	*********
40 to 50	*******
50 to 60	***
60 to 70	***
70 to 80	*
80 to 90	
90 to 100	

Empirical distribution	Theoretical distribution
Mean = 19.6745	Mean = 19.2160
Standard error = 16.9519	Standard error = 16.2835

The empirical distribution of 100 sample means based on a sample size of $n = 5$ is as follows:

Class boundaries	Sample means
0 to 10	***************************
10 to 20	**********************************
20 to 30	*************************
30 to 40	********
40 to 50	***
50 to 60	*
60 to 70	
70 to 80	
80 to 90	
90 to 100	

Empirical distribution	Theoretical distribution
Mean = 17.7009	Mean = 19.2160
Standard error = 10.0414	Standard error = 10.2986

The empirical distribution of 100 sample means based on a sample size of $n = 25$ is as follows:

Class boundaries	Sample means
9 to 12	*********
12 to 15	**********************
15 to 18	*****************
18 to 21	*********************
21 to 24	******************
24 to 27	********
27 to 30	***
30 to 33	***

Empirical distribution	Theoretical distribution
Mean = 18.6566	Mean = 19.2160
Standard error = 5.1883	Standard error = 4.6057

As another illustration of the central limit theorem, suppose ordered samples of size 2 are drawn with replacement from a population consisting of the values 0, 2, 4, 6, and 8. The 25 samples listed in Table 8-5 can be drawn.

Table 8-5 Samples of Size 2 from Population of Size 5, with Values of Sample Mean

Sample	$\bar{x}$	Sample	$\bar{x}$	Sample	$\bar{x}$
{0, 0}	0	{4, 0}	2	{8, 0}	4
{0, 2}	1	{4, 2}	3	{8, 2}	5
{0, 4}	2	{4, 4}	4	{8, 4}	6
{0, 6}	3	{4, 6}	5	{8, 6}	7
{0, 8}	4	{4, 8}	6	{8, 8}	8
{2, 0}	1	{6, 0}	3		
{2, 2}	2	{6, 2}	4		
{2, 4}	3	{6, 4}	5		
{2, 6}	4	{6, 6}	6		
{2, 8}	5	{6, 8}	7		

The mean of the population is

$$\mu = \frac{0 + 2 + 4 + 6 + 8}{5} = 4$$

The variance of the population is

$$\sigma^2 = \frac{(0-4)^2 + (2-4)^2 + (4-4)^2 + (6-4)^2 + (8-4)^2}{5}$$

$$= \frac{40}{5} = 8$$

and the standard deviation of the population is

$$\sigma = \sqrt{8} \simeq 2.83$$

A graph of the frequency distribution for the population is shown in Figure 8-13. The graph could not be considered bell-shaped or normal.

FIGURE 8-13 Frequency distribution for population of values 0, 2, 4, 6, 8

The 25 sample means can be grouped into the following frequency table:

$\bar{x}$	f
0	1
1	2
2	3
3	4
4	5
5	4
6	3
7	2
8	1

A graph of the frequency distribution for $\bar{x}$ is shown in Figure 8-14. From the bell-shaped appearance of the distribution of means, we conclude that it is

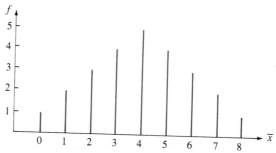

FIGURE 8-14 Frequency distribution for $\bar{x}$

reasonable to approximate the sampling distribution of $\bar{x}$ by a normal distribution, once the mean and standard deviation of the sampling distribution are known. Table 8-6 makes it convenient to compute $\sigma_{\bar{x}}$, the standard error of the mean, using the formulas

$$\mu_{\bar{x}} = \frac{\sum \bar{x}f}{N} \quad \text{and} \quad \sigma_{\bar{x}} = \sqrt{\frac{\sum (\bar{x} - \mu_{\bar{x}})^2 f}{N}}$$

Table 8-6 Computations for Determining Standard Error of Mean

$\bar{x}$	f	$f\bar{x}$	$\bar{x} - \mu_{\bar{x}}$	$(\bar{x} - \mu_{\bar{x}})^2$	$(\bar{x} - \mu_{\bar{x}})^2 f$
0	1	0	$0 - 4 = -4$	16	16
1	2	2	$1 - 4 = -3$	9	18
2	3	6	$2 - 4 = -2$	4	12
3	4	12	$3 - 4 = -1$	1	4
4	5	20	$4 - 4 = 0$	0	0
5	4	20	$5 - 4 = 1$	1	4
6	3	18	$6 - 4 = 2$	4	12
7	2	14	$7 - 4 = 3$	9	18
8	1	8	$8 - 4 = 4$	16	16
	$N = 25$	100			100

The mean of the sampling distribution of the mean is

$$\mu_{\bar{x}} = \frac{100}{25} = 4$$

and the variance of the sampling distribution of the mean is

$$\sigma_{\bar{x}}^2 = \frac{100}{25} = 4$$

Hence, the standard error of the mean is

$$\sigma_{\bar{x}} = 2$$

Notice that the value of the standard error of the mean found by using formula (8-2) agrees with the above result, since

$$\sigma_{\bar{x}} = \frac{\sigma}{\sqrt{n}} = \frac{\sqrt{8}}{\sqrt{2}} = 2$$

We continue this illustration of the central limit theorem in the following example.

EXAMPLE 8-12   If in the previous illustration, an ordered sample of size 60 is drawn with replacement from the population, what is the probability that the sample mean will be between 3.5 and 4.3? That is, determine $P(3.5 \le \bar{x} \le 4.3)$.

Solution   In this case, $\sigma_{\bar{x}} = \sigma/\sqrt{n} = \sqrt{8}/\sqrt{60} \simeq .37$. Since $n \ge 30$ and as a consequence of the central limit theorem, the sampling distribution of the mean is well approximated by a normal distribution. The z scores for 3.5 and 4.3 are found by using the z-score formula

$$z = \frac{\bar{x} - \mu_{\bar{x}}}{\sigma_{\bar{x}}}$$

The z score for 3.5 is

$$z = \frac{3.5 - 4}{.37} = -1.35$$

and the z score for 4.3 is

$$z = \frac{4.3 - 4}{.37} = .81$$

Thus,

$$P(3.5 \leq \bar{x} \leq 4.3) = P(-1.35 \leq z \leq .81) = .4115 + .2910 = .7025 \qquad \blacksquare$$

It is important to note once again that as n increases, the variation among sample means decreases. This is due to the fact that $\sigma_{\bar{x}} = \sigma/\sqrt{n}$, which decreases as n increases. Recall that the area under a normal curve is equal to 1. So as the variance of the sampling distribution decreases, the sampling distribution of the mean narrows and becomes taller to contain an area of 1. This decrease in variance and the effects on the sampling distribution of the mean as n increases are illustrated in Figure 8-15.

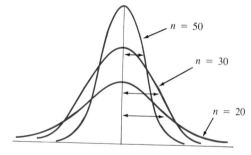

FIGURE 8-15 Effects of increasing n on sampling distribution of mean

Sampling Distribution of Sample Sums

Since the sum of the measurements in a sample is directly related to the sample mean, one might suspect that if the sampling is from a nonnormal population, then the sampling distribution of the sample sums is approximately normal, and that the approximation improves as the sample size increases. This is indeed the case, and it can be shown that the mean of the sampling distribution of the sample sums has a mean equal to $n\mu$ and a standard deviation equal to $\sqrt{n}\sigma$, where μ and σ are the mean and standard deviation, respectively, of the sampled population and n is the sample size. That is,

$$\mu_{\Sigma x} = n\mu$$

$$\sigma_{\Sigma x} = \sqrt{n}\sigma$$

In Chapter 6 we saw that the number of successes in a binomial experiment is a binomial random variable with $\mu = np$ and $\sigma = \sqrt{np(1-p)}$. For a binomial experiment, if we let the random variable X_i take on the value 1 if a success is obtained on the ith trial and 0 if a failure is obtained on the ith trial, then $\sum X_i$ represents the number of successes. By the preceding remarks, we know that the sampling distribution of $\sum X_i$, the number of successes, is approximately normal. This observation reinforces the result of Section 7.3 that normal distributions can be used to approximate binomial distributions.

Applications of the Central Limit Theorem

EXAMPLE 8-13   A traffic study shows that the average number of occupants in a car is 1.75 and the standard deviation is .65. In a sample of 50 cars, find the probability that the mean number of occupants is more than 2.

Solution   We are given that $\mu = 1.75$ and $\sigma = .65$.

1. As a consequence of the central limit theorem, the sampling distribution of the mean is approximately normal. The approximation is considered good since $n = 50$.
2. The mean of the sampling distribution of the mean is $\mu_{\bar{x}} = \mu = 1.75$ (see Figure 8-16).
3. The standard error of the mean is $\sigma_{\bar{x}} = \sigma/\sqrt{n} = .65/\sqrt{50} = .092$.
4. The z value for a mean of 2 is

$$z = \frac{2 - 1.75}{.092} = 2.72$$

Thus, $P(\bar{x} > 2) = P(z > 2.72) = .5 - .4967 = .0033$.

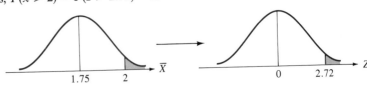

FIGURE 8-16 Sampling distribution of mean for Example 8-13   ■

EXAMPLE 8-14   The manager of a credit bureau has calculated that the average amount of money borrowed by customers for a new car is $4685.54 and the standard deviation is $748.72. During the next month, if 40 people receive car loans, what is the probability that the average amount borrowed is between $4500 and $4800?

Solution We are given that $\mu = \$4685.54$ and $\sigma = \$748.72$.

1. As a consequence of the central limit theorem and the fact $n = 40 \geq 30$, a normal distribution provides a good approximation to the sampling distribution of the mean.

2. The mean of the sampling distribution of the mean is

$$\mu_{\bar{x}} = \mu = \$4685.54$$

3. The standard error of the mean is

$$\sigma_{\bar{x}} = \frac{\sigma}{\sqrt{n}} = \frac{\$748.72}{\sqrt{40}} = \$118.38$$

4. The z values for $\$4500$ and $\$4800$ are (see Figure 8-17):

$$z = \frac{4500 - 4685.54}{118.38} = -1.57$$

$$z = \frac{4800 - 4685.54}{118.38} = .97$$

Thus, $P(4500 < \bar{x} < 4800) = P(-1.57 < z < .97) = .4418 + .3340 = .7758$. Hence, the probability that 40 people borrow an average amount of money for a car between $\$4500$ and $\$4800$ is .7758.

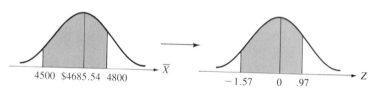

FIGURE 8-17 Sampling distribution of mean for Example 8-14

■

EXAMPLE 8-15 The mean breaking point for a batch of 1950 special fuses is 4.5 amperes and the standard deviation is 1.5 amperes. If 200 of these special fuses are sampled (without replacement), find the probability that the mean breaking point of the sample is less than 4.3 amperes.

Solution We are given that $N = 1950$, $n = 200$, $\mu = 4.5$, and $\sigma = 1.5$. As a consequence of the central limit theorem, the sampling distribution of the mean is approximately normal. The mean of the sampling distribution of the mean is $\mu_{\bar{x}} = 4.5$. Since sampling is done without replacement from a finite population and $20n > N$, we should calculate $\sigma_{\bar{x}}$ using the following formula:

$$\sigma_{\bar{x}} = \left(\frac{\sigma}{\sqrt{n}}\right)\sqrt{\frac{N - n}{N - 1}}$$

Thus, the standard error of the mean is

$$\sigma_{\bar{x}} = \left(\frac{1.5}{\sqrt{200}}\right)\sqrt{\frac{(1950-200)}{(1950-1)}} = .10$$

The z value for a sample mean of 4.3 is

$$z = \frac{4.3 - 4.5}{.10} = -2$$

Thus, $P(\bar{x} < 4.3) = P(z < -2) = .5 - .4772 = .0228$. Hence, the probability that the mean breaking point for a sample of 200 fuses is less than 4.3 amperes is .0228. ∎

EXAMPLE 8-16   A collection of 1000 test scores is bimodal with a mean of $\mu = 76$ and a standard deviation of 15. If a sample of 40 of these scores is selected without replacement, find the probability that the sample mean is larger than 79.

Solution   We are given that $N = 1000$, $n = 40$, $\mu = 76$, and $\sigma = 15$. The sampling distribution of the mean is approximately normal, as a consequence of the central limit theorem, with mean $\mu_{\bar{x}} = 76$. Since sampling is done without replacement from a finite population and $N > 20n$, we can calculate $\sigma_{\bar{x}}$ by using

$$\sigma_{\bar{x}} = \frac{\sigma}{\sqrt{n}}$$

Thus, the standard error of the mean is

$$\sigma_{\bar{x}} = \frac{15}{\sqrt{40}} = 2.37$$

The z value for a mean of 79 is

$$z = \frac{79 - 76}{2.37} = 1.27$$

The probability that the sample mean is larger than 79 is thus

$$P(\bar{x} > 79) = P(z > 1.27)$$
$$= .5 - .3980 = .1020$$ ∎

Problem Set 8.4

A

1. If ordered samples of size 30 are selected with replacement from the population with values 0, 3, 6, 9, and 12, find $\mu_{\bar{x}}$, $\sigma_{\bar{x}}$, and the approximate shape of the sampling distribution of the mean.

2. Suppose ordered samples of size 2 are selected with replacement from the population 1, 3, 5, and 7.
   a. Find μ and σ.
   b. Find $\mu_{\bar{x}}$ and $\sigma_{\bar{x}}$.

c. Construct a frequency distribution for $\bar{x}$.

d. Verify that the results of the central limit theorem hold.

3. Repeat Problem 2, parts (a) and (b), for ordered samples of size 3.

4. If in Problem 1, a sample is selected from the sampling distribution of the mean, what is the probability that the sample mean is

a. less than 6.6?

b. between 5 and 7?

c. greater than 5.5?

5. A service station located in a large city has found that its gasoline sales average 12.4 gallons per customer and have a standard deviation of 2.9 gallons. For a random sample of 40 gasoline customers, find the probability that the average gasoline purchase:

a. is less than 13 gallons.

b. exceeds 12 gallons.

c. is between 11.5 gallons and 13.1 gallons.

d. exceeds 12.6 gallons.

6. The average life of a certain brand of light bulbs is 1000 hours and the standard deviation is 50 hours. What is the probability that the average life for a sample of 36 such light bulbs is greater than 1010 hours?

7. Identify a practical situation where a variable of interest is probably not normal, but where the central limit theorem could be used for calculating probabilities associated with a sample mean.

8. An automatic machine fills bottles with root beer. The mean amount of fill is 16 ounces and the standard deviation is .5 ounce. What is the probability that a sample of 35 bottles will have a mean fill:

a. greater than 16.1 ounces?

b. between 15.9 ounces and 16.1 ounces?

c. less than 15.9 ounces?

9. If the sampling distribution of the mean is normal for all sample sizes n, what do you know about the population sampled? Explain.

10. The birth weights of 5000 newborn babies have a mean of 7.3 pounds and a standard deviation of 1.5 pounds. If a sample of 100 newborn babies is selected and weighed, what is the probability that the mean weight is

a. between 7 and 7.5 pounds?

b. less than 7.1 pounds?

c. greater than 7.2 pounds?

11. Five thousand students took a statistics examination. The mean score was 75 and the standard deviation was 10. If 300 random samples of size 40 are drawn from this population of test scores, find:

a. $\mu_{\bar{x}}$ and $\sigma_{\bar{x}}$.

b. the approximate number of sample means that fall between 73 and 77.

c. the approximate number of sample means that fall above 72.

12. Suppose samples of size 2 are drawn with replacement from the population having values 1, 7, and 13.

a. Construct a frequency distribution for the 9 sample sums.

b. Verify that the mean of the sampling distribution of sample sums is given by $\mu_{\Sigma x} = n\mu$.

c. Verify that the standard error of the sample sums is given by $\sigma_{\Sigma X} = \sqrt{n}\sigma$.

B

13. The sampling distribution of the sample variance s^2 is not normal. But if sampling is from a normal population with mean μ and variance σ^2, the sampling distribution of s^2 has a mean equal to σ^2 and a variance equal to $2\sigma^4/(n-1)$. For large values of n, the sampling distribution of s^2 is approximately normal. If a sample of size 75 is taken from a normal population with $\sigma^2 = 85$, find the probability that s^2 will be between 60 and 100.

8.5 *SAMPLING DISTRIBUTIONS OF PROPORTIONS* _____

A **binomial** or **dichotomous population** is a population consisting of qualitative data that can be classified into one of two distinct classes, successes or failures. In Figure 8-18, successes are labeled *a* and failures are labeled *b*.

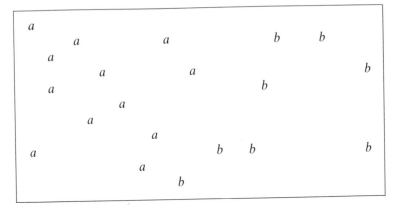

FIGURE 8-18 Dichotomous or binomial population

The following are examples of binomial populations:

Population	Classes
a. Sex classifications of 100 people	Males, females
b. Marital status of 40 adults	Single, married
c. 70 whole numbers	Even numbers, odd numbers
d. 100 tosses of a coin	Heads, tails
e. 30 test scores	Passing scores, failing scores
f. Programs of study for 50 students	Math majors, not math majors
g. Classifications of 80 senior citizens	Retired, not retired
h. Classifications of 500 adult males	Employed, unemployed

Any binomial population can be associated with a population of 0s and 1s. A population with a members labeled A and b members labeled B can be associated with a population having a 0s and b 1s or a 1s and b 0s. Let's assume that the observations in one of the two classes in a binomial population are assigned the value 1 and the observations in the other class are assigned the value 0. If X is a variable that can assume only the values 0 and 1, then $\sum x$ is just the number of observations for which X equals 1 in the binomial population. For example, suppose we have a binomial population with 4 As and 6 Bs and each A is associated with 1 and each B is associated with 0, as shown here:

A	A	1	1
A	A	1	1
B	B	0	0
B	B	0	0
B	B	0	0

Dichotomous   **Population of**
population   **0s and 1s**

The proportion p of As in the binomial population is $p = \frac{4}{10}$, and the mean of the population of 0s and 1s is

$$\mu = \frac{\Sigma x}{N}$$

$$= \frac{1 + 1 + 1 + 1 + 0 + 0 + 0 + 0 + 0 + 0}{10} = \frac{4}{10}$$

Thus, if a binomial population consists of As and Bs, each A is associated with 1, and each B is associated with 0, then the proportion of As, p, in the population is equal to the mean of the population of 0s and 1s, μ. As a result, we have

$$\mu = p$$

The variance σ^2 of a population of 0s and 1s can be found by using the formula

$$\sigma^2 = p(1 - p) \qquad (8\text{-}4)$$

where p is the proportion of 1s. For the above population of 4 As and 6 Bs, let's suppose X is a variable that assigns 1 to A and 0 to B. The following is a frequency table for the binomial population:

Observation	x	f
B	0	6
A	1	4
		$\overline{10}$

To find σ^2 for the binomial population of 0s and 1s, we complete the following table, sum the last column, and divide the sum by $N = 10$:

x	f	xf	$x - \mu$	$(x - \mu)^2$	$(x - \mu)^2 f$
0	6	0	$-.4$	.16	.96
1	4	4	.6	.36	1.44
	$\overline{10}$	$\overline{4}$			$\overline{2.40}$

The population variance is therefore equal to

$$\sigma^2 = \frac{\Sigma f(x - \mu)^2}{N} = \frac{2.40}{10} = .24$$

This same result can be obtained by using formula (8-4):

$$\sigma^2 = p(1 - p) = (.6)(.4) = .24$$

Estimating Population Proportions

Frequently we will have occasion to estimate or determine an unknown proportion for a binomial population. For example, we may want to estimate the proportion of college students who live off campus, or we may be interested in the percentage of women who favor abortion. To estimate the population proportion, we will use a random sample from the binomial population and use the sample proportion $\hat{p} = x/n$ to estimate p, where x is the number of successes (the observations of interest) and n is the size of the sample. For example, suppose a sample from a binomial population consists of 2 As and 3 Bs, as shown in Figure 8-19. For the sample of As and Bs, the proportion of As is $\hat{p} = \frac{2}{5}$ and the mean of the sample of 0s and 1s is $\bar{x} = \frac{2}{5}$. Thus, we have $\hat{p} = \bar{x}$.

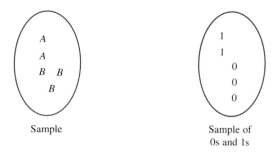

FIGURE 8-19 Binomial population of 2 As and 3 Bs

Sampling Distribution of Sample Proportions

If random samples of size n are drawn from a large binomial population and the sample proportion $\hat{p}$ is computed for each sample, the collection of sample proportions is called the **sampling distribution of sample proportions**. For example, suppose we have a population of five transistors, consisting of two defective ones (D) and three good ones (G). If we associate a 1 with a defective transistor, then we have the following population of 0s and 1s:

Transistor	Transistor quality	Population value
1	*D*	1
2	*G*	0
3	*G*	0
4	*D*	1
5	*G*	0

If unordered samples of size 3 are drawn without replacement from this population, there are ten possible samples. They are listed in Table 8-7 along with the proportion of defectives $\hat{p}$ in each sample.

Table 8-7 Samples of Size 3 from Binomial Population

Sample number	Sample of transistors	Proportion of defectives	Sample number	Sample of transistors	Proportion of defectives
1	1, 2, 3	$\frac{1}{3}$	6	3, 4, 5	$\frac{1}{3}$
2	1, 2, 4	$\frac{2}{3}$	7	1, 4, 5	$\frac{2}{3}$
3	1, 2, 5	$\frac{1}{3}$	8	2, 4, 5	$\frac{1}{3}$
4	2, 3, 4	$\frac{1}{3}$	9	1, 3, 5	$\frac{1}{3}$
5	2, 3, 5	0	10	1, 3, 4	$\frac{2}{3}$

The mean for the population of ten sample proportions is

$$\mu = \frac{6(\frac{1}{3}) + (0) + 3(\frac{2}{3})}{10} = \frac{2}{5}$$

Note that the proportion of defectives in the population is also $\frac{2}{5}$. A frequency graph for the sampling distribution of $\hat{p}$ is as follows:

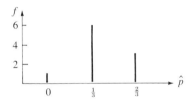

Since samples from a binomial population can be identified with samples of 0s and 1s and the proportion of successes in a sample from the binomial population is equal to the mean of the corresponding sample of 0s and 1s, the sampling distribution of the mean can assist us in answering the following three questions concerning the sampling distribution of sample proportions:

1. What distribution does the sampling distribution of $\hat{p}$ have?
2. What is the mean of the sampling distribution of $\hat{p}$?
3. What is the standard deviation of the sampling distribution of $\hat{p}$?

Recall the following facts concerning the sampling distribution of the mean:

1. The sampling distribution of the mean is approximately normal; the approximation is considered good for $n \geq 30$.
2. $\mu_{\bar{x}} = \mu$
3. $\sigma_{\bar{x}} = \sigma/\sqrt{n}$

As a consequence of these facts, we have the following properties concerning the sampling distribution of the sample proportion:

1. The sampling distribution of sample proportions is approximately normal; the approximation is considered good for $n \geq 30$.
2. $\mu_{\hat{p}} = \mu_{\bar{x}} = \mu = p$, where μ is the mean of the population of 0s and 1s and p is the proportion of 1s in the population.

3. $\sigma_{\hat{p}} = \sigma_{\bar{x}} = \sigma/\sqrt{n}$, where σ is the standard deviation of the population of 0s and 1s.

Because the standard deviation for a population of 0s and 1s is given by $\sigma = \sqrt{p(1-p)}$, the standard error of $\hat{p}$ (the standard deviation of the distribution of sample proportions) is given by

$$\sigma_{\hat{p}} = \frac{\sigma}{\sqrt{n}}$$

$$= \frac{\sqrt{p(1-p)}}{\sqrt{n}}$$

$$= \sqrt{\frac{p(1-p)}{n}}$$

Hence, the standard error of the sample proportion is given by

$$\sigma_{\hat{p}} = \sqrt{\frac{p(1-p)}{n}} \qquad (8\text{-}5)$$

If sampling is done from a finite binomial population with $20n > N$, then the standard error of $\hat{p}$ is given by

$$\sigma_{\hat{p}} = \sqrt{\frac{p(1-p)}{n}} \sqrt{\frac{N-n}{N-1}} \qquad (8\text{-}6)$$

EXAMPLE 8-17   For the previous illustration of selecting samples of size 3 from a population of five transistors, two of which are defective, find the standard deviation of the sampling distribution of $\hat{p}$:

a. without using formula (8-6).
b. using formula (8-6).

Solution   **a.** We shall use the formula for calculating the standard deviation of a population. Recall that $\mu_{\hat{p}} = \frac{2}{5}$. By the definition of population standard deviation, we have

$$\sigma_{\hat{p}} = \sqrt{\frac{\sum(\hat{p} - \mu_{\hat{p}})^2 f}{N}}$$

$$= \sqrt{\frac{(\frac{1}{3} - \frac{2}{5})^2 6 + (0 - \frac{2}{5})^2 1 + (\frac{2}{3} - \frac{2}{5})^2 3}{10}} = .2$$

b. By using formula (8-6):

$$\sigma_{\hat{p}} = \sqrt{\frac{p(1-p)}{n}} \sqrt{\frac{N-n}{N-1}}$$

$$= \sqrt{\frac{(\frac{2}{5})(\frac{3}{5})}{3}} \sqrt{\frac{5-3}{5-1}}$$

$$= \sqrt{.08}\sqrt{.5} = \sqrt{.04} = .2$$

■

In summary, a sample proportion can be thought of as a special case of a sample mean for a binomial population consisting of 0s and 1s, where a success is identified with 1. Thus, by using the population of 0s and 1s as the sampled population and as a consequence of the central limit theorem for the sampling distribution of the mean, we have the following properties for the sampling distribution of the sample proportion $\hat{p}$:

1. The sampling distribution of $\hat{p}$ is approximately normal.
2. The mean of the sampling distribution of $\hat{p}$ is equal to p, the population proportion.
3. The standard deviation of $\hat{p}$ (the standard error of $\hat{p}$) is equal to $\sqrt{p(1-p)/n}$, where n is the size of the sample and p is the proportion of successes in the population.

In addition, if $n \geq 30$, the above normal approximation is considered good.

As a result of these properties for the sampling distribution of $\hat{p}$, it follows that the z score for $\hat{p}$ is given by

$$z = \frac{\hat{p} - p}{\sqrt{p(1-p)/n}} \tag{8-7}$$

and the sampling distribution of z is approximately the standard normal.

EXAMPLE 8-18   It has been determined that 60% of the students at a large university smoke cigarettes. If a random sample of 800 students polled indicates that 440 smoke cigarettes, locate the sample proportion $\hat{p}$ within the sampling distribution of $\hat{p}$ by finding what percentage of the sample proportions fall below it.

Solution   Since $n = 800$, the sampling distribution of $\hat{p}$ is approximately normal. The mean of the sampling distribution of $\hat{p}$ is

$$\mu_{\hat{p}} = p = .60$$

The standard error of $\hat{p}$ is

$$\sigma_{\hat{p}} = \sqrt{\frac{p(1-p)}{n}} = \sqrt{\frac{(.6)(.4)}{800}} = .0713$$

The sample proportion $\hat{p}$ is

$$\hat{p} = \frac{x}{n} = \frac{440}{800} = .55$$

Therefore, the z value for $\hat{p}$ is

$$z = \frac{\hat{p} - p}{\sqrt{p(1-p)/n}}$$
$$= \frac{.55 - .60}{.0173} = -2.89$$

Thus, $P(\hat{p} < .55) = P(z < -2.89) = .5 - .4981 = .0019 = .19\%$ of the sampling distribution of $\hat{p}$ falls to the left of $\hat{p} = .55$, a rare event indeed (see Figure 8-20).

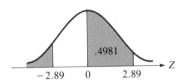

FIGURE 8-20 Sampling distribution of $\hat{p}$ for Example 8-18   ■

EXAMPLE 8-19   A medication for an upset stomach carries a warning that some users may have an adverse reaction to it. Further, it is thought that about 3% of the users have such a reaction. If a random sample of $n = 150$ people with upset stomachs use the medication, find the probability that $\hat{p}$, the proportion of the users who actually have an adverse reaction, will be greater than 5%.

Solution   To determine the probability that more than 5% of the users will have an adverse reaction to the medication, we find the area to the right of $\hat{p} = .05$ under the following normal curve:

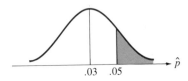

The z score for $\hat{p} = .05$ is

$$z = \frac{\hat{p} - p}{\sqrt{p(1-p)/n}}$$

$$= \frac{.05 - .03}{\sqrt{(.03)(.97)/150}}$$

$$= 1.44$$

By examining the z table (see the front endpaper), we determine the area under the standard normal curve between $z = 0$ and $z = 1.44$ to be

$$P(0 < z < 1.44) = .4251$$

Thus, the probability that more than 5% of the users will have an adverse reaction is

$$P(\hat{p} > .05) = P(z > 1.44)$$
$$= .5 - .4251 = .0749 \qquad ■$$

Binomial Probability Distributions

A binomial population is closely associated with a binomial probability distribution. A binomial population is a collection of successes and failures, whereas a binomial probability distribution contains the probabilities or proportions of all possible numbers of successes in a binomial experiment. As a consequence of this relationship, probability statements involving the sample proportion $\hat{p}$ can be evaluated by using the normal approximation to the binomial, provided that $np \geq 5$ and $n(1 - p) \geq 5$. Recall that we worked with counts of successes, instead of proportions, when dealing with binomial experiments in Chapter 6. Any count can be converted to a proportion by dividing the count by the number of trials, n; that is, $x/n = \hat{p}$.

For Example 8-19, let's calculate the probability that the proportion of users who actually have an adverse reaction will be greater than 5% by using the normal approximation to the binomial. Recall that $n = 150$ and $p = .03$. Since $\hat{p} = x/n$ and x represents a count and must be a whole number, we have

$$P(\hat{p} > .05) = P\left(\frac{x}{150} > .05\right)$$
$$= P(x > 7.5)$$
$$= P(x \geq 8)$$

The binomial distribution of the number of users who have an adverse reaction has a mean of

$$\mu = np = (150)(.03) = 4.5$$

and a standard deviation of

$$\sigma = \sqrt{np(1 - p)} = \sqrt{(4.5)(.97)} = 2.089$$

To approximate the area of all the bars labeled $x = 8$ or higher in a probability graph for the binomial distribution, we use the normal distribution with a mean of 4.5 and a standard deviation of 2.089. The z score for $x = 7.5$ is

$$z = \frac{7.5 - 4.5}{2.089} = 1.44$$

Hence, the probability that $\hat{p}$ is greater than .05 is

$$P(\hat{p} > .05) = P(x \geq 8)$$
$$\simeq P(z > 1.44)$$
$$= .5 - .4251 = .0749$$

which agrees with the result obtained in Example 8-19. Note that $np \geq 5$ and $n(1 - p) \geq 5$.

Problem Set 8.5

A

1. Five percent of all diodes manufactured by a certain firm are defective. If a random sample of 50 diodes is selected from 1500 newly manufactured diodes and $\hat{p}$ represents the percentage of defective diodes, find:
   a. $P(\hat{p} < .08)$
   b. $P(.01 < \hat{p} < .10)$
   c. $P(\hat{p} > .04)$

2. A particular county in Pennsylvania has a 12% unemployment rate. A survey of 500 individuals was made.
   a. Describe the sampling distribution of $\hat{p}$ and find $\mu_{\hat{p}}$ and $\sigma_{\hat{p}}$.
   b. If $\hat{p}$ represents the percentage unemployed in the sample, find $P(\hat{p} \geq .11)$.
   c. If $\hat{p} = .13$, locate $\hat{p}$ within the sampling distribution of $\hat{p}$ by finding what percentage of sample proportions will fall below .13.

3. If a drug is 80% effective in treating a certain disease and a random sample of 500 patients is treated with the drug, find the following probabilities if $\hat{p}$ represents the percentage of effective treatments:
   a. $P(\hat{p} > .81)$
   b. $P(.79 < \hat{p} < .81)$
   c. $P(\hat{p} < .84)$

4. An executive officer for a large brokerage firm surveyed 120 of its clients and learned that 72 of them were extremely satisfied with the firm's service.
   a. Find $\hat{p}$, the proportion of customers in the sample who were extremely satisfied with the firm's service.
   b. Estimate the standard error of the sample proportion of customers who are extremely satisfied by using $\hat{p}$ as an estimate for p, the true proportion of customers who are extremely satisfied.

5. For Problem 4, find the percentage of sample proportions that fall within 2 standard deviations of the population proportion p.

6. If it is known that the true proportion of all manufactured components made by a firm that are defective is 4%, find the probability that a random sample of size 60 will have fewer than 3% defective components.

7. In Problem 6, suppose the sample was drawn from a population of size 1000. Find:
   a. $\sigma_{\hat{p}}$

   b. $P(\hat{p} < .03)$
   c. $P(.02 < \hat{p} < .07)$

8. For Problem 3, find $\sigma_{\hat{p}}$ if it was determined that the drug was 80% effective for a population of size 50,000.

9. According to a recent survey of 500 doctors in the United States, 75% approve of requiring second opinions before nonemergency surgery. Use this value as the true proportion of U.S. doctors who approve of requiring second opinions before nonemergency surgery.
   a. Find $\sigma_{\hat{p}}$.
   b. Find an interval centered at .75 that contains 80% of the sample proportions based on samples of size 500.

10. For Problem 9, find $\sigma_{\hat{p}}$ if the random sample was drawn from a population of $N = 8000$ doctors.

B

11. Suppose an experiment consists of tossing a fair coin 50 times. If $\hat{p}$ represents the proportion of heads obtained, find:
    a. $\mu_{\hat{p}}$ and $\sigma_{\hat{p}}$
    b. $P(\hat{p} > .6)$
    c. $P(\hat{p} < .04)$
    d. $P(.4 < \hat{p} < .6)$

12. Show that formula (8-7) can be written as

$$z = \frac{x - np}{\sqrt{np(1 - p)}}$$

What type of discrete random variable is x? Explain.

13. Suppose a large binomial population consists of 0s and 1s, where the proportion of 1s is p and the proportion of 0s is $(1 - p)$.
    a. Prove that the mean of the population is p.
    b. Prove that the variance of the population is $p(1 - p)$.

14. Refer to Problem 13 and recall that $\hat{p} = x/n$ can be regarded as the mean of a sample of size n drawn from a population of 0s and 1s.
    a. Prove that the mean of the sampling distribution of $\hat{p}$ is p.
    b. Prove that the standard error of $\hat{p}$ is $\sqrt{p(1 - p)/n}$.

CHAPTER SUMMARY

In this chapter we learned about sampling distributions of sample statistics, a concept that is basic to understanding inferential statistics. Random samples are used in order to ensure a representative subcollection from the whole population. We learned several methods for obtaining a random sample from a population of interest. We learned that the standard error of the mean depends on the size of the sample relative to the population and decreases as the sample size increases. The central limit theorem states that the sampling distribution of the mean is approximately normal when sampling is from nonnormal populations. If the population is normal, then the sampling distribution of the mean is normal. Finally, we studied the sampling distribution of the sample proportion, where sampling is done from a binomial population. We learned that the distribution of a sample statistic has a certain shape, mean, variance, and standard deviation. When such information is known, as in the case for the sampling distributions of the mean and sample proportions, probability statements can be made about the sampled population and the sampling distribution.

C H A P T E R   R E V I E W

IMPORTANT TERMS

For each of the following terms, provide a definition in your own words. Then check your responses against the definitions given in the chapter.

binomial population
census
central limit theorem
cluster sample
correction factor
degrees of freedom
dichotomous population
estimate
estimator
hextat

nonsampling error
randomization
random number table
random sample
random variable
sample bias
sample proportion
sampling distribution
 of the sampling mean
sampling distribution
 of sample proportions

sampling distribution
 of a statistic
sampling distributions
sampling error
simple random sample
standard error of
 the mean
standard error of the
 statistic

statistical bias
strata
stratified sample
systematic sample
t distribution
t statistic
unbiased
zero bias

IMPORTANT SYMBOLS

e, sampling error

$\mu_{\bar{x}}$, mean of the sampling distribution of the mean

$\sigma_{\bar{x}}$, standard deviation of the sampling distribution of the

mean (also the standard error of the mean)

t, t statistic

df, degrees of freedom

$\hat{p}$, sample proportion

p, population proportion

μ, mean of the sampling distribution of the sample proportion

$\sigma_{\hat{p}}$, standard deviation of the sampling distribution of the sample proportion (also the standard error of the sample proportion)

IMPORTANT FACTS AND FORMULAS

1. a. If $20n > N$ and sampling is without replacement, then

$$\sigma_{\bar{x}} = \frac{\sigma}{\sqrt{n}} \sqrt{\frac{N-n}{N-1}}$$

b. If $20n \leq N$ or sampling is from an infinite population, then

$$\sigma_{\bar{x}} = \frac{\sigma}{\sqrt{n}}$$

2. $\mu_{\bar{x}} = \mu$, mean of sampling distribution of $\bar{x}$

3. $z = \dfrac{\bar{x} - \mu}{(\sigma/\sqrt{n})}$, z value for $\bar{x}$

4. $t = \dfrac{\bar{x} - \mu}{(s/\sqrt{n})}$, t value for $\bar{x}$

5. df $= n - 1$, degrees of freedom for t

6. $\hat{p} = \dfrac{x}{n}$, sample proportion

7. $\mu_{\hat{p}} = p$, mean of the sampling distribution of $\hat{p}$

8. $\sigma_{\hat{p}} = \sqrt{p(1-p)/n}$, standard error of $\hat{p}$

9. $z = \dfrac{\hat{p} - p}{\sqrt{p(1-p)/n}}$, z value for $\hat{p}$

10. *Central limit theorem*   The sampling distribution of the mean is approximately normal. The approximation is considered good for $n \geq 30$.

REVIEW PROBLEMS

1. Suppose it is desired to obtain a random sample of size 5 from the numbers 1, 2, 3, 4, 5, 6, 7, 8, 9, and 10. To select the sample a coin is tossed. If it lands heads up, then the numbers 1, 3, 5, 7, and 9 are chosen. If the coin lands tails up, then the numbers 2, 4, 6, 8, and 10 are chosen.
   a. Does every number between 1 and 10 have the same chance of being chosen?
   b. Will the resulting sample be a simple random sample? Explain.

2. Use the random number table (Table 8-1) to select a sample of size 5 from the 26 letters of the alphabet and explain your procedure.

3. Is the sample {1, 2} a random sample from the population of values {1, 2, 3, 4, 5, 6, 7, 8, 9, 10}? Explain.

4. Suppose ordered samples of size 2 are drawn with replacement from the population of values 1, 3, 5, and 7. For the sampling distribution of the mean:
   a. find $\mu_{\bar{x}}$.
   b. find $\sigma_{\bar{x}}$.
   c. construct a graph for the frequency distribution.

5. Suppose ordered samples of size 2 are drawn without replacement from the population of values 1, 3, 5, and 7. For the sampling distribution of the mean:
   a. find $\mu_{\bar{x}}$.
   b. find $\sigma_{\bar{x}}$.
   c. construct a graph for the frequency distribution.

6. Suppose ordered samples of size 2 are drawn with replacement from the population of values 0, 3, 5, and 7. If the sample mean is computed for each sample, find:
   a. the sampling error for each sample mean.
   b. the mean of the sampling errors.
   c. the standard deviation of the sampling errors.

7. Ordered samples of size 3 are selected with replacement from the population of values 1, 2, 3, and 6. The following is a frequency table for the sample medians:

Median	f
1	10
2	22
3	22
6	10

Suppose the median of a sample is used to measure or estimate the population mean.
   a. Find the sampling error for each median.
   b. Construct a graph for the sampling distribution of the median.
   c. Find the mean of the sampling errors.
   d. Find the mean of the sampling distribution of the median.
   e. Find the standard deviation of the sampling errors.

f. Find the standard error of the median.

g. Compare your results for parts (e) and (f).

8. Consider the following population of six voters, where "yes" means that the voter will vote for candidate A and "no" indicates that the voter will not vote for candidate A.

Voter	Response
1	yes
2	no
3	yes
4	yes
5	yes
6	no

a. Selecting unordered samples of size 4 without replacement provides a total of 15 possible samples. List the 15 samples.

b. Compute the proportion of "yes" votes for each sample.

c. Find the mean of the sampling distribution of the proportion of "yes" votes.

d. Find the standard deviation of the sampling distribution of the proportion of "yes" votes.

e. Construct a graph for the sampling distribution of the proportion of "yes" votes.

9. Suppose an automatic machine used to fill cans of soup has $\mu = 16$ ounces and $\sigma = .5$ ounce. If a sample of 50 cans is obtained, find the probability that the sample mean fill $\bar{x}$ is

a. greater than 15.88 ounces.

b. greater than 15.9 ounces and less than 16.09 ounces.

c. less than 16.2 ounces.

10. Suppose a sample of size 50 is selected from a population of 5000 employees in order to estimate the average age for the population. If the standard deviation of the population is 7.5 years, find $\sigma_{\bar{x}}$, the standard error of the mean.

11. What is the probability that the sample mean age of the employees in Problem 10 will be within 2 years of the population mean age?

12. It is known that 10% of all items produced on an assembly line are defective. If a sample of 50 parts are tested and $\hat{p}$ denotes the percentage of tested items that are defective, find:

a. $P(\hat{p} > 10.8\%)$

b. $P(9.5\% < \hat{p} < 10.5\%)$

c. $P(\hat{p} < 9.6\%)$

13. It is claimed that a new drug is 80% effective in treating patients with a particular disease. If the drug is used on a sample of 64 patients and $\hat{p}$ denotes the sample proportion of patients effectively treated, find:

a. $P(.78 < \hat{p} < .82)$

b. $P(.82 < \hat{p} < .83)$

c. $P(\hat{p} < .81)$

14. Suppose the daily water usage by customers is normally distributed with a mean of 250 gallons. On a particular day, a sample of 20 meter readings showed a standard deviation of 45 gallons. If $P(\bar{x} > c) = .95$, find the value of c.

15. The number of passengers carried each day by a certain bus company is normally distributed with a mean of 220 passengers and a standard deviation of 50 passengers. If a sample of 12 days is used to estimate the average number of passengers using a company bus each day, find the probability that a 12-day sample will produce an average of less than 300 customers.

CHAPTER ACHIEVEMENT TEST

(5 points) **1.** If a random sample has a mean $\bar{x} = 20$ and the sampled population has a mean $\mu = 25$, determine the sampling error if $\bar{x}$ is used to estimate μ.

(15 points) **2.** Describe how to obtain a random sample of three letters from the alphabet using a random number table.

(10 points) **3.** If random samples of size 36 are obtained from a population with mean $\mu = 10$ and standard deviation $\sigma = 9$, determine $\sigma_{\bar{x}}$.

(15 points)   **4.** If a random sample with $\bar{x} = 7$, $s = 2.5$, and $n = 10$ was obtained from a normal population with $\mu = 5$ and $\sigma = 3$, find:

    a. the value of the t statistic associated with $\bar{x} = 7$.

    b. the degrees of freedom associated with t.

    c. the value of the z statistic associated with $\bar{x} = 7$.

(15 points)   **5.** If a random sample of size 100 is obtained from a population with $\mu = 40$ and $\sigma = 12$, find:

    a. $P(38 < \bar{x} < 41)$

    b. $P(40 < \bar{x} < 55)$

    c. the value c so that $P(\bar{x} > c) = .40$

(10 points)   **6.** Eighty percent of the students at a large university favor pass/fail grades for elective courses. If a random sample of 100 students is selected to determine the proportion who favor pass/fail grades and the sample proportion $\hat{p}$ is calculated, find:

    a. $P(\hat{p} > .85)$ **b.** $P(.78 < \hat{p} < .83)$

(10 points)   **7.** If a random sample with $\bar{x} = 5$, $s = 2$, and $n = 10$ is selected from a normal population with $\mu = 8$ and $\sigma = 3$, find the value of the z statistic associated with $\bar{x} = 5$.

(10 points)   **8.** Sixty percent of a large number of people polled indicated that they prefer brand X toothpaste to other brands. If in a sample of 40 people, 65% indicated they prefer brand X toothpaste to other brands, find the value of the z statistic associated with $\hat{p} = .65$.

(10 points)   **9.** If ordered samples of size 4 are obtained with replacement from the population of values 2, 3, 4, 5, and 6, find:

    a. $\mu_{\bar{x}}$ **b.** $\sigma_{\bar{x}}$

9

ESTIMATION

Chapter Objectives

In this chapter you will learn:

- *the difference between an estimate and an estimator*
- *two kinds of estimates for an unknown parameter*
- *what a biased estimator is*
- *what a critical value is*
- *how to find critical values*
- *point estimates of μ, σ^2, σ, and p*
- *what the maximum error of estimate is for a point estimate*
- *how to make probability statements concerning the maximum error of estimate*
- *how to construct confidence intervals for the mean of a population using large samples*

- *how to construct confidence intervals for the mean of a normal population using small samples*
- *how to determine sample sizes in order to be confident that the error of estimate is at most a specified value*
- *what the chi-square distributions are*
- *how to construct confidence intervals for the variance and standard deviation of a normal population*

Chapter Contents

Estimation is a major objective of inferential statistics. By studying one sample from a population, we want to generalize our findings to the entire population. As we have seen in Chapter 8, statistics vary greatly within their sampling distributions, and the smaller the standard error of the statistic, the closer the values of the statistic are to each other. While studying **statistical inference** (generalizing from a sample to a population), we will concern ourselves with two elements: the inference and how good it is. For example, if we estimate the annual average income of all families living within 1 mile of a new shopping center to be $28,200, we would like to know how good this estimate is. Since the population mean is unknown, the sampling error, $\bar{x} - \mu = 28,200 - \mu$, is unknown. In this chapter we will learn how to make probability statements concerning the size of the error.

There are two general processes for making inferences about unknown population parameters: **estimation** and **hypothesis testing**. While the two processes share common elements (for example, both procedures involve probability theory), the process of estimation is more direct and conceptually easier to understand; we shall start with it. The logic of hypothesis testing involves an indirect method for making inferences and, as a result, is conceptually more involved; we save it for Chapter 10. It should be kept in mind that the method used for making an inference is usually a matter of choice for the experimenter. Since both processes are used in experimental research, it is important to understand both approaches.

9.1 POINT ESTIMATES FOR μ

As an illustration of the ideas involved in estimation, suppose a biologist is interested in determining the average number of eggs laid per nest per season for the Eastern Phoebe bird. A random sample of 50 nests was examined and the following results were obtained:

Number of eggs/nest	1	2	3	4	5	6
f	2	1	1	8	36	2

The mean number of eggs per nest for the sample of 50 nests is 4.62. This figure can serve as an estimate of μ, the true average number of eggs laid per nest by the Eastern Phoebe bird during one season. If this experiment were to be repeated a large number of times, values of the sample means would vary, providing different estimates. As a result of the central limit theorem, we know the sampling distribution of these means is approximately normal with $\mu_{\bar{x}} = \mu$ and $\sigma_{\bar{x}} = \sigma/\sqrt{50}$. By using these facts, we will be able to determine the goodness of our estimate.

There are two types of estimates for parameters: point and interval estimates. A **point estimate** is a single value of a statistic that is used to estimate a parameter. The statistic that we use is called an **estimator**. For our example above, the sample mean $\bar{x}$ is the estimator and the value 4.62 is the point estimate. An **interval estimate**, on the other hand, is an interval (usually of finite width) that is expected to contain the parameter. For example, the interval (4.57, 4.67) could be an interval estimate for the true average number of eggs per nest. The width of this interval is $4.67 - 4.57 = .10$. This interval either contains the population mean μ or it does not.

For estimation purposes, we can think of the sample mean $\bar{x}$ as a measurement of the value of the population mean μ. Any measurement has an associated degree of precision, which indicates how accurate it is. If the interval (4.57, 4.67) is an interval estimate for μ, based on $\bar{x} = 4.62$, we feel reasonably confident that $4.57 < \mu < 4.67$. Stated differently, we have a certain degree of confidence that the value of μ is strictly within $4.62 \pm .05$. The value .05 indicates a probable degree of precision involved in using $\bar{x} = 4.62$ to estimate μ. Knowing that a measurement is precise is of limited value if it is not reliable. **Reliability** is the probability that an estimate is correct. As a result of the central limit theorem, the sampling distribution of $\bar{x}$ is approximately normal for large samples. Therefore, when we estimate μ using $\bar{x}$, we can increase the reliability and precision by using larger samples. Since $\mu_{\bar{x}} = \mu$ and $\sigma_{\bar{x}} = \sigma/\sqrt{n}$, the values for $\bar{x}$ will cluster more tightly around μ as the size of the sample increases.

Population values are commonly estimated by their corresponding sample values. A sample mean $\bar{x}$ is an estimate of the population mean μ. A sample proportion $\hat{p} = x/n$ is an estimate of the population proportion p. The population variance σ^2 is estimated by s^2, the sample variance, while the standard deviation σ of a population is estimated by the standard deviation s of a sample drawn from the population.

An estimator is **unbiased** if the mean of its sampling distribution is the parameter being estimated. If the sample mean $\bar{x}$ is used to estimate the population μ, we learned in Chapter 8 that $\bar{x}$ is an unbiased estimator—that is, $\mu_{\bar{x}} = \mu$. Similarly, we learned that $\hat{p}$ is an unbiased estimator of p.

The following is a list of parameters and corresponding estimators:

Parameter	Estimator	Unbiased
μ	$\bar{x}$	Yes
σ^2	s^2	Yes
p	$\hat{p} = x/n$	Yes
σ	s	No

While $\bar{x}$, s^2, and $\hat{p}$ are unbiased estimators of μ, σ^2, and p, respectively, s is a biased estimator of σ. But for most practical applications, the error resulting is negligible, provided the size of the sample is at least 30.

It should be clear that the precision of an estimator increases with the size of the sample. If the sample is the entire population, then $\bar{x} = \mu$. The absolute value of the difference between an estimate and the parameter being estimated is called the **error of estimate**. When $\bar{x}$ is used to estimate μ, $|\bar{x} - \mu|$ is the error of estimate. Recall from Section 8.1 that $\bar{x} - \mu$ is called **sampling error**. The difference between *sampling error* and *error of estimate* is that the error of estimate is always greater than or equal to 0. The average error of estimate decreases as the sample size increases. Unfortunately, we never know the error of estimate, since we do not know the value of the parameter. But we can make probability statements concerning the error of estimate.

Suppose we use $\bar{x}$ to estimate μ, the mean number of eggs per nest for the Eastern Phoebe bird data, and want to determine the probability that the error of estimate is less than .05. That is, we want to determine the probability $P(-.05 < \bar{x} - \mu < .05)$. Further, suppose we use the sample standard deviation, $s = .99$, to estimate the true standard deviation σ. Since $n = 50$, the standard error of the mean is given by

$$\sigma_{\bar{x}} = \frac{\sigma}{\sqrt{n}}$$

$$\simeq \frac{s}{\sqrt{n}}$$

$$= \frac{.99}{\sqrt{50}} = .14$$

Note that since

$$z = \frac{\bar{x} - \mu}{\sigma/\sqrt{n}}$$

it follows that

$$P(|\bar{x} - \mu| < .05) = P(-.05 < \bar{x} - \mu < .05)$$
$$= P\left(-\frac{.05}{.14} < z < \frac{.05}{14}\right)$$
$$= P(-.36 < z < .36)$$
$$= 2P(0 < z < .36) = 2(.1406) = .2812$$

If $\bar{x}$ is used to estimate μ, we have determined that the probability of the error of estimate being less than .05 is .2812. The probability value .2812 is called the **confidence level** or **reliability level**. Thus, we can say that we are 28.12% confident that the error of estimate is at most .05. Notice that in order to evaluate the probability that the error of estimate $|\bar{x} - \mu|$ is at most some value E, we need know only $\sigma_{\bar{x}}$, the standard error of the mean.

EXAMPLE 9-1 A study was conducted to determine the mean amount of cola dispensed from a cola machine. In a previous study, it was determined that the standard deviation of the amounts dispensed is $\sigma = .40$ ounce. If a sample of size 40 is used to estimate μ, find the confidence level for the error of estimate to be at most .01 ounce.

Solution   We want to determine the probability $P(|\bar{x} - \mu| < .01)$. As a consequence of the central limit theorem and the fact that $n = 40$, the sampling distribution of $\bar{x}$ is approximately normal. The standard error of the mean is given by

$$\sigma_{\bar{x}} = \frac{\sigma}{\sqrt{n}} = \frac{.40}{\sqrt{40}} = .063$$

Thus,

$$\begin{aligned} P(|\bar{x} - \mu| < .01) &= P\left(|z| < \frac{.01}{.063}\right) \\ &= P(|z| < .16) \\ &= 2P(0 < z < .16) = 2(.0636) = .1272 \end{aligned}$$

Hence, we can be 12.72% confident that the error of estimate is at most .01.   ∎

For each of the illustrations thus far, the maximum error of estimate was stipulated in advance, and the confidence level was found. We now want to develop the ideas involved in determining the maximum error of estimate if the confidence level has been stipulated in advance. But in order to make probability statements about the error of estimate, we must first examine the concept of critical value. In Figure 9-1, the value z_α is called a critical value. The **critical value** z_α is that standard normal value such that the area under the standard normal curve to the right of it is α.

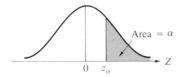

FIGURE 9-1 Critical value z_α

For example, $z_{.025}$ is the z value that has an area of .025 to the right of it under the standard normal curve (see Figure 9-2). Since the standard normal table (z table; see front endpaper) provides areas under the standard normal curve between 0 and specified z values, to determine the value of $z_{.025}$ we need to identify the z value that corresponds to a tabled area of .475 ($= .5 - .025$). [If the exact area cannot be found, we will determine the area that comes closest to

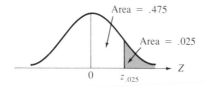

FIGURE 9-2 Determination of $z_{.025}$

it and identify its corresponding z value as the critical value.] Checking the z table on the front endpaper, we can locate .475 in the table; its corresponding z value is 1.96. Hence, the desired critical value is $z_{.025} = 1.96$.

A distribution can have two critical values. In Figure 9-3, $-z_{\alpha/2}$ and $z_{\alpha/2}$ are both called critical values; one is the negative of the other. The area between the two critical values is called the **confidence level**. The confidence level is equal to $1 - \alpha$, since the sum of three areas, $\alpha/2$, $1 - \alpha$, and $\alpha/2$, is 1.

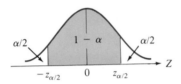

FIGURE 9-3 Critical values for confidence level $1 - \alpha$

The following examples illustrate the concepts of confidence level and critical value.

EXAMPLE 9-2   If the confidence level is 85%, find the positive critical value.

Solution   Since $1 - \alpha = 85\%$, $\alpha = .15$ and $\alpha/2 = .075$. To find the critical value $z_{.075}$ we use the standard normal table. Since the area under the curve to the right of $z_{.075}$ is .075 (see Figure 9-4), the area between 0 and $z_{.075}$ is $.5 - .075 = .425$. We look in the z table (see the front endpaper) for the area that comes closest to .425 and find the corresponding z value. Since the area .4251 comes closest to the area .425, the corresponding z value is 1.44. Thus, the positive critical value $z_{.075}$ is 1.44.

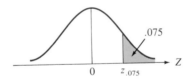

FIGURE 9-4 Critical value $z_{.075}$ for Example 9-2   ■

EXAMPLE 9-3   If the positive critical value is 1.65, find the confidence level $1 - \alpha$.

Solution   The area under the standard normal curve between 0 and 1.65 is .4505, as shown in the accompanying figure. We double this to obtain the confidence level. Thus, the confidence level is $1 - \alpha = .9010$.

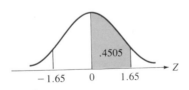

■

Point Estimates for μ Using Large Samples

Suppose we want our estimate of μ to be $(1 - \alpha)100\%$ reliable. What can we say about the level of precision or error associated with the estimate? The maximum error E that corresponds to a confidence level of $1 - \alpha$ can be found as follows. Since

$$P(-z_{\alpha/2} < z < z_{\alpha/2}) = 1 - \alpha \qquad (9\text{-}1)$$

and

$$z = \frac{\bar{x} - \mu}{\sigma/\sqrt{n}} \qquad (9\text{-}2)$$

we substitute the value of z given in (9-2) into the left-hand side of the equation given in (9-1) to get

$$P\left(-z_{\alpha/2} < \frac{\bar{x} - \mu}{\sigma/\sqrt{n}} < z_{\alpha/2}\right) = 1 - \alpha$$

If we now apply some relatively simple algebra, we can rewrite this last statement as

$$P\left(|\bar{x} - \mu| < z_{\alpha/2}\left(\frac{\sigma}{\sqrt{n}}\right)\right) = 1 - \alpha \qquad (9\text{-}3)$$

Thus, we see that the maximum error of estimate E is given by

$$E = z_{\alpha/2}\left(\frac{\sigma}{\sqrt{n}}\right) \qquad (9\text{-}4)$$

One should be careful not to conclude that E is the maximum *possible* error. The error of estimate $|\bar{x} - \mu|$ can exceed the "maximum error E." What the probability statement (9-3) indicates is that in repeated sampling, $(1 - \alpha)100\%$ of the sample means will fall no more than a distance of E from the population mean μ. Since our estimate $\bar{x}$ of μ is based on a single random sample, we feel $(1 - \alpha)100\%$ confident that the error of estimate $|\bar{x} - \mu|$ is less than the maximum error of estimate E. In other words, the reliability of our estimate $\bar{x}$ having precision E is $(1 - \alpha)100\%$. Equation (9-4) defines the maximum error as the product of the critical value and the standard error of the mean.

Consider the following examples.

EXAMPLE 9-4   In an effort to estimate the average resting pulse rate of adults 40 years of age, a random sample of pulse-rate measurements from 50 40-year-old individuals was used. If $\bar{x}$ is used to estimate μ and it is desired to be 95% confident that the population mean μ differs by no more than E from $\bar{x}$, find the maximum error E, assuming that $\sigma = 10$ beats per minute.

Solution   Since $1 - \alpha = .95$, $\alpha = .05$ and $\alpha/2 = .025$. From the z table (see the front end-paper) we find $z_{.025} = 1.96$. By using (9-4), we have

$$E = z_{\alpha/2}\left(\frac{\sigma}{\sqrt{n}}\right) = 1.96\left(\frac{10}{\sqrt{50}}\right) = 2.77$$

Thus, if $\bar{x}$ is used to estimate μ, we can be 95% confident that the maximum error of estimate is at most 2.77 beats per minute.   ■

EXAMPLE 9-5   Suppose a random sample of 40 newborn baby boys is to be used to estimate the mean weight of all newborn baby boys. If it is found that $\bar{x} = 7.5$ pounds and $s = 1.2$ pounds, find the maximum error E so that we can be 95% confident that the error of estimate is at most E.

Solution   Since $n \geq 30$ and σ is unknown, we can use s to estimate σ. Thus, by using (9-4) we have

$$E = z_{\alpha/2}\left(\frac{\sigma}{\sqrt{n}}\right) \simeq z_{\alpha/2}\left(\frac{s}{\sqrt{n}}\right) = 1.96\left(\frac{1.2}{\sqrt{40}}\right) = .37 \text{ pound}$$

Thus, we can be 95% sure that μ is within $\pm.37$ pound of $\bar{x} = 7.5$ pounds.   ■

The maximum error E depends, in part, on the confidence level. If we want to be more confident, then the maximum error E must increase, as illustrated in the following examples.

EXAMPLE 9-6   For Example 9-5, find the maximum error E so that we can be 99% confident that the error of estimate is at most E.

Solution   Since $1 - \alpha = .99$, $\alpha = .01$ and $\alpha/2 = .005$. We find that $z_{.005} = 2.58$ by locating the z score corresponding to the area which comes closest to the area .495. By applying (9-4), we obtain the maximum error of estimate:

$$E = z_{\alpha/2}\left(\frac{\sigma}{\sqrt{n}}\right) = 2.58\left(\frac{1.2}{\sqrt{40}}\right) = .49$$

Thus, we can be 99% confident that the error of estimate $|\bar{x} - \mu|$ is at most .49.   ■

EXAMPLE 9-7   For Example 9-5, find E so that we can be 90% confident that the maximum error of estimate is at most E.

Solution   Since $1 - \alpha = .90$, $\alpha = .10$ and $\alpha/2 = .05$. By examining the z table (see the front endpaper), we determine the positive critical value to be 1.65. By using (9-4), we have

$$E = z_{\alpha/2}\left(\frac{\sigma}{\sqrt{n}}\right) = 1.65\left(\frac{1.2}{\sqrt{40}}\right) = .31$$ ■

The results of Examples 9-5, 9-6, and 9-7 are summarized in the following table:

Confidence level	Maximum error E
.90	.31
.95	.37
.99	.49

Notice that as the confidence level increases, the maximum error also increases.

There is a direct relationship between the confidence level and the maximum error of estimate. The maximum error of estimate varies directly with the confidence level. When one increases so does the other, and when one decreases so does the other. Note also that the maximum error E varies directly with the standard error of the mean, $\sigma_{\bar{x}}$, since $E = z_{\alpha/2}\sigma_{\bar{x}}$. This means that E gets smaller as the standard error $\sigma_{\bar{x}}$ gets smaller and E gets larger when $\sigma_{\bar{x}}$ gets larger. Since $\sigma_{\bar{x}} = \sigma/\sqrt{n}$, $\sigma_{\bar{x}}$ decreases as n increases. Consequently, the maximum error of estimate E will also decrease as the sample size increases. It is usually desirable to choose as large a sample as possible when using the sample mean $\bar{x}$ to estimate the population mean μ.

Point Estimates for μ Using Small Samples

If $\bar{x}$ is used to estimate μ and a random sample of size $n < 30$ is selected from a normal population with σ unknown, then the sampling distribution of the sample mean is not normal. Recall from Chapter 8 that the sampling distribution of the statistic $(\bar{x} - \mu)/(s/\sqrt{n})$ is a t distribution with df $= (n - 1)$.

Critical values for t are identified with the t distributions in a fashion similar to the standard normal distribution. The values $-t_{\alpha/2}(\text{df})$ and $t_{\alpha/2}(\text{df})$ in Figure 9-5 are called **critical values**. The value $t_{\alpha/2}(\text{df})$ is such that the area under the t curve (with df $= n - 1$) to the right of it is $\alpha/2$.

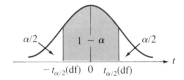

FIGURE 9-5 Critical values for t distribution

The following probability statement can be made for a t distribution.

$$P[-t_{\alpha/2}(\text{df}) < t < t_{\alpha/2}(\text{df})] = 1 - \alpha$$

Since the t value for $\bar{x}$ is given by

$$t = \frac{\bar{x} - \mu}{s/\sqrt{n}}$$

we can substitute this value for t in the probability statement and use some simple algebra to deduce the maximum error of estimate E when $\bar{x}$ is used to estimate μ for a small sample ($n < 30$) drawn from a normal population with unknown variance:

$$E = t_{\alpha/2}(\mathrm{df})\left(\frac{s}{\sqrt{n}}\right) \tag{9-5}$$

EXAMPLE 9-8 In an effort to determine the average noise level for large trucks, the Environmental Protection Agency (EPA) obtained a random sample of noise-level readings (in decibels) for eight large trucks. The sample mean and standard deviation were found to be 85.6 decibels and .65 decibel, respectively. If the sample mean is used to estimate μ, determine the maximum error E so that the EPA can be 95% confident that μ is within $\pm E$ of $\bar{x}$. Assume that the truck noise levels are normally distributed.

Solution By reading from the t table (see the back endpaper) under the column labeled .025 for df = 7, we find that $t_{.025}(7) = 2.365$. Hence, using (9-5), we have

$$E = t_{\alpha/2}(\mathrm{df})\left(\frac{s}{\sqrt{n}}\right) = (2.365)\left(\frac{.65}{\sqrt{8}}\right) = .54$$

Thus, if $\bar{x} = 85.6$ is used to estimate μ, we can be 95% confident that μ is within $\pm.54$ decibel of $\bar{x} = 85.6$ decibels. Stated differently, we can be 95% confident that the maximum error of estimate is at most .54 when $\bar{x} = 85.6$ is used to estimate μ. ∎

EXAMPLE 9-9 It is known that the time (in minutes) required to hand assemble a certain electronic module is normally distributed. If the following sample data were obtained, find the maximum error of estimate so that we can be 99% confident that the population mean assembly time μ is within $\pm E$ of the sample mean $\bar{x}$.

6.2   7.1   5.7   6.8   5.4

Solution **1.** We first find $\bar{x}$ and s for the sample. The following table organizes the computations:

x	x^2
6.2	38.44
7.1	50.41
5.7	32.49
6.8	46.24
5.4	29.16
Sums 31.2	196.74

The sample mean is

$$\bar{x} = \frac{\sum x}{n} = \frac{31.2}{5} = 6.24$$

The sum of squares is

$$SS = \sum x^2 - \frac{(\sum x)^2}{n} = 196.74 - \frac{(31.2)^2}{5} = 2.052$$

Therefore, the sample standard deviation is

$$s = \sqrt{\frac{SS}{n-1}} = \sqrt{\frac{2.052}{4}} = .716$$

The confidence level is $(1 - \alpha) = .99$. Thus, $\alpha = .01$ and $\alpha/2 = .005$.

2. The critical value is found in the t table (see the back endpaper) with df $=$ $n - 1 = 5 - 1 = 4$: $t_{.005}(4) = 4.604$.

3. We use (9-5) to find E:

$$E = t_{\alpha/2}(\text{df})\left(\frac{s}{\sqrt{n}}\right) = (4.604)\left(\frac{.716}{\sqrt{5}}\right) = 1.47$$

Thus, we can be 99% confident that μ is within ± 1.47 of $\bar{x}$. In other words, we can be 99% confident that the error of estimate is at most 1.47 minutes when $\bar{x} = 6.24$ is used to estimate μ.   ■

Problem Set 9.1

A

1. Find the confidence level for the following critical values:
   a. $z_{\alpha/2} = 1.28$
   b. $-z_{\alpha/2} = -1.44$
   c. $z_{\alpha/2} = 2.58$
   d. The positive critical z value is 2.76.

2. Find the positive critical z value if
   a. the confidence level is .94
   b. $\alpha = 10\%$
   c. the confidence level is 88%

3. If the confidence level is 95%, find the t critical value for $n = 15$.

4. If the confidence level is 90%, find $t_{\alpha/2}(19)$.

5. If $1 - \alpha = .98$, find $t_{\alpha/2}(29)$.

6. In an effort to determine the average amount spent per customer for lunch at a large Chicago restaurant, data were collected for 64 customers over a 1-month period. If $\sigma = \$2.25$ and the sample mean $\bar{x}$ is used to estimate μ, find the maximum error E so that we can be 95% confident that the error of estimate is at most E.

7. Fifteen employees of a large manufacturing company were involved in a study to test a new pro-

duction method. The mean production rate for the sample of 15 employees was 63 components per hour and the standard deviation was 8 components per hour. If $\bar{x}$ is used to estimate μ, determine the maximum error E in order to be 99% confident that μ is within $\pm E$ of $\bar{x}$. Assume that the production rates are normally distributed.

8. A random sample of 50 entrance examination scores at a large university is used to estimate the true mean of all entrance examination scores. If $\bar{x} = 98.2$, $s = 17$, and the confidence level is 99%, find the maximum error if $\bar{x}$ is used to estimate μ and interpret your result.

9. Each member of a random sample of 60 college students was asked how many hours he or she studies per week. If $\bar{x} = 18$ and $s = 2$, and $\bar{x}$ is to be used to estimate μ, find the maximum error so that the probability that the error of estimate is at most E is .95.

10. If a random sample of 50 cups of cola dispensed from an automatic vending machine showed that the mean amount of cola dispensed was 7.9 ounces with $s = .35$ ounce, find the confidence level if $E = .06$.

11. The drying times (in hours) for newly painted parts are normally distributed. Using the following sample data, find the maximum error of estimate E so that we can be 95% confident that the mean drying time of all newly painted parts, μ, is within $\pm E$ of $\bar{x}$.

| 6.4 | 7.2 | 5.9 | 6.8 | 7.1 | 5.5 |

12. A car manufacturer tested 100 cars to determine the mileage traveled before a motor overhaul was needed. He obtained $\bar{x} = 81,250$ miles and $s = 6325$ miles. Find the maximum error of estimate so that we can be 95% confident that μ is within $\pm E$ of $\bar{x}$.

B

13. Using the statement $P(-z_{\alpha/2} < z < z_{\alpha/2}) = 1 - \alpha$ and the fact that $z = (\bar{x} - \mu)/(\sigma/\sqrt{n})$, prove that

$$P\left[|\bar{x} - \mu| < z_{\alpha/2}\left(\frac{\sigma}{\sqrt{n}}\right)\right] = 1 - \alpha$$

and

$$E = z_{\alpha/2}\left(\frac{\sigma}{\sqrt{n}}\right)$$

14. Using the statement $P[-t_{\alpha/2}(df) < t < t_{\alpha/2}(df)] =$ $1 - \alpha$ and the fact that $t = (\bar{x} - \mu)/(s/\sqrt{n})$, prove that

$$P\left[|\bar{x} - \mu| < t_{\alpha/2}(df)\left(\frac{s}{\sqrt{n}}\right)\right] = 1 - \alpha$$

and

$$E = t_{\alpha/2}(df)\left(\frac{s}{\sqrt{n}}\right)$$

15. If μ is within $\pm E$ of $\bar{x}$, show that $\bar{x}$ is within $\pm E$ of μ. That is, if $\bar{x}$ is close to μ, then μ is close to $\bar{x}$.

16. Suppose we wanted to estimate the mean number of children per family. There are at least two methods we could use to arrive at this estimate:

Method 1. Select a random sample of families, count the number of children per family, and determine the average number of children per family.

Method 2. Select a random sample of people, count the number of children each has in his or her family, and determine the average number of children per family.

Do these methods provide unbiased estimates of μ, the average number of children per family? Explain.

17. Which of the two methods listed in Problem 16 would you use to estimate the mean number of passengers per car? Explain.

9.2 CONFIDENCE INTERVALS FOR μ

Suppose we want to estimate a population mean μ by using an interval estimate and that either the population is normal or the sample size n is large ($n \geq 30$). For the present, we shall assume that sampling is done without replacement. If the population is normal, then the sampling distribution of the sample mean is normal; if the sample size is large, then, as a consequence of the central limit theorem, the sampling distribution of the sample mean is approximately normal. In Section 9.1 we saw that if $\bar{x}$ is used to estimate μ, then the maximum error E can be determined so that we can be $(1 - \alpha)100\%$ confident that μ is within $\pm E$ of $\bar{x}$ or that $\bar{x} - E \leq \mu \leq \bar{x} + E$. That is, we can determine the maximum error E so that $P(\bar{x} - E < \mu < \bar{x} + E) = 1 - \alpha$. The interval extending from $\bar{x} - E$ to $\bar{x} + E$ is called a $(1 - \alpha)100\%$ *confidence interval for* μ:

$$\begin{array}{ccc} \uparrow & \uparrow & \uparrow \\ \bar{x} - E & \bar{x} & \bar{x} + E \end{array}$$

The value $\bar{x} - E$ is called the **lower confidence limit** and is denoted by L_1; the value $\bar{x} + E$ is called the **upper confidence limit** and is denoted by L_2.

The confidence interval is written using interval notation as (L_1, L_2). Consequently, the limits for a $(1 - \alpha)100\%$ confidence interval for μ are given by the following three equivalent forms:

$$\bar{x} \pm E \quad \text{or} \quad \bar{x} \pm z_{\alpha/2}\left(\frac{\sigma}{\sqrt{n}}\right) \quad \text{or} \quad \bar{x} \pm z_{\alpha/2}\sigma_{\bar{x}} \tag{9-6}$$

The maximum error E represents the precision when we use $\bar{x}$ to estimate μ. That is, when $\bar{x}$ is used as a measurement of μ, we can be $(1 - \alpha)100\%$ confident that μ is somewhere between $\bar{x} - E$ and $\bar{x} + E$:

$$\underset{\substack{\uparrow \\ Measurement}}{\bar{x}} \quad \pm \quad \underset{\substack{\uparrow \\ Precision}}{E}$$

EXAMPLE 9-10   Type B light bulbs supplied by a firm to a theater for use in its display signs have useful lives that have a standard deviation of 35 hours. If a random sample of 45 type B bulbs has an average life of 750 hours, find a 95% confidence interval for the population mean life of all type B light bulbs manufactured by the firm.

Solution   We use (9-6). Since $1 - \alpha = .95$, $\alpha = .05$ and $\alpha/2 = .025$. The positive critical value is $z_{.025} = 1.96$, and the maximum error E is found by using (9-4):

$$E = z_{\alpha/2}\left(\frac{\sigma}{\sqrt{n}}\right) = (1.96)\left(\frac{35}{\sqrt{45}}\right) = 10.23$$

From (9-6), the confidence interval limits are

$$\bar{x} \pm E$$
$$750 \pm 10.23$$

The 95% confidence interval for μ is $(750 - 10.23, 750 + 10.23)$ or $(739.77, 760.23)$. The width of this confidence interval is found by subtracting the lower confidence limit from the upper confidence limit:

$$L_2 - L_1 = 760.23 - 739.77 = 20.46 \qquad \blacksquare$$

It is important to realize that the confidence interval specified by (9-6) has L_1 and L_2 as variables. Their values depend on the values for $\bar{x}$ and α once the sample size n is determined. Once L_1 and L_2 are fixed, the confidence interval (L_1, L_2) either contains μ, with a probability of 1, or it does not, with a probability of 0. That is, if μ is contained in (L_1, L_2), then the probability that the interval contains μ is 1, and if μ is not contained in the interval (L_1, L_2), then the probability that the interval contains μ is 0.

In Example 9-10, we obtained $(739.77, 760.23)$ as a 95% confidence interval for μ. How should we interpret this interval? It either contains μ or it does not. The statement that "the probability that the interval $(739.77, 760.23)$ contains μ is 95%" is *false*. Since the values of L_1 and L_2 depend on the sample mean

$\bar{x}$ once α and n are specified, we can think of the variable confidence interval (L_1, L_2) as generating a collection of confidence intervals. Every confidence interval is centered at a sample mean $\bar{x}$ and has a width w equal to

$$
\begin{aligned}
w &= L_2 - L_1 \\
&= (\bar{x} + E) - (\bar{x} - E) \\
&= 2E \\
&= 2z_{\alpha/2}\left(\frac{\sigma}{\sqrt{n}}\right)
\end{aligned}
$$

which is a fixed value. Some of the intervals generated in this manner contain μ and some do not. The value $1 - \alpha$ is the proportion of the intervals that contain μ, while α is the proportion of the intervals that do not contain μ. For Example 9-10, since 95% of the confidence intervals constructed with a fixed width of 20.46 contain μ, we can say we are 95% confident that our interval (739.77, 760.23) contains μ. A good point to remember is

> Being 95% confident that a confidence interval contains μ cannot be interpreted to mean that the probability the interval contains μ is 95%.

Table 9-1 contains twenty 95% confidence intervals for μ. Twenty random samples of size 50 were drawn from a normal population with $\mu = 100$ and $\sigma = 10$. For each sample, a 95% confidence interval was constructed using (9-6). Note that 19 intervals contain μ. We would expect 95% of *all* such intervals

Table 9-1 95% Confidence Intervals for μ for $n = 50$, $\mu = 100$, $\sigma = 10$, and Sampling from a Normal Population

Sample	L_1	L_2	Contains μ
1	97.61	103.15	Yes
2	96.60	102.14	Yes
3	94.34	99.88	No
4	99.18	104.72	Yes
5	95.17	100.71	Yes
6	98.64	104.18	Yes
7	98.89	104.43	Yes
8	98.91	104.45	Yes
9	99.12	104.66	Yes
10	94.81	100.35	Yes
11	95.77	101.31	Yes
12	97.09	102.63	Yes
13	95.40	100.94	Yes
14	95.60	101.14	Yes
15	99.62	105.16	Yes
16	99.47	105.01	Yes
17	97.47	103.01	Yes
18	96.08	101.62	Yes
19	97.00	102.54	Yes
20	98.17	103.71	Yes

constructed to contain μ. Of course, if we selected another 20 samples and computed the 95% confidence intervals for the mean, there is no guarantee that exactly 19 of them would contain μ.

Sampling from Small Populations without Replacement

Sometimes it is desirable to construct a confidence interval for μ when the population is not large and sampling is done without replacement. Recall from Chapter 8 that whenever the population size is less than 20 times the sample size ($N < 20n$), then the standard error of the mean $\sigma_{\bar{x}}$ is found by using the following formula:

$$\sigma_{\bar{x}} = \left(\frac{\sigma}{\sqrt{n}}\right)\sqrt{\frac{N-n}{N-1}} \tag{9-7}$$

The maximum error of estimate E is then

$$E = z_{\alpha/2}\sigma_{\bar{x}}$$

And the confidence interval limits are then found by using (9-6):

$$\bar{x} \pm z_{\alpha/2}\sigma_{\bar{x}}$$

The following example illustrates the process.

EXAMPLE 9-11   A sample of 45 students is selected from a population of 800 students and given a test to determine their reaction times to respond to a given stimulus. If the mean reaction time is determined to be $\bar{x} = .75$ second and the standard deviation is $s = .15$ second, find a 95% confidence interval for μ, the mean reaction time for the population of 800 students.

Solution   Since $20n = (20)(45) = 900 > N = 800$, we use (9-7) to find $\sigma_{\bar{x}}$. Also, since $n \geq 30$ and σ is unknown, we can use s to estimate σ. Thus,

$$\sigma_{\bar{x}} = \left(\frac{\sigma}{\sqrt{n}}\right)\sqrt{\frac{N-n}{N-1}}$$

$$= \left(\frac{.15}{\sqrt{45}}\right)\sqrt{\frac{800-45}{800-1}} = .022$$

The positive critical value is $z_{.025} = 1.96$, and the limits for the 95% confidence interval are given by

$$\bar{x} \pm z_{\alpha/2}\sigma_{\bar{x}}$$
$$.75 \pm (1.96)(.022)$$
$$.75 \pm .04$$

Thus, the 95% confidence interval is

$$(.75 - .04, .75 + .04)$$
$$(.71, .79)$$

We can be 95% confident that the interval $(.71, .79)$ contains μ; 95% of all such intervals constructed contain μ and 5% of them do not.   ■

Confidence Intervals Using Small Samples

If we want to construct a $(1 - \alpha)100\%$ confidence interval for μ using a sample of size $n < 30$ selected from a normal population with an unknown standard deviation σ, the maximum error E is determined by using a t distribution. Recall from Section 9.1 that the maximum error of estimate E is given by $E = t_{\alpha/2}(\text{df})(s/\sqrt{n})$. As a result, the limits for a $(1 - \alpha)100\%$ confidence interval for μ are given by

$$\bar{x} \pm E \quad \text{or} \quad \bar{x} \pm t_{\alpha/2}(\text{df})\left(\frac{s}{\sqrt{n}}\right) \tag{9-8}$$

EXAMPLE 9-12   Five similar bags of potatoes weigh 14.8, 16.1, 15.3, 15.2, and 15.4 pounds. Find a 95% confidence interval for the mean weight of all such bags of potatoes. Assume that the population of weights is approximately normal.

Solution   The sample mean and standard deviation are found to be $\bar{x} = 15.36$ and $s = .47$, respectively. For df $= 5 - 1 = 4$, we find $t_{.025}(4) = 2.776$ from the t table (see the back endpaper). Thus, the maximum error E is given by

$$E = t_{.025}(\text{df})\left(\frac{s}{\sqrt{n}}\right) = (2.776)\left(\frac{.47}{\sqrt{5}}\right) = .58$$

and the limits for the 95% confidence interval are

$$\bar{x} \pm E$$
$$15.36 \pm .58$$

Thus, the required interval is $(15.36 - .58, 15.36 + .58)$, or $(14.78, 15.94)$. With this result we feel 95% confident that the interval contains μ, since 95% of all such intervals constructed contain μ. In addition, we are 95% confident that μ is within $\pm.58$ of $\bar{x} = 15.36$. ∎

If (L_1, L_2) is a $(1 - \alpha)100\%$ confidence interval for μ found by using (9-8), then the width of the interval is given by

$$w = L_2 - L_1$$
$$= (\bar{x} + E) - (\bar{x} - E)$$
$$= 2E$$
$$= 2t_{\alpha/2}(\text{df})\left(\frac{s}{\sqrt{n}}\right)$$

Note that the width of the intervals constructed in repeated sampling is not constant since each is dependent on the value of s, which varies from sample to sample. Recall that this was not the case when we constructed a confidence interval for μ using a random sample drawn from a large population with a known standard deviation; when σ is known, all $(1 - \alpha)100\%$ confidence intervals constructed using a fixed sample size have the same width.

Problem Set 9.2

A

1. A sample of 100 observations is taken from a population with unknown mean μ and standard deviation $\sigma = 4.5$. If the mean of the sample is 28.3, construct confidence intervals for μ for each of the following confidence levels. In addition, for each interval, find the width and compare your results for parts (a)–(d).
   a. 90%   **b.** 95%   **c.** 99%   **d.** 99.7%

2. A mathematical skills test given to 12 randomly selected eighth-grade students showed an average score of 77.8 and a standard deviation of 11.1. Assuming that the population of scores is normally distributed, find a 95% confidence interval for the mean mathematical skills score for eighth-graders.

3. A college student measured the heights of 35 male colleagues in her statistics class. Assuming this sample to be representative of all males attending college, find a 95% confidence interval for the true average height of all male college students, if the sample yielded a mean of $\bar{x} = 70.4$ inches and a standard deviation of $s = 2.45$ inches.

4. A study of the health records of a large group of deceased males who smoked at least one pack of cigarettes daily over a 5-year period was conducted to determine the mean life span for all such individuals. A random sample of 16 health records for deceased smokers indicated an average life span of 65.7 years and a standard deviation of 3.4 years. Using these statistics, construct a 99% confidence interval for the true average life span μ for the population of male smokers who smoke at least one pack of cigarettes daily over a 5-year period. Assume that the life spans for such males are normally distributed.

5. The management of a large national chain of 20 motels, each with 100 rooms, decided to estimate the mean cost per room of repairing damages made by its customers. A random sample of 150 vacated rooms was inspected by the management and indicated a mean repair cost of $\bar{x} = \$28.10$ and a sample standard deviation of $s = \$12.40$. Construct a 95% confidence interval for the mean repair cost μ for the 2000 motel rooms.

6. Repeat Problem 5 if the national chain consists of 50 motels, each with 100 rooms. Compare the width of this interval to the width of the interval obtained in Problem 5.

7. Assume that the useful lifetimes (in minutes) of a certain brand of type D batteries are normally distributed. A random sample of seven batteries was tested to determine how long the batteries would adequately function in a certain electronic device. The sample yielded a mean of $\bar{x} = 152$ minutes and a standard deviation of $s = 5$ minutes. Construct a 95% confidence interval for the true mean lifetime of all type D batteries operating the electronic device.

8. Assuming that the useful lifetimes (in minutes) of type C batteries are normally distributed, use the following sample of battery lives to construct a 90% confidence interval for the true mean lifetime of all type C batteries:

   $$150 \quad 162 \quad 178 \quad 158 \quad 162 \quad 171$$

9. **a.** For df = 11, find a value for c such that
   $$P(-c < t < c) = .90.$$
   b. Find a value for α such that
   $$P(-z_\alpha < z < z_\alpha) = .40.$$
   c. Find $P(-z_{.05} < z < z_{.10})$.

10. A health-care organization conducted a study to estimate the average number of days a surgical patient is hospitalized. A random sample of 150 surgical patients yielded an average stay of 4.8 days and a standard deviation of 2.6 days. Find a 90% confidence interval for the mean number of days a surgical patient remains in the hospital.

11. Repeated assessments on a chemical determination of human blood during a laboratory analysis are known to be normally distributed. A sample of ten assessments on a given sample of blood yielded the values 1.002, .958, 1.014, 1.009, 1.041, .962, 1.058, 1.024, 1.019, and 1.020. Find a 99% confidence interval for a true chemical determination in the blood for repeated assessments on the sample.

12. A special type of hybrid corn was planted on eight different plots. The plots produced yield values (in bushels) of 140, 70, 39, 110, 134, 104, 100, and 125. Assuming the yields follow a normal distribution, find a 95% confidence interval for the true average yield of this type of hybrid corn.

B

13. Refer to Problem 3. If $L_1 = \bar{x} - z_{.02}(s/\sqrt{n})$ and $L_2 = \bar{x} + z_{.03}(s/\sqrt{n})$, is (L_1, L_2) a 95% confidence

interval for μ? Why? How does the width of this interval compare to the width of the interval found in Problem 3?

14. Prove that the inequalities $\bar{x} - E < \mu < \bar{x} + E$ are equivalent to the inequality $|\bar{x} - \mu| < E$.

15. If $P(-z_{\alpha/2} < z < z_{\alpha/2}) = 1 - \alpha$ and $z = (\bar{x} - \mu)/(\sigma/\sqrt{n})$, prove that

$$P\left[\bar{x} - z_{\alpha/2}\left(\frac{\sigma}{\sqrt{n}}\right) < \mu < \bar{x} + z_{\alpha/2}\left(\frac{\sigma}{\sqrt{n}}\right)\right] = 1 - \alpha$$

16. If $P(-t_{\alpha/2}(df) < t < t_{\alpha/2}(df)) = 1 - \alpha$ and $t = (\bar{x} - \mu)/(s/\sqrt{n})$, prove that

$$P\left[\bar{x} - t_{\alpha/2}(df)\left(\frac{s}{\sqrt{n}}\right) < \mu < \bar{x} + t_{\alpha/2}(df)\left(\frac{s}{\sqrt{n}}\right)\right]$$
$$= 1 - \alpha$$

17. A one-sided $(1 - \alpha)100\%$ confidence interval for μ is an interval with an area equal to α located either in the left tail or the right tail of the distribution,

but not both. A $(1 - \alpha)100\%$ one-sided confidence interval for μ has one of the following two forms:

$$\mu > \bar{x} - z_\alpha\left(\frac{\sigma}{\sqrt{n}}\right) \quad \text{or} \quad \mu < \bar{x} + z_\alpha\left(\frac{\sigma}{\sqrt{n}}\right)$$

A random sample of thirty statistics scores was chosen from a large group of scores to determine the mean μ for the entire group. From previous tests, σ was determined to be 13. If the sample mean was $\bar{x} = 72$, construct an appropriate one-sided 95% confidence interval for μ.

18. Based on a random sample of 80 Jersey cows, the confidence interval (38.6, 42.3) was constructed for estimating the true average yield (in ounces) per milking for Jersey cows. If the amounts of milk from Jersey cows per milking are normally distributed and $\sigma = 6$, determine the level of confidence for the interval. Also, determine the mean of the sample used to construct the interval.

9.3 ESTIMATES OF POPULATION PROPORTIONS ───────────

In Section 9.1 we learned that the sample proportion $\hat{p}$ can be used to estimate the population proportion p. For example, in 1984 a Baltimore newspaper, *The Sun*, was interested in determining, among other things, the proportion of Maryland residents who smoke. It hired a Baltimore agency to conduct a survey of Maryland adults. The survey revealed a total of 250 smokers from a random sample of 806 adults (see *The Sun*, Nov. 11, 1984). By using this information, the polling agency determined the proportion of smokers in the sample to be .31. The estimate $\hat{p} = .31$ serves as a point estimate for the true proportion of all Maryland adults who smoke. The maximum error of estimate was reported by *The Sun* to be no more than 3.5% with a confidence level of 95%. This means that if $\hat{p} = .31$ is used to estimate p, the true proportion of Maryland residents who smoke, then *The Sun* is 95% confident that p is within $\pm .035$ of $\hat{p} = .31$.

Given a certain level of confidence, we would like to determine the **maximum error of estimate** E when the sample proportion $\hat{p}$ is used as a point estimate for the population proportion p. Toward this goal, recall the following facts from Section 8.5 concerning the sampling distribution of $\hat{p}$:

1. When the sample size n is large, the sampling distribution of $\hat{p}$ is approximately normal.
2. The mean of the sampling distribution of $\hat{p}$ is $\mu_{\hat{p}} = p$.
3. The standard error of $\hat{p}$ is $\sigma_{\hat{p}} = \sqrt{p(1 - p)/n}$.

For those applications where p is to be estimated, the standard error of $\hat{p}$ will be unknown. Fortunately, it can be estimated by using the sample proportion $\hat{p}$. The estimate $\hat{\sigma}_{\hat{p}}$ of $\sigma_{\hat{p}}$ is given by

$$\hat{\sigma}_{\hat{p}} = \sqrt{\frac{\hat{p}(1 - \hat{p})}{n}}$$

For *The Sun*'s poll mentioned above, the standard error of $\hat{p}$ can be estimated as

$$\hat{\sigma}_{\hat{p}} = \sqrt{\frac{\hat{p}(1 - \hat{p})}{n}} = \sqrt{\frac{(.31)(.69)}{806}} = .0163$$

The sampling distribution of the statistic $(\hat{p} - \mu_{\hat{p}})/\hat{\sigma}_{\hat{p}}$ is approximately the standard normal. Thus, we have

$$z \simeq \frac{(\hat{p} - p)}{\sqrt{p(1 - p)/n}} \tag{9-9}$$

Toward finding the maximum error of estimate E, we begin with the following probability statement:

$$P(-z_{\alpha/2} < z < z_{\alpha/2}) = 1 - \alpha$$

By substituting the z value from (9-9) into this probability statement and using some simple algebra, we can obtain the following probability statement:

$$P\left(\underset{\underset{\textit{Error of estimate, E}}{\uparrow}}{|\hat{p} - p|} < z_{\alpha/2}\sqrt{\frac{\hat{p}(1 - \hat{p})}{n}}\right) \simeq \underset{\underset{\textit{Confidence level}}{\uparrow}}{1 - \alpha} \tag{9-10}$$

Therefore, the maximum error of estimate E can be approximated by

$$E = z_{\alpha/2}\sqrt{\frac{\hat{p}(1 - \hat{p})}{n}} = z_{\alpha/2}\hat{\sigma}_{\hat{p}}$$

Thus, if $\hat{p}$ is used to estimate p with a confidence level of $1 - \alpha$, the maximum error of estimate E can be approximated by using

$$E \simeq z_{\alpha/2}\sqrt{\frac{\hat{p}(1 - \hat{p})}{n}} = z_{\alpha/2}\hat{\sigma}_{\hat{p}} \tag{9-11}$$

For *The Sun*'s poll of Maryland adults, if $\hat{p} = .31$ is used to estimate the true proportion of Maryland adults who smoke, *The Sun* can be 95% confident that the error of estimate is approximately

$$E \simeq z_{\alpha/2}\hat{\sigma}_{\hat{p}} = (1.96)(.0163) = .032$$

Hence, *The Sun* can be 95% confident that p is within approximately $\pm 3.2\%$ of $\hat{p} = .31$. Notice that our maximum error of estimate differs from *The Sun*'s by .3% or .003.

EXAMPLE 9-13   In a sample of 400 type B batteries manufactured by Everlast Co., 20 defective batteries were found. If the proportion $\hat{p}$ of defective batteries in the sample is used to estimate p, the true proportion of all defective type B batteries manufactured by Everlast Co., find the maximum error of estimate E so that one can be 95% confident that p is within $\pm E$ of $\hat{p}$.

Solution   Since $1 - \alpha = .95$, $\alpha = .05$ and $\alpha/2 = .025$. Hence, $z_{\alpha/2} = 1.96$. Since $\hat{p} = \frac{x}{n} = \frac{20}{400} = .05$, an estimate for $\sigma_{\hat{p}}$ is given by

$$\hat{\sigma}_{\hat{p}} = \sqrt{\frac{\hat{p}(1 - \hat{p})}{n}} = \sqrt{\frac{(.05)(.95)}{(400)}} = .0109$$

Hence, by using (9-11), we have

$$E \simeq z_{\alpha/2}\sqrt{\frac{\hat{p}(1 - \hat{p})}{n}} = (1.96)(.0109) = .021$$

If $\hat{p} = .05$ is used to estimate p, we can be 95% confident that p is within approximately $\pm.021$ of $\hat{p}$. In other words, if $\hat{p} = .05$ is used to estimate p, the maximum error of estimate will be approximately .021 with a confidence level of 95%.   ■

If $\hat{p}$ is used to estimate p and we are 95% confident that p is within approximately $\pm E$ of $\hat{p}$, then the expression $\hat{p} \pm E$ can be used to establish an approximate 95% confidence interval for p. For Example 9-13, an approximate 95% confidence interval for the true proportion of defective type B batteries is found by using the limits

$$\hat{p} \pm E$$
$$.05 \pm .021$$

Thus, an approximate 95% confidence interval for p, the true proportion of defective type B batteries, is

$$(.029, .071)$$

In summary, the limits for an approximate $(1 - \alpha)100\%$ confidence interval for the population proportion p are given by the following three equivalent expressions:

$$\hat{p} \pm E \quad \text{or} \quad \hat{p} \pm z_{\alpha/2}\sqrt{\frac{\hat{p}(1 - \hat{p})}{n}} \quad \text{or} \quad \hat{p} \pm z_{\alpha/2}\hat{\sigma}_{\hat{p}} \tag{9-12}$$

EXAMPLE 9-14   In a study of 300 automobile accidents in a particular city, 60 resulted in fatalities. Based on this sample, construct an approximate 90% confidence interval for the proportion of all auto accidents in the city that result in fatalities.

Solution   We first note that the best point estimate of p is $\hat{p} = \frac{60}{300} = .2$. The critical value $z_{.05}$ is found in the standard normal table (z table; see the front endpaper) to be 1.65. Since an approximation for the standard error of $\hat{p}$ can be found by using

$$\hat{\sigma}_{\hat{p}} = \sqrt{\frac{\hat{p}(1 - \hat{p})}{n}}$$

we have

$$\hat{\sigma}_{\hat{p}} = \sqrt{\frac{(.2)(.8)}{300}} = .0231$$

Approximate confidence interval limits are found from (9-12) as follows:

$$\hat{p} \pm z_{\alpha/2}\hat{\sigma}_{\hat{p}}$$
$$.20 \pm (1.65)(.0231)$$
$$.20 \pm .038$$

Thus, an approximate 90% confidence interval for p is (.162, .238). In addition, using $\hat{p} = .2$ to estimate p, we can be 90% confident that our error of estimate is approximately $E = .038$.   ■

When sampling without replacement from a small population (relative to the sample size) and estimating p, we can approximate the standard error of $\sigma_{\hat{p}}$ by using the following formula:

$$\hat{\sigma}_{\hat{p}} = \sqrt{\frac{\hat{p}(1 - \hat{p})}{n}} \sqrt{\frac{N - n}{N - 1}} \tag{9-13}$$

Note that formula (9-13) is used whenever $N < 20n$, as illustrated in the following example.

EXAMPLE 9-15   A political candidate is planning his campaign strategy and would like to determine the extent to which he is known. In a random sample of 3000 of the county's 25,000 registered voters, 1800 indicated that they recognized the candidate's name. Construct a 95% confidence interval for the true proportion of voters in the county who are familiar with the candidate.

Solution   The critical value is $z_{.025} = 1.96$. Since $25,000 < (20)(3000)$, formula (9-13) should be used to compute $\hat{\sigma}_{\hat{p}}$ (note that $\hat{p} = 1800/3000 = .6$):

$$\hat{\sigma}_{\hat{p}} = \sqrt{\frac{(.6)(.4)}{3000}} \sqrt{\frac{25,000 - 3000}{25,000 - 1}}$$
$$= .00839$$

Hence, by using (9-12), we obtain the approximate confidence limits:

$$\hat{p} \pm z_{\alpha/2} \sqrt{\frac{\hat{p}(1 - \hat{p})}{n}}$$
$$.6 \pm (1.96)(.008)$$
$$.6 \pm .02$$

Thus, an approximate 95% confidence interval for p is (.58, .62). We can be approximately 95% confident that the interval (.58, .62) contains p, since approximately 95% of all such intervals constructed contain p.   ■

Problem Set 9.3

A

1. Construct the appropriate confidence intervals for p using the given information:
   a. $n = 400$, $x = 100$, $\alpha = .05$
   b. $n = 500$, $x = 125$, $1 - \alpha = .90$
   c. $n = 1500$, $x = 900$, $1 - \alpha = .80$
   d. $n = 250$, $N = 2000$, $x = 100$, $1 - \alpha = .95$

2. A survey of 672 audited tax returns showed that 448 resulted in additional payments. Construct an approximate 95% confidence interval for the true proportion of all audited tax returns that result in additional payments to the Internal Revenue Service. Also, find E so that we can be 95% confident that p is within approximately $\pm E$ of $\hat{p}$.

3. A coin was tossed 500 times and 255 heads appeared.
   a. Find E so that we can be 90% sure that p is within approximately $\pm E$ of $\hat{p}$.
   b. Construct an approximate 90% confidence interval for the probability of getting a head on one toss and interpret your result.

4. An intrauterine device used by 500 women for the purpose of preventing pregnancy failed in 20 women. Construct an approximate 99% confidence interval for the true proportion of failures for this device and interpret your result.

5. If the device in Problem 4 is experimental and used by only 4000 women, construct an approximate 95% confidence interval for the true proportion of failures out of 4000 and interpret your result. Compare this interval to the one obtained in Problem 4.

6. Suppose a random sample of 25 mining caps was tested from a population of 300 mining caps and 20 exploded properly. Construct an approximate 95% confidence interval for the proportion of mining caps that will explode properly and interpret your result.

7. When sprayed with a certain type of insecticide, 38 of 60 Japanese beetles died. Construct an approximate 99% confidence interval for the true proportion of beetles that will die when exposed to the insecticide.

8. In a random poll taken in a large city, 428 of 975 people indicated they drink at least one cup of coffee a day. Construct an approximate 80% confidence interval for the true proportion of people who drink at least one cup of coffee per day.

9. A sample of voters was polled to determine the support for candidate A. Out of 140 voters surveyed, 74 expressed plans to vote for candidate A during the election. Construct an approximate 90% confidence interval for the proportion of the voters in the population who will vote for candidate A.

10. In a city of size 25,000, a random sample of 700 voters revealed that 420 opposed the reelection of their mayor. Construct an approximate 92% confidence interval for the proportion of all city voters who oppose the mayor's reelection.

11. In a sample of 350 students on a university campus who were questioned regarding their participation in athletics, 161 said they participate. Construct an approximate 96% confidence interval for the true proportion of students who participate in athletics.

12. Of 125 people who were given a vaccine for type A flu virus, 100 developed immunity to the virus. Construct an approximate 95% confidence interval for the true proportion of those vaccinated who develop immunity.

13. A sample of 1168 rabbis revealed that 315 experience their work as very stressful. Construct an approximate 99% confidence interval for the true proportion of rabbis who find their work very stressful.

14. In a random sample of 249 Maryland smokers, it was found that 87% believe smoking is harmful to them. Construct an approximate 92% confidence interval for the true proportion of Maryland smokers who believe smoking is harmful to them.

15. In a random sample of 249 Maryland smokers, it was found that 147 believe smoking is hazardous to nonsmokers. Construct an approximate 96% confidence interval for the true proportion of Maryland

smokers who believe that smoking is hazardous to nonsmokers.

16. On a certain day at the New York Stock Exchange, 2500 different stocks were traded. Of a random sample of 100 stocks, 70 declined in price. Construct an approximate 99% confidence interval for the proportion of all stocks that declined.

17. Researchers at The Centers for Disease Control (CDC) believe that the flu vaccine does not work as well for nursing-home residents as it does for younger, healthier people. The flu vaccine is usually 50%–80% effective in warding off disease in the general population. Reasons for the lower effectiveness in the nursing-home population include a decline with age in the body's immunity levels and the close contact of residents in a nursing home. In an effort to estimate the effectiveness of the flu vaccine in nursing-home residents, CDC researchers conducted a survey of 1068 nursing-home residents who received the flu vaccine and found that 269 came down with flu or flu-like illness. Construct a 95% confidence interval for the effectiveness of the flu vaccine in nursing-home residents.

18. An unconventional new treatment for advanced liver cancer, a generally fatal disease, shows promise in shrinking tumors and has even produced a handful of apparent cures. Dr. Stanley Order of Johns Hopkins Hospital has indicated that the treatment, which involves injections of antibodies carrying radioactive isotopes, has significantly shrunk inoperable tumors in 50 of 104 patients (*The Cumberland News*, Aug. 15, 1985). Construct a 99% confidence interval for the true proportion of inoperable tumors that are significantly shrunk by Dr. Order's new treatment.

19. As one of the criteria for job employment as a teacher in Baltimore City, newly-hired teachers must take a writing test to determine a teacher's ability to spell, punctuate, and construct a sentence. Of 158 newly-hired teachers in 1985, 32 were unable to score a passing grade on the test (*The Cumberland News*, Sept. 13, 1985). Construct a 95% confidence interval for the true proportion of all newly-hired teachers who are unable to pass the test.

B

20. By using the statement $P(-z_{\alpha/2} < z < z_{\alpha/2}) = 1 - \alpha$ and the fact that $z = (\hat{p} - p)/\sigma_{\hat{p}}$, prove that

$$P(\hat{p} - z_{\alpha/2}\sigma_{\hat{p}} < p < \hat{p} + z_{\alpha/2}\sigma_{\hat{p}}) = 1 - \alpha$$

21. Since $P(|\hat{p} - p|/\sigma_{\hat{p}} < z_{\alpha/2}) = 1 - \alpha$, the inequality $|\hat{p} - p| < z_{\alpha/2}\sigma_{\hat{p}}$ can be solved for p to provide a $(1 - \alpha)100\%$ confidence interval for p. Let $f(p) = |\hat{p} - p| - z_{\alpha/2}\sigma_{\hat{p}}$ and let z represent $z_{\alpha/2}$.
  a. Show that $f(p) = (n + z^2)p^2 - (2n\hat{p} + z^2)p + n\hat{p}^2$.
  b. Show that the zeros p_0 of the quadratic function $f(p)$ are given by

$$p_0 = \frac{2n\hat{p} + z^2 \pm z\sqrt{4n\hat{p}(1 - \hat{p}) + z^2}}{2n + 2z^2}$$

  c. Show that the two zeros found in part (b) determine a $(1 - \alpha)100\%$ confidence interval for p.
  d. Show that if n is large, the interval obtained in part (c) is approximately equal to the interval determined by (9-12). [*Hint:* Divide the numerator and denominator of the expression in part (b) by $2n$ and observe the limiting values for p_0 as n gets large.]

22. Using the results in Problem 21, construct a 95% confidence interval for the true proportion of all defective type B batteries in Example 9-13 and compare your result with that obtained in Example 9-13.

9.4 DETERMINING SAMPLING SIZES FOR ESTIMATES _____

Population Mean

How large should a sample be if the sample mean is to be used to estimate the population mean? The answer depends on the standard error of the mean. If the standard error of the mean is 0, then only one measurement would be needed; this measurement must necessarily equal the unknown mean μ, since $\sigma = 0$. This extreme case is not encountered in practice, but it supports the fact that the smaller the standard error of the mean, the smaller the sample size needed to achieve a desired degree of accuracy.

It was stated earlier that one way to decrease the error of estimate is to increase the size of the sample. If the sample included the whole population, then $\bar{x}$ would equal μ and the error of estimate $|\bar{x} - \mu|$ would equal 0. With this in mind, it seems reasonable that for a fixed confidence level, it should be possible to determine a sample size so that the error of estimate is as small as we want. To be more precise, given a fixed confidence level $(1 - \alpha)$ and a fixed error of estimate E, we can choose a sample size n so that $P(|\bar{x} - \mu| < E) = 1 - \alpha$. Toward the task of determining n, recall from Section 9.1 that the maximum error of estimate is given by

$$E = z_{\alpha/2} \left(\frac{\sigma}{\sqrt{n}} \right)$$

If we square both sides of this equation and solve the resulting equation for n, we have

$$n = \left(\frac{z_{\alpha/2}\sigma}{E} \right)^2 \tag{9-14}$$

Since n must be a whole number, we round up all fractional parts. For example, if formula (9-14) produced a value of $n = 89.13$, we would round up to $n = 90$. If instead of rounding this value up to 90, we used $n = 89$, we could not be $(1 - \alpha)100\%$ confident that μ is within $\pm E$ of $\bar{x}$. With the value $n = 90$, we can be *at least* $(1 - \alpha)100\%$ confident that μ is within $\pm E$ of $\bar{x}$.

Formula (9-14) requires that the population standard deviation σ be known. Since this is seldom the case, a preliminary sampling must be conducted with $n \geq 30$ in order to obtain an estimate s for σ. The following example illustrates the use of formula (9-14).

EXAMPLE 9-16   A biologist wants to estimate the mean weight of deer killed in the state of Maryland. A preliminary study of ten deer killed showed the standard deviation of deer weights to be 12.2 pounds. How large a sample should be taken so that the biologist can be 95% confident that the error of estimate is at most 4 pounds?

Solution   By using formula (9-14) we have

$$n = \left(\frac{z_{\alpha/2}\sigma}{E} \right)^2$$
$$= \left[\frac{(1.96)(12.2)}{4} \right]^2$$
$$= 35.736$$

Thus, if the size of the sample is $n = 36$, we can be 95% confident that μ is within ± 4 of $\bar{x}$.   ∎

Population Proportions

We next want to determine the sample size n that is necessary in order to be $(1 - \alpha)100\%$ confident that the error of estimate is at most E when $\hat{p}$ is used to estimate p. Stated more precisely, given the maximum error E that can be tolerated and a fixed confidence level $1 - \alpha$, we want to find the sample size n so that $P(|\hat{p} - p| < E) = 1 - \alpha$. Recall from Section 9.3 that the maximum error of estimate E is given by

$$E = z_{\alpha/2} \sqrt{\frac{p(1-p)}{n}}$$

Squaring both sides and solving for n, we have

$$n = \left(\frac{z_{\alpha/2}}{E}\right)^2 p(1-p)$$

Since we are trying to estimate p, which is unknown, we need a preliminary estimate $\hat{p}$ for p in order to use the above formula. If a preliminary estimate $\hat{p}$ for p is available, we can use the following formula to determine the sample size n:

$$n = \left(\frac{z_{\alpha/2}}{E}\right)^2 \hat{p}(1-\hat{p}) \tag{9-15}$$

Again, we round up the value for n obtained by using (9-15) to the next larger integer value.

    If a preliminary estimate $\hat{p}$ for p is not available, we could substitute the maximum possible value of $p(1-p)$ in formula (9-15) to obtain the sample size n. In most cases this strategy would probably make the sample size unnecessarily large for a given level of confidence. The value of n would then be a conservative estimate in the sense that it is larger than what it needs to be; in this case, we can say that we are at least $(1 - \alpha)100\%$ confident that p is within $\pm E$ of $\hat{p}$. The maximum value of $p(1-p)$ is easily found to be $\frac{1}{4}$ (see Problem 15 at the end of this section). By using this maximum value, we obtain the following sample size:

$$n = \left(\frac{z_{\alpha/2}}{2E}\right)^2 \tag{9-16}$$

    For *The Sun*'s poll to determine the proportion of adult smokers in Maryland (discussed in Section 9.3), we can use (9-16) to determine the size of sample to be polled in order to be 95% confident that the error of estimate is no more than 3.5 percentage points. The size of the random sample to be polled is

$$n = \left(\frac{z_{\alpha/2}}{2E}\right)^2 = \left(\frac{1.96}{.07}\right)^2 = 784$$

Since *The Sun*'s poll involved 806 adults, which is 22 more than the required sample size we calculated, we can be at least 95% confident that the true proportion of adult smokers in Maryland is within $\pm.035$ of $\hat{p} = .31$.

EXAMPLE 9-17   A large teachers' organization wants to estimate the percentage of its membership who are in favor of collective bargaining. The union wants to be certain that the error of estimate is at most 1.5% with a confidence level of 95%.

a. If no preliminary estimate for p is available, how large a sample must be polled?

b. If a preliminary sample of 200 teachers indicated that 65% favored collective bargaining, how many more teachers should be polled?

Solution   **a.** When no preliminary estimate of p is available, we use formula (9-16). Thus,

$$n = \left(\frac{z_{\alpha/2}}{2E}\right)^2$$

$$= \left[\frac{1.96}{(2)(.015)}\right]^2$$

$$= 4268.44$$

$$\simeq 4269$$

b. In this case, since we have a preliminary estimate, $\hat{p} = .65$, we can use (9-15). Hence, the sample size is

$$n = \left(\frac{z_{\alpha/2}}{E}\right)^2 \hat{p}(1 - \hat{p})$$

$$= \left(\frac{1.96}{.015}\right)^2 (.65)(.35)$$

$$= 3884.28$$

$$\simeq 3885$$

Thus, we need to poll $3885 - 200 = 3685$ more teachers in order to achieve the desired accuracy. Note that by having a preliminary estimate for p, we can sample 384 fewer teachers than required for the same nominal level of confidence when no preliminary estimate for p is available. ∎

Problem Set 9.4

A

1. A biometrician wants to estimate the percentage of oxygen in the blood of all newborn babies, and prior research has suggested that this value is $p = 68\%$. If it is desired to be off by no more than 2 percentage points with a confidence level of 95%, find the sample size needed. If no prior estimate of p is available, how large should the sample be?

2. It is desired to determine the mean weight of a certain type of fish. If it is also desired to estimate μ to within .05 gram with a confidence level of .80, how large must the sample be if σ is approximately 2.5?

3. If it is desired in Problem 2 to estimate μ so that the width of the confidence interval is no more than .12 gram with a confidence level of 95%, how large must the sample be?

4. In an area where water is not fluoridated, a random sample of citizens is to be surveyed to estimate the percentage of citizens who favor having the water fluoridated. How large a sample is needed to be 99% confident that the true percentage is within 1 percentage point of the sample percentage?

5. Answer Problem 4 if it was previously estimated that 85% of the residents favor fluoridation.

6. The amount of time that a doctor spends with patients has a standard deviation of approximately 7.8 minutes. If it is desired to estimate the mean time doctors spend with their patients, find the sample size that is needed to be 95% confident that the true mean is within 2.5 minutes of the estimate.

7. A student group wants to estimate the percentage of students living in dormitories who are dissatisfied with cafeteria food. Determine the number to be surveyed if the student group wants to be 95% confident that the true proportion p is within 1 percentage point of the estimate $\hat{p}$.

8. A seed company wants to estimate the proportion of its cucumber seeds that will germinate. How large a sample of cucumber seeds should be used if the seed company wants to be 94% confident that the error of estimate is at most .05?

9. A random sample of 50 lawyers indicated that the average length of law experience is 9.6 years and the standard deviation is 4.4 years. How many more lawyers should be included in the sample to be 99% confident that the true mean work experience is within 6 months of the sample mean work experience?

10. A random sample of 75 medical doctors in Chicago indicated they worked an average of 51.2 hours per week; the standard deviation of the number of hours worked was 6.8. If it is desired to gather another random sample of doctors in Chicago to estimate the true average number of hours worked per week, how large a sample is needed in order to be 95% confident that the true mean is within .5 hour of the sample mean?

11. Doctors at Sinai Hospital in Baltimore have been studying the effects of doses of vitamin E in the treatment of the common ailment of noncancerous breast lumps in females. When treating 26 patients with 600 i.u. of vitamin E daily for 8 weeks, they found that 22 were good or fair responders to the treatment and 4 were nonresponders. How many more patients should be tested in order for the doctors to be 95% confident that the sample proportion of female patients taking 600 i.u. of vitamin E daily for 8 weeks who are good or fair responders to the treatment differs no more than .10 from the true proportion?

B

12. An opinion survey conducted by a college reported that 1067 students were surveyed and that the error factor was 3%. Verify that an error factor of 3% is approximately correct. [*Hint:* Use $\alpha = .05$ and find E.]

13. Suppose sampling is done without replacement from a finite population of size N. Show that in order to be $(1 - \alpha)100\%$ confident that p will differ from $\hat{p}$ by at most E, the sample size should be at least

$$n = \frac{Nz_{\alpha/2}^2}{z_{\alpha/2}^2 + 4(N-1)E^2}$$

when no prior information concerning p is available.

14. A labor organization that has 4000 members wants to conduct a survey to estimate the percentage of its members favoring a strike. What is the minimum number of members that should be included in a sample so that the organization can be 98% confident that the true proportion p will be within 4 percentage points of the sample estimate $\hat{p}$?

15. Show that the maximum value of $p(1 - p)$ is $\frac{1}{4}$.

9.5 *THE CHI-SQUARE DISTRIBUTIONS*

In order to estimate the population variance or standard deviation, we need to be familiar with the χ^2 statistic. If a sample of size n is selected from a normal population with variance σ^2, the statistic

$$\frac{(n-1)s^2}{\sigma^2}$$

has a sampling distribution which is a chi-square distribution with degrees of freedom (df) equal to $(n - 1)$ and is denoted by χ^2 (χ is the lowercase Greek letter chi). The chi-square statistic is given by

$$\chi^2 = \frac{(n-1)s^2}{\sigma^2}$$

where n is the sample size, s^2 is the sample variance, and σ^2 is the variance of the sampled population. Note that since $s^2 = SS/(n-1) = \sum(x - \bar{x})^2/(n-1)$, the chi-square statistic can also be written as either of the following two expressions:

$$\chi^2 = \frac{SS}{\sigma^2} \quad \text{or} \quad \chi^2 = \frac{\sum(x - \bar{x})^2}{\sigma^2}$$

The chi-square distributions have the following properties:

1. The values of χ^2 are greater than or equal to 0.
2. The shapes of the χ^2 distributions depend on df. As a result, there are an infinite number of χ^2 distributions.
3. The area under a chi-square curve is 1.
4. The χ^2 distributions are not symmetric. They have narrow tails extending to the right; that is, they are skewed to the right.
5. When $n > 2$, the mean of a χ^2 distribution is $(n-1)$ and the variance is $2(n-1)$.

Figure 9-6 illustrates three χ^2 distributions. Notice that the modal value for a χ^2 distribution occurs at the value (df $-$ 2).

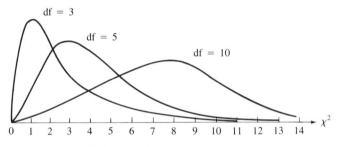

df = 3

df = 5

df = 10

0 1 2 3 4 5 6 7 8 9 10 11 12 13 14 χ^2

FIGURE 9-6 Chi-square distributions

Table 2 in Appendix B gives critical values $\chi_\alpha^2(df)$ for ten selected values of α. The symbol $\chi_\alpha^2(df)$ is used to denote the critical value for a chi-square distribution with df degrees of freedom. This critical value specifies an area of α under the χ^2 curve to the right of it (see Figure 9-7). For example, to find $\chi_{.05}^2(6)$ in Table 4, we locate df $= 6$ at the left side of the table and $\alpha = .05$ across the top of the table. The critical value is found where the row labeled by df $= 6$ intersects the column labeled by $\alpha = .05$:

$$\alpha = .05$$
$$\downarrow$$

$$df = 6 \longrightarrow \quad 12.592 \left[= \chi_{.05}^2(6) \right]$$

Thus, the critical value is $\chi_{.05}^2(6) = 12.592$.

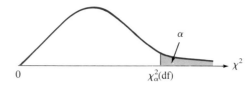

FIGURE 9-7 Critical value of chi-square distribution

EXAMPLE 9-18   Find the critical value $\chi^2_{.10}(13)$ in Table 2 of Appendix B.

Solution   Locate the row corresponding to df = 13 and the column corresponding to $\alpha = .10$. Their intersection in the table determines the critical value; it is found to be $\chi^2_{.10}(13) = 19.812$. We have the following diagram for the χ^2 distribution:

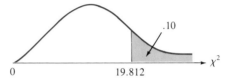

EXAMPLE 9-19   Suppose the times required for a particular bus to arrive at one of its destinations in a large city form a normal distribution with a standard deviation of $\sigma = .5$ minute. If a sample of 17 bus arrival times is selected at random, find the probability that the sample variance is greater than .50; that is, find $P(s^2 > .50)$.

Solution   We find the chi-square value corresponding to $s^2 = .50$ as follows:

$$\chi^2 = \frac{(n-1)s^2}{\sigma^2}$$

$$= \frac{(17-1)(.50)}{(.5)^2} = 32$$

Thus,

$$P(s^2 > .50) = P(\chi^2 > 32)$$

By observing Table 2 and reading across the row labeled by df = 17 − 1 = 16, we locate 32.000 under the .01 column. Hence, the probability value is $P(s^2 > .50) = .01$.   ■

Confidence Intervals for σ^2 and σ

Often it is required to estimate the variance of a given population of measurements. For example, an industry that manufactures nuts and bolts must have close tolerances on all produced parts. A bolt that is 8.4 mm in diameter will not screw into an 8-mm hole. As an example, suppose a bolt manufacturer is producing 8-mm diameter bolts and the bolt diameters are normally distributed. For quality control purposes, a random sample of 25 bolts is obtained from a production line to estimate the variance of all 8-mm bolt diameters. The variance for the sample of bolt diameters is $s^2 = .009$ square mm. The sample variance $s^2 = .009$ provides a point estimate for σ^2, the variance of the diameters of all 8-mm bolts produced.

Toward obtaining a 95% confidence interval for σ^2, consider Figure 9-8. We want to determine two critical values so that the area between them is .95 and the "tail" regions each contain an area of .025. Note that the critical value $\chi^2_{.975}(24)$ has an area of .975 to its right and an area of .025 to its left (.975 + .025 = 1). In general, if $\chi^2_{\alpha/2}(\text{df})$ is the right-tail critical value, then $\chi^2_{1-\alpha/2}(\text{df})$ will be the left-tail critical value. By using Table 2, we find the critical values to be

$$\chi^2_{.975}(24) = 12.401$$
$$\chi^2_{.025}(24) = 39.364$$

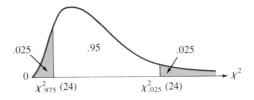

FIGURE 9-8 Critical values for 95% confidence interval for σ^2

We will use these values to construct a 95% confidence interval for σ^2. The confidence interval (L_1, L_2) will be such that

$$P(L_1 < \sigma^2 < L_2) = .95$$

L_1 represents the smallest value that σ^2 can be and L_2 represents the largest value that σ^2 can be. Since the χ^2 statistic is defined by

$$\chi^2 = \frac{(n-1)s^2}{\sigma^2}$$

we can solve this equation for σ^2 in order to obtain the confidence limits L_1 and L_2. By doing so, we have

$$\sigma^2 = \frac{(n-1)s^2}{\chi^2} \tag{9-17}$$

To find the confidence limits L_1 and L_2, we substitute the chi-square critical values for χ^2 in formula (9-17). The limits become

$$\sigma^2 = \frac{(n-1)s^2}{\chi^2}$$
$$= \frac{(25-1)(.009)}{39.364} = .0055$$

and

$$\sigma^2 = \frac{(n-1)s^2}{\chi^2}$$

$$= \frac{(25-1)(.009)}{12.401} = .0174$$

The smaller value obtained by using (9-17) is the lower confidence limit L_1 and the larger value is the upper confidence limit L_2. Thus, $L_1 = .006$, $L_2 = .017$, and a 95% confidence interval for σ^2 is (.006, .017). If this is acceptable to the plant engineer and if at some future date a sample of bolt diameters produces a sample variance of $s^2 = .020$, this may be an indication that something is wrong, since the value .020 is not contained in the 95% confidence interval (.006, .017).

In general, to find a $(1 - \alpha)100\%$ confidence interval (L_1, L_2) for the variance σ^2 of a normal population, the confidence limits are found by using L_1 and L_2 given by

$$L_1 = \frac{(n-1)s^2}{\chi^2_{\alpha/2}(\mathrm{df})}$$

$$L_2 = \frac{(n-1)s^2}{\chi^2_{1-\alpha/2}(\mathrm{df})}$$

(9-18)

where n is the sample size and s^2 is the sample variance.

To obtain a confidence interval for the population standard deviation σ of a normal population, we use the positive square roots of the limits found by (9-18). A confidence interval for σ is therefore given by

$$(\sqrt{L_1}, \sqrt{L_2})$$

(9-19)

where L_1 and L_2 are defined by (9-18).

EXAMPLE 9-20   The variation of the potency of a drug must be small; otherwise, the drug could be harmful or ineffective. A pharmaceutical company was interested in determining the variance of potency measurements for a new batch of a certain drug. Toward this end, a random sample of 20-cc vials of the drug produced a variance equal to .0018. Construct a 99% confidence interval for the variance of the drug's potency measurements.

Solution   **1.** Since $1 - \alpha = .99$, we have $\alpha = .01$, $\alpha/2 = .005$, and $1 - \alpha/2 = .995$.
2. $\mathrm{df} = n - 1 = 20 - 1 = 19$
3. The critical values $\chi^2_{.005}(19)$ and $\chi^2_{.995}(19)$ are found in Table 2 of Appendix B:

$$\chi^2_{.005}(19) = 38.582$$
$$\chi^2_{.995}(19) = 6.844$$

4. By dividing the value $(n - 1)s^2$ by the critical values, we obtain the confidence interval limits L_1 and L_2. Thus, we have

$$\frac{(n - 1)s^2}{\chi^2_{.005}(19)} = \frac{(19)(.0018)}{38.582} = 0.000886$$

and

$$\frac{(n - 1)s^2}{\chi^2_{.995}(19)} = \frac{(19)(.0018)}{6.844} = .0050$$

Hence, $L_1 = .00089$ and $L_2 = .005$. Thus, a 99% confidence interval for σ^2 is (.00089, .005).

The confidence interval limits for σ are found by taking the positive square roots of the limits L_1 and L_2:

$$(\sqrt{L_1}, \sqrt{L_2})$$
$$(\sqrt{.00089}, \sqrt{.005})$$
$$(.030, .071)$$

Thus, (.030, .071) is a 99% confidence interval for σ.   ■

Problem Set 9.5 ─────────────────────────────────────

A

1. Find the critical values $\chi^2_{1-\alpha/2}(df)$ and $\chi^2_{\alpha/2}(df)$ if:
  a. $n = 17$ and $1 - \alpha = 99\%$
  b. $n = 25$ and $1 - \alpha = 95\%$
  c. $n = 11$ and $1 - \alpha = 95\%$
  d. $n = 18$ and $1 - \alpha = 90\%$

2. Construct 90% confidence intervals for σ^2 and σ in each of the following cases. Assume that the statistics resulted from samples drawn from normal populations.
  a. $s = .15$; $n = 25$
  b. $s = 28.7$; $n = 14$
  c. $s = 1.01$; $n = 38$
  d. $s = 842$; $n = 29$

3. Assume that $n = 41$, $\bar{x} = 72.4$, and $s^2 = 2.6$ result from a random sample taken from a normal population.
  a. What is the best point estimate for σ?
  b. Construct a 90% confidence interval for σ.

4. A sample of 20 castings is taken from a production run known to produce flaws that are normally distributed with a standard deviation of 5 flaws. If the sample has a mean of 4 flaws and a standard deviation of 2.83 flaws, find the value of the χ^2 statistic.

5. A sample of 12 cans of soup produced by XYZ Soup Company yielded the following net weights, measured in ounces:

11.9	12.2	11.6	12.1
12.1	11.8	11.9	11.8
12.0	12.3	11.8	12.0

Assuming the net weights are normally distributed, construct 95% confidence intervals for the variance and standard deviation of the population of net weights of all cans of soup produced by the company.

6. The daily arrival times for a certain train at one of its destinations are normally distributed. A sample of 12 train arrival times indicated $s = 1.789$ minutes. Construct a 99% confidence interval for the variance of the population of arrival times.

7. The weights of a certain species of fish are known to be normally distributed with a standard deviation of 2 grams. If a sample of 17 fish is selected from the population, find the probability that the sample variance is less than 8.

8. Suppose the number of hours that teenagers spend watching television per week has a normal distribution with a variance of 3. If a sample of 17 teenagers is selected, and the number of hours per week of viewing television is recorded for each, find the probability that the sample standard deviation of viewing times is greater than $\sqrt{6}$.

9. A precision meter is guaranteed to be accurate within 2% of full scale. A sample of five meter readings on the same object yielded the following measurements:

| 350 | 348 | 348 | 352 | 351 |

Construct a 95% confidence interval for σ, assuming the population of measurements is normally distributed.

10. The concentrations of artificial food coloring in six lots were recorded as follows:

| .010 | .013 | .018 |
| .014 | .015 | .013 |

If the food coloring concentrations are assumed to be normally distributed, construct a 95% confidence interval for the variance of concentrations for the population of lots.

11. For the data in Problem 10, construct a 90% confidence interval for the standard deviation of the concentrations for the population of lots.

B

12. If $P[\chi^2_{1-\alpha/2}(\mathrm{df}) < \chi^2 < \chi^2_{\alpha/2}(\mathrm{df})] = 1 - \alpha$ and $\chi^2 = (n-1)s^2/\sigma^2$, prove that

$$P\left[\frac{(n-1)s^2}{\chi^2_{\alpha/2}(\mathrm{df})} < \sigma^2 < \frac{(n-1)s^2}{\chi^2_{1-\alpha/2}(\mathrm{df})}\right] = 1 - \alpha$$

13. When df > 30, the statistic $\sqrt{2\chi^2} - \sqrt{2(\mathrm{df}) - 1}$ has a sampling distribution that is approximately the standard normal. That is,

$$z \simeq \sqrt{2\chi^2} - \sqrt{2(\mathrm{df}) - 1}$$

Use this result to find $\chi^2_{.05}(40)$ and compare the result to that contained in the chi-square table. [*Hint:* Let $z = 1.65$, df = 40, and solve for the value of χ^2.]

14. For large values of n, the sampling distribution of

$$\frac{\chi^2 - (n-1)}{\sqrt{2(n-1)}}$$

is approximately the standard normal. Use this fact to find $\chi^2_{.05}(40)$ and compare the result to that contained in the chi-square table.

15. For samples of size $n \geq 30$, the sampling distribution of s is approximately normal with $\mu_s = \sigma$ and $\sigma_s = \sigma/\sqrt{2n}$. Use this result to solve Problem 3(b) and compare the results obtained by using the two procedures.

CHAPTER SUMMARY

In this chapter we learned how to construct point and interval estimates for μ, σ^2, σ, and p. The estimates result from random samples taken from populations and are used to estimate characteristics of the populations, such as the mean and variance. A confidence interval shows a range of values within which the population parameter might fall. A given parameter is either contained in a confidence interval or it is not. Being 95% confident that a confidence interval contains the population mean does *not* mean that the probability the interval contains μ is 95%; rather, it means that 95% of all such intervals constructed will contain μ. All confidence intervals for μ and p have the following form:

$$\text{Estimate} \pm [z_{\alpha/2} \text{ or } t_{\alpha/2}(\mathrm{df})]$$
$$\cdot (\text{Standard error of the estimate})$$

We learned also how to determine the sample size needed to construct confidence intervals of a given width. Finally, we learned how to construct confidence intervals for the variance and standard deviation of a normal population.

The accompanying chart organizes the ideas used in Chapter 9.

To estimate	Employ sampling distribution	Conditions	Use	Formula number
A. μ	$\bar{x}$	$n \geq 30$ or the population is normal	$\bar{x} \pm z\left(\dfrac{\sigma}{\sqrt{n}}\right)$	(9-6)
			or	
			$\bar{x} \pm z\left(\dfrac{s}{\sqrt{n}}\right)$	
		$n < 30$, σ is unknown, and the population is normal	$\bar{x} \pm t(\mathrm{df})\left(\dfrac{s}{\sqrt{n}}\right)$	
B. p	$\hat{p}$	The population is large	$\hat{p} \pm z\sqrt{\dfrac{\hat{p}(1 - \hat{p})}{n}}$	(9-12)
C. σ^2	χ^2	The population is normal	(L_1, L_2) where $L_1 = \dfrac{(n-1)s^2}{\chi^2_{\alpha/2}(\mathrm{df})}$ and $L_2 = \dfrac{(n-1)s^2}{\chi^2_{1-\alpha/2}(\mathrm{df})}$	(9-18)
σ	χ^2	The population is normal	$(\sqrt{L_1}, \sqrt{L_2})$	(9-19)

D. Determining the sample size n so that $P(|\bar{x} - \mu| < E) = 1 - \alpha$

Conditions	Use	Formula number
σ is known	$n = \left(\dfrac{z\sigma}{E}\right)^2$	(9-14)
σ is unknown, $n \geq 30$, and have an estimate for σ	$n = \left(\dfrac{zs}{E}\right)^2$	

E. Determining the sample size n so that $P(|\hat{p} - p| < E) = 1 - \alpha$

Conditions	Use	Formula number
Have an estimate for p	$n = \left(\dfrac{z}{E}\right)^2 \hat{p}(1 - \hat{p})$	(9-15)
Have no estimate for p	$n = \left(\dfrac{z}{2E}\right)^2$	(9-16)

C H A P T E R   R E V I E W

IMPORTANT TERMS

For each of the following terms, provide a definition in your own words. Then check your responses against the definitions given in the chapter.

chi-square distribution	estimation	maximum error of estimate	sampling error
confidence level	estimator	point estimate	statistical inference
critical value	hypothesis testing	reliability	unbiased estimator
degrees of freedom	interval estimate	reliability testing	upper confidence limit
error of estimate	lower confidence limit		

IMPORTANT SYMBOLS

E, maximum error of estimate

$z_{\alpha/2}$, critical value for z

$t_{\alpha/2}(\text{df})$, critical value for t

$1 - \alpha$, confidence level

$\sigma_{\bar{x}}$, the standard error of $\bar{x}$

$\sigma_{\hat{p}}$, the standard error of $\hat{p}$

L_1, lower confidence limit

$\hat{\sigma}_{\hat{p}}$, an estimate of $\sigma_{\hat{p}}$

L_2, upper confidence limit

$|\bar{x} - \mu|$, error of estimate when estimating μ

$|\hat{p} - p|$, error of estimate when estimating p

χ^2, chi-square statistic

df, degrees of freedom

$\chi^2_{1-\alpha/2}(\text{df})$, lower critical value for χ^2

$\chi^2_{\alpha/2}(\text{df})$, upper critical value for χ^2

IMPORTANT FACTS AND FORMULAS

1. $\bar{x} \pm z_{\alpha/2}\left(\dfrac{\sigma}{\sqrt{n}}\right)$, confidence limits for μ

2. $\bar{x} \pm t_{\alpha/2}(\text{df})\left(\dfrac{s}{\sqrt{n}}\right)$, confidence limits for μ

3. $\hat{p} \pm z_{\alpha/2}\sqrt{\dfrac{\hat{p}(1 - \hat{p})}{n}}$, approximate confidence limits for p

4. $\chi^2 = \dfrac{(n-1)s^2}{\sigma^2}$, df $= (n-1)$

5. (L_1, L_2) where $L_1 = \dfrac{(n-1)s^2}{\chi^2_{\alpha/2}(\text{df})}$ and $L_2 = \dfrac{(n-1)s^2}{\chi^2_{1-\alpha/2}(\text{df})}$, confidence limits for σ^2

6. $n = \left(\dfrac{z_{\alpha/2}\sigma}{E}\right)^2$, sample size needed in order to estimate μ

7. $n = \left(\dfrac{z_{\alpha/2}}{2E}\right)^2$, sample size needed in order to estimate p when no prior estimate of p is available

8. $n = \left(\dfrac{z_{\alpha/2}}{E}\right)^2 \hat{p}(1 - \hat{p})$, sample size needed in order to estimate p when a prior estimate of p is available

CHAPTER REVIEW PROBLEMS

1. a. In a study to estimate the true proportion of U.S. citizens who are in favor of gun-control legislation, a random sample of 5000 people yielded 2800 in favor of gun-control legislation. Construct an approximate 90% confidence interval for the true proportion of U.S. citizens who are in favor of gun-control legislation.

b. How large a sample must be selected if we want to be 95% confident that the error of estimate is at most .01? Use the value of $\hat{p}$ obtained in part (a) to estimate p.

2. A random sample of 125 college students was taken to determine the percentage of students who receive financial aid at a certain college. It was found that 88 received some form of financial aid. Construct an approximate 99% confidence interval for the true proportion of college students receiving financial aid.

3. How large a sample is needed in Problem 2 if we want to estimate the true proportion to within .05 with a confidence level of 99% and no preliminary estimate for p is available?

4. In order to determine if ball bearings are conforming to manufacturing specifications, a sample of 25 ball-bearing diameters was inspected, yielding a variance of .0015. Construct a 95% confidence interval for the true variance of all ball-bearing diameters manufactured. Assume that the ball-bearing diameters are normally distributed.

5. In order to determine the proportion of students at a large university who are in default on their government loans, administrators selected a sample of 380 former students. The investigation indicated that 150 loans were in default. Construct an approximate 95% confidence interval for the true proportion of university students who are in default on their loans. What is the maximum error of estimate?

6. A new cigarette was just marketed. In a study to determine the mean nicotine content, 35 cigarettes yielded $\bar{x} = 25.4$ mg and $s = 1.9$ mg.
   a. Find E so that we can be 95% confident that μ is within $\pm E$ of $\bar{x}$.
   b. Construct a 95% confidence interval for the true average nicotine content.
   c. Using $s = 1.9$ to estimate σ, find the sample size needed in order to be 95% confident that the error of estimate is at most .01 mg.

7. In order to estimate the variance in the diameters of screw-top lids for medicine bottles, a sample of 20 lids was examined, yielding a sample variance of .18 square mm. Assuming the lid diameters are normally distributed, construct a 99% confidence interval for the true variance of lid diameters.

8. A large university wants to estimate the mean number of days that students are sick each school year. A sample of 500 students indicated that $\bar{x} = 2.3$ days and $s = 10.2$ days.
   a. Find E so that we can be 90% confident that when $\bar{x}$ is used to estimate μ, the error of estimate is at most E.
   b. Estimate μ by constructing a 90% confidence interval for μ.
   c. By using $s = 10.2$ as a point estimate for σ, find the sample size needed in order to be 90% confident that the maximum error of estimate is .25 day.

9. On a final examination for a calculus course at a large university, a sample of 45 tests was graded early to estimate the average grade. From previous experience, it had been determined that $\sigma = 14.5$. If the 45 tests produced a mean of 62 and a standard deviation of 17.1, construct a 95% confidence interval for the true average test grade for all calculus students at the university.

10. A sample of 75 customers at a certain gasoline station indicated that the mean number of gallons purchased was 14.3 gallons and the standard deviation was 2.7 gallons.
   a. By using $\bar{x}$ to estimate μ, find the maximum error of estimate E so that μ is within $\pm E$ of $\bar{x}$ with a confidence level of 95%.
   b. Construct a 95% confidence interval for the true average number of gallons of gasoline purchased.
   c. By using $s = 2.7$ to estimate σ, find the sample size needed in order to be 95% confident that the maximum error of estimate is at most .2 gallon.

11. Shown below are the durations (in minutes) for a sample of telephone orders at a large department store. Assuming the lengths of telephone calls to place orders are normally distributed, construct a 95% confidence interval for the true mean time needed to place telephone orders.

10.1	5.0	4.3
4.2	3.6	6.7
3.1	2.8	4.2

12. Sales personnel at a large company are required to submit weekly reports concerning the number of customer contacts. A random sample of 78 weekly reports showed an average of 27.2 customer contacts per week; the standard deviation was 7.2 contacts.

a. By using the value 27.2 to estimate the true average, find E so that we can be 90% confident that μ is within $\pm E$ of $\bar{x} = 27.2$.

b. Construct a 90% confidence interval for μ.

c. By using $s = 7.2$ to estimate σ, find the sample size necessary to be 90% confident that E is at most .5 customer.

13. The diameters of a certain type of bolt produced by a large firm are normally distributed. Suppose a sample of 20 bolts yields $\bar{x} = 8.02$ mm and $s = .06$ mm.

a. Find E so that we can be 95% confident that μ is within $\pm E$ of $\bar{x}$.

b. Estimate μ by constructing a 95% confidence interval.

14. In a survey to determine the proportion of viewers for a new television program, a sample of 575

households indicated that 238 were watching the program.

a. Construct an approximate 99% confidence interval for the true proportion of households watching the program.

b. If no preliminary estimates are available, how large a sample must be polled in order to be 95% confident that E is at most 3%?

15. A random sample of monthly rents for one-bedroom apartments in a large city was taken to estimate the average rent. Suppose $n = 800$, $\bar{x} = \$235.10$, and $s = \$47.55$.

a. Find E so that we can be 85% confident that μ is within $\pm E$ of $\bar{x}$.

b. Estimate μ by constructing an 85% confidence interval.

c. If s is used as a point estimate for σ, find the sample size needed in order to be 85% confident that E is at most $\$5.00$.

CHAPTER ACHIEVEMENT TEST

(10 points) **1.** Find the following critical values:
    a. $t_{.05}(\mathrm{df})$ if $n = 18$
    b. $\chi^2_{.01}(\mathrm{df})$ if $n = 25$

(15 points) **2.** Find the value of the following statistics:
    a. z; $\bar{x} = 17$, $s = 4.2$, $n = 80$, and $\mu = 15$
    b. t; $\bar{x} = 75$, $s = 1.8$, $n = 12$, and $\mu = 80$
    c. χ^2; $n = 16$, $s = 18$, and $\sigma^2 = 9$

(10 points) **3.** Construct a 95% confidence interval for μ given $\bar{x} = 17.2$, $s = 2.5$, and $n = 45$.

(10 points) **4.** Construct an approximate 90% confidence interval for p given $x = 25$ and $n = 150$.

(10 points) **5.** Construct a 99% confidence interval for σ given that a sample from a normal population yielded $n = 18$, $\bar{x} = 2.1$, and $s = 1.4$.

(10 points) **6.** If $\bar{x}$ is used to estimate μ and $\sigma = 1.7$, find the value of n so that we can be 90% confident that E is at most .12.

(10 points) **7.** If $\hat{p}$ is used to estimate p, find the value of n so that we can be 95% confident that $\hat{p}$ is within $\pm.11$ of p.

(25 points) **8.** Find the following:
    a. $P(\chi^2_{.70} < \chi^2 < \chi^2_{.10})$
    b. $P[t_{.90}(\mathrm{df}) < t < t_{.05}(\mathrm{df})]$
    c. $P(-z_{.05} < z < z_{.15})$
    d. the confidence level $1 - \alpha$ if $\chi^2_{\alpha/2}(17) = 7.564$
    e. the confidence level $1 - \alpha$ if $t_{\alpha/2}(18) = 2.552$

10

HYPOTHESIS TESTING

Chapter Objectives

In this chapter you will learn:

- *what hypothesis testing is*

- *the logic and language of hypothesis testing*

- *the types of errors associated with hypothesis testing*

- *the steps in the hypothesis testing procedure*

- *how to test hypotheses concerning a population mean*

- *how to test hypotheses concerning a population proportion*

- *how to test hypotheses concerning a variance or standard deviation of a normal population*

Chapter Contents

In Chapter 9 we studied estimation procedures for making inferences about unknown population parameters, such as μ, σ, and p, by using single values and intervals. In this chapter we want to develop an alternate procedure for making inferences—**hypothesis testing**. With the hypothesis testing procedure, we follow a formal set of steps or ordered procedures that lead to decisions concerning statements (called **statistical hypotheses**) about the value of a parameter.

10.1 THE LOGIC OF HYPOTHESIS TESTING

Unlike estimation procedures, hypothesis testing is not an exploratory procedure. With hypothesis testing, we are more interested in confirming the relationship of the parameter of interest to a fixed, known value, than in exploring its unknown value. For example, according to the 1978 Energy Act passed by Congress, a tax is imposed on the manufacturer of any new car that fails to average at least 19.5 miles per gallon of gasoline. (This cost gets passed on, of course, to the consumer.) As a result, a new-car manufacturer may not want to estimate the average gas mileage, but instead would want to determine if the average mileage exceeds 19.5 miles per gallon. The car manufacturer would want to test the hypothesis

> A: The mean gas mileage does not exceed 19.5 miles per gallon

against the hypothesis

> B: The mean gas mileage exceeds 19.5 miles per gallon

with the hope that it can gather sufficient information to support hypothesis B.

Some time ago it was accidentally discovered that minoxidil, a drug manufactured by Upjohn Pharmaceutical Company and prescribed for severe high blood pressure, causes hair growth. The drug is usually given with two other drugs to control blood pressure. It has been estimated that 80% of patients taking minoxidil experience thickening, elongation, and darkening of body hair within 3 to 6 weeks after starting the drug. As a result of these side effects of the drug, doctors have been researching the possibilities of using minoxidil to treat baldness, a condition that is widespread in the population and is generally untreatable by drugs. An experiment was conducted by a researcher to test the effects of minoxidil on baldness. The experiment was conducted over a 6-month period to test the hypothesis

> A: Minoxidil has no therapeutic benefits for treating baldness

against the hypothesis

> B: Minoxidil has therapeutic benefits for treating baldness

The experiment was conducted under the assumption that hypothesis A was correct, but with the hope of gathering evidence to the contrary. Two groups of bald people were used. One group (the treatment group) received a fixed dosage of minoxidil, and the other group (the control group) received a placebo—a substance having an appearance identical to minoxidil but having no known effects for treating baldness. The medical staff administering the drug was not made aware of which subjects were in the treatment and control groups; this was done to control any possible effects that the persons administering the drug might have. (An experiment such as this one, in which neither the experimental subject nor the person administering the treatment knows who is in the treatment or control groups, is called a **double-blind experiment**.) After the experimental period of 6 months, the researcher found no evidence to suggest that minoxidil has any real benefits for treating baldness in humans. She thus failed to reject hypothesis *A* in favor of hypothesis *B*. The process that the researcher used to arrive at her decision is called **hypothesis testing**.

As a more involved example of a hypothesis testing situation, consider the following: A leading drug in the treatment of hypertension has an advertised therapeutic success rate of 84%. A medical researcher believes he has found a new drug for treating hypertensive patients that has a higher therapeutic success rate than the leading drug, but with fewer side effects. To test his assertion, the researcher gets approval from the Food and Drug Administration to conduct an experimental study involving a random sample of 60 hypertensive patients. The proportion of patients in the sample receiving therapeutic success will be used to determine if the success rate, p, for the population of hypertensive patients is higher than .84. The researcher arbitrarily decides that he will conclude that the therapeutic success rate of the new drug is higher than .84 if the proportion of hypertensive patients in the sample having therapeutic success after taking the drug is .86 or higher. Otherwise, he will conclude that his drug is no more effective than the currently used drug. It appears, at first glance, that the decision should be clear-cut—just compute the sample proportion, compare it to .84, then make a decision. But the decision rests on the outcome of a single sample proportion, which will vary from sample to sample. If the true proportion in the population receiving therapeutic success is less than or equal to .84 (i.e., if $p \leq .84$), there is the possibility that $\hat{p}$ could be greater than .86, by chance; similarly, there is the possibility that the sample proportion could be less than .84 even if the true proportion is, say, .87. The uncertainty involving the researcher's decision can be attributed, in part, to the standard error of the sample proportion.

For this example, the two statements

$p \leq .84$

$p > .84$

are called statistical hypotheses. A **statistical hypothesis** is an assertion concerning one or more parameters involving one or more populations.

Null Hypothesis and Alternative Hypothesis

The logic of hypothesis testing is based on the simple fact that a statement and its negation cannot both be true (or false) at the same time. Consider the following two statements concerning the population proportion p:

> A: $p \leq .84$
>
> B: $p > .84$

If statement A is true, then statement B is false, and if statement B is true, then statement A is false. Both statements cannot be true (or false) at the same time.

Hypothesis testing involves a pair of statistical hypotheses, such that acceptance of one means rejection of the other. The hypothesis that the researcher would like to establish is called the **experimental hypothesis** or **alternative hypothesis** and is denoted by H_1, while the other hypothesis is called the **null hypothesis** and is denoted by H_0. *Null* means none, so it is sometimes convenient to think of the null hypothesis as the hypothesis of "no difference." The null hypothesis is commonly written as a statement involving the equality relation when a population parameter is being compared with a numerical value. For example, H_0: $\mu = 20$, H_0: $\mu \geq 20$, and H_0: $\mu \leq 20$ could be null hypotheses. For the above illustration, the null hypothesis is H_0: $p \leq .84$ and the alternative hypothesis is H_1: $p > .84$. If the researcher rejects the null hypothesis, he will have support for his drug's having superior therapeutic benefits over the leading drug that has widespread use.

Note that the relation expressed by the null hypothesis must include one of the three symbols "$\leq$," "$\geq$," or "$=$." The statistical hypothesis $p < .84$ is not an acceptable null hypothesis, but it is an acceptable alternative hypothesis. The following examples further illustrate pairs of statistical hypotheses. For each, H_0 is under test; if H_0 is rejected, the researcher has produced evidence that the alternative hypothesis H_1 is true.

> a. H_0: $\mu = 25$ (a population mean equals 25)
> H_1: $\mu \neq 25$ (a population mean does not equal 25)
> b. H_0: $\sigma \leq 4$ (a population standard deviation is less than or equal to 4)
> H_1: $\sigma > 4$ (a population standard deviation is greater than 4)
> c. H_0: $p \geq .5$ (a population proportion is greater than or equal to .5)
> H_1: $p < .5$ (a population proportion is less than .5)

Note that in each pair, H_1 is the negation (opposite) of H_0.

In order for an experimental or alternative hypothesis to be established, the null hypothesis must be rejected. This occurs when the evidence obtained from one sample (for the hypotheses to be tested in this chapter) is inconsistent with the statement of the null hypothesis. If the sample evidence is not inconsistent with the null hypothesis, it does not necessarily imply that the null hypothesis is true; it simply means we do not have sufficient evidence to reject it. The statistician or researcher generally labels the statistical hypothesis that he hopes to reject as the null hypothesis. The process of choosing between H_0 and H_1 is called **hypothesis testing**. Consider the following examples:

1. A doctor believes that drug A is better than drug B in treating a certain disease. This becomes the alternative hypothesis H_1. The null hypothesis H_0 is that drug A is not better than drug B. If the doctor gathers sufficient evidence to reject H_0, there is a strong likelihood that H_1 is true.

2. A consumer who purchased a new car 4 months ago does not believe the advertised claim that the average gas mileage is at least 45 miles per gallon on the highway; he suspects it is less than 45 miles per gallon. Thus, the statement of the alternative hypothesis becomes $H_1: \mu < 45$, and the statement of the null hypothesis is $H_0: \mu \geq 45$. Note that if H_0 is rejected, then the consumer will have support for H_1 being true.

3. A prosecuting attorney believes the accused is guilty. His job is to convict the accused party. The null hypothesis is the statement that the accused is innocent, while the alternative hypothesis is the statement that the accused is guilty. Failure to reject the null hypothesis does not mean the accused is innocent; it means only that there is insufficient evidence to convict.

4. A business executive believes sales for this year will be higher than last year. The null hypothesis becomes H_0: sales for this year will be no higher than sales last year. The alternative hypothesis becomes H_1: sales this year will be higher than sales last year.

5. An agronomist would like to determine if a new brand of fertilizer will increase grain production. The null hypothesis is H_0: the new fertilizer will not increase grain production. The alternative hypothesis is H_1: the new brand of fertilizer will increase grain production.

6. A pollster would like to determine if the proportion of voters in Maryland who are Democrats is different from the proportion of voters in Pennsylvania who are Democrats. The null hypothesis is H_0: the proportion of Democrats in Pennsylvania is equal to the proportion of Democrats in Maryland. The alternative hypothesis is H_1: the proportion of Democrats in Pennsylvania is different from the proportion of Democrats in Maryland.

As another example, suppose a drug company conducted an experiment to determine which of two hypertension drugs, A or B, is more effective on patients with mild hypertension. Drug A was given to one group of patients and drug B was given to another group of patients. On the whole, the drug company found that the patients receiving drug B had more therapeutic benefits than those receiving drug A. There are two possible explanations for the observed difference:

1. Drug B is superior to drug A in terms of therapeutic benefits.
2. The observed difference in therapeutic benefits is due to chance factors or sampling error.

Suppose the drug company was able to calculate the plausibility of the second explanation and found it to be 5%. The company might then conclude, based on the evidence at hand, that it is reasonably certain that drug B is superior to drug A.

The truth of a statistical hypothesis is never known with certainty. The only way to determine the unknown value of the parameter is to examine the whole population. Since this is not always practical or possible, a decision involving the null hypothesis could involve an error.

Types of Errors with Hypothesis Testing

When any null hypothesis H_0 is tested, there are four possible outcomes that can occur. The following diagram contains the four possibilities:

	H_0 is true	H_0 is false
Reject H_0	Incorrect decision (Type I error)	Correct decision
Do not reject H_0	Correct decision	Incorrect decision (Type II error)

Two of the outcomes involve correct decisions and two involve incorrect decisions. Rejecting H_0 when it is true and not rejecting H_0 when it is false are incorrect decisions. Rejecting the null hypothesis when it is true is called a **Type I error**, and not rejecting the null hypothesis when it is false is called a **Type II error**. The probability of committing a Type I error is denoted by the Greek letter α and is called the **level of significance**. For the earlier example of the hypertension drug study, deciding on the basis of the test data that the new drug is more effective than the old one (i.e., deciding there is greater than .84 therapeutic success) when in fact the drug is not more effective for the population as a whole would be a Type I error. The probability of making this error can be arrived at using techniques we will develop shortly. The probability of committing a Type II error is denoted by the Greek letter β. In the same hypertension study, deciding from the test data that the new drug is only as effective or less effective than the old one (i.e., deciding that the therapeutic success is not greater than 84%), when in reality it is more effective for the entire population, would be a Type II error.

Testing a null hypothesis H_0 is similar to trying an accused individual under our judicial system. The defendant is on trial and H_0 is the statement that the accused is innocent. H_1 is the statement that the defendant is guilty. The prosecution lawyers try to prove H_1 is true (or H_0 is false). In arriving at their verdict, the jury may render a correct decision or an incorrect decision. The correct decisions are the jury votes not guilty when the accused is innocent or votes guilty when the accused is guilty. The incorrect decisions are the jury votes to convict the accused when the accused is innocent (Type I error) or votes to acquit the accused when the accused is guilty (Type II error).

As another example to illustrate the types of errors that can result when testing a null hypothesis, suppose a new and more expensive diagnostic procedure for detecting breast cancer in women is being tested for superiority over the currently used method. The statistical hypotheses are:

H_0: The new method is no better than the currently used method
H_1: The new method is better than the currently used method

A Type I error results when the null hypothesis H_0 is rejected when it is true. The consequences of this error would be increased medical costs. A Type II error would result when H_0 is not rejected when it is false. Consequences of this error would be inferior testing and possibly a higher death rate from breast cancer. The more serious type of error in this case is the Type II error. If H_0 is true and it is not rejected or if H_0 is false and it is rejected, then no errors are committed.

The probability associated with making a correct decision—not rejecting H_0 when it is true or rejecting H_0 when it is false—can be determined. The probability of not rejecting H_0 when it is true is equal to $(1 - \alpha)$. This can be shown by noting that

$$P(\text{rejecting } H_0 \text{ when it is true}) + P(\text{not rejecting } H_0 \text{ when it is true}) = 1$$

Since $P(\text{rejecting } H_0 \text{ when it is true}) = \alpha$, we have

$$P(\text{not rejecting } H_0 \text{ when } H_0 \text{ is true}) = 1 - \alpha$$

Note that the probability of not rejecting H_0 when it is true is the confidence level $(1 - \alpha)$ studied in Chapter 9.

The probability of rejecting H_0 when it is false is equal to $(1 - \beta)$. This can be shown by noting that

$$P(\text{rejecting } H_0 \text{ when it is false}) + P(\text{not rejecting } H_0 \text{ when it is false}) = 1$$

But since $P(\text{not rejecting } H_0 \text{ when it is false}) = \beta$, we have

$$P(\text{rejecting } H_0 \text{ when it is false}) = 1 - \beta$$

The probability of rejecting the null hypothesis H_0 when it is false is called the **power of the test**.

The probabilities associated with the four possible outcomes of a hypothesis test are summarized in Table 10-1.

Table 10-1 Outcomes and Probabilities for Hypothesis Testing

Probability symbol	Definition
α	Level of significance: probability of a Type I error
β	Probability of a Type II error
$1 - \alpha$	Level of confidence: probability of not rejecting H_0 when it is true
$1 - \beta$	Power of the test: probability of rejecting H_0 when it is false

Proof-by-contradiction in mathematics provides a reasonable parallel to hypothesis testing. The only difference is that proof-by-contradiction does constitute a proof in the mathematical sense, but hypothesis testing does not. With hypothesis testing, a statement is rejected when probabilities derived from the data based on the statement of the null hypothesis cast serious doubt on the truth of the null hypothesis. A mathematician does not conclude that $A = B$

just because he cannot show otherwise. The reasoning with hypothesis testing is no less logical in that failure to reject H_0 does not constitute a proof that H_0 is true or that it has been verified.

Hypothesis testing can never be used to "establish" absolute truths. With any decision, there is always the possibility of error. When the null hypothesis is rejected, we have statistical evidence which indicates that the alternate hypothesis is plausible, but not necessarily true. In addition, failure to reject H_0 should not imply that one should accept H_0. Rather, judgment should be reserved unless the probability of committing a Type II error is known. If β is small, then one might conclude that H_0 is plausible, but not necessarily true.

Types of Hypothesis Tests

Hypothesis tests are classified as directional or nondirectional, depending on whether the alternative hypothesis H_1 involves the "$\neq$" sign. If the statement of H_1 involves the unequal sign "$\neq$", then the test is called **nondirectional**, while if the statement of H_1 does not involve "$\neq$" (that is, if it involves "$<$" or "$>$"), then the test is called **directional**. Nondirectional tests are also called **two-tailed tests**, and directional tests are also called **one-tailed tests**. For tests involving one sample of data, if the statement of H_1 involves the symbol "$<$", then the test is called a **left-tailed** test, and if the statement of H_1 involves the symbol "$>$", then the test is called a **right-tailed** test. The following table summarizes the language used. Terms listed in each column are frequently used interchangeably.

	Sign in the statement of H_1	
<	**$\neq$**	**>**
Directional	Nondirectional	Directional
One-tailed	Two-tailed	One-tailed
Left-tailed		Right-tailed

The following are examples of the different types of hypothesis tests:

Directional	Nondirectional	Left-tailed	Right-tailed
a. H_1: $\mu > 2$	H_1: $\mu \neq 2$	H_1: $\mu < 2$	H_1: $\mu > 2$
b. H_1: $p < .2$	H_1: $p \neq .2$	H_1: $p < .2$	H_1: $p > .2$
c. H_1: $\sigma < 4$	H_1: $\sigma \neq 4$	H_1: $\sigma < 4$	H_1: $\sigma > 4$

Determining H_1

The form of the alternative hypothesis H_1 will be dictated by the practical aspects of the problem. It will reflect the direction of the desired result as formulated by the experimenter. Frequently it is possible to select H_0 so that an erroneous rejection of H_0 is the more serious consequence. As we shall see later, this is due to the fact that the probability of committing a Type I error can be controlled by the experimenter (since it is the level of significance and should be stipulated in advance).

As an illustration, suppose a medical researcher is interested in testing the effects of a new drug. Consider the following two pairs of statistical hypotheses:

a. H_0:   The drug is safe
   H_1:   The drug is harmful
b. H_0:   The drug is harmful
   H_1:   The drug is safe

For situation (a), a Type I error is to conclude that a safe drug is harmful and a Type II error is to conclude that a harmful drug is safe. For situation (b), a Type I error is to conclude that a harmful drug is safe, and a Type II error is to conclude that a safe drug is harmful. Since the Type I error for situation (b) (concluding that a harmful drug is safe) is more serious and α can be controlled, the researcher would select the hypotheses in (b) for the experiment and would formulate the null hypothesis to reflect that the drug is harmful.

In verbal problems written to provide experience in using the hypothesis testing procedure (such as the exercises in this chapter), the type of test (one-tailed or two-tailed) can be determined by careful reading of the problem. A one-sided test can usually be identified by use of such terms as *more than, less than, better than, worse than, at least, at most, too much,* and so forth. Two-sided tests are usually identified in verbal problems by such terms as *not equal to, different from, changed for better or worse, unequal, a difference,* and so on.

EXAMPLE 10-1   For each of the following situations, identify the alternative hypothesis H_1, the type of test (one-tailed or two-tailed), and the identifying term(s).

a. The national average salary for state and local government employees was $19,044 for the year 1982. Employees of the state of Maryland suspected that they were below the national average in terms of annual salary.

b. The mean yield of cotton grown by a particular farmer was 1225 pounds per acre for a recent year. In the next year, the farmer tried a new type of insecticide on a random sample of plots and was interested in determining if the average yield of cotton per acre would increase.

c. A sample of Camel cigarettes shows the average tar content to be 12 mg. Does this indicate that the average tar content for Camel cigarettes exceeds 11 mg?

d. A pharmaceutical company claims that its best-selling painkiller has a mean effective period of at least 3 hours. Do the sample data indicate that the company's claim is too high?

e. In 1979, the average age of students enrolled in college was 19.8 years. Based on a sample taken last year, has the average age of college students changed?

f. John Smith, a Democrat, is running against Jane Doe, a Republican, for the office of governor. Last month's poll showed that 65% of the voters supported Smith, but then Doe started mudslinging and now Smith is interested in whether the proportion who support him has dropped.

Solution

Alternative hypothesis	Type of test	Identifying terms
a. H_1: $\mu < \$19{,}044$	Left-tailed test	below
b. H_1: $\mu > 1225$	Right-tailed test	increase
c. H_1: $\mu > 11$	Right-tailed test	exceeds
d. H_1: $\mu < 3$	Left-tailed test	too high
e. H_1: $\mu \neq 19.8$	Two-tailed test	changed
f. H_1: $p < .65$	Left-tailed test	dropped

■

Problem Set 10.1

A

1. For each pair of statistical hypotheses, indicate which one is the null hypothesis, H_0.
   a. A: $\mu > 21$; B: $\mu \leq 21$
   b. A: $p = .7$; B: $p \neq .7$
   c. A: $\sigma \neq 1.2$; B: $\sigma = 1.2$
   d. A: $\sigma^2 > 8$; B: $\sigma^2 \leq 8$

2. Identify which of the hypotheses in Problem 1 are associated with one-tailed tests.

3. For each of the following situations, identify the hypotheses and the type of error when appropriate.
   a. The drug cibenzoline is being compared to propranolol for possible use in controlling cardiac arrhythmia. The claim is that cibenzoline is better than propranolol. In fact, this is not the case. However, research concluded that cibenzoline is better.
   b. Two schools are equally effective in training students. An evaluation team concluded that there was no significant difference between the effectiveness of the two schools in training students.
   c. A greater proportion of RCA televisions need repairs than Zenith televisions. A study done by a consumer advocate group concluded the same thing.
   d. Two people are compared for efficiency in performing a particular task. Individual B is really more efficient than individual A. An evaluation concluded that there is no difference in efficiency ratings between A and B.

4. For each of the following research situations, indicate whether a one-tailed or two-tailed test is appropriate.
   a. The Food and Drug Administration wants to test a new prescription drug manufactured by Upjohn to determine if it contains 5 mg of codeine as claimed by Upjohn.
   b. A survey was done by mechanics to determine if

there is a difference between the proportion of Fords and the proportion of Chevrolets that need repairs. They think there is a difference, but are not sure.
   c. An accrediting association was convinced that college A's grading of students was different from college B's; however, the association was not sure from which college the students got better grades.
   d. Dr. Jones believes that the discovery method of teaching is more effective than the expository method. She is interested in showing this is the case.
   e. The National Academy of Sciences has recommended that the mean daily sodium intake for a person should be between 1100 mg and 3300 mg. A medical researcher thinks the mean daily sodium intake for the average person exceeds 4500 mg.

5. Refer to Problem 4(a).
   a. Why did you choose your answer over the other two possibilities for stating the alternative hypothesis?
   b. Describe Type I and Type II errors and discuss the possible consequences of making each type of error.

6. For a statistical test, what do $(1 - \alpha)$ and $(1 - \beta)$ represent?

7. Suppose it is desired to test the following hypotheses:

   H_0: Smoking is not harmful to your health
   H_1: Smoking is harmful to your health

   In terms of the null hypothesis, state in words what is represented by
   a. a Type I error
   b. a Type II error
   c. a good decision
   Which type of error is more serious?

8. For each of the following null hypotheses, state what actions would constitute Type I and Type II errors:

a. H_0:   The discovery method of teaching statistics is at least as good as the expository method.

b. H_0:   At most 2% of the machines are defective.

c. H_0:   At least 95% of Americans are against war.

d. H_0:   The new production process is at least as good as the old process.

e. H_0:   Ninety percent of doctors recommend preparation A.

9. Suppose we conduct a hypothesis test with a level of significance equal to .05. Determine whether each of the following statements is true or false:

a. The probability of committing a Type I error is .95.

b. The probability of rejecting H_0 is .05.

c. The probability of committing a Type I error is .05.

d. The probability of not rejecting H_0 is .95.

e. The probability of not rejecting H_0 when it is true is .95.

10. When we fail to reject H_0, why do we not automatically accept H_0? Explain.

11. Is the probability of making a Type I error equal to 1 minus the probability of making a Type II error? Explain.

12. Consider each of the following situations. For each outcome, state whether it is a correct decision and (if applicable) indicate the type of error.

a. H_0:   The lifetime of battery A does not exceed the lifetime of battery B.

(1) Change to A when B lasts as long or longer.

(2) Keep B when A lasts longer.

(3) Keep B when B lasts at least as long.

(4) Change to A when A lasts at least as long.

b. H_0:   The new drug is safe.

(1) Approve the drug when it is unsafe.

(2) Disapprove the drug when it is unsafe.

(3) Disapprove the drug when it is safe.

(4) Approve the drug when it is safe.

B

13. If you have rejected H_0, is it possible that you have made a Type II error?

14. If it has been determined to reject the null hypothesis H_0 when $\alpha = .05$, does $\beta = 0$? Explain.

10.2 AN INTRODUCTION TO THE HYPOTHESIS TESTING PROCEDURE

The hypothesis testing procedure involves an indirect approach for making an inference about an unknown population parameter. The procedure is based on the assertion that the null hypothesis H_0 is true. A statistical test is then conducted to determine the plausibility of H_0. A null hypothesis is plausible if a researcher finds insufficient evidence to suggest that it is false. If the results of the test are judged unlikely to occur (or indicate a rare event), then the original assumption that H_0 is true is rejected; in such a case, there is support for the alternative hypothesis H_1 to be plausible. Of course, with any decision, we may make an error. By using the hypothesis testing procedure, we will make generalizations about population parameters based on the evidence gathered from only one sample (for the tests in this chapter).

The hypothesis testing procedure involves the following steps:

1. Formulate and state H_1.
2. Formulate and state H_0.
3. Choose an appropriate test statistic.
4. Formulate and state a decision rule.

5. Compute a statistic from a test sample.
6. Make one of the following decisions: (a) fail to reject H_0, or (b) reject H_0.

The hypothesis testing procedure involves gathering a random sample of test data relating to the hypotheses and computing a statistic, called the **test statistic**. This test statistic belongs to a sampling distribution with a particular shape, mean, and standard deviation. If a value of the test statistic (such as the sample mean $\bar{x}$) is located too far to the left of the mean of its sampling distribution (in the left tail of the sampling distribution) or too far to the right of the mean (in the right tail of its sampling distribution), then the null hypothesis H_0 is rejected. How far is "too far" is determined by a decision rule. The **decision rule** is stated in terms of α, the level of significance. Recall that the level of significance is the probability of committing a Type I error—that is, rejecting H_0 when it is true. The decision rule specifies those values in the sampling distribution of the test statistic that result in H_0 being rejected. These values are the values of the test statistic that are unlikely to occur if H_0 is true, and are commonly referred to as the **rejection region**. Consider the following example.

Suppose a certain coin-operated soft-drink machine was designed to dispense, on the average, 8 ounces of beverage per cup. Over a long period of use, we suspect that the machine is dispensing an average of less than 8 ounces per cup. As a result, we decide to test the following statistical hypotheses:

Step 1

H_0: $\mu \geq 8$
H_1: $\mu < 8$

If we can reject H_0, we conclude that our suspicions are correct.

Step 2   We next stipulate the level of significance α. The level of significance can be any value between 0 and 1, but is usually .05 or .01. Let's use $\alpha = .05$. This means that the probability of rejecting H_0 when it is true is .05. Since we are interested in producing evidence that supports H_1 being true, we shall assume that H_0 is true and hope to reject H_0. The statement that H_0 is true is only an assumption and must be tested for truth or plausibility. As a result, we proceed under the assumption that H_0 is true.

Step 3   Next we decide to take a random sample of 30 cups of beverage and compute $\bar{x}$ and s. The sample mean is chosen as the test statistic. Suppose $\bar{x} = 7.6$ ounces and $s = .75$ ounce. We want to locate the value of the sample mean $\bar{x} = 7.6$ within its sampling distribution. As a consequence of the central limit theorem and the fact that $n = 30$, the sampling distribution of the mean is approximately normal. Its mean is μ, the mean of the sampled population. Since we assumed H_0 to be true, the mean of the sampling distribution is greater than or equal to 8 ounces. Which value for μ should we use? We always use the equality value expressed by H_0 (called the **null value**). Thus, $\mu = 8$ is the null value in this illustration. If a value of μ greater than 8 is used, it can be shown that the probability of rejecting H_0 when H_0 is true is less than the stipulated level of significance $\alpha = .05$. That is, the null value presents the worst possible case in terms of providing the largest possible probability of rejecting H_0 when it is assumed to be true. As a result, throughout the remainder of the text, we shall always employ the null value as the mean of the sampling distribution of the test statistic.

Since the level of significance is $\alpha = .05$ and the test is a left-tailed test (since $H_1: \mu < 8$), a rejection region can be established. This is done by locating the value in the sampling distribution that has 5% of the distribution below it. In other words, a value can be determined that cuts off an area of .05 in the left tail of the sampling distribution. This value is called the **critical value** and it separates the sampling distribution into two regions, the rejection region and the nonrejection region. We shall denote a critical value by the symbol C.

Step 4   All values to the left of the critical value make up the rejection region. If $\bar{x} = 7.6$ falls in the rejection region, we reject H_0; otherwise, we fail to reject H_0, as indicated in the figure:

$$\text{Reject } H_0 \qquad \text{Do not reject } H_0$$
$$\underline{\hspace{4cm}|\hspace{4cm}} \ \bar{x}$$
$$C$$

Since the sampling distribution of the mean is approximately normal, to find the critical value C, we find the z value for C and refer to the standard normal distribution (see Figure 10-1). Since σ is unknown and $n \geq 30$, s is an adequate point estimate for σ. Thus,

$$z = \frac{C - \mu}{\sigma/\sqrt{n}}$$

$$-1.65 = \frac{C - 8}{.75/\sqrt{30}} = \frac{C - 8}{.137}$$

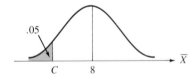

FIGURE 10-1 Critical value for soft-drink example

The value 1.65 was found in the z table (see the front endpaper) corresponding to an area of .45 $(=.5 - .05)$. Multiplying both sides of the last equation by .137, we have

$$-.23 = C - 8$$

And adding 8 to both sides, we find that $C = 7.77$. Since $\bar{x} = 7.6 < 7.77$, we reject H_0 in favor of H_1. With this decision we risk committing a Type I error, rejecting H_0 when it is true. The probability of this type of error is $\alpha = .05$. When a test results in rejecting a null hypothesis under a specified level of significance, the test is said to be **significant** and the result is referred to as a **significant finding**. For our example, the result $\mu < 8$ is classified as a significant finding. Thus, we have significant statistical evidence to indicate that the machine is dispensing, on the average, less than 8 ounces per cup of soft-drink beverage.

A test procedure is considered good if both its Type I and Type II error probabilities are small. We have control over the Type I error probability α, since it is stipulated before the data are gathered. We generally have little control over β, the probability of a Type II error. The probability of committing a Type II error varies, depending on the true value of the population parameter. For example, if we want to test the null hypothesis H_0: $\mu = 25$ and it is assumed that the true value of μ is 26, then a value for β can be calculated. The value for β (for a fixed sample size) depends on the difference between the asserted true value and the hypothesized value.

It is desirable to minimize the probabilities for both types of error. For a fixed α, β can generally be kept to a minimum by choosing the sample size as large as possible. Recall that the probability of rejecting the null hypothesis when it is false is called the power of the test and is denoted by $(1 - \beta)$. Power can be increased by increasing n.

Let's examine the logic underlying the hypothesis test procedure introduced in this section. The following events occurred in the order listed:

1. The statistical hypotheses H_0 and H_1 were formulated such that rejection of the null hypothesis, H_0: $\mu \geq 8$ ounces, would establish that the alternative hypothesis, H_1: $\mu < 8$ ounces, is plausible.

2. The null hypothesis was assumed to be true and the null value $\mu = 8$ was chosen to be the mean of the sampling distribution of the sample mean.

3. The maximum risk of committing a Type I error, that of rejecting H_0 when it is true, was stipulated to be $\alpha = .05$.

4. To test the null hypothesis, a random sample of 30 cups of beverage was obtained and the sample mean and standard deviation were computed ($\bar{x} = 7.6$ and $s = .75$). The sample mean $\bar{x}$ is the test statistic and 7.6 is the value of the test statistic.

5. Next, we located $\bar{x} = 7.6$ in its sampling distribution to determine the likelihood of $\bar{x}$ being at most as large as 7.6 under the assumption that H_0: $\mu = 8$ was true. Since the level of significance was stipulated to be .05, a cutoff point C (called the critical value) was established in the left tail of the distribution such that $P(\bar{x} < C) = .05$. All values in the sampling distribution to the left of the critical value were identified as the rejection region. If the value of the test statistic falls within the rejection region, we reject H_0; otherwise, we fail to reject H_0. For this example, $\bar{x} = 7.6$ fell in the rejection region, so H_0 was rejected in favor of H_1 being true. If H_0 is true and our sample is random, we would not expect the sample mean to be located so far into the left tail of its sampling distribution. A sample mean falling in the lower left portion of the distribution would signal a rare or unlikely event, in which case the null hypothesis is not plausible and should be rejected.

Problem Set 10.2

A

For each of the applications in Problems 1–4, find the critical value(s) and determine the rejection region.

1. The mean of SAT math scores is 500 and the standard deviation is 100. If a sample of 64 students received an average score of 480, do we have any evidence to suggest that the mean score is less than 500? Use $\alpha = .05$.

2. A soft-drink machine is set to dispense 7.0 ounces per cup. If 35 cups yielded $\bar{x} = 7.23$ ounces and $s = .14$ ounce, is this evidence that the machine is overfilling cups? Use $\alpha = .01$.

3. A new type of tire is tested to determine if it can average 30,000 miles under normal highway driving conditions. A sample of 100 tire sets was tested and averaged 30,200 miles; the standard deviation was 850 miles. Do the data indicate that the tires wear significantly more than 30,000 miles? Use $\alpha = .05$.

4. A toy company manufactures snow sleds having a pull rope. The mean breaking strength of the rope is advertised to be at least 150 pounds. Suspecting that the mean breaking strength of the rope is less than 150 pounds, company engineers chose a random sample of 50 ropes from a new shipment and found a mean breaking strength of 145 pounds and a standard deviation equal to 4 pounds. Should the toy manufacturer revise its advertisement? Use $\alpha = .05$.

For Problems 5–8, identify the following information:
  a. Null hypothesis
  b. Alternative hypothesis
  c. Test statistic and corresponding value
  d. Significance level
  e. Decision
  f. Type of error associated with the decision in part (e)

5. Problem 1

6. Problem 2

7. Problem 3

8. Problem 4

B

9. In nine test runs, a new large diesel truck was operated with 1 gallon of special fuel oil, and obtained the following mileages: 7, 9, 9, 6, 8, 11, 9, 7, and 6. Is this evidence that the truck is not operating at an average of 10.4 miles per gallon of fuel oil? Use $\alpha = .05$. [*Hint:* Use the t statistic.]

10. For Problem 9, identify each of the following:
  a. Null hypothesis
  b. Alternative hypothesis
  c. Value of the test statistic
  d. Significance level
  e. Critical value(s)
  f. Decision
  g. Type of error associated with the decision in part (f)

11. For large sample sizes, small differences tend to be significant, but not always of practical importance. Occasionally, a small sample yields a difference that is of practical importance. For example, a weight gain of 5 pounds would not be considered of practical importance for a 20-year-old male football player but it would be for a 1-year-old baby. The following situations illustrate that small differences become significant for large sample sizes. In which of the following situations would you reject H_0 for a one-tailed test?
  a. $n = 30$, $\sigma = 5$, $\alpha = .05$, $H_0: \mu = 70$, and $\bar{x} = 71$
  b. $n = 100$, $\sigma = 5$, $\alpha = .05$, $H_0: \mu = 70$, and $\bar{x} = 71$
  c. $n = 1000$, $\sigma = .5$, $\alpha = .05$, $H_0: \mu = 70$, and $\bar{x} = 71$
  d. $n = 2500$, $\sigma = 2$, $\alpha = .05$, $H_0: \mu = 1.70$, and $\bar{x} = 1.80$
  e. $n = 3600$, $\sigma = .5$, $\alpha = .05$, $H_0: \mu = .60$, and $\bar{x} = .61$

12. Suppose it is desired to test the null hypothesis $H_0: \mu \le 10$ versus the alternative hypothesis $H_1: \mu > 10$. If the decision rule is to reject H_0 if $\bar{x} > 10$, what is the probability of committing a Type I error?

13. For Problem 12, state a different rule that has a probability of 0 of leading to a Type I error. With such a rule, what is the probability of committing a Type II error?

14. Can a null hypothesis be of the form $H_0: \mu > \mu_0$, where μ_0 is a fixed value? Explain.

15. If H_0 cannot be rejected at $\alpha = .05$, can it be rejected at $\alpha = .01$? Explain.

10.3 HYPOTHESIS TESTS INVOLVING μ

Let's examine another hypothesis test in detail. Suppose, as members of a consumer protection group, we are interested in determining if the average weight of a certain brand of chocolate chip morsels, packaged in 15-ounce packages, is less than the advertised weight of 15 ounces. We formulate the following statistical hypotheses and decide to test the null hypothesis H_0 against the alternative hypothesis H_1:

$$H_0: \quad \mu \geq 15$$
$$H_1: \quad \mu < 15$$

As the level of significance, we decide on $\alpha = .05$. Since we are dealing with the parameter μ and desire information concerning its magnitude, we choose a random sample of 50 bags and compute the sample mean $\bar{x}$. Suppose we find that $\bar{x} = 14.4$ ounces and $s = 1.2$ ounces. Our best point estimate of μ is then 14.4. Recall from Chapter 8 that $\bar{x}$ can be represented as a sum of two components, the population mean μ and sampling error e:

$$\bar{x} = \mu + e$$

Testing the null hypothesis $H_0: \mu \geq 15$ is equivalent to determining whether the observed difference between $\bar{x}$ and μ can be attributed to sampling error alone if we assume that $\mu = 15$. To aid us in our determination, we use the sampling distribution of the mean.

If the two assumptions

1. $H_0: \quad \mu = 15$
2. the sample is random

are true, we would expect the sample mean ($\bar{x} = 14.4$) to be representative of the population for which $\mu = 15$ and to fall somewhere near the center of the sampling distribution of the mean. On the other hand, if $\bar{x} = 14.4$ falls too far into the lower tail of the distribution, this would indicate an unusual (or rare) event and signal that perhaps something was wrong.

How can we determine if $\bar{x} = 14.4$ is located "too far" into the left-tail area? As we learned in Section 10.2, the level of significance $\alpha = .05$ assists us in this determination (see Figure 10-2). Recall that the sample mean C that has 5% of the sample means to the left of it is called the critical value for the test and determines the rejection region. Thus, if $\bar{x} = 14.4$ is less than C, we will reject H_0;

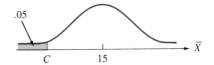

FIGURE 10-2 Critical value for hypothesis test

otherwise, we will fail to reject H_0. What remains is to determine the critical value C, located within the normal distribution having $\mu = 15$ and $\sigma = 1.2/\sqrt{50}$.

The critical value C can be found by using the z score formula:

$$z = \frac{C - \mu}{\sigma_{\bar{x}}}$$

$$= \frac{C - 15}{1.2/\sqrt{50}}$$

$$= \frac{C - 15}{.17}$$

By examining the normal table, we find $z_{.05} = 1.65$. Thus, we have the following equation:

$$-1.65 = \frac{C - 15}{.17}$$

Multiplying both sides of the above equation by .17 and then adding 15 to both sides, we get $C = 14.72$. The rejection region becomes all values of $\bar{x}$ less than 14.72:

Rejection region	Fail to reject region	
	$C = 14.72$	$\bar{x}$

Since the value of our test statistic, $\bar{x} = 14.4$, falls in the rejection region, we reject $H_0: \mu \geq 15$ in favor of $H_1: \mu < 15$ being true. It thus follows that we have statistically significant information that supports the alternative hypothesis $H_1: \mu < 15$ as being the true state of affairs. As a result, we can conclude that the average weight of a 15-ounce package of chocolate chips is less than 15 ounces. Of course, with this decision we may have committed a Type I error, the error of rejecting H_0 when it is true. But the chance of doing this is only 5%, or 1 chance in 20.

If the test is predicated on the assumptions that H_0 is true and the sample is random, then to have the test statistic $\bar{x} = 14.4$ fall in the rejection region signals that something may be wrong—either the sample is not random or H_0 is not true. If we are confident concerning the randomness of the sample, then it is probably true that $H_0: \mu \geq 15$ is false. That is, $\mu \geq 15$ is false and $\mu < 15$ is true. If H_0 is true, a sample mean as small as 14.4 (i.e., 14.4 or smaller) would occur only 5% of the time.

In the hypothesis test we have just conducted, if H_0 is true and $n \geq 30$, then the sampling distribution of the mean is approximately normal and has a mean equal to 15. In Figure 10-3, this distribution is identified with the normal curve drawn with the solid line. If the sample was random, we would expect the value of $\bar{x}$ to fall near 15, the mean of its distribution. In our illustration, the observed value of the sample mean was $\bar{x} = 14.4$, which suggests that $\mu_{\bar{x}} \leq 15$ in order for the observed sample mean $\bar{x} = 14.4$ to be located near the mean of

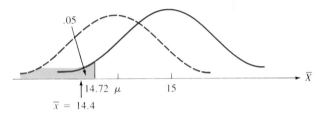

FIGURE 10-3 Sampling distribution of mean for
hypothesis test

its sampling distribution. Such a distribution is indicated by the dashed normal
curve in Figure 10-3. Reflecting back on the error term $e = \bar{x} - \mu = -.6$ (if H_0
is true), we see that the data suggest this difference cannot be attributed solely
to sampling error alone.

In summary, the following important terminology was used in our hypothesis testing procedure for the above illustration:

- Null hypothesis, H_0: $\mu \geq 15$
- Alternative hypothesis, H_1: $\mu < 15$
- Level of significance, $\alpha = .05$
- Value of test statistic, $\bar{x} = 14.4$
- Critical value, $C = 14.72$
- Rejection region, all values of $\bar{x}$ less than $C = 14.72$
- Sampling distribution of the mean, the normal distribution with $\mu_{\bar{x}} = 15$
  and $\sigma_{\bar{x}} = 1.2/\sqrt{50} = .17$

Equivalent Testing Procedures

There are three equivalent procedures for deciding whether the null hypothesis
H_0 is false. They are described in terms of the previous example as follows:

1. Compare the value of the test statistic $\bar{x}$ (calculated from the sample) to
   the critical value obtained from the sampling distribution of the mean
   with $\mu_{\bar{x}} = 15$ and $\sigma_{\bar{x}} = .17$.
2. Find the z value for the value of the test statistic and compare it with
   $-z_{.05}$ (the z value corresponding to an area equal to $\alpha = .05$ in the left
   tail of the standard normal distribution).
3. Find the probability of obtaining a value at most as large as the test
   statistic $\bar{x} = 14.4$ and compare this with the significance level $\alpha = .05$.
   If this probability is less than α, we will reject H_0; otherwise, we will
   fail to reject H_0.

We have already used the first procedure with the illustration in this section.
We now illustrate the second procedure, which will be our preference in later
work dealing with hypothesis testing.

For the second procedure, we find the z value for $\bar{x} = 14.4$ and compare it with $-z_{.05}$. The z value for $\bar{x} = 14.4$ is

$$z = \frac{\bar{x} - \mu}{\sigma_{\bar{x}}}$$

$$= \frac{14.4 - 15}{.17}$$

$$= -3.53$$

Within the standard normal distribution, $-z_{.05}$ is found to be -1.65. Since $z = -3.53 < -z_{.05} = -1.65$, we reject H_0 in favor of H_1 being true. When we use the second procedure to arrive at a decision regarding H_0, the z value for the sample mean $\bar{x}$ becomes the **test statistic** and the tabulated value of z corresponding to α in the standard normal table becomes the **critical value**. The only difference between the first and second approaches is the normal distribution that is used—the standard normal or some other normal distribution (see Figure 10-4 comparing the normal distributions used for procedures 1 and 2).

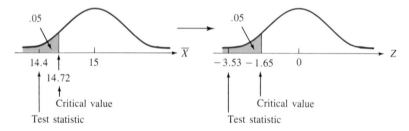

FIGURE 10-4 Normal distributions of test statistics

Before we examine the third procedure for testing a null hypothesis, let's consider several examples illustrating the details of the second procedure.

EXAMPLE 10-2   A cold tablet is supposed to contain 10 grains of aspirin. A random sample of 100 tablets yielded a mean of 10.2 grains and a standard deviation of 1.4 grains. Can we conclude at the .05 level of significance that μ is different from 10?

Solution   Note that in this case, the alternative hypothesis indicates a nondirectional test. As such, $\mu < 10$ or $\mu > 10$ and the rejection region consists of values in the left and right tails of the distribution. Since the probability of committing a Type I error, rejecting H_0 when it is true, is .05 and both tails constitute the rejection region, we place $\alpha/2 = .025$ of the distribution in each of the two tail regions, as indicated in the accompanying figure.

1. H_0: $\mu = 10$
2. H_1: $\mu \neq 10$ (two-tailed test)

3. Decision rule: Reject H_0 if $z < -z_{.025}$ or $z > z_{.025}$.

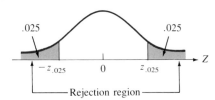

4. Sampling distribution: The sampling distribution of the mean. The test statistic is the z value for $\bar{x}$. Since σ is unknown and $n = 100$, the sample standard deviation s provides a good estimate for σ. The test statistic is computed as

$$z = \frac{\bar{x} - \mu}{\sigma/\sqrt{n}} = \frac{10.2 - 10}{1.4/\sqrt{100}} = 1.43$$

5. The critical values are $\pm z_{.025} = \pm 1.96$, as shown in the figure.

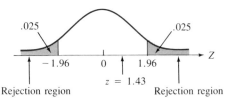

6. Decision: We cannot reject H_0. We thus conclude that there is no statistical support for μ being different from 10. Our failure to reject H_0: $\mu = 10$ is interpreted to mean that our finding is not statistically significant at $\alpha = .05$.

7. Type of error possible: Type II, since H_0 may be false and we fail to reject it. The probability β in this case is unknown. As a result, the experimenter should reserve judgment on H_0 until more data are gathered. In this case, the decision is "H_0 cannot be rejected." As we pointed out earlier, this statement does not imply that H_0 is accepted as being true or plausible.   ■

EXAMPLE 10-3   It is known that the voltages of a certain brand of size C batteries are normally distributed. A random sample of 15 batteries was tested and it was found that $\bar{x} = 1.4$ volts and $s = .21$ volt. At the .01 level of significance, does this indicate that $\mu < 1.5$ volts?

Solution   **1.** H_0: $\mu \geq 1.5$
2. H_1: $\mu < 1.5$ (one-tailed test, left tail)
3. $\alpha = .01$
4. Sampling distribution: Distribution of the sample mean. The test statistic is the t value for $\bar{x}$:

$$t = \frac{\bar{x} - \mu}{s/\sqrt{n}}, \quad df = 14$$

Note that t is used instead of z, since $n < 30$ and σ is unknown. In addition, we use the normality assumption here, since the t statistic requires sampling from a normal population. The value of the test statistic is

$$t = \frac{1.4 - 1.5}{21/\sqrt{15}} = -1.84$$

5. Critical value: $-t_{.01} = -2.624$. This value was obtained from the t table (see the back endpaper) and represents the t value corresponding to $\alpha = .01$ and df = 14 (see the accompanying figure).

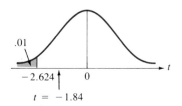

6. Decision: Since $t = -1.84 > -2.624 = -t_{.01}$, we cannot reject H_0. Hence, there is no evidence to support that $\mu < 1.5$. In this case the difference $e = \bar{x} - \mu_{\bar{x}} = 1.4 - 1.5 = -.1$ can be attributed to sampling error alone.
7. Type of error possible: Type II; the value of β is unknown.   ∎

EXAMPLE 10-4   A commercial school advertises that its students can type an average of 80 words per minute (wpm) upon graduation. A sample of 60 recent graduates was tested and the results showed $\bar{x} = 78$ wpm and $s = 6.2$ wpm. Can this difference of $78 - 80 = -2$ be explained by sampling error alone, or is $\mu < 80$? Test at the .05 level of significance.

Solution   1. H_0: $\mu \geq 80$
2. H_1: $\mu < 80$ (one-tailed test, left tail)
3. $\alpha = .05$
4. Sampling distribution: Distribution of the mean. The test statistic is the z value for $\bar{x}$:

$$z = \frac{\bar{x} - \mu}{\sigma_{\bar{x}}}$$

$$= \frac{78 - 80}{6.2/\sqrt{60}} = -2.50$$

5. Critical value: $-z_{.05} = -1.65$
6. Decision: Reject H_0, since $-2.50 < -1.65$. The difference -2 is not likely due to sampling error alone. We conclude that $\mu < 80$. The finding is significant at the .05 level.
7. Type of error possible: Type I. The probability of rejecting H_0 when it is true is .05.   ∎

p-Values

The third procedure for deciding the fate of a null hypothesis H_0 involves the concept of p-value.

When computer programs involving library packages (such as SPSS, BMDP, and SAS) are used to analyze statistical data, it is common for the resulting printouts to contain p-values. The **p-value** (or **probability value**) associated with a test statistic is the smallest level of significance that would have resulted in H_0 being rejected. Thus, for a right-tailed test, the p-value is the probability that the test statistic is at least as large as the value of the statistic calculated from the sample under the assumption that H_0 is true. For a left-tailed test, the p-value corresponds to the probability that the test statistic is at most as large as the value of the statistic calculated from the sample when H_0 is true. And for a two-tailed test, the p-value is the probability of getting a sample result at least as extreme in either direction as the one obtained when H_0 is true. If the level of significance of a test is greater than or equal to the reported p-value, then H_0 is rejected. Thus, the following rule can be used for deciding whether to reject the null hypothesis:

Reject H_0 if p-value $< \alpha$

The p-value provides an indication of how strongly the data disagree with H_0. Consider the following illustrations. In each case, the p-value is indicated by the shaded area in the figure.

    a. Left-tailed test: We would reject H_0 if p-value $< \alpha$.

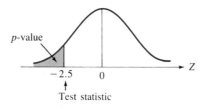

    b. Right-tailed test: We would reject H_0 if p-value $< \alpha$.

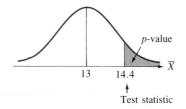

c. Two-tailed test: We would reject H_0 if p-value $< \alpha$.

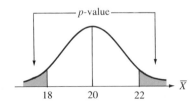

The following printout was obtained by using Minitab, a general-purpose statistical computing system developed at Pennsylvania State University. The ZTEST command was used to test the null hypothesis $H_0: \mu = .06$ versus the alternative hypothesis $H_1: \mu \neq .06$.

```
ZTEST MU = 0.0600, SIGMA = .0005, DATA IN C1
C1    N = 15   MEAN = 0.060020 ST. DEV. = .000576
TEST OF MU = 0.0600 VS. MU N.E.  0.0600
THE ASSUMED SIGMA = 0.0005
Z = 0.155
THE TEST IS SIGNIFICANT AT 0.8769
CANNOT REJECT AT ALPHA = 0.05
```

For this example the p-value is .8769. It is common practice by some researchers to report that the hypothesis test is significant at the p-value. Recall that if the p-value is greater than the significance level α, then the null hypothesis cannot be rejected.

EXAMPLE 10-5   For the left-tailed test in Example 10-4, compute the p-value.

Solution   To compute the p-value, we need to find the area under the standard normal curve to the left of the test statistic, as indicated in the accompanying figure. By examining the normal table, we find the p-value to be $.5 - .4938 = .0062$. Thus, p-value $= .0062$. Since p-value $< \alpha = .05$, we reject H_0. This test would have been significant at any level of significance less than .0062.

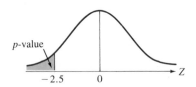

EXAMPLE 10-6   For the left-tailed test in Example 10-3, compute the p-value.

Solution   The t value for the test statistic is -1.84. We need to find the area under the t curve with df $= 14$ to the left of $t = -1.84$ (see the accompanying figure). We scan along the df $= 14$ row of the t table (see the back endpaper) and find that $t = 1.84$ lies between $t_{.05}(14) = 1.761$ and $t_{.025}(14) = 2.145$. Thus, $.025 <$ p-value $< .05$. Since $\alpha = .01$ and p-value $> \alpha$, we cannot reject H_0.

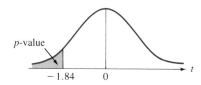

EXAMPLE 10-7   For the two-tailed test in Example 10-2, compute the p-value.

Solution   Since this test is two-tailed, we need to compute the p-value using both tails. Since the p-value is the smallest level of significance that would have resulted in H_0 being rejected and since the distribution is symmetric, we just double the p-value for one side. The value of the test statistic was $z = 1.43$. We thus need to find the area under the z curve to the right of $z = 1.43$; this value will represent $\frac{1}{2}(p\text{-value})$, as illustrated in the figure. From the z table (see the front endpaper), we find $\frac{1}{2}(p\text{-value}) = .5 - .4236 = .0764$. Hence, p-value $= 2(.0764) = .1528$. This test would have been significant at the 15.3% significance level. Since p-value $> \alpha = .05$, we cannot reject H_0.

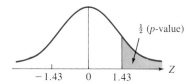

Comparing Confidence Intervals and Two-Tailed Hypothesis Tests

There are two standard techniques for making an inference about the value of an unknown parameter: estimation and hypothesis testing. We discussed estimation in Chapter 9. A comparison of an unknown parameter with a known constant involving a two-tailed test at a significance level equal to α can be made by constructing a $(1 - \alpha)100\%$ confidence interval for the parameter. If the hypothesized value of the parameter is contained in the confidence interval, then we cannot conclude that the parameter is different from the known constant.

As an illustration, refer again to the cold-tablet data in Example 10-2, and let's construct a 95% confidence interval for the average amount of aspirin contained in the cold tablet. Recall that the confidence interval limits are found by using

$$\bar{x} \pm z_{\alpha/2}\left(\frac{\sigma}{\sqrt{n}}\right)$$

The critical value is $z_{.025} = 1.96$. Since $n = 100$ and σ is unknown, s provides a good estimate of σ. Hence, the limits are

$$10.2 \pm (1.96)\left(\frac{1.4}{\sqrt{100}}\right)$$

$$10.2 \pm .27$$

A 95% confidence interval for μ is (9.93, 10.47). Since the hypothesized value 10 is contained in the interval, we cannot conclude that $\mu \neq 10$. Note that this result is the same conclusion we arrived at by using the hypothesis testing procedure in Example 10-2.

A confidence interval provides more information than does a hypothesis test. Based on the data, we could reject the null hypothesis and find that the result is of no practical significance. But by using the corresponding confidence interval and some common sense, we can determine if the significant results of the hypothesis test are of any practical importance. For example, let's consider a hypothetical situation in which a new study guide for increasing SAT mathematics scores is being tested. The mean and standard deviation for the SAT mathematics scores are known to be 500 and 100, respectively. A sample of 1,000,000 students is used to determine whether the new study guide produces a mean SAT mathematics score different from 500. The null hypothesis is that the mean SAT mathematics test score for students using the new study guide is 500, the same as the group not using the new study guide. That is, the null hypothesis is

$$H_0: \quad \mu = 500$$

Let's suppose that the group using the new study guide has a mean SAT mathematics score of 500.4. By using the formula

$$\bar{x} \pm z_{.025}\left(\frac{\sigma}{\sqrt{n}}\right)$$

we find that a 95% confidence interval for μ is

$$500.4 \pm (1.96)\left(\frac{100}{\sqrt{1,000,000}}\right)$$

$$500.4 \pm (1.96)(.1)$$

$$(500.2, 500.6)$$

Since 500 is not contained in the interval, we would reject the null hypothesis and conclude that the new study guide does have a statistically significant effect on the SAT mathematics test scores. The average test score for the group, $\bar{x} = 500.4$, is 4 standard deviations above the mean of its sampling distribution. But does this significant finding have any practical importance? Surely, few people (if any) would use the new study guide in order to raise their SAT mathematics score by .4 point. This example, while hypothetical, does serve to illustrate two points:

1. A hypothesis test may yield a significant finding but have no *practical* importance.
2. A large sample size increases the chance of rejecting the null hypothesis.

We learned in Chapter 9 that as the sample size increases, the width of the confidence interval shrinks to 0. When the sample is the entire population, $\bar{x} = \mu$ and the width of the confidence interval is 0. In this situation, any null hypothesis $H_0: \mu = \mu_0$ would be rejected, except for the case where μ_0 is the true value of μ. From a theoretical point of view, any null hypothesis can be rejected by choosing a large enough sample. One might then conclude that failure to reject a null hypothesis is a result of the sample not being large enough. Of course, in most practical applications involving hypothesis tests, the amount of data collected is based on economic considerations, as well as the nature of the experiment. For some experiments, such as in the study of rare diseases, it is not possible to gather a large amount of data.

Problem Set 10.3

A

For each problem, conduct a hypothesis test and determine the following information:

a. H_0
b. H_1 and type of test (one-tailed or two-tailed)
c. α
d. Sampling distribution
e. Test statistic
f. Critical value(s)
g. Decision
h. Type of error possible and associated probability (if a Type I error)
i. *p*-value

1. A newspaper article stated that college students at a state university spend an average of $95 a year on beer. A student investigator who believed the stated average was too high polled a random sample of 50 students and found that $\bar{x} = \$92.25$ and $s = \$10$. Use these results to test at the .05 level of significance the statement made by the newspaper and find the *p*-value.

2. A survey of 50 homemakers selected at random showed that they watch television an average of 15 hours per week; the standard deviation was 12.5 hours. Test $H_0: \mu \geq 20$ against $H_1: \mu < 20$ at the .02 level of significance and find the *p*-value.

3. Truckloads of coal arriving at a power plant are contracted to carry 10 tons of coal per load. A

sample of 15 loads showed $\bar{x} = 9.5$ tons and $s = .9$ ton. If the distribution of weights is assumed to be normal, test the null hypothesis $H_0: \mu \geq 10$ versus the alternative hypothesis $H_1: \mu < 10$ by using a level of significance of .01. Find the *p*-value.

4. A random sample of ten high-school seniors took a standardized mathematics test and made the following scores:

88	86	90	84	85
89	91	86	83	87

Past scores at the same high school have been normally distributed with $\sigma = 4$ and $\mu = 85$. Test $H_0: \mu \leq 85$ against $H_1: \mu > 85$. Use a level of significance of .05 and find the *p*-value.

5. The scores of ten students on a statistics examination were as follows:

43	61	67	70	74
76	79	85	94	81

Assuming these scores are from a normal population, test $H_0: \mu = 70$ against $H_1: \mu \neq 70$ at the .05 level of significance and find the *p*-value.

6. A certain restaurant advertises that it puts .25 pound of beef in its burgers. A customer who frequents the restaurant thinks the burgers contain less than .25

pound of beef. With permission from the owner, the customer selected a random sample of 50 burgers and found that $\bar{x} = .23$ and $s = .12$. Test the restaurant's claim using a level of significance of .10 and find the p-value.

7. A dairy advertises that a tub of its ice cream produces an average of 84 scoops. An ice-cream parlor that buys wholesale from the dairy found that its clerks obtained an average of 83.7 scoops per tub from 72 tubs. The standard deviation was 11.43 scoops. Test the null hypothesis $H_0: \mu \geq 84$ against the alternative hypothesis $H_1: \mu < 84$ by using a level of significance of .05 and find the p-value. To what do you attribute the results?

8. The lengths (in cm) of a random sample of 30 trout caught at Piney Dam were determined; it was found that $\bar{x} = 37.8$ and $s^2 = 27.04$. Do the data indicate that $\mu > 35$? Test by using the .05 level of significance and find the p-value.

9. The weights (in pounds) of a random sample of 6-month-old babies are 14.6, 12.5, 15.3, 16.1, 14.4, 12.9, 13.7, and 14.9. Test at the 5% level of significance to determine if the average weight of all 6-month-old babies is different from 14 pounds and find the p-value. Assume that the weights of 6-month-old babies are normally distributed.

10. For Problem 5, construct a 95% confidence interval for μ. Does the interval contain $\mu = 70$? Compare your results with those obtained in Problem 5.

11. For Problem 9, construct a 95% confidence interval for μ. Does the interval contain $\mu = 14$? Compare your results with those obtained in Problem 9.

12. A sugar manufacturer uses an automatic machine to fill 5-pound bags with sugar. The machine was initially set to dispense an average of 5 pounds, and to have a standard deviation of .12 pound. To determine if the machine is out of adjustment, a sample of 35 bags was weighed, and it was found that the average weight was 4.9 pounds. Use the 5% level of significance to determine if the machine needs readjusting.

13. For Problem 12, construct a 95% confidence interval for μ. Does the interval contain $\mu = 5$? Compare your results with those obtained in Problem 12.

B

14. Refer to Problem 8. If the true value of μ is 37, find β, the probability of committing a Type II error.

15. Refer to Problem 6. If the true value of μ is .24, find the probability of committing a Type II error.

10.4 TESTING PROPORTIONS AND VARIANCES ────────────

Testing Proportions

The hypothesis testing procedure for testing proportions is similar to the procedure for testing means. The logic is the same; the only difference is that we work with sampling distributions of proportions rather than means. Recall that if n is large, the sampling distribution of $\hat{p} = x/n$ is approximately normal with $\mu_{\hat{p}} = p$ and $\sigma_{\hat{p}} = \sqrt{p(1-p)/n}$. A sample is considered large if both $np \geq 5$ and $n(1-p) \geq 5$. Care should be taken not to confuse the p-value (introduced in Section 10.3) with p, the population proportion, since both are numbers between 0 and 1, inclusive.

Consider the following examples.

EXAMPLE 10-8   A doctor claims that 12% of all his appointments are canceled. Over a 6-week period, 21 of the doctor's 200 appointments were canceled. Test at $\alpha = .05$ to determine if the true proportion of all appointments that are canceled is different from 12%.

Solution   **1.** H_0: $p = .12$

2. H_1: $p \neq .12$ (two-tailed test)

3. $\alpha = .05$

4. Sampling distribution: Distribution of sample proportion. The test statistic is the z value for $\hat{p}$, where $\hat{p} = \frac{21}{200} = .105$.

Note that $np = (200)(.12) = 24$ and $n(1 - p) = (200)(.88) = 176$. Thus, n can be considered large and the sampling distribution of $\hat{p}$ is approximately normal. The z value for $\hat{p}$ is

$$z = \frac{\hat{p} - \mu_{\hat{p}}}{\sigma_{\hat{p}}}$$

$$= \frac{\hat{p} - p}{\sqrt{p(1 - p)/n}}$$

$$= \frac{.105 - .12}{\sqrt{(.12)(.88)/200}} = -.65$$

Note that we used the hypothesized value of p in the denominator of z since we assumed that H_0: $p = .12$ is true. In Chapter 9, we used $\hat{p}$ in the construction of a confidence interval for p.

5. Critical values: $\pm z_{.025} = \pm 1.96$ (see the accompanying figure).

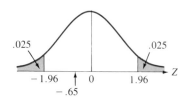

6. Decision: We cannot reject H_0. Thus, we have no statistical evidence to reject the doctor's claim.

7. Type of error possible: Type II; the probability β is unknown.

8. p-value: We have the situation illustrated in the figure. Thus, the p-value is $2(.2422) = .4844$.

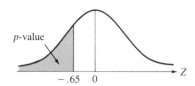

EXAMPLE 10-9   Suppose it has been suggested that pregnant women who once took birth-control pills are more likely to have girls than boys. A random sample of 30 mothers who once used the pill and subsequently gave birth to one child produced 7

boys and 23 girls. Do the data indicate that girls are more likely to be born than boys to mothers who once used the pill? Use the .05 level of significance.

Solution   Suppose p represents the proportion of girls born to mothers who once used the pill. The value of the sample proportion is $\hat{p} = \frac{23}{30} = .767$.

1. H_0: $p \le .5$
2. H_1: $p > .5$ (one-tailed, right tail)
3. $\alpha = .05$
4. Sampling distribution: Distribution of $\hat{p}$. Note that $np = n(1 - p) = 30(.5) = 15 \ge 5$. The test statistic is the z value for $\hat{p}$:

$$z = \frac{\hat{p} - p}{\sqrt{p(1 - p)/n}}$$

$$= \frac{.767 - .5}{\sqrt{(.5)(.5)/30}} = 2.92$$

5. Critical value: $z_{.05} = 1.65$
6. Decision: Since $2.92 > 1.65$, we reject H_0. We thus conclude that the proportion of girls born to mothers who once used the pill is greater than .5. In other words, we can conclude that more girls than boys are born to mothers who once used the pill.
7. Type of error possible: Type I. The probability of a Type I error is $\alpha = .05$.
8. p-value: We have the situation shown in the figure. Therefore, the p-value is $.5 - .4982 = .0018 = .18\%$. Thus, the finding is significant at the .2% level.

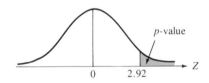

Testing Variances

Many practical applications require making inferences about population variances. In quality control studies, the concern is consistency of the product. For example, a bolt manufacturing company may be interested in testing its manufacturing process to determine whether bolt diameters have a variance smaller than an allowable specification.

Recall that if a random sample is drawn from a normal population, then the statistic $(n - 1)s^2/\sigma^2$ has a sampling distribution which is a chi-square distribution with df $= (n - 1)$. That is,

$$\chi^2 = \frac{(n - 1)s^2}{\sigma^2}$$

serves as our test statistic, and the critical values are read directly from the chi-square table of probabilities (Table 2 of Appendix B). Consider the following example.

EXAMPLE 10-10   An engineer believes that the variance in waiting times (in seconds) by machinists at a tool cage is greater than 25. To test this, he selects a random sample of 25 machinists' times logged in at the cage and found that $s^2 = 41.4$ square seconds. Assuming that the waiting times are normally distributed, test the null hypothesis $H_0: \sigma^2 \leq 25$ using the .05 level of significance.

Solution   **1.** H_0: $\sigma^2 \leq 25$
2. H_1: $\sigma^2 > 25$ (one-tailed, right tail)
3. $\alpha = .05$
4. Sampling distribution: χ^2, df $= 25 - 1 = 24$. The value of the χ^2 test statistic is

$$\chi^2 = \frac{(n-1)s^2}{\sigma^2}$$

$$= \frac{(24)(41.4)}{25}$$

$$= 39.74$$

5. Critical value: $\chi^2_{.05}(24) = 36.415$ (found in Table 2, Appendix B)
6. Decision: Since $39.74 > 36.415$, we reject H_0. (See the accompanying figure.)

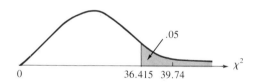

The engineer can thus conclude that the variance in waiting times is greater than 25.
7. Type of error possible: Type I, $\alpha = .05$
8. *p*-value: Reading across the row labeled df $= 24$ in Table 2 (Appendix B), we see that the chi-square test statistic 39.74 falls between $\chi^2_{.025}(24) = 39.364$ and $\chi^2_{.01}(24) = 42.980$. Thus, $.01 < p\text{-value} < .025$.   ■

Sometimes we are interested in testing a null hypothesis involving the population standard deviation. Since the variance is equal to the square of the standard deviation, the procedure used to test a null hypothesis involving the variance can be used. Consider the following example.

EXAMPLE 10-11   An automatic sugar packaging machine is used to fill 5-pound bags of sugar. A random sample of 15 bags indicated that $\bar{x} = 4.94$ pounds and $s = .02$ pound. If the weights are assumed to be normal and from past experience it is known that the standard deviation of the weights is .015 pound, can the apparent increase in variability be explained by sampling error alone? Use the .05 level of significance.

Solution 1. H_0: $\sigma \le .015$
2. H_1: $\sigma > .015$ (right-tailed test)
3. $\alpha = .05$
4. Note that if we assume that the null hypothesis H_0: $\sigma = .015$ is true, then the hypothesis H_0: $\sigma^2 = (.015)^2$ is also true and we can use the chi-square distribution. The test statistic is χ^2. Its value is

$$\chi^2 = \frac{(n-1)s^2}{\sigma^2}$$

$$= \frac{(15-1)(.02)^2}{(.015)^2}$$

$$= 24.89$$

5. Critical value: $\chi^2_{.05}(14) = 23.685$
6. Decision: Reject H_0, since $\chi^2 = 24.89 > \chi^2_{.05}(14) = 23.685$.
7. Type of error possible: Type I, $\alpha = .05$
8. *p*-value: By reading across the row labeled by df $= 14$ in Table 2 (Appendix B), we locate $\chi^2 = 24.89$ between $\chi^2_{.05}(14) = 23.685$ and $\chi^2_{.025}(14) = 26.119$. Thus, $.025 < p\text{-value} < .05$.

Thus, the increase in variability should not be attributed to sampling error alone, but to some other factor; perhaps the machine is not being adequately calibrated. ∎

Problem Set 10.4

1. Conduct each of the following hypothesis tests at $\alpha = .01$:
 a. H_0: $p = .5$, H_1: $p \ne .5$; $\hat{p} = .45$, $n = 500$
 b. H_0: $p = .5$, H_1: $p \ne .5$; $\hat{p} = .6$, $n = 100$
 c. H_0: $p \ge .6$, H_1: $p < .6$; $\hat{p} = .5$, $n = 200$
 d. H_0: $p \le .3$, H_1: $p > .3$; $\hat{p} = .34$, $n = 50$

2. Conduct the following hypothesis tests:
 a. H_0: $\sigma^2 \ge 100$, H_1: $\sigma^2 < 100$; $s^2 = 50$,
  $n = 28$; $\alpha = .05$
 b. H_0: $\sigma^2 \le 200$, H_1: $\sigma^2 > 200$; $s^2 = 250$,
  $n = 25$; $\alpha = .01$
 c. H_0: $\sigma^2 = 150$, H_1: $\sigma^2 \ne 150$; $s^2 = 120$,
  $n = 37$; $\alpha = .05$

3. From a random sample of 500 males interviewed, 125 indicated that they watch professional football on Monday night television. Does this evidence indicate that more than 20% of the male TV viewers watch professional football on Monday evenings? Use the .01 level of significance, and find the *p*-value.

4. A random sample of 300 shoppers in a shopping mall is selected and 182 are found to favor longer shopping hours. Is this sufficient evidence to conclude that less than 65% of the shoppers favor longer shopping hours? Use $\alpha = .05$ and find the *p*-value.

5. A candidate for the local board of education thinks that at least 55% of the voters will vote for her. If a random sample of 50 voters indicated that 46% would vote for her, test H_0: $p \ge .55$ against H_1: $p < 55$ at $\alpha = .05$. Find the *p*-value.

6. If 60% of a sample of 100 new-car buyers indicated that they want air conditioning, does this indicate that less than 70% of all new-car buyers desire air conditioning? Use $\alpha = .02$ and find the *p*-value.

7. A national survey indicated that 60% of all new homes constructed have four bedrooms. If a random sample of 100 newly constructed homes indicated 52 with four bedrooms, does this suggest that the national survey is wrong? Use $\alpha = .05$ and find the p-value.

8. The following sample was taken from a normal population:

12	10	13	12	11	13
14	13	14	12	10	

Test $H_0: \sigma \geq 1.5$ against $H_1: \sigma < 1.5$ using the .05 level of significance.

9. The following sample was drawn from a normal population:

41	10	25	5	10	10
30	19	6	10	14	14
41	25	14	30	25	14
30	25				

By using $\alpha = .01$, test $H_0: \sigma^2 = 121$ versus $H_1: \sigma^2 \neq 121$.

10. The following sample is thought to have come from a normal population with variance $\sigma^2 = 25$:

2.1	3.6	3.8	4.2	4.7	15.3

Test at the .05 level of significance to determine if $\sigma^2 \neq 25$.

11. The Metro Bus Company in a large city claims to have a variance in bus-arrival times (arrival times measured in minutes) at its various bus stops of no more than 5. A bus company executive ordered that arrival data be collected at various bus stops in order to determine if bus drivers are maintaining consistent schedules. If a sample of 12 bus arrivals at a particular bus stop produced a variance of 5.7 and the population of arrival times is assumed to be normal, test the null hypothesis $H_0: \sigma^2 \leq 5$ versus $H_1: \sigma^2 > 5$. Use $\alpha = .05$.

12. The variance in the diameters of roller bearings during production is of critical importance. Large variances in bearing diameters promote wear and bearing failure. Industry standards call for a variance of no more than .0001 when the bearing diameters are measured in inches. A bearing manufacturer selected a random sample of 25 bearings and found that $s = .015$ inch. Does this indicate that $\sigma^2 > .0001$? Use $\alpha = .01$ and assume that the bearing diameters are normally distributed.

13. For Problem 7, construct a 95% confidence interval for the true proportion of all new homes constructed that have four bedrooms. Compare your results with those obtained in Problem 7.

14. For Problem 10, construct a 95% confidence interval for the true population variance σ^2. Compare your results with those obtained in Problem 10.

CHAPTER SUMMARY

In this chapter we discussed hypothesis testing, an alternative to estimation for making inferences. When using hypothesis testing, we are interested in determining whether a parameter is different from a specified value. The logic of hypothesis testing is indirect; that is, we assume the null hypothesis is true, and based on this assumption, we determine if our test statistic is likely to be from the hypothesized population. If not, we conclude that the null hypothesis should be rejected in favor of the alternative hypothesis. Since for each type of decision—reject H_0 or fail to reject H_0—there is a probability of committing an error, we never "prove" anything using hypothesis testing. Failure to reject the null hypothesis does not necessarily mean that we will "accept" the null hypothesis. Instead, we shall reserve judgment on H_0 since the probability of committing a Type II error is unknown. We discussed tests for μ, p, σ^2, and σ. The test procedures are summarized in Table 10-2.

Table 10-2 Summary of Hypothesis Tests Covered in Chapter 10

Parameter	Null hypothesis	Assumptions	Sampling distribution
μ	$H_0: \mu = \mu_0$	Normal population or $n \geq 30$	Normal
μ	$H_0: \mu = \mu_0$	Normal population $n < 30$ σ unknown	t distribution
p	$H_0: p = p_0$	$n \geq 30$	Normal
σ^2	$H_0: \sigma^2 = \sigma_0^2$	Normal population	χ^2 distribution
σ	$H_0: \sigma = \sigma_0$		

C H A P T E R   R E V I E W

IMPORTANT TERMS

For each of the following terms, provide a definition in your own words. Then check your responses against the definitions given in the chapter.

alternative hypothesis	hypothesis testing	one-tailed test	significant finding
critical value	left-tailed test	power of the test	statistical hypothesis
decision rule	level of significance	*p*-value	test statistic
directional test	nondirectional test	rejection region	two-tailed test
double-blind experiment	null hypothesis	right-tailed test	Type I error
experimental hypothesis	null value	significant	Type II error

IMPORTANT SYMBOLS

H_0, null hypothesis

H_1, alternative hypothesis

α, probability of Type I error, level of significance

β, probability of Type II error

C, critical value

IMPORTANT FACTS AND FORMULAS

If the null hypothesis cannot be rejected, judgment should be reserved concerning the truth of the null hypothesis. One should conclude that the data offer no significant evidence to warrant that the null hypothesis be rejected in favor of the alternative hypothesis.

Table 10-2 (*continued*)

Test statistic	Critical value(s)	
	1-tailed test	**2-tailed test**
$z = \dfrac{\bar{x} - \mu_0}{\sigma/\sqrt{n}}$	z_α or $-z_\alpha$	$\pm z_{\alpha/2}$
$z = \dfrac{\bar{x} - \mu_0}{s/\sqrt{n}}$		
$t = \dfrac{\bar{x} - \mu}{s/\sqrt{n}}$	$t_\alpha(\mathrm{df})$ or $-t_\alpha(\mathrm{df})$	$\pm t_{\alpha/2}(\mathrm{df})$
$z = \dfrac{\hat{p} - p_0}{\sqrt{p_0(1 - p_0)/n}}$	z_α or $-z_\alpha$	$\pm z_{\alpha/2}$
$\chi^2 = \dfrac{(n-1)s^2}{\sigma_0^2}$	$\chi_\alpha^2(\mathrm{df})$ or $\chi_{1-\alpha}^2(\mathrm{df})$	$\chi_{\alpha/2}^2(\mathrm{df})$ and $\chi_{1-\alpha/2}^2(\mathrm{df})$

REVIEW PROBLEMS

1. A large television repair service claims that its average repair charge is $24. Suspecting that the average repair charge is higher, a consumer group randomly chose a sample of 35 statements for TV repairs done by the repair service and found $\bar{x} = \$25.50$ and $s = \$2.25$. Test the claim at the 5% level of significance and find the *p*-value.

2. A basketball team claims that their opponents scored an average of at most 87 points against them over the past 10 years. Suspecting that the true average was somewhat higher, a sports reporter gathered a random sample of 30 game summaries for the period. He found $\bar{x} = 89.6$ points and $s = 6.2$ points. Test the team's claim at $\alpha = .05$.

3. A newspaper in a large city advertised that 62% of the registered voters were opposed to abortions. A social service agency, believing the estimate was too high, polled a random sample of 500 registered voters and found that 290 were opposed to abortions. Using the 5% level of significance, test to determine if the newspaper's estimate is too high and find the *p*-value.

4. A coffee dispenser at a local cafeteria is supposed to dispense an average of 7 ounces of coffee. Suspecting that the average amount dispensed is somewhat lower, a customer obtained a random sample of 15 cups over a period of 2 weeks and found $\bar{x} = 6.4$ ounces and $s = .71$ ounce. Assuming that the amounts of coffee dispensed are normally distributed, test at $\alpha = .01$ to determine if the average amount of coffee dispensed is less than 7 ounces.

5. The weights (in pounds) of a random sample of 16-year-old male high-school students are 146, 149, 137, 153, 125, 219, and 161. Assuming that the weights of 16-year-old boys are normally distributed, test the hypothesis that the average weight of 16-year-old boys is different from 140 pounds. Use the .05 level of significance.

6. The statistics department at a certain university has never been able to achieve a failure rate less than 11% for its introductory statistics course. During an experimental semester, all students enrolled in introductory statistics were required to attend a 1-hour laboratory in addition to classes with the hope of lowering the failure rate. At the end of the semester, 171 students out of 1800 students failed. Test using the .05 level of significance to determine if there has been a significant decrease in failure rate since the lab was instituted.

7. A beer company uses dispensing machines to fill beer cans that provide a maximum variance of .05

(amount of beer is measured in ounces) so that cans are not overfilled or underfilled. A sample of fills for 25 cans yielded $s^2 = .07$. If the amounts of beer dispensed are normally distributed, test the null hypothesis $\sigma^2 \le .05$ against the alternative hypothesis $\sigma^2 > .05$. Use $\alpha = .05$.

8. A moped manufacturer advertises that its moped gets an average of at least 127.6 miles per gallon of gasoline. The standard deviation σ is known to be 9.8 miles per gallon and the gas mileage ratings are normally distributed. Suspecting the manufacturer's claim is too high, a consumer took a sample of ten mopeds and found a sample mean of 120.6 miles per gallon and a sample standard deviation of 8.4 miles per gallon. Test the manufacturer's claim using $\alpha = .05$.

9. A new drug, Cyclosporin-A, is claimed to have been 86% successful in increasing the success rate in 30 organ transplant operations. Before this new drug was available, a success rate of 60% had been obtained with organ transplant patients. By using $\alpha = .05$, determine if the success rate has improved as a result of the new drug.

10. For Problem 8, test using $\alpha = .05$ to determine if $\sigma \ne 9.8$.

11. One-pound cans of nuts are to contain a net weight of 16 ounces, but there is considerable variability. A random sample of six cans of brand A nuts revealed the following net weights (in ounces):

$$16.1 \quad 15.8 \quad 15.1 \quad 15.4 \quad 16.1 \quad 16.2$$

Using $\alpha = .01$, determine if the true net weight is different from 16 ounces.

12. A new alloy is claimed to have a tensile strength of 120 pounds. A sample of seven independent tests provided the following readings (in pounds):

$$116.5 \quad 118.7 \quad 122.3 \quad 118.7 \quad 122.3 \quad 122.6 \quad 121.6$$

At $\alpha = .05$, do the data indicate that $\mu \ne 120$?

CHAPTER ACHIEVEMENT TEST

(20 points)   **1.** A claim is made that the mean height of male college teachers is 71 inches. In an investigation to test the claim, a random sample of 12 male teachers yielded $\bar{x} = 72$ inches and $s = 3$ inches. Assuming that the heights of male teachers are normally distributed, test the claim at the .05 level of significance and provide the following information:
a. Test statistic   **b.** Critical value(s)   **c.** Decision   **d.** p-value

(20 points)   **2.** A particular medicine is claimed to be 85% effective in alleviating a certain type of allergic reaction. A consumer group believes the medicine is less than 85% effective and gathered a sample of 60 people who experience the type of allergic reaction. Of this group who used the medicine, 48 people got relief. Test the claim at the .01 level of significance and provide the following information:
a. Test statistic   **b.** Critical value(s)   **c.** Decision   **d.** p-value

(20 points)   **3.** A manufacturer claims that the diameters of its 8-mm bolts have a variance of at most .02. The diameters of a sample of 20 8-mm bolts yielded a variance of .025. Test at $\alpha = .05$ to determine if $\sigma^2 > .02$, and provide the following information:
a. Test statistic   **b.** Critical value   **c.** Decision

(20 points)   **4.** A new type of radial tire is tested to determine if it can average at least 60,000 miles of road wear. A sample of 35 tires was experimentally tested and it was determined that $\bar{x} = 59,600$ miles and $s = 968$ miles. At $\alpha = .05$, do the data indicate that $\mu < 60,000$ miles?

(20 points)   **5.** In Problem 4, assume that tread wear is normally distributed. Test, using $\alpha = .01$, to determine if $\sigma < 1200$ miles.

11

INFERENCES COMPARING TWO PARAMETERS

Chapter Objectives

In this chapter you will learn:

- *about two types of samples used in making inferences about two populations*

- *the properties of the sampling distribution of the differences between sample means*

- *how to construct confidence intervals for $(\mu_1 - \mu_2)$*

- *how to test the null hypothesis $H_0: \mu_1 - \mu_2 = 0$*

- *the properties of the sampling distribution of the differences between sample proportions*

- *how to construct confidence intervals for $(p_1 - p_2)$*

- *how to test the null hypothesis $H_0: p_1 - p_2 = 0$*

- *the properties of the F distributions*

- *how to construct confidence intervals for σ_1^2/σ_2^2*

- *how to test the null hypothesis $H_0: \sigma_1^2 = \sigma_2^2$*

Chapter Contents

Many practical applications involve the comparison of two populations. We commonly compare two populations by comparing their corresponding parameters, such as μ_1 and μ_2, p_1 and p_2, σ_1 and σ_2, or σ_1^2 and σ_2^2. For example, population means μ_1 and μ_2 might be compared when one is deciding which brand of toothpaste, A or B, is more effective in preventing cavities. Or population percentages p_1 and p_2 might be compared when trying to decide if a greater proportion of males than females favor lowering the minimum wage. Two population variances σ_1^2 and σ_2^2 might be compared when a drug manufacturer is interested in comparing the consistencies of two different methods of producing a heart drug containing 25 mg of digoxin.

When researchers compare two parameters, such as μ_1 and μ_2, it is common practice to consider the difference of the parameters $(\mu_1 - \mu_2)$. By determining whether the difference $(\mu_1 - \mu_2)$ equals zero, we can make comparisons concerning μ_1 and μ_2. That is, if $(\mu_1 - \mu_2) = 0$, then $\mu_1 = \mu_2$; if $(\mu_1 - \mu_2) > 0$, then $\mu_1 > \mu_2$; and if $(\mu_1 - \mu_2) < 0$, then $\mu_1 < \mu_2$. To estimate the difference $(\mu_1 - \mu_2)$, we usually use the point estimator $(\bar{x}_1 - \bar{x}_2)$; to estimate $(p_1 - p_2)$, we use the point estimator $(\hat{p}_1 - \hat{p}_2)$, the difference between sample proportions. When certain assumptions are satisfied, inferences comparing two parameters can be made by considering the sampling distribution of the difference between sample statistics, such as $(\bar{x}_1 - \bar{x}_2)$ and $(\hat{p}_1 - \hat{p}_2)$. We begin by drawing a distinction between two basic types of samples used in comparing two population parameters.

11.1 INDEPENDENT AND DEPENDENT SAMPLES

In order to make statistical inferences about two populations, we need to have a sample from each population. The two samples will be independent or dependent, according to how they are selected. If the selection of sample data from one population is unrelated to the selection of sample data from the other population, the samples are called **independent samples**. If the samples are chosen in such a way that each measurement in one sample can be naturally paired with a measurement in the other sample, the samples are called **dependent samples**. Each piece of data results from some source. A **source** is anything—a person or an object—that produces a piece of data. If two measurements result from the same source, then the measurements can be thought of as being paired. As a result, two samples resulting from the same set of sources are dependent. The following illustrations will help clarify the two types of samples:

1. Ten overweight adults were randomly selected to evaluate a particular diet. Each person was weighed before beginning the diet and again after being on the diet 12 weeks. The sample of weights before dieting and the sample of weights after dieting are dependent samples. A source is a person, who provides two measurements, one for each sample. The paired observations provide the sample measurements.

2. A farmer from the midwest conducted an experiment to determine if the use of a special chemical additive with the fertilizer he has been using to grow

soybeans will accelerate plant growth. Fifteen locations were randomly chosen for the study. At each location, two soybean plants located close to one another were treated, one with the standard fertilizer and one with the standard fertilizer with the chemical additive. Plant growth was measured (in inches) for each plant after a 4-week period. The plant measurements associated with each type of fertilizer constitute two dependent samples. A source is a location; each location produces a pair of measurements.

3. A medical researcher compared two flu vaccines, A and B, for localized side effects. Ten subjects were randomly selected and each subject was injected twice, with vaccine A in the left arm and vaccine B in the right arm. Sufficient time was allowed between injections to serve as a *washout* period for vaccine A. The side effects of each vaccine were measured using a special numerical index. The set of numerical indices for each vaccine constitutes a sample, and the two samples are dependent. A source is a person.

4. In an experiment to determine whether persons afflicted with glaucoma have abnormally thick corneas, eight persons with glaucoma in only one eye were examined. The cornea thickness (in microns) was measured for each person's eyes. The eight thickness measurements for the glaucomatous eyes constitute one sample, and the eight other eye measurements constitute the other sample. The samples are dependent, and each person serves as a source.

5. In an experiment to determine whether persons afflicted with glaucoma have abnormally thick corneas, 16 subjects were involved. Eight had glaucoma and eight did not. Cornea thickness was measured (in microns) for each subject. The measurements for the glaucoma patients comprise one sample and the cornea measurements for the subjects not afflicted with glaucoma comprise the other sample. The two samples are independent; no pairing is involved.

6. Twelve 1-week old white male infants from middle- and upper-middle-class families were involved in an experiment to determine whether special walking exercises in the newborn can lower the average age at which infants first walk alone. Two groups were randomly chosen from the twelve infants. Group A received special training for an 8-week period, while group B received no special training. After the 8-week period, the ages (in months) when the children first walked were recorded. The ages for the two groups comprise independent samples.

EXAMPLE 11-1   A medical researcher wants to determine whether drug therapy can improve the IQs of children with learning and behavioral problems. An experiment will be conducted in which two groups of children, A and B, are used. Both groups are to be the same size. Group A will receive a placebo for 6 weeks, while group B will receive a widely used anticonvulsant drug for 6 weeks. A verbal IQ test will be used to assess results at the end of the 6-week period.

a. Describe how you would obtain two independent samples of IQ scores to evaluate.

b. Describe how you would obtain two dependent samples of IQ scores to evaluate.

Solution **a.** Randomly divide 20 children with learning and behavioral problems into two groups of ten. One group will receive the placebo (no therapy) and the other group will receive the drug therapy. The two samples of IQ test scores will be independent.

b. Choose ten children with learning and behavioral difficulties. Each child will be administered the placebo for 3 weeks and the drug for 3 weeks. After each 3-week period, all children will be administered an IQ test. Because a child might do better the second time he or she takes the IQ test, the order in which the placebo and drug are administered will be randomized; some children will be given the placebo first and some will be given the drug first. IQ scores for each child will be recorded following the 3-week placebo period and following the 3-week drug period. The two samples of ten IQ scores form dependent samples. A source in each case is a student. ∎

EXAMPLE 11-2 Ten pairs of identical twins are used in an experiment to determine which of two experimental methods is better for teaching statistics. One method involves a small-group discovery method and the other method involves a lecture/discussion format with computer-assisted instruction. One member of each pair of twins is assigned to each method. Following the instruction, an examination is administered to each group. Are the two samples of test scores independent? Explain.

Solution Identical twins come from the same fertilized egg and therefore have the same set of genes, which determine their physical and mental characteristics. As a result, the samples can be considered dependent. The pairing is natural. ∎

It should be noted that if two samples are dependent, then they necessarily have the same size. Therefore, if two samples have different sizes, they cannot be dependent.

Why Use Dependent Samples?

The procedures for making inferences about two population means involve the sample means calculated from two dependent or two independent samples. Dependent samples are used to control the effects of certain "nuisances," thus reducing the effects of unwanted factors. For example, suppose we were given the task of determining the wearing quality of two different brands of tires, A and B. To do this, we must road-test the two brands of tires. Automobiles must be selected, equipped with both brands of tires, and then driven for a fixed distance under similar conditions. Tread wear will then be measured (in thousandths of an inch) for each brand of tire. If we want the tread-wear measurements to reflect only the quality of the tires, then we want to control as many of the following factors as possible that could also affect tread wear:

1. Type of car, including size and weight
2. Mechanical condition of the car
3. Driver's driving habits

4. Type of roads, including terrain
5. Distance traveled
6. Location of tires on the car
7. Weather conditions

If we randomly choose, say, ten cars, randomly place one tire of each brand on the rear wheels of each car, and then drive the cars a fixed distance under similar conditions, we can control or limit the influence that the above factors can have on the results. Those factors we cannot control, such as driver habits and course traveled, should have an equal effect on both brands of tires due to the way we designed the experiment. Thus, the tread-wear measurements should reflect the wearing qualities of the two brands of tires.

Inferences concerning two populations using dependent samples are covered in Section 11.5. The remaining sections of this chapter deal with the methods for making inferences about population parameters using independent samples.

Problem Set 11.1

1. An experiment to test a new variety of sugar cane against an old variety was conducted on ten different 1-acre plots. The new variety was expected to yield higher sugar content than the variety currently used. The ten different 1-acre plots were chosen where soil and general climate conditions vary. Each acre was divided into two $\frac{1}{2}$-acre plots containing similar soil, and the new variety was planted in one $\frac{1}{2}$-acre plot and the old type was planted in the other $\frac{1}{2}$-acre plot. After harvest, the average sugar content for each $\frac{1}{2}$-acre plot was determined. Do the samples of data for the two varieties of sugar cane represent dependent or independent samples? Explain.

2. An experiment is to be designed to compare two competing headache remedies, A and B, for fast-acting relief. Sixteen sets of identical twins are available for the study. Eight subjects are to be assigned to each of two groups. Each subject in one group is to be administered 500 mg of remedy A and each subject in the other group is to be administered 500 mg of remedy B. The length of time (in minutes) for each drug to reach a specified level in the blood will be recorded for each individual.
   a. Describe how you could obtain two independent samples of absorption times to evaluate.
   b. Describe how you could obtain two dependent samples of absorption times to evaluate.

3. To test whether honeybees show a preference for stinging objects that have already been stung, an experiment was repeated ten times. Each time eight cotton balls wrapped in muslin were dangled up and down in front of a beehive entrance. Four of the balls were exposed to a swarm of angry bees and were filled with stingers, while the other four were fresh. After a specified length of time, the number of new stingers in each ball was counted. Are the two sets of bee-sting counts independent or dependent? Explain.

4. An experiment was conducted in Milwaukee, Wisconsin, to test the effectiveness of artificial food flavoring in rat poison. For each survey, approximately 1600 poison baits were placed around garbage storage bins; half of the poison baits were plain cornmeal and half were butter-vanilla flavored cornmeal. To ensure that a rat would have equal access to both kinds of poison bait, the baits were always placed in pairs. After 2 weeks, the sites were inspected and the percentage of poison baits that were gone was recorded. A different set of locations in the same general vicinity was selected and the experiment was repeated again for another 2-week period. This same procedure was repeated three more times. For each of the five surveys, the percentage of baits accepted was recorded for each poison. Are the two samples of five percentages independent or dependent? Explain.

5. An experiment was designed to determine if fluoride helps to prevent tooth cavities. Fifteen children had their teeth cleaned and treated with flouride. Another 15 had their teeth cleaned, but received no fluoride. Six months later, the number of cavities was recorded

for each child. Are the two sets of cavity counts independent or dependent? Explain.

6. Describe how you would redesign the experiment of Problem 5 to obtain samples of the opposite type.

7. A manufacturer of waterproofing for footwear claims its product is superior to the leading brand. Ten pairs of shoes are available for a test.

   a. Explain how you would conduct a test using dependent samples. How would you make your assignments of waterproofing?

   b. Explain how you would conduct a test using independent samples.

8. To determine what effect a rat's environment early in life has on his behavior late in life, 16 rats from the same litter were separated from their mothers when they were 3 weeks old. One-half (group A) were put into individual cages and the other half (group B) were put into the same cage. After living under these conditions for a period of approximately 8 months, each rat was individually put through a series of swimming trials to determine the length of time required to swim a specified distance underwater. Based on the swimming times, the rats within each group were ranked, and the rats in the two groups were paired according to respective ranks: the rat ranking #1 in group A was paired with the rat ranking #1 in group B, and so on. Rats in each pair had approximately the same swimming rates. Pair by pair, the rats were subjected to competitive underwater swimming competitions. Average times were recorded for each rat under competition. Each of the 8 pairs of rats competed 60 times. The average times for each group constitute a sample. Are the two samples independent or dependent? Explain.

11.2 INFERENCES CONCERNING $(\mu_1 - \mu_2)$ USING LARGE, INDEPENDENT SAMPLES

In each of the following situations, one might be interested in comparing two population means, μ_1 and μ_2:

- *Medicine:* Determining which of two treatments is more effective in treating a certain disease
- *Education:* Comparing test scores to determine the better mode of instruction
- *Manufacturing:* Comparing two automobiles for the better average gas mileage
- *Sports:* Comparing two defensive strategies in basketball by examining opponents' scores
- *Agriculture:* Comparing two types of fertilizer on crop yields
- *Law:* Comparing two municipal courts in a large city for the average length of time it takes to try a case
- *Science:* Comparing a new synthetic material for strength against an old material
- *Business:* Comparing two methods of marketing a product
- *Mining:* Comparing two methods of mining coal
- *Religion:* Comparing attitudes between youth and adults concerning the Bible

There are few new ideas covered in this section that were not presented in Chapter 10; apart from involving a new sampling distribution, the techniques and logic involved in comparing two parameters are the same as those presented in Chapter 10.

Before we can begin to make inferences concerning the difference between two population means, we need to develop the background needed for making such comparisons. We begin by examining the sampling distribution of the differences between sample means.

The Sampling Distribution of the Differences between Sample Means

Suppose we want to decide which of two brands of toothpaste, A or B, is more effective in preventing cavities in teenagers. How do we decide whether a difference exists, or how can we estimate the difference? We would agree that if there is a difference in therapeutic benefits, then this difference should be reflected in the difference between the average numbers of cavities for teenagers using the two brands. A starting point is to gather some data from two different groups of teenagers, one using brand A and the other using brand B. The best estimate for the difference between population means is the corresponding difference between the means for the two samples. The difference in sample means $(\bar{x}_1 - \bar{x}_2)$ becomes our estimator for the difference in population means $(\mu_1 - \mu_2)$. Therefore, we need to examine and understand the sampling distribution of $(\bar{x}_1 - \bar{x}_2)$; we will need to use it to determine if there is a difference in the corresponding population parameters by using a hypothesis test or estimation.

In general, suppose we have two distinct populations, the first with mean μ_1 and standard deviation σ_1, and the second with mean μ_2 and standard deviation σ_2. Further, suppose a random sample of size n_1 is selected from the first population and an independent random sample of size n_2 is selected from the second population. The sample mean is computed for each sample and the difference between sample means is calculated. The collection of all such differences is called the **sampling distribution of the differences between means** or **the sampling distribution of the statistic** $(\bar{x}_1 - \bar{x}_2)$. Figure 11-1 illustrates the sampling distribution of $(\bar{x}_1 - \bar{x}_2)$.

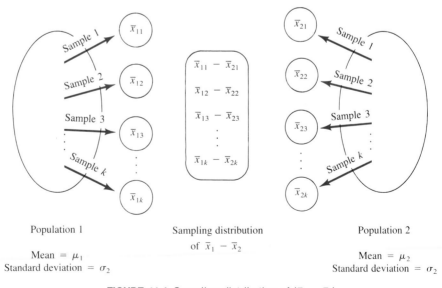

FIGURE 11-1 Sampling distribution of $(\bar{x}_1 - \bar{x}_2)$

The sampling distribution of $(\bar{x}_1 - \bar{x}_2)$ has the following properties:

1. The distribution is approximately normal for $n_1 \geq 30$ and $n_2 \geq 30$. If the populations are normal, then the sampling distribution is normal regardless of the sample sizes.

2. $\mu_{\bar{x}_1 - \bar{x}_2} = \mu_1 - \mu_2$ (11-1)

3. $\sigma_{\bar{x}_1 - \bar{x}_2} = \sqrt{\dfrac{\sigma_1^2}{n_1} + \dfrac{\sigma_2^2}{n_2}}$

The following examples illustrate the sampling distribution of the differences between sample means.

EXAMPLE 11-3   In a study to compare the average weights of sixth-grade boys and girls at a large middle school, a random sample of 20 boys and a random sample of 25 girls are to be used. All sixth-grade boys at the school have an average weight of 100 pounds and a standard deviation of 14.142 pounds, while all sixth-grade girls have an average weight of 85 pounds and a standard deviation of 12.247 pounds. If $\bar{x}_1$ represents the average weight of a sample of 20 boys and $\bar{x}_2$ represents the average weight of a sample of 25 girls, find $P(\bar{x}_1 - \bar{x}_2 > 20)$, the probability that the average weight for the 20 males is at least 20 pounds more than the average weight for the 25 females.

Solution   A value for the difference between sample means $(\bar{x}_1 - \bar{x}_2)$ is an element in the sampling distribution of the differences between sample means. As a result of (11-1), we have:

1. The sampling distribution of $(\bar{x}_1 - \bar{x}_2)$ is approximately normal.

2. $\mu_{\bar{x}_1 - \bar{x}_2} = \mu_1 - \mu_2 = 100 - 85 = 15$

3. $\sigma_{\bar{x}_1 - \bar{x}_2} = \sqrt{\dfrac{\sigma_1^2}{n_1} + \dfrac{\sigma_2^2}{n_2}}$

$= \sqrt{\dfrac{(14.142)^2}{20} + \dfrac{(12.247)^2}{25}} = 4.0$

Thus, to determine $P(\bar{x}_1 - \bar{x}_2 > 20)$, we find the area under the following standard normal curve:

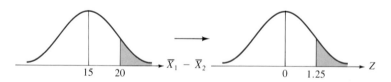

The z value for a difference of 20 is given by

$z = \dfrac{20 - \mu_{\bar{x}_1 - \bar{x}_2}}{\sigma_{\bar{x}_1 - \bar{x}_2}}$

$= \dfrac{20 - 15}{4} = 1.25$

Hence, we have

$$P(\bar{x}_1 - \bar{x}_2 > 20) = P(z > 1.25)$$
$$= .5 - .3944 = .1056$$

Therefore, the probability that the average weight of the sample of boys is at least 20 pounds greater than the average weight of the sample of girls is .1056.

■

EXAMPLE 11-4   A leading television manufacturer purchases picture tubes from two companies, A and B. The tubes from company A have a mean lifetime of 7.2 years and a standard deviation of .8 year, while those from company B have a mean lifetime of 6.7 years and a standard deviation of .7 year. Determine the probability that a random sample of 34 picture tubes from company A will have a mean lifetime that is at least 1 year more than the mean lifetime of a random sample of 40 picture tubes from company B.

Solution   We summarize the given information as follows:

Company A	Company B
$\mu_1 = 7.2$	$\mu_2 = 6.7$
$\sigma_1 = .8$	$\sigma_2 = .7$
$n_1 = 34$	$n_2 = 40$

As a consequence of (11-1), the mean of the sampling distribution of differences between sample means is

$$\mu_{\bar{x}_1 - \bar{x}_2} = \mu_1 - \mu_2 = 7.2 - 6.7 = .5$$

and the standard error of the difference between sample means is

$$\sigma_{\bar{x}_1 - \bar{x}_2} = \sqrt{\frac{\sigma_1^2}{n_1} + \frac{\sigma_2^2}{n_2}}$$
$$= \sqrt{\frac{(.8)^2}{34} + \frac{(.7)^2}{40}} = .176$$

If the mean of the sample from company A is at least 1 year more than the mean of the sample from company B, then $\bar{x}_1 - \bar{x}_2 \geq 1$. The z value for $(\bar{x}_1 - \bar{x}_2) = 1$ is given by

$$z = \frac{1 - .5}{.176} = 2.84$$

Thus, the required probability is

$$P(\bar{x}_1 - \bar{x}_2 > 1) = P(z > 2.84)$$
$$= .5 - .4977 = .0023$$

Hence, the probability that the mean lifetime of a random sample of 34 picture tubes from company A will exceed the mean lifetime of a random sample of 40 picture tubes from company B by at least 1 year is .0023. ∎

It is interesting to note that if independent samples of sizes n_1 and n_2 are drawn from the same normal population with mean μ and variance σ^2, then by (11-1) we have

1. The sampling distribution of $(\bar{x}_1 - \bar{x}_2)$ is normal.
2. $\mu_{\bar{x}_1 - \bar{x}_2} = \mu - \mu = 0$
3. $\sigma_{\bar{x}_1 - \bar{x}_2} = \sqrt{\dfrac{\sigma^2}{n_1} + \dfrac{\sigma^2}{n_2}}$

$$= \sigma \sqrt{\dfrac{1}{n_1} + \dfrac{1}{n_2}}$$

(11-2)

Example 11-5 illustrates sampling from the same normal population.

EXAMPLE 11-5   Suppose two independent samples of size 20 are drawn from a normal population with mean 10 and variance 2.5. Find $P(\bar{x}_1 - \bar{x}_2 < 1)$.

Solution   Since the sampling distribution of the differences between sample means is normal, we will find the z score for 1 and then determine the area under the standard normal curve below this z value. From (11-2), we know that the mean of the sampling distribution of the differences between sample means is

$$\mu_{\bar{x}_1 - \bar{x}_2} = 10 - 10 = 0$$

and the standard deviation of the sampling distribution of the differences between means is

$$\sigma_{\bar{x}_1 - \bar{x}_2} = \sqrt{2.5} \sqrt{\frac{1}{20} + \frac{1}{20}}$$

$$= \sqrt{(2.5)(.10)} = .5$$

The z score for 1 thus becomes

$$z = \frac{1 - \mu_{\bar{x}_1 - \bar{x}_2}}{\sigma_{\bar{x}_1 - \bar{x}_2}}$$

$$= \frac{1 - 0}{.5} = 2$$

Thus, $P(\bar{x}_1 - \bar{x}_2 < 1) = P(z < 2) = .5 + .4772 = .9772.$ ∎

If two independent samples of the same size are selected from the same population, one would expect the difference between any pair of sample means to be close to 0 for large sample sizes. This can be seen by examining the variance of the sampling distribution of the difference between sample means:

$$\sigma^2_{\bar{x}_1 - \bar{x}_2} = \frac{\sigma^2}{n} + \frac{\sigma^2}{n} = \frac{2\sigma^2}{n}$$

As n gets large, the values of $2\sigma^2/n$ will approach 0. Thus, for large values of n, the variability of the differences between means is small, and the differences between sample means are close to their mean, which is 0.

Confidence Intervals for $(\mu_1 - \mu_2)$

Confidence intervals can be used to compare two population means, μ_1 and μ_2. To do so, a $(1 - \alpha)100\%$ confidence interval for $(\mu_1 - \mu_2)$ is constructed. If the interval contains 0, then μ_1 could equal μ_2, or μ_1 could be smaller than μ_2, or μ_1 could be larger than μ_2. If the confidence interval contains only negative values, then one can be $(1 - \alpha)100\%$ confident that $(\mu_1 - \mu_2) < 0$ or $\mu_1 < \mu_2$. Or, if the confidence interval contains only positive values, one can be $(1 - \alpha)100\%$ confident that $(\mu_1 - \mu_2) > 0$ or $\mu_1 > \mu_2$. Recall from Chapter 9 that the limits for a confidence interval can be obtained by employing a point estimator and its standard error. Consequently, the limits for a $(1 - \alpha)100\%$ confidence interval for $(\mu_1 - \mu_2)$ can be obtained by using the following:

$$(\bar{x}_1 - \bar{x}_2) \pm z_{\alpha/2}\sigma_{\bar{x}_1 - \bar{x}_2} \qquad (11\text{-}3)$$

Point estimate   *Standard error*

When a confidence interval is constructed for the difference between population means, there are two possibilities for expressing the difference between means: $(\mu_1 - \mu_2)$ and $(\mu_2 - \mu_1)$. The confidence intervals for $(\mu_1 - \mu_2)$ and $(\mu_2 - \mu_1)$ are not identical, but they are related in a special way. If (L_1, L_2) is a $(1 - \alpha)100\%$ confidence interval for $(\mu_1 - \mu_2)$, then $(-L_2, -L_1)$ is a $(1 - \alpha)100\%$ confidence interval for $(\mu_2 - \mu_1)$. This can be seen by observing that if

$$L_1 < \mu_1 - \mu_2 < L_2$$

then by multiplying both inequalities by -1, we have $-L_1 > \mu_2 - \mu_1 > -L_2$. And rewriting these inequalities, we have

$$-L_2 < \mu_2 - \mu_1 < -L_1$$

The choice of which interval to construct is arbitrary, but we need to be consistent when interpreting the interval. As a general rule, when we are requested to determine a confidence interval for the difference between population means and the difference is not specified, we shall use the interval corresponding to a positive difference between sample means.

Hypothesis Tests for $(\mu_1 - \mu_2)$

Hypothesis tests can also be used to compare μ_1 and μ_2. The logic and procedures for testing are identical to those presented in Chapter 10. Two samples of data are collected and the value of the test statistic $(\bar{x}_1 - \bar{x}_2)$ is located in its sampling distribution under the assumption that H_0 is true. If the value of the test statistic falls in the tail area(s) determined by α, then the value of the statistic is determined to be unlikely from the assumed normal population, and the null hypothesis H_0 is rejected in favor of the alternative hypothesis H_1. And if the value of the statistic falls near the center of its sampling distribution, there is no statistical evidence to suggest that H_0 is not true. When $H_0: \mu_1 - \mu_2 = 0$ is tested, the z value for the test statistic $(\bar{x}_1 - \bar{x}_2)$ is computed using the following formula:

$$z = \frac{(\bar{x}_1 - \bar{x}_2) - 0}{\sqrt{\dfrac{\sigma_1^2}{n_1} + \dfrac{\sigma_2^2}{n_2}}} \tag{11-4}$$

When directional hypotheses (e.g., $H_1: \mu_1 - \mu_2 < 0$) are tested, an upper-tailed test for one researcher might be a lower-tailed test for another. For example, if one is interested in establishing that $\mu_1 > \mu_2$, there are two possibilities for stating H_1:

$$H_1: \quad \mu_1 - \mu_2 > 0 \qquad \text{or} \qquad H_1: \quad \mu_2 - \mu_1 < 0$$

This is because the inequality $\mu_1 > \mu_2$ is equivalent to the inequality $\mu_2 < \mu_1$; hence, $(\mu_1 - \mu_2) > 0$ or $(\mu_2 - \mu_1) < 0$. If the alternative hypothesis is $H_1: \mu_1 - \mu_2 > 0$, then the test statistic becomes $(\bar{x}_1 - \bar{x}_2)$, and if the alternative hypothesis is $H_1: \mu_2 - \mu_1 < 0$, then the test statistic becomes $(\bar{x}_2 - \bar{x}_1)$. These observations are summarized below:

H_1	Test statistic
$\mu_1 - \mu_2 < 0$	$\bar{x}_1 - \bar{x}_2$
$\mu_2 - \mu_1 > 0$	$\bar{x}_2 - \bar{x}_1$
$\mu_1 - \mu_2 \neq 0$	$(\bar{x}_1 - \bar{x}_2)$ or $(\bar{x}_2 - \bar{x}_1)$

It is generally a good practice to use the test statistic that preserves uniformity of "match" between the subscripts used for the sample means and the subscripts used for the corresponding population means. For example, we would choose $(\bar{x}_1 - \bar{x}_2)$ as our test statistic for the one-tailed test involving $H_1: \mu_1 - \mu_2 > 0$. It is best not to label directional tests involving two parameters as right-tailed or left-tailed.

The following examples illustrate the procedures of estimation and hypothesis testing for making inferences concerning a comparison of two population means, μ_1 and μ_2. In practice, a researcher would choose only one procedure, based on personal preference, for drawing an inference concerning $(\mu_1 - \mu_2)$. We will use both procedures, side-by-side, to illustrate that both procedures lead to consistent results.

EXAMPLE 11-6   The state department of education in a southern state compared high school seniors' knowledge of the basic skills in mathematics at two different high schools, one located in the northern part of the state and one located in the southern part of the state. Random samples of 50 seniors from each high school were obtained and given a standardized mathematics achievement examination. An analysis of the examination scores yielded the following results:

Northern school	Southern school
$n_1 = 50$	$n_2 = 50$
$\bar{x}_1 = 81.4$	$\bar{x}_2 = 84.5$
$s_1 = 4.6$	$s_2 = 4.0$

Determine whether μ_1 is significantly different from μ_2 using $\alpha = .05$.

Solution   *Estimation procedure.*   Since $\alpha = .05$, $1 - \alpha = .95$. Also, since $\bar{x}_2 > \bar{x}_1$, a 95% confidence interval will be constructed for ($\mu_2 - \mu_1$). The best point estimator for the difference between population means ($\mu_2 - \mu_1$) is the difference between sample means ($\bar{x}_2 - \bar{x}_1$). The value of ($\bar{x}_2 - \bar{x}_1$) is $84.5 - 81.4 = 3.1$. Since σ_1 and σ_2 are unknown and $n_1 = n_2 = 50$, the values of s_1 and s_2 can be used as point estimates for σ_1 and σ_2, respectively. In addition, since n_1 and n_2 are both greater than 30, the sampling distribution of the differences between sample means is approximately normal. The standard error of the difference between means is found as follows:

$$\sigma_{\bar{x}_2 - \bar{x}_1} = \sqrt{\frac{\sigma_2^2}{n_2} + \frac{\sigma_1^2}{n_1}}$$

$$= \sqrt{\frac{(4.0)^2}{50} + \frac{(4.6)^2}{50}} = .862$$

The positive critical value is $z_{.025} = 1.96$. Limits for the confidence interval are found using (11-3):

$$(\bar{x}_2 - \bar{x}_1) \pm z_{\alpha/2}\sigma_{\bar{x}_2 - \bar{x}_1}$$
$$3.1 \pm (1.96)(.862)$$
$$3.1 \pm 1.69$$

The 95% confidence interval for ($\mu_2 - \mu_1$) is $(1.41, 4.79)$. Thus, we can be 95% confident that ($\mu_2 - \mu_1$) is contained in the interval $(1.41, 4.79)$; 95% of all such intervals constructed will contain ($\mu_2 - \mu_1$). Since the confidence interval $(1.41, 4.79)$ contains only positive values, we can be 95% confident that $\mu_2 - \mu_1 > 0$ or $\mu_2 > \mu_1$.

Hypothesis testing procedure.   We carry out the following steps:

1. The null hypothesis is H_0: $\mu_1 - \mu_2 = 0$.
2. The alternative hypothesis is H_1: $\mu_1 - \mu_2 \neq 0$ (two-tailed test).
3. The level of significance is $\alpha = .05$.
4. The sampling distribution is the distribution of the differences between sample means. The standard error of the difference was found above to be .862. The

z value for −3.1 is then

$$z = \frac{\bar{x}_1 - \bar{x}_2}{\sqrt{\dfrac{\sigma_1^2}{n_1} + \dfrac{\sigma_2^2}{n_2}}}$$

$$= \frac{-3.1}{.862} = -3.60$$

5. The critical values for *z* are $\pm z_{.025}$ or ± 1.96.

6. Decision: Since $-3.60 < -1.96$, we reject H_0.

7. Type of error possible: Type I; $\alpha = .05$.

8. *p*-value: $p = 2(.5 - .49984) = .00032$ (a value very close to 0).

  Note that with hypothesis testing we cannot conclude that $(\mu_1 - \mu_2) < 0$ or $\mu_1 < \mu_2$, since the alternative hypothesis $H_1: \mu_1 - \mu_2 \neq 0$ is a two-tailed test. We can conclude only that $\mu_1 \neq \mu_2$. Another directional hypothesis test would have to be used to statistically establish (at $\alpha = .05$) that $\mu_1 < \mu_2$. ■

  Although one-sided confidence intervals can be constructed to parallel directional hypothesis tests, we shall not develop these concepts since they are not consistent with the general objectives of this text.

  As another illustration of comparing two population means, let's consider the following example.

EXAMPLE 11-7 A large company wants to hire a secretary. Two private secretarial schools are available to recruiters. The personnel officer of the company gave a typing test to independent random samples of 50 recent graduates from each school. The test was developed to determine the correct number of words typed per minute. The following results were obtained:

School A	School B
$n_1 = 50$	$n_2 = 50$
$\bar{x}_1 = 67$	$\bar{x}_2 = 70$
$s_1 = 15$	$s_2 = 11$

  Based on the data, can we conclude at the 95% level of confidence that there is a significant difference between the average scores of students from the two schools? Which school should be chosen to recruit from, if all other factors are equal?

Solution Again, we shall use both procedures for drawing inferences to illustrate how the procedures compare. In practice, a researcher would use only one procedure.

  Hypothesis testing procedure. A personnel officer not trained in statistics might think that school B is better, since it has the higher sample average score.

But do the mean scores differ because of sampling error or do the means differ because graduates of school B actually have a greater average typing score than graduates of school A? Inferential statistics can help us answer this question. The difference between means $(\bar{x}_2 - \bar{x}_1)$ is $70 - 67 = 3$. Can this difference of 3 be attributed to random error alone? If the schools were identically effective and if two more samples of size 50 were obtained, we would not expect to obtain a difference in sample means of exactly 3.

Hypothesis testing.

1. The null hypothesis is H_0: $\mu_2 - \mu_1 = 0$.
2. The alternative hypothesis is H_1: $\mu_2 - \mu_1 \neq 0$ (two-tailed test).
3. The level of significance is $\alpha = .05$.
4. The sampling distribution is the distribution of differences between sample means, and the test statistic is the z value for $(\bar{x}_2 - \bar{x}_1)$:

$$z = \frac{(\bar{x}_2 - \bar{x}_1) - 0}{\sqrt{\dfrac{\sigma_1^2}{n_1} + \dfrac{\sigma_2^2}{n_2}}}$$

$$= \frac{3}{\sqrt{\dfrac{15^2}{50} + \dfrac{11^2}{50}}} = \frac{3}{2.63} = 1.14$$

5. The critical values are $\pm z_{.025}$ or ± 1.96.
6. Decision: Since $-1.96 < 1.14 < 1.96$, we cannot reject H_0.
7. Type of error possible: Type II; β is unknown.
8. *p*-value: $p = 2(.5 - .3729) = 2(.1271) = .2542$.

As a result of this test, we conclude that the observed difference of 3 between sample means can be accounted for by random error in sampling; there is no statistical evidence that one school is better than the other. Note that we should not conclude that there is no difference in school averages, since β is unknown.

Estimation.   We shall construct a $(1 - \alpha)100\% = (1 - .05)100\% = 95\%$ confidence interval for $(\mu_2 - \mu_1)$ (note that $\bar{x}_2 - \bar{x}_1$ is positive). We find the limits for the confidence interval by using (11-3):

$$(\bar{x}_2 - \bar{x}_1) \pm z_{\alpha/2}\sigma_{\bar{x}_2 - \bar{x}_1}$$
$$3 \pm (1.96)(2.63)$$
$$3 \pm 5.15$$

Hence, we find that $(-2.15, 8.15)$ is a 95% confidence interval for $(\mu_2 - \mu_1)$. Since the interval contains 0, any of the three relationships may be true: $\mu_1 = \mu_2$, $\mu_1 < \mu_2$, or $\mu_1 > \mu_2$. Thus, we would not want to conclude that the null hypothesis H_0: $\mu_2 - \mu_1 = 0$ is true.   ■

Problem Set 11.2

A

1. Two brands of golf balls, A and B, are to be compared with respect to driving distance. The balls are tested on an automatic driving device known to give normally distributed distances with a standard deviation of 15 yards. It is known that $\mu_A = 285$ yards and $\mu_B = 280$ yards. If random samples of 25 of each type of ball are hit, determine the probability that $(\bar{x}_A - \bar{x}_B)$ is greater than 11 yards.

2. At a certain eastern college, the average score for freshmen on an entrance examination is 450 and the standard deviation is 40. If two groups of freshman students are selected at random, one of size 40 and the other of size 45, what is the probability that the two groups of students will differ in their mean scores by more than 10 points?

3. Independent random samples of grades for males and females were selected from the student population of a large university in an effort to determine which sex had the higher grade-point average (GPA). The results are as follows:

Males	Females
$n_1 = 50$	$n_2 = 75$
$\bar{x}_1 = 2.1$	$\bar{x}_2 = 2.3$
$s_1 = .8$	$s_2 = .7$

a. Using hypothesis testing, determine if there is a difference in average college male and female GPAs. Use $\alpha = .01$.

b. Determine if there is a difference between average male and female GPAs by constructing a 99% confidence interval for the difference in averages.

4. A study was conducted to determine the difference between salaries of college science teachers and industrial employees who were once college science teachers. Two random samples of salary information were gathered and the results follow:

Science teachers	Industrial employees
$n_1 = 50$	$n_2 = 60$
$\bar{x}_1 = \$34,960$	$\bar{x}_2 = \$35,440$
$s_1 = \$1,200$	$s_2 = \$1,000$

a. Construct a 90% confidence interval for $(\mu_2 - \mu_1)$ and interpret it.

b. At the .05 level of significance, do the data support that $\mu_2 > \mu_1$? Find the p-value.

5. Samples of hourly wages of truck drivers in cities A and B yielded the following data:

City A	City B
$n_1 = 30$	$n_2 = 30$
$\bar{x}_1 = \$5.30$	$\bar{x}_2 = \$5.40$
$s_1 = \$.16$	$s_2 = \$.15$

Test the null hypothesis $H_0: \mu_2 - \mu_1 \leq 0$ against the alternative hypothesis $H_1: \mu_2 - \mu_1 > 0$ using $\alpha = .01$. Find the p-value.

6. The director of athletics at a large university was interested in determining whether male students who participate in college athletics are taller than other male students. Two independent random samples of height data (in inches) were collected and the following results were obtained:

Participants	Nonparticipants
$n_1 = 50$	$n_2 = 75$
$\bar{x}_1 = 68.2$	$\bar{x}_2 = 67.4$
$s_1 = 5.2$	$s_2 = 2.9$

Test the hypothesis at the .01 significance level that male students who participate in college athletics are taller than nonparticipants. Find the p-value.

7. Of 80 recently hired employees for a large firm, half were assigned to a special 1-day orientation class and half received no special orientation. After 3 months on the job, evaluations were conducted, producing the following information:

Received orientation	No orientation
$n_1 = 40$	$n_2 = 40$
$\bar{x}_1 = 84.1$	$\bar{x}_2 = 81.8$
$s_1 = 3.6$	$s_2 = 4.1$

At the .05 level of significance, do the data indicate that employees receiving special orientation perform better on the job than those who do not? Find the *p*-value.

8. The length of time (in days) to complete recovery for hernia patients randomly assigned to two different surgical procedures is as follows:

Procedure 1	Procedure 2
$n_1 = 30$	$n_2 = 35$
$\bar{x}_1 = 7.50$	$\bar{x}_2 = 8.25$
$s_1 = 1.12$	$s_2 = 1.38$

At $\alpha = .05$, do the data indicate a difference in mean recovery times for the two surgical procedures? Use estimation and hypothesis testing to arrive at your conclusion. What is the *p*-value for the test?

9. Two drugs, A and B, are compared for duration of pain relief in postoperative patients. Records are kept on the number of hours of pain relief for 40 randomly selected patients using drug A and 50 randomly selected patients using drug B. The results are as follows:

Drug A	Drug B
$n = 40$	$n = 50$
$\bar{x} = 5.14$	$\bar{x} = 4.53$
$s = 1.20$	$s = 1.79$

By using $\alpha = .05$, determine if drug A has a significantly longer duration of pain relief than drug B.

10. Two brands of cigarettes, C and D, are compared for their nicotine contents (in milligrams). Random samples of 40 brand C cigarettes and 50 brand D cigarettes yielded the following results:

Brand C	Brand D
$n = 40$	$n = 50$
$\bar{x} = 14.3$	$\bar{x} = 15.7$
$s = 2.9$	$s = 3.8$

a. At the 1% level of significance, do the two brands of cigarettes differ in their mean nicotine contents?
b. Construct a 99% confidence interval for the difference between mean nicotine contents for the two brands of cigarettes.

B

11. Two random samples of the same size are taken from normal populations with variances of 5 and 10. If $(\bar{x}_1 - \bar{x}_2)$ is used to estimate $(\mu_1 - \mu_2)$, determine the sample size needed in order to be 95% confident that $(\bar{x}_1 - \bar{x}_2)$ differs by no more than 1 unit from $(\mu_1 - \mu_2)$.

12. Two methods of teaching ninth-grade science are being compared, an old method and a new method. The following data on test scores resulted:

Old method	New method
$n_1 = 60$	$n_2 = 75$
$\bar{x}_1 = 68.1$	$\bar{x}_2 = 72.9$
$s_1 = 5.1$	$s_2 = 5.5$

At the .05 level of significance, test the hypothesis that the mean science score of students taught under the new method is over 3 points above the mean science score of students taught by the old method. Find the *p*-value.

11.3   INFERENCES COMPARING
POPULATION PROPORTIONS OR PERCENTAGES ⎯⎯⎯⎯⎯⎯⎯⎯⎯

Many applications involve populations of qualitative data that need to be compared by using proportions or percentages. The following are examples:

- *Politics:* Is there a difference between the percentages of Democrats and Republicans favoring SALT talks?

- *Education:* Is the proportion of students who pass mathematics greater than the proportion who pass English?
- *Medicine:* Is the percentage of users of drug A who have an adverse reaction less than the percentage of users of drug B who have an adverse reaction?
- *Management:* Is there a difference between the percentages of men and women in management positions?
- *Marketing:* Is the percentage of beer-drinkers who prefer Budweiser greater than the percentage who prefer Miller?

Both estimation and hypothesis testing procedures can be used to draw inferences comparing two population proportions or percentages. The techniques are similar to those involved with comparing population means using large samples. The major exception is that a sampling distribution of the differences between sample proportions is used instead of a sampling distribution of differences between sample means.

Sampling Distribution of $(\hat{p}_1 - \hat{p}_2)$

Many practical applications involve qualitative data that can be placed into one of two categories. In the case of one-sample problems encountered in Chapter 10, we saw that as a consequence of the central limit theorem, the sampling distribution of the sample proportion $\hat{p} = x/n$ is approximately normal with $\mu_{\hat{p}} = p$ and $\sigma_{\hat{p}} = \sqrt{p(1 - p)/n}$. When sampling is from two binomial populations and two sample proportions are involved, the sampling distribution of $(\hat{p}_1 - \hat{p}_2)$ is approximately normal for large sample sizes, again a consequence of the central limit theorem. The following properties of the sampling distribution of $(\hat{p}_1 - \hat{p}_2)$ also hold:

1. $\mu_{\hat{p}_1 - \hat{p}_2} = p_1 - p_2$

2. $\sigma_{\hat{p}_1 - \hat{p}_2} = \sqrt{\dfrac{p_1(1 - p_1)}{n_1} + \dfrac{p_2(1 - p_2)}{n_2}}$

$$(11\text{-}5)$$

where p_1 is the first population proportion, p_2 is the second population proportion, n_1 is the size of the first sample, and n_2 is the size of the second sample.

Consider the following example illustrating the sampling distribution of the difference between sample proportions.

EXAMPLE 11-8   Adult males and females living in a large northern city differ in their views concerning the issue of the death penalty for persons found guilty of murder. It is believed that 12% of the adult males favor the death penalty, while only 10% of the adult females favor the death penalty. If a random sample of 150 males and a random sample of 100 females are polled concerning their views on the issue of the death penalty for persons found guilty of murder, determine the probability that the percentage of males who favor the death penalty is at least 3% higher than the percentage of females who favor the death penalty.

Solution Let p_1 represent the percentage of males who favor the death penalty and p_2 represent the percentage of females who favor the death penalty. As a consequence of (11-5), the mean of the sampling distribution of the differences between sample proportions is

$$\mu_{\hat{p}_1 - \hat{p}_2} = p_1 - p_2 = .12 - .10 = .02$$

and the standard error of the differences between sample proportions is

$$\sigma_{\hat{p}_1 - \hat{p}_2} = \sqrt{\frac{p_1(1 - p_1)}{n_1} + \frac{p_2(1 - p_2)}{n_2}}$$

$$= \sqrt{\frac{(.12)(.88)}{150} + \frac{(.10)(.90)}{100}} = .04$$

Thus, the z value for $\hat{p}_1 - \hat{p}_2 = .03$ is given by

$$z = \frac{(\hat{p}_1 - \hat{p}_2) - \mu_{\hat{p}_1 - \hat{p}_2}}{\sigma_{\hat{p}_1 - \hat{p}_2}}$$

$$= \frac{.03 - .02}{.04} = .25$$

Hence,

$$P(\hat{p}_1 - \hat{p}_2 \geq .03) = P(z \geq .25) = .5 - .0987 = .4013$$

Thus, the probability that the percentage of males who favor the death penalty for persons found guilty of murder is at least 3% higher than the percentage of females who favor the death penalty is .4013. ∎

Confidence Intervals for $(p_1 - p_2)$

By (11-5), the standard error of the difference between sample proportions is given by

$$\sigma_{\hat{p}_1 - \hat{p}_2} = \sqrt{\frac{p_1(1 - p_1)}{n_1} + \frac{p_2(1 - p_2)}{n_2}}$$

Since p_1 and p_2 are unknown, we can use $\hat{p}_1$ and $\hat{p}_2$ to estimate p_1 and p_2, respectively. As a result, $\hat{\sigma}_{\hat{p}_1 - \hat{p}_2}$ is used to estimate $\sigma_{\hat{p}_1 - \hat{p}_2}$, where

$$\hat{\sigma}_{\hat{p}_1 - \hat{p}_2} = \sqrt{\frac{\hat{p}_1(1 - \hat{p}_1)}{n_1} + \frac{\hat{p}_2(1 - \hat{p}_2)}{n_2}}$$

Hence, the limits for an approximate $(1 - \alpha)100\%$ confidence interval for $(p_1 - p_2)$ can be found by using

$$(\hat{p}_1 - \hat{p}_2) \pm z_{\alpha/2}\hat{\sigma}_{\hat{p}_1 - \hat{p}_2} \tag{11-6}$$

where the sample estimates $\hat{p}_1$ and $\hat{p}_2$ involve large samples ($n_1 \geq 30$ and $n_2 \geq 30$).

Hypothesis Tests for $(p_1 - p_2)$

Whenever the null hypothesis is of the form $H_0: p_1 - p_2 = 0$ and we assume that $p_1 = p_2 (= p)$, we will not know the common value p. In order to compute an approximate value of $\sigma_{\hat{p}_1 - \hat{p}_2}$, we must have an estimate for p, since

$$\sigma_{\hat{p}_1 - \hat{p}_2} = \sqrt{\frac{p(1 - p)}{n_1} + \frac{p(1 - p)}{n_2}}$$

$$= \sqrt{p(1 - p)\left[\frac{1}{n_1} + \frac{1}{n_2}\right]}$$

Which estimate for p should we use, $\hat{p}_1$ or $\hat{p}_2$ or neither? An average $(\hat{p}_1 + \hat{p}_2)/2$ would be fair only when $n_1 = n_2$. Since $\hat{p}_1 = x_1/n_1$ and $\hat{p}_2 = x_2/n_2$, we can form a pooled or weighted estimate for p defined by

$$\hat{p} = \frac{x_1 + x_2}{n_1 + n_2} \tag{11-7}$$

The value of $\hat{p}$ must necessarily lie between $\hat{p}_1$ and $\hat{p}_2$. Then the standard error of the difference between sample proportions can be estimated by using the following formula:

$$\hat{\sigma}_{\hat{p}_1 - \hat{p}_2} = \sqrt{\hat{p}(1 - \hat{p})\left[\frac{1}{n_1} + \frac{1}{n_2}\right]} \tag{11-8}$$

where $\hat{p}$ is given by (11-7).

The above techniques are illustrated using the following examples.

EXAMPLE 11-9 A college student has heard that professor X gives a higher percentage of A grades in Math 101 than professor Y. Suspecting that professor X is the easier "A grader," the student compares published grades for both professors the past semester. Assuming that students are randomly assigned to Math 101 sections by the registrar, do the following data indicate that professor X gives a higher percentage of A grades than professor Y?

Professor X	Professor Y
$x_1 = 60$	$x_2 = 72$
$n_1 = 150$	$n_2 = 195$

Use $\alpha = .01$ and find the *p*-value.

Solution The sample proportion of A grades given by professor X is $\hat{p}_1 = \frac{60}{150} = .40$, and the sample proportion of A grades given by professor Y is $\hat{p}_2 = \frac{72}{195} = .37$. From (11-7), the pooled estimate for p is

$$\hat{p} = \frac{x_1 + x_2}{n_1 + n_2}$$

$$= \frac{60 + 72}{150 + 195} = .38$$

The estimated standard error of $\hat{p}_1 - \hat{p}_2$ is found using (11-8):

$$\hat{\sigma}_{\hat{p}_1 - \hat{p}_2} = \sqrt{\hat{p}(1 - \hat{p})\left[\frac{1}{n_1} + \frac{1}{n_2}\right]}$$

$$= \sqrt{(.38)(.62)\left[\frac{1}{150} + \frac{1}{195}\right]} = .053$$

By using the hypothesis testing procedure, we have:

1. The null hypothesis is H_0: $p_1 - p_2 \leq 0$.
2. The alternative hypothesis is H_1: $p_1 - p_2 > 0$ (one-tailed test).
3. The level of significance is $\alpha = .01$.
4. The sampling distribution is the distribution of the differences between sample proportions. The test statistic is the z value for $\hat{p}_1 - \hat{p}_2$:

$$z = \frac{(\hat{p}_1 - \hat{p}_2) - 0}{\hat{\sigma}_{\hat{p}_1 - \hat{p}_2}}$$

$$= \frac{.40 - .37}{.053} = .57$$

5. The critical value is $z_{.01} = 2.33$.
6. We cannot reject H_0, since the test statistic does not fall in the rejection region; that is, $z = .57 < 2.33$.
7. Type of error possible: Type II; β is unknown.
8. The p-value is $.5 - .2157 = .2843$.

Thus, the observed difference $(\hat{p}_1 - \hat{p}_2) = .03$ can be attributed to sampling error, and one cannot conclude that professor X gives a higher percentage of A grades in Math 101 than professor Y.   ∎

EXAMPLE 11-10   To landscape its campus, a large university experimented with two varieties of plants. Four hundred plants of variety A and 600 plants of variety B were initially planted. If 42 of the variety A plants failed to grow, while 48 of the variety B plants failed to grow, test at the .05 level of significance whether there is a difference between the percentages of varieties A and B that will fail to grow on campus.

Solution   We will illustrate hypothesis testing and estimation procedures for solving the problem. Let p_1 represent the proportion of variety A plants that fail to grow on campus and let p_2 represent the proportion of variety B plants that fail to grow.

Hypothesis testing.   We first compute the sample proportions and the pooled estimate for p. The sample proportions are

$$\hat{p}_1 = \frac{42}{400} = .105$$

$$\hat{p}_2 = \frac{48}{600} = .08$$

The pooled estimate for p is

$$\hat{p} = \frac{x_1 + x_2}{n_1 + n_2}$$

$$= \frac{42 + 48}{1000} = .09$$

The approximate standard error of $\sigma_{\hat{p}_1 - \hat{p}_2}$ is

$$\hat{\sigma}_{\hat{p}_1 - \hat{p}_2} = \sqrt{\hat{p}(1 - \hat{p})\left[\frac{1}{n} + \frac{1}{n_2}\right]}$$

$$= \sqrt{(.09)(.91)\left[\frac{1}{400} + \frac{1}{600}\right]} = .018$$

1. The null hypothesis is H_0: $p_1 - p_2 = 0$.
2. The alternative hypothesis is H_1: $p_1 - p_2 \neq 0$ (two-tailed test).
3. The level of significance is $\alpha = .05$.
4. The sampling distribution is the distribution of differences between sample percentages. The test statistic is the z value for $(\hat{p}_1 - \hat{p}_2)$:

$$z = \frac{\hat{p}_1 - \hat{p}_2}{\hat{\sigma}_{\hat{p}_1 - \hat{p}_2}} = \frac{.105 - .08}{.018} = 1.39$$

5. The critical values are $\pm z_{.025} = \pm 1.96$.
6. The decision is that we cannot reject H_0. There is no statistical evidence to reject H_0.
7. Type of error possible: Type II; β is unknown.
8. The p-value is $2(.5 - .4177) = 2(.0823) = .1646$.

Thus, the difference $(\hat{p}_1 - \hat{p}_2) = .025$ can be accounted for as sampling error and not due to the variety of the plant.

Estimation. We first find the estimated value of $\sigma_{\hat{p}_1 - \hat{p}_2}$ by using (11-8):

$$\hat{\sigma}_{\hat{p}_1 - \hat{p}_2} = \sqrt{\frac{\hat{p}_1(1 - \hat{p}_1)}{n_1} + \frac{\hat{p}_2(1 - \hat{p}_2)}{n_2}}$$

$$= \sqrt{\frac{(.105)(.895)}{400} + \frac{(.08)(.92)}{600}} = .019$$

Note that we do not use the pooled estimate, since we have not assumed that $p_1 = p_2$ is true.

By (11-6), the limits for the 95% confidence interval for $(p_1 - p_2)$ are given by

$$(\hat{p}_1 - \hat{p}_2) \pm z_{.025}\hat{\sigma}_{\hat{p}_1 - \hat{p}_2}$$

$$.025 \pm (1.96)(.019)$$

$$.025 \pm .037$$

Thus, a 95% confidence interval for $(p_1 - p_2)$ is $(-.012, .062)$. Since 0 is contained in the interval, $(p_1 - p_2)$ may equal 0. Hence, we have no statistical evidence to suggest that there is a difference in the percentages of varieties A and B that fail to grow on campus.   ∎

If it is desired to test a null hypothesis of the form $H_0: p_1 - p_2 = p_0$, where p_0 is a nonzero value, there is no need to obtain a pooled estimate for p, since $p_1 \neq p_2$. In this case $\hat{p}_1$ is used to estimate p_1 and $\hat{p}_2$ is used to estimate p_2.

EXAMPLE 11-11   An administrator of a large college claims that at least 10% more male students than female students have an automobile on campus. A statistics professor takes issue with the claim and randomly polls 100 males and 100 females. He found that 34 males have cars on campus and 27 females have cars on campus. Can he conclude at the 5% level of significance that the administrator's claim is false?

Solution   Let p_1 represent the proportion of male students who have cars on campus and let p_2 represent the proportion of female students who have cars on campus.

1. The null hypothesis is $H_0: p_1 - p_2 \geq .1$
2. The alternative hypothesis is $H_1: p_1 - p_2 < .1$ (one-tailed test).
3. The level of significance is $\alpha = .05$.
4. The sampling distribution is the sampling distribution of $(\hat{p}_1 - \hat{p}_2)$. The test statistic is the z value for $\hat{p}_1 - \hat{p}_2 = .34 - .27 = .07$:

$$z = \frac{(\hat{p}_1 - \hat{p}_2) - (p_1 - p_2)}{\sqrt{\dfrac{\hat{p}_1(1 - \hat{p}_1)}{n_1} + \dfrac{\hat{p}_2(1 - \hat{p}_2)}{n_2}}}$$

$$= \frac{.07 - .1}{\sqrt{\dfrac{(.34)(.66)}{100} + \dfrac{(.27)(.73)}{100}}} = -.46$$

5. The critical value is $-z_{.05} = -1.65$.
6. We cannot reject H_0, since $-.46 > -1.65$.
7. Type of error possible: Type II; β is unknown.
8. The p-value is $P(z < -.46) = .5 - .1772 = .3228$.

Hence, there is no statistical evidence to reject the administrator's claim that at least 10% more male students have cars on campus than female students.   ∎

Problem Set 11.3

A

1. In a study to estimate the proportions of residents in a small city and its suburbs who subscribe to the newspaper, it was found that 48 of 120 residents subscribe while only 30 of 125 suburban residents subscribe to the newspaper. If the true proportion of the residents who subscribe to the newspaper is .5 and the true proportion of the suburbanites who subscribe is .3, find:

a. $\mu_{\hat{p}_r - \hat{p}_s}$
b. $\sigma_{\hat{p}_r - \hat{p}_s}$
c. Corresponding z score for $\hat{p}_r - \hat{p}_s$

2. A study was conducted to determine whether home-makers from Boston have the same preference as homemakers from St. Louis for one of two brands of floor wax, A and B. It was found that among 300 randomly selected homemakers from Boston, 171 preferred brand A to brand B, while among 400 homemakers from St. Louis, 236 preferred A to B.
   a. At $\alpha = .05$ determine if there is a significant dif-ference between the percentages of homemakers who prefer brand A for the two groups. Find the p-value.
   b. Construct a 95% confidence interval for the dif-ference in percentages who prefer brand A for the two groups. Does the interval contain 0?

3. In a study to determine the effects of color in com-mercial television advertising, two groups of 500 persons were exposed to a program, including com-mercials. One group watched the program in color and the other group watched the identical program in black and white. Two hours after the program, each person was asked which products were adver-tised. Two hundred of those who watched the pro-gram in color remembered the products, while 180 of those who watched the program without color remembered the products. At $\alpha = .05$ test the null hypothesis that there is no difference in retention between persons who watch TV in color as opposed to those who watch TV in black and white. Find the p-value.

4. For the situation described in Problem 3, construct a 95% confidence interval for the true difference between proportions. Is 0 in the interval?

5. Independent random samples of 50 males and 75 females were selected to study GPAs of students at a large university. The group of males included 15 who had GPAs less than 2.0, while the group of females had 24 with GPAs below 2.0. Construct a 95% confidence interval for the true difference between the proportions of all male and female students with GPAs below 2.0.

6. For Problem 5, test the null hypothesis that there is no difference between the true proportion of univer-sity males with GPAs below 2.0 and the true pro-portion of university females with GPAs below 2.0. Use $\alpha = .05$.

7. A manufacturer of microprocessors buys its proces-sor chips from two suppliers. A sample of 300 chips from supplier A resulted in 50 defective chips, where-as a sample of 400 chips from supplier B resulted

in 70 defective chips. Construct a 95% confidence interval for the difference between the proportions of defective chips received from the two suppliers.

8. It is common knowledge that not everyone coop-erates with answering questions from door-to-door poll-takers. For an experiment to determine whether women are more cooperative than men, the follow-ing results indicate the number of each sex who co-operated with an interviewer:

Men	Women
$n_1 = 175$	$n_2 = 250$
$x_1 = 97$	$x_2 = 143$

Construct a 99% confidence interval for the differ-ence in the true proportions of men and women who cooperate with poll-takers.

9. For Problem 8, test whether there is a difference be-tween the proportions of men and women who co-operate with door-to-door poll-takers. Use $\alpha = .01$ and find the p-value.

10. Test the null hypothesis $H_0: p_1 - p_2 = .01$ versus the alternative hypothesis $H_1: p_1 - p_2 \neq .01$ using $\alpha = .05$ if the following data were obtained:

Sample 1	Sample 2
$n_1 = 200$	$n_2 = 300$
$x_1 = 120$	$x_2 = 168$

Find the p-value.

11. The western Maryland branch of the American Heart Association studied cardiovascular-disease-related deaths in Allegheny and Garrett counties for 1983. It reported that of the 969 deaths in Allegheny County, 510 were attributed to cardiovascular dis-ease, while of the 250 deaths reported in Garrett County, 150 were due to cardiovascular disease. Do the data suggest that the percentages of deaths re-sulting from cardiovascular disease differ for the two western Maryland counties? Use $\alpha = .05$.

B

12. Construct a 95% confidence interval for $(p_1 - p_2)$ in Problem 10. Does the interval contain .01?

13. For Problem 7, test the null hypothesis $H_0: p_2 - p_1 = .5\%$ against $H_1: p_2 - p_1 \neq .5\%$ using the 5% level of significance. Find the p-value.

14. A study is to be conducted to estimate the difference between the proportions of defective parts shipped by two suppliers, A and B. If samples of the same size are to be used, how large must the samples be in order to be 95% confident that the maximum error of estimate is at most .01?

11.4 COMPARING POPULATION VARIANCES _____

Another method of comparing two populations is to compare their variances. Many statistical applications arise in which population variances must be compared. In industrial applications involving two different methods or machines for producing the same product, variances are often used and compared for quality control purposes. Many statistical tests, such as those to be presented in Section 11.5, require population variances to be equal in order for the tests to be appropriate.

In order to be able to compare two population variances, we need to be familiar with a new sampling distribution, the *F* distribution.

F Distributions

If two independent random samples are taken from normal populations with equal population variances ($\sigma_1^2 = \sigma_2^2$), then the sampling distribution of the statistic s_1^2/s_2^2 is an **F distribution**. Thus, the **F statistic** is defined by

$$F = \frac{s_1^2}{s_2^2}$$

The exact shape of an *F* distribution depends on two degrees-of-freedom parameters, df_1 and df_2. These are defined by:

$df_1 = n_1 - 1$, where n_1 is the size of the sample yielding s_1^2
$df_2 = n_2 - 1$, where n_2 is the size of the sample yielding s_2^2

We will identify the degrees of freedom as an ordered pair, (df_1, df_2). The *F* distributions, like the χ^2 distributions, are not symmetric distributions, but all elements of both distributions are greater than or equal to 0. (see Figure 11-2).

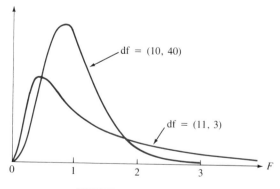

FIGURE 11-2 *F* distributions

EXAMPLE 11-12   In a test of the effectiveness of two different kinds of sleeping pills, A and B, two independent groups of insomniacs are to be used. One group of size 40 will be administered pill A and the other group of size 60 will be administered pill B. The number of hours of sleep will be recorded for each individual used in the study. If the number of hours of sleep for individuals using each pill is assumed to be normally distributed and $\sigma_A^2 = \sigma_B^2$, calculate the value of the F statistic and determine df_A and df_B if $s_A^2 = 9$, $s_B^2 = 5$, $n_A = 40$, and $n_B = 60$.

Solution   The value of the F statistic is

$$F = \frac{s_A^2}{s_B^2} = \frac{9}{5} = 1.8$$

The degrees of freedom are

$$df_A = n_A - 1 = 40 - 1 = 39$$
$$df_B = n_B - 1 = 60 - 1 = 59$$

■

Note that if the statistic s_1^2/s_2^2 has a sampling distribution which is an F distribution with $df_1 = n_1 - 1$ and $df_2 = n_2 - 1$, then the statistic s_2^2/s_1^2 (the reciprocal of F) has a sampling distribution which is an F distribution with $df_1 = n_2 - 1$ and $df_2 = n_1 - 1$. That is,

> If F has an F distribution with $df = (n_1 - 1, n_2 - 1)$,
> then $1/F$ has an F distribution with $df = (n_2 - 1, n_1 - 1)$.

Since both statistics have F distributions, it is common practice to place the larger sample variance in the numerator of the F ratio.

Hypothesis Tests Comparing σ_1^2 and σ_2^2

To compare population variances we need to use the quotient of the variances, rather than the difference of the variances. If the quotient of the population variances is equal to 1, then we can say that the variances are equal; if the quotient of the population variances is not equal to 1, then the population variances are not equal. That is,

$\sigma_1^2/\sigma_2^2 = 1$ is equivalent to $\sigma_1^2 = \sigma_2^2$

If $\sigma_1^2 = \sigma_2^2$, we would expect the value of the F statistic to be close to 1; the difference from 1 may be attributed to sampling error. The more removed the value of F is from 1, the more unlikely that the F value belongs to a particular F distribution.

As we have previously noted, the F distributions are not symmetric; as a result, to form the rejection region for a two-tailed test, we can simplify the computations by making certain that the right tail of the F distribution is used. A right-tailed test is used because only these areas are given in the F tables. If the larger sample variance is placed in the numerator of the F statistic, then a right-tailed test will always be indicated. By doing this, we are in effect doubling the tabled value for α by making the test a two-tailed test and using only the right tail. Thus, for a two-tailed test employing a significance level of α, a right-tailed test is used that has a tail area of α/2. The F statistic becomes

$$F = \frac{\text{Larger sample variance}}{\text{Smaller sample variance}}$$

For directional hypotheses, we place the sample variance corresponding to the larger population variance, as expressed by the alternative hypothesis H_1, in the numerator of the F ratio. For example, if the alternative hypothesis is $\sigma_1^2 > \sigma_2^2$, then the sample variance corresponding to the population variance σ_1^2 is placed in the numerator of the F ratio. One should be careful that df_1 represents the degrees of freedom corresponding to the variance in the numerator of the F ratio.

To determine a critical value in the F table (Table 3 of Appendix B) we locate the correct table for α (.01 or .05) and intersect the *row* identified by df_2 with the *column* identified by df_1:

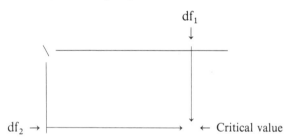

We shall denote the F distribution with $df_1 = (n_1 - 1)$ and $df_2 = (n_2 - 1)$ as $F(n_1 - 1, n_2 - 1)$. As a result, the F statistic has a sampling distribution which is an $F(n_1 - 1, n_2 - 1)$ distribution. Note that the df corresponding to the numerator of the F ratio is always located at the top of the F table, and the df corresponding to the denominator of the F ratio is always located at the left side of the F table.

The following examples illustrate testing hypotheses to determine if two population variances differ.

EXAMPLE 11-13   Two different processes are used to produce 100-ohm, 1% resistors. A random sample of resistors is selected for each process. Assume that the distribution of resistor values for each process is normally distributed. If $n_1 = 21$, $s_1^2 = 39$, $n_2 = 31$, and $s_2^2 = 33$, test using α = .05 to determine if the variance of the resistors manufactured by process 1 is greater than the variance of the resistors manufactured by process 2.

Solution   **1.** The null hypothesis is H_0: $\sigma_1^2 \leq \sigma_2^2$.
2. The alternative hypothesis is H_1: $\sigma_1^2 > \sigma_2^2$ (one-tailed test).
3. The level of significance is α = .05.

4. The sampling distribution is $F(20, 30)$, the F distribution with $df_1 = 20$ and $df_2 = 30$. The value of the test statistic is

$$F = \frac{s_1^2}{s_2^2} = \frac{39}{33} = 1.18$$

5. The critical value is $F_{.05}(20, 30) = 1.93$. This value was found by locating the F distribution table corresponding to $\alpha = .05$ (Table 3b of Appendix B) and determining the value where the column labeled by 20 intersects the row labeled by 30.

6. The decision is that we cannot reject H_0 (see the accompanying figure).

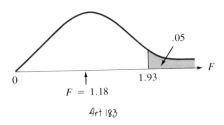

$F = 1.18$

Art 183

7. With this decision, we may have committed a Type II error; β is unknown.

The value of the F statistic differs from 1 by .18; we conclude that this difference can be explained as sampling error. We have no statistical evidence to indicate that the variance of process 1 is greater than the variance of process 2. ∎

EXAMPLE 11-14   Assume that the weights (in kg) of items produced by two different manufacturing processes are normally distributed. If the weights of a sample of $n_1 = 15$ items produced by one process have a variance of 33 and the weights of a sample of $n_2 = 41$ items produced by the other process have a variance of 15, test using $\alpha = .10$ to determine if $\sigma_1^2 \neq \sigma_2^2$.

Solution   **1.** The null hypothesis is $H_0: \sigma_1^2 = \sigma_2^2$.
2. The alternative hypothesis is $H_1: \sigma_1^2 \neq \sigma_2^2$ (two-tailed test).
3. The level of significance is $\alpha = .10$.
4. The sampling distribution is $F(14, 40)$. The test statistic is F. Note that the larger variance is placed in the numerator of the F statistic. The value of the F statistic is

$$F = \frac{33}{15} = 2.2$$

5. The critical value is $F_{.05}(14, 40) = 1.95$, as indicated in the figure. Note that it is not necessary to obtain a critical value in the lower tail of the F distribution, since we placed the larger sample variance in the numerator of the F statistic. The larger critical value corresponds to a right-tail area of $\alpha/2$ for a two-tailed test.

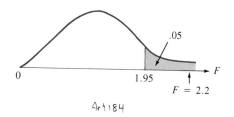

4ሩ∤184

6. The decision is to reject H_0.
7. With this decision we may have committed a Type I error; $\alpha = .10$.

Thus, we can conclude that the population variances are significantly different at the 10% significance level. ∎

Comparing Population Standard Deviations

To determine whether $\sigma_1 \neq \sigma_2$, we determine whether $\sigma_1^2 \neq \sigma_2^2$. That is, $\sigma_1 \neq \sigma_2$ if $\sigma_1^2 \neq \sigma_2^2$. Thus, to test the null hypothesis $H_0: \sigma_1 = \sigma_2$, we test the equivalent hypothesis $H_0: \sigma_1^2 = \sigma_2^2$.

EXAMPLE 11-15   In an attempt to determine whether the standard deviation in weights among first-grade boys is greater than the standard deviation in weights among first-grade girls, two random samples of weights were obtained. The following data were recorded:

Boys	Girls
$s_1 = 6.3$	$s_2 = 3.9$
$n_1 = 10$	$n_2 = 25$

Do the data indicate that $\sigma_1 > \sigma_2$? Use $\alpha = .05$.

Solution   Note that $\sigma_1 > \sigma_2$ if $\sigma_1^2 > \sigma_2^2$. Thus, we test to determine if $\sigma_1^2 > \sigma_2^2$.

1. The null hypothesis is $H_0: \sigma_1^2 \leq \sigma_2^2$.
2. The alternative hypothesis is $H_1: \sigma_1^2 > \sigma_2^2$.
3. The level of significance is $\alpha = .05$.
4. The sampling distribution is $F(9, 24)$. The test statistic is

$$F = \frac{(6.3)^2}{(3.9)^2} = 2.61$$

5. The critical value is $F_{.05}(9, 24) = 2.30$.
6. The decision is that we cannot reject H_0.
7. A Type II error is possible with this decision; β is unknown.

Hence, we have no statistical evidence to suggest that the standard deviation of the weights of first-grade boys is greater than the standard deviation of the weights of first-grade girls. ∎

Left-Tail Critical Values for F

In order to construct a confidence interval for σ_1^2/σ_2^2, we need to be able to find both critical values associated with the F statistic. To find the critical value for the left tail in an F distribution, we use the following fact:

$$F_{1-\alpha/2}(n_2 - 1, n_1 - 1) = \frac{1}{F_{\alpha/2}(n_1 - 1, n_2 - 1)} \tag{11-9}$$

For example, to find $F_{.95}(14, 40)$ for the F distribution in Example 11-14, we use (11-9):

$$F_{.95}(14, 40) = \frac{1}{F_{.05}(40, 14)} = \frac{1}{2.27} = .441$$

Recall that the critical value 1.95 represents $F_{.05}(14, 40)$, which was found in the F table (Appendix B).

Confidence Intervals for the Quotient of Two Population Variances

When constructing a confidence interval for the quotient of two population variances, we shall follow the convention adopted earlier of placing the larger sample variance in the numerator of the F statistic. In the notation that follows, s_1^2 will represent the larger sample variance. The confidence intervals for σ_1^2/σ_2^2 and σ_2^2/σ_1^2 are not identical, but they are related to one another in a special way. If (L_1, L_2) is a **$(1 - \alpha)100\%$ confidence interval for σ_1^2/σ_2^2**, then $(1/L_2, 1/L_1)$ is a $(1 - \alpha)100\%$ confidence interval for σ_2^2/σ_1^2.
A $(1 - \alpha)100\%$ confidence interval for σ_1^2/σ_2^2 is given by (L_1, L_2) where

Larger sample variance

$$L_1 = \frac{s_1^2}{s_2^2} \cdot F_{1-\alpha/2}(n_2 - 1, n_1 - 1)$$

and

―Note―

$$L_2 = \frac{s_1^2}{s_2^2} \cdot F_{\alpha/2}(n_2 - 1, n_1 - 1) \tag{11-10}$$

If the confidence interval for σ_1^2/σ_2^2 does not contain 1, then we can be $(1 - \alpha)100\%$ confident that $\sigma_1^2 \neq \sigma_2^2$. And we can reject $H_0: \sigma_1^2 = \sigma_2^2$ using a level of significance equal to α.
A **$(1 - \alpha)100\%$ confidence interval for σ_1/σ_2** is given by $(\sqrt{L_1}, \sqrt{L_2})$, where L_1 and L_2 are given by (11-10).

The following four steps can be used to construct a $(1 - \alpha)100\%$ confidence interval for σ_1^2/σ_2^2:

1. Find the value of s_1^2/s_2^2, making sure the larger sample variance is placed in the numerator of the F ratio.
2. Find the value of $F_{1-\alpha/2}(n_2 - 1, n_1 - 1)$ by using (11-9). Be sure that n_2 represents the size of the sample from the population having variance σ_2^2.
3. Find the value of $F_{\alpha/2}(n_2 - 1, n_1 - 1)$, making sure n_2 represents the size of the sample from the population having variance σ_2^2.
4. Find the confidence interval limits L_1 and L_2 by using (11-10).

EXAMPLE 11-16   For Example 11-14 construct a 90% confidence interval for σ_1^2/σ_2^2. Recall that $n_1 = 15$, $s_1^2 = 33$, $n_2 = 41$, and $s_2^2 = 15$.

Solution   **1.** The value of the F statistic is

$$\frac{s_1^2}{s_2^2} = \frac{33}{15} = 2.2$$

Note that the larger sample variance is placed in the numerator of the F statistic.

2. The left-hand critical value for F is

$$F_{1-\alpha/2}(n_2 - 1, n_1 - 1) = F_{.95}(40, 14)$$
$$= \frac{1}{F_{.05}(14, 40)}$$
$$= \frac{1}{1.95} = .51$$

3. The right-hand critical value for F is

$$F_{\alpha/2}(n_2 - 1, n_1 - 1) = F_{.05}(40, 14)$$
$$= 2.27$$

From (11-10), the limits for the 90% confidence interval are

$$L_1 = \left(\frac{s_1^2}{s_2^2}\right) \cdot F_{1-\alpha/2}(n_2 - 1, n_1 - 1)$$
$$= (2.2)(.51) = 1.122$$

and

$$L_2 = \left(\frac{s_1^2}{s_2^2}\right) \cdot F_{\alpha/2}(n_2 - 1, n_1 - 1)$$
$$= (2.2)(2.27) = 4.994$$

Thus, a 90% confidence interval for σ_1^2/σ_2^2 is (1.12, 4.99). Note that if a $(1 - \alpha)100\%$ confidence interval for σ_1^2/σ_2^2 contains 1, then we cannot be

$(1 - \alpha)100\%$ confident that the population variances are different. In this example, we can be 90% confident that the population variances are different, since the interval (1.12, 4.99) does not contain 1. Note that the 90% confidence interval for σ_2^2/σ_1^2 is $(1/4.99, 1/1.12) = (.20, .89)$. ∎

EXAMPLE 11-17   Two new motor assembly methods are tested by an automobile manufacturer for variance in assembly times (in minutes). The results follow:

Method 1	Method 2
$n_1 = 31$	$n_2 = 25$
$s_1^2 = 50$	$s_2^2 = 24$

a. Construct a 90% confidence interval for σ_1^2/σ_2^2.
b. Conduct the corresponding hypothesis test to determine if there is a difference in the variances of the assembly times for the two methods.

Solution   **a.** Construction of a 90% confidence interval
   (1) The value of s_1^2/s_2^2 is $\frac{50}{24} = 2.08$.
   (2) The left-hand critical value is

$$F_{1 - \alpha/2}(n_2 - 1, n_1 - 1) = F_{.95}(24, 30)$$

$$= \frac{1}{F_{.05}(30, 24)}$$

$$= \frac{1}{1.94} = .52$$

   (3) The right-hand critical value is

$$F_{\alpha/2}(n_2 - 1, n_1 - 1) = F_{.05}(24, 30) \simeq 1.93$$

Note that the F table does not have a column for df $= 24$, but that it does contain df $= 20$ and df $= 30$. For these adjacent values, the critical values for F are $F(20, 30) = 1.93$ and $F(30, 30) = 1.84$. Let's use the larger and more conservative value of 1.93. [In this text, whenever the tables do not contain a df value entry, we shall follow the practice of choosing the larger F value corresponding to the adjacent values of df.]
   The limits for the 90% confidence interval for σ_1^2/σ_2^2 are given by (11-10):

$$L_1 = \left(\frac{s_1^2}{s_2^2}\right) \cdot F_{1 - \alpha/2}(n_2 - 1, n_1 - 1)$$

$$= (2.08)(.52) = 1.08$$

and

$$L_2 = \left(\frac{s_1^2}{s_2^2}\right) \cdot F_{\alpha/2}(n_2 - 1, n_1 - 1)$$

$$= (2.08)(1.93) = 4.01$$

Thus, a 90% confidence interval for σ_1^2/σ_2^2 is $(1.08, 4.01)$. Since this interval does not contain 1, we can be 90% confident that $\sigma_1^2 \neq \sigma_2^2$.

b. Hypothesis test

(1) The null hypothesis is $\sigma_1^2 = \sigma_2^2$.

(2) The alternative hypothesis is $\sigma_1^2 \neq \sigma_2^2$ (two-tailed test).

(3) The level of significance is 10%.

(4) The sampling distribution is the F distribution with $df_1 = 30$ and $df_2 = 24$, $F(30, 24)$. The value of the F statistic is

$$F = \frac{50}{24} = 2.08$$

(5) The critical value is $F_{.05}(30, 24) = 1.94$. Since we put the larger sample variance in the numerator of the F statistic, we only need to compare the value of the F statistic against the larger critical value, $F_{.05}(30, 24)$.

(6) The decision is to reject H_0, since $F = 2.08 > F_{.05}(30, 24) = 1.94$. Note that this result is consistent with our interpretation of the confidence interval found in part (a).   ∎

Problem Set 11.4

A

1. To determine whether the grades of instructor A are more variable than those of instructor B, two independent random samples of grades on equivalent examinations were selected. The following results were obtained:

Instructor A	Instructor B
$s_A^2 = 81.24$	$s_B^2 = 97.86$
$n_A = 15$	$n_B = 20$

If it is assumed that $\sigma_A = \sigma_B$, calculate the value of the F statistic and find its associated df values.

2. Find the following critical values for F:

a. $F_{.95}(21, 31)$ **b.** $F_{.95}(15, 10)$
c. $F_{.95}(10, 10)$ **d.** $F_{.99}(31, 21)$
e. $F_{.99}(10, 15)$ **f.** $F_{.99}(10, 10)$

3. Find 90% confidence intervals for σ_1^2/σ_2^2 if:

a. $n_1 = 15$, $s_1^2 = 15.1$, $n_2 = 31$, and $s_2^2 = 6.8$
b. $n_1 = 21$, $s_1^2 = 4.8$, $n_2 = 11$, and $s_2^2 = 10.4$

4. For Problem 3, find 90% confidence intervals for σ_1/σ_2.

5. Perform the following hypothesis tests. Assume sampling is from normal populations.

a. $H_0: \sigma_1^2 \geq \sigma_2^2$, $H_1: \sigma_1^2 < \sigma_2^2$; $\alpha = .05$; $n_1 = 16$, $s_1^2 = 12.83$, $n_2 = 26$, $s_2^2 = 21.75$
b. $H_0: \sigma_1^2 = \sigma_2^2$, $H_1: \sigma_1^2 \neq \sigma_2^2$; $\alpha = .10$; $n_1 = 21$, $s_1^2 = 10.89$, $n_2 = 11$, $s_2^2 = 22.87$

c. $H_0: \sigma_1 = \sigma_2$, $H_1: \sigma_1 \neq \sigma_2$; $\alpha = .10$; $n_1 = 26$, $s_1^2 = 3.84$, $n_2 = 11$, $s_2^2 = 16.4$

6. The following data represent statistics from independent samples from two normal populations with unknown parameters. For each, test at $\alpha = .02$ to determine whether $\sigma_1^2 \neq \sigma_2^2$.

a. $n_1 = 25$, $s_1 = 14$, $n_2 = 31$, $s_2 = 18$
b. $n_1 = 21$, $s_1 = 5.2$, $n_2 = 24$, $s_2 = 3.2$
c. $n_1 = 10$, $s_1 = 4.1$, $n_2 = 12$, $s_2 = 2.1$
d. $n_1 = 13$, $s_1 = 3.1$, $n_2 = 14$, $s_2 = 1.6$

7. Assume that an industrial plant produces 70-mm bolts on two different shifts. Samples of bolts are taken from those made by both shifts and their diameters are measured (in mm). The results follow:

First shift	Second shift
$n_1 = 31$	$n_2 = 21$
$s_1^2 = .045$	$s_2^2 = .080$

a. By using $\alpha = .10$ test the null hypothesis
$$H_0: \sigma_1 = \sigma_2$$
against the alternative hypothesis
$$H_1: \sigma_1 \neq \sigma_2.$$

b. Construct a 90% confidence interval for σ_1/σ_2 to determine whether $\sigma_1 \neq \sigma_2$.

8. The following data represent two samples from different normal populations:

Sample 1		Sample 2	
9	9	13	14
9	7	12	16
11	8	11	10
11	6	13	8
12	5	12	9

a. Construct a 90% confidence interval for σ_1^2/σ_2^2.

b. Test the null hypothesis $H_0: \sigma_1^2 = \sigma_2^2$ against the alternative hypothesis $H_1: \sigma_1^2 \neq \sigma_2^2$ using $\alpha = .10$.

9. A new machine is being considered by a sugar packaging firm to replace its present machine. The weights of a sample of 21 5-pound bags of sugar packaged by the old machine produced a variance of $s_1^2 = .16$, while the weights for 20 5-pound bags of sugar packaged by the new machine produced a variance of $s_2^2 = .09$. Based on this data, would you advise management to consider the new machine as being superior to the old machine? Use $\alpha = .05$.

10. A criminologist is interested in comparing the consistency of the lengths of sentences given to persons convicted of robbery by two county judges. A random sample of 17 people convicted of robbery by one judge showed a standard deviation of $s_1 = 1.34$ years, while a random sample of 21 persons convicted by the other judge showed a standard deviation of $s_2 = 2.53$ years.

a. Do the data suggest that the variances of the lengths of sentences for the two judges differ? Use $\alpha = .10$.

b. Construct a 90% confidence interval for σ_1/σ_2.

c. Construct a 90% confidence interval for σ_2/σ_1.

B

11. If χ_1^2 and χ_2^2 are independent random variables, then $(\chi_1^2/df_1)/(\chi_2^2/df_2)$ has an F distribution with df $= (df_1, df_2)$. Use this fact to derive formulas (11-10).

12. A variance-reduction process is being considered by a plant to cut rising costs. The new process will not be implemented unless it can be verified statistically at $\alpha = .10$ that the new process will reduce the standard deviation by more than 20%. Suppose a study resulted in the following data:

Old process	New process
$n_1 = 17$	$n_2 = 11$
$s_1 = 3.61$	$s_2 = 1.72$

Should the new process be implemented? Explain.

11.5 INFERENCES CONCERNING $(\mu_1 - \mu_2)$ USING SMALL SAMPLES

If the sample sizes are large enough ($n_1 \geq 30$ and $n_2 \geq 30$), the techniques presented in Section 11.2 can be used to compare almost any two populations. This is a consequence of the central limit theorem. In situations where the population variances are unknown, sample variances provide satisfactory substitutes for the population variances in most situations when large samples are used.

In this section, we shall develop methods for comparing two populations by using the difference between sample means obtained from small samples ($n_1 < 30$ and/or $n_2 < 30$) to estimate $(\mu_1 - \mu_2)$. The samples can be either independent of one another or related, as would be the case with paired data on the same subjects. We shall treat the case of independent samples first.

Independent Samples

The assumptions underlying the methods for making inferences concerning $(\mu_1 - \mu_2)$ based on small, independent samples from the two populations are:

1. Both populations are normal.
2. The samples are independent and random.
3. The population variances are equal ($\sigma_1^2 = \sigma_2^2$).

Assumption 3 can be tested using the F statistic, explained in Section 11.4. If we cannot reject the null hypothesis $H_0: \sigma_1^2 = \sigma_2^2$, we cannot necessarily conclude that $\sigma_1^2 = \sigma_2^2$. Nevertheless, whenever we cannot reject the null hypothesis of equal population variances, we will proceed under the assumption that the population variances are equal.

Inferences comparing μ_1 and μ_2 using small samples require certain parameters to be estimated. If the population variances are equal, we will denote the common population variance by σ^2. Point estimates for μ_1, μ_2, and σ must be known. The best point estimate for μ_1 is $\bar{x}_1$ and the best estimate for μ_2 is $\bar{x}_2$. The difference between sample means $(\bar{x}_1 - \bar{x}_2)$ provides an estimate for the difference between population means $(\mu_1 - \mu_2)$. How should we estimate σ? If the sample sizes are equal $(n_1 = n_2)$, then the average sample variance $(s_1^2 + s_2^2)/2$ is appropriate. But if $n_1 \neq n_2$, we need a weighted average that takes into account both sample sizes. This weighted average (sometimes called the **pooled estimate**), denoted by s_p, is defined as follows:

$$s_p = \sqrt{\frac{(n_1 - 1)s_1^2 + (n_2 - 1)s_2^2}{n_1 + n_2 - 2}} \qquad (11\text{-}11)$$

Recall that the standard error of the difference between sample means is defined by

$$\sigma_{\bar{x}_1 - \bar{x}_2} = \sqrt{\frac{\sigma_1^2}{n_1} + \frac{\sigma_2^2}{n_2}} \qquad (11\text{-}12)$$

Since $\sigma_1^2 = \sigma_2^2 = \sigma^2$, we can rewrite (11-12) as

$$\sigma_{\bar{x}_1 - \bar{x}_2} = \sqrt{\frac{\sigma_1^2}{n_1} + \frac{\sigma_2^2}{n_2}}$$

$$= \sqrt{\sigma^2 \left[\frac{1}{n_1} + \frac{1}{n_2}\right]}$$

$$= \sigma \sqrt{\left[\frac{1}{n_1} + \frac{1}{n_2}\right]}$$

If σ were known, the following statistic would have the standard normal as its sampling distribution:

$$\frac{(\bar{x}_1 - \bar{x}_2) - (\mu_1 - \mu_2)}{\sigma\sqrt{(1/n_1) + (1/n_2)}}$$

But since σ is unknown, we can use s_p to estimate σ, and it can be shown that the following statistic has a sampling distribution, which is a t distribution with df $= n_1 + n_2 - 2$:

$$t = \frac{(\bar{x}_1 - \bar{x}_2) - (\mu_1 - \mu_2)}{s_p\sqrt{(1/n_1) + (1/n_2)}} \qquad (11\text{-}13)$$

The following example illustrates the use of formula (11-13) for calculating the value of the t statistic.

EXAMPLE 11-18   Two methods for teaching calculus, A and B, were compared using two random groups of students. At the conclusion of the experimental instruction, each group was given the same achievement examination. The scores for each group comprise random samples from two normal populations with equal variances. Find the value of the t statistic and its associated degrees of freedom if we assume that $\mu_1 = \mu_2$ and $n_1 = 16$, $n_2 = 12$, $\bar{x}_1 = 80.7$, $\bar{x}_2 = 73.2$, $s_1^2 = 12.8$, and $s_2^2 = 8.5$.

Solution   We first compute the value of s_p using (11-11):

$$s_p = \sqrt{\frac{(n_1 - 1)s_1^2 + (n_2 - 1)s_2^2}{n_1 + n_2 - 2}}$$

$$= \sqrt{\frac{(16 - 1)(12.8) + (12 - 1)(8.5)}{16 + 12 - 2}} \simeq 3.31$$

From (11-13), the value of the t statistic is

$$t = \frac{(\bar{x}_1 - \bar{x}_2) - (\mu_1 - \mu_2)}{s_p\sqrt{[(1/n_1) + (1/n_2)]}}$$

$$= \frac{(80.7 - 73.2) - 0}{3.31\sqrt{\frac{1}{16} + \frac{1}{12}}} \simeq 5.93$$

The df associated with t is

$$\begin{aligned} \text{df} &= n_1 + n_2 - 2 \\ &= 16 + 12 - 2 = 26 \end{aligned}$$ ∎

Inferences Concerning $(\mu_1 - \mu_2)$

Hypothesis testing and estimation procedures can be used to draw inferences concerning $(\mu_1 - \mu_2)$ using small samples. If we are testing the null hypothesis $H_0: \mu_1 - \mu_2 = 0$ under the assumptions of normality and equal variances, then the test statistic becomes

$$t = \frac{\bar{x}_1 - \bar{x}_2}{s_p\sqrt{(1/n_1) + (1/n_2)}} \qquad (11\text{-}14)$$

with df $= n_1 + n_2 - 2$.

The limits for a $(1 - \alpha)100\%$ confidence interval for $(\mu_1 - \mu_2)$ are given by

$$(\bar{x}_1 - \bar{x}_2) \pm t_{\alpha/2}(\text{df}) \cdot s_p\sqrt{\frac{1}{n_1} + \frac{1}{n_2}} \qquad (11\text{-}15)$$

The following examples illustrate the methods used to draw inferences concerning $(\mu_1 - \mu_2)$ for small, independent samples from two normal populations with unknown, but equal, variances.

EXAMPLE 11-19   An experiment was conducted to compare the mean length of time required for the human body to absorb two drugs, A and B. Assume that the length of time it takes either drug to reach a specified level in the bloodstream is normally distributed. Twelve people were randomly assigned to test each drug, and the length of time (in minutes) for each drug to reach a specified level in the blood was recorded. The following results were obtained:

Drug A	Drug B
$n_1 = 12$	$n_2 = 12$
$\bar{x}_1 = 26.8$	$\bar{x}_2 = 32.6$
$s_1^2 = 15.57$	$s_2^2 = 17.54$

Determine using $\alpha = .05$ if $(\mu_2 - \mu_1) \neq 0$.

Solution   **Hypothesis testing.**   We first test the assumption $\sigma_1^2 = \sigma_2^2$ by using hypothesis testing with $\alpha = .10$.

1. The null hypothesis is $H_0: \sigma_1^2 = \sigma_2^2$.
2. The alternative hypothesis is $H_1: \sigma_1^2 \neq \sigma_2^2$ (two-tailed test).
3. The level of significance is $\alpha = .10$.
4. The sampling distribution is $F(11, 11)$. The value of the test statistic is:

$$F = \frac{17.54}{15.57} = 1.13$$

5. Recall that we can use the right-hand critical value if we put the larger sample variance in the numerator of the F ratio. Although we cannot find $F_{.05}(11, 11)$ in Table 3 of Appendix B, the following values can be found:

$$F_{.05}(10, 11) = 2.86$$
$$F_{.05}(12, 11) = 2.79$$

The critical value for F is some value between 2.86 and 2.79. We shall follow our convention and choose the larger value.
6. The decision is that we cannot reject H_0, since $F = 1.13$ is less than the larger of the two critical F values, $F_{.05}(10, 11) = 2.86$. Thus, there is no statistical evidence to conclude that the population variances are different.

We next determine if $\mu_2 - \mu_1 \neq 0$ by using hypothesis testing:

1. The null hypothesis is $H_0: \mu_2 - \mu_1 = 0$.
2. The alternative hypothesis is $H_1: \mu_2 - \mu_1 \neq 0$ (two-tailed test).
3. The level of significance is $\alpha = .05$.
4. The sampling distribution is the distribution of the differences between sample means. The test statistic is the t value for $(\bar{x}_2 - \bar{x}_1)$.
  We first find the value for s_p from (11-11):

$$s_p = \sqrt{\frac{(n_1 - 1)s_1^2 + (n_2 - 1)s_2^2}{n_1 + n_2 - 2}}$$

$$= \sqrt{\frac{(11)(15.57) + (11)(17.54)}{22}} = 4.07$$

Note that since $n_1 = n_2$, s_p^2 represents the average variance:

$$s_p^2 = \sqrt{\frac{15.57 + 17.54}{2}} = \sqrt{16.555} = 4.07$$

which agrees with the above result.

From (11-14), the value of the t statistic is

$$t = \frac{\bar{x}_2 - \bar{x}_1}{s_p \sqrt{\frac{1}{n_1} + \frac{1}{n_2}}}$$

$$= \frac{32.6 - 26.8}{4.07 \sqrt{\frac{1}{12} + \frac{1}{12}}} = 3.49$$

The degrees of freedom are

$$df = n_1 + n_2 - 2$$
$$= 12 + 12 - 2 = 22$$

5. The critical values are $\pm t_{.025}(22)$ or ± 2.074.

6. The decision is to reject H_0, since $t = 3.49 > 2.074$.

7. With this decision, we risk committing a Type I error, and $\alpha = .05$.

8. The p-value associated with this test is less than .005.

Hence, we can conclude that the mean length of time for drug A to reach a specified level in the bloodstream is different from the mean length of time it takes drug B to reach the specified level in the bloodstream.

Estimation. We next demonstrate the use of estimation to determine if $(\mu_1 - \mu_2) \neq 0$. Since $\alpha = .05$, $1 - \alpha = .95$. The limits for a 95% confidence interval for $(\mu_1 - \mu_2)$ are found using (11-15):

$$(\bar{x}_1 - \bar{x}_2) \pm t_{\alpha/2}(df)s_p \sqrt{\frac{1}{n_1} + \frac{1}{n_2}}$$

$$5.8 \pm (2.074)(4.07) \sqrt{\frac{1}{12} + \frac{1}{12}}$$

$$5.8 \pm 3.45$$

$$(2.35, 9.25)$$

Since 0 is not contained in this interval, we can be 95% confident that $\mu_1 - \mu_2 \neq 0$, or that $\mu_1 \neq \mu_2$. ∎

EXAMPLE 11-20 A common method to check for pollution in a river is to measure the dissolved oxygen content in the river water at different locations. A reduction of the oxygen content in the water from one location to another indicates the presence of pollution somewhere between the two locations. The Environmental Protection Agency (EPA) suspected that a small town was dumping untreated sewage into a river running through the town. Six random specimens of river water were taken at a location above town and five random specimens of river water were taken at a location below the town. The dissolved oxygen readings, in parts per million, are as follows:

Above town	Below town
4.7	5.1
5.1	4.6
4.9	4.8
4.8	4.9
5.2	5.0
5.0	

Determine whether there is sufficient evidence to indicate that the mean oxygen content above the town is greater than the mean oxygen content below the town. Use $\alpha = .05$.

Solution   The following statistics were calculated:

Above town	Below town
$n_1 = 6$	$n_2 = 5$
$\bar{x}_1 = 4.95$	$\bar{x}_2 = 4.88$
$s_1 = .19$	$s_2 = .19$

Note that since the two sample variances are equal, there is no reason to suspect that the two population variances are unequal. In addition, with an F value of 1, we cannot reject a null hypothesis of equal population variances, since all the tabulated critical values of F are greater than 1.

We now proceed to test the hypothesis of interest to the EPA.

1. The null hypothesis is $H_0: \mu_1 - \mu_2 \leq 0$.
2. The alternative hypothesis is $H_1: \mu_1 - \mu_2 > 0$ (one-tailed test).
3. The level of significance is $\alpha = .05$.
4. The sampling distribution is the distribution of the differences between sample means. The test statistic is the t value for $(\bar{x}_1 - \bar{x}_2)$. In order to calculate the value of t, we first need to calculate the value of s_p:

$$s_p = \sqrt{\frac{(n_1 - 1)s_1^2 + (n_2 - 1)s_2^2}{n_1 + n_2 - 2}}$$

$$= \sqrt{\frac{5(.19)^2 + 4(.19)^2}{9}} = .19$$

The value of the t statistic is

$$t = \frac{\bar{x}_1 - \bar{x}_2}{s_p \sqrt{(1/n_1) + (1/n_2)}}$$

$$= \frac{4.95 - 4.88}{.19 \sqrt{\frac{1}{6} + \frac{1}{5}}} = .61$$

The value of df is

$$df = n_1 + n_2 - 2$$
$$= 6 + 5 - 2 = 9$$

5. The critical value is $t_{.05}(9) = 1.833$.

6. The decision is not to reject H_0, since $t = .61 < 1.833$. Thus, there is no statistical evidence to indicate that the community is polluting the stream. The difference between sample means $(\bar{x}_1 - \bar{x}_2) = .07$ can be explained by sampling error and should not be attributed to pollution.

7. With this decision we may be committing a Type II error; β is unknown.

8. The *p*-value for the test is greater than .1.

For illustrative purposes, let's construct a 95% confidence interval for $(\mu_1 - \mu_2)$. The limits are given by (11-15):

$$(\bar{x}_1 - \bar{x}_2) \pm t_{\alpha/2}(df)s_p \sqrt{\frac{1}{n_1} + \frac{1}{n_2}}$$

$$.07 \pm (2.262)(.19)\sqrt{\frac{1}{6} + \frac{1}{5}}$$

$$.07 \pm .26$$

Thus, a 95% confidence interval for $(\mu_1 - \mu_2)$ is $(-.19, .33)$. Since this interval contains 0, we cannot conclude that there is a difference in mean oxygen contents for the two locations.   ■

Dependent Samples

Many practical applications involve making comparisons between two populations based on paired data or dependent samples. The following are examples of applications that could involve dependent samples:

- *Medicine:* Testing the effects of a diet by obtaining weight measurements on the same people before and after dieting
- *Teaching:* Testing the effectiveness of a teaching strategy by giving pretests and posttests to the same individuals
- *Agriculture:* Testing the effects of two fertilizers on soybean production by comparing the yields from similar plots under similar conditions
- *Business:* Comparing repair estimates of two different auto-body shops for the same wrecked automobiles
- *Industry:* Testing two brands of tires for tread wear by placing one of each brand on the rear wheels of a sample of the same type of car

Reducing Two Samples of Data to One Sample

If we have two dependent random samples of size *n*, where each element in the first sample is paired with exactly one element in the second sample, then the two samples give rise to a sample of pairs or a sample of differences as indicated in Figure 11-3.

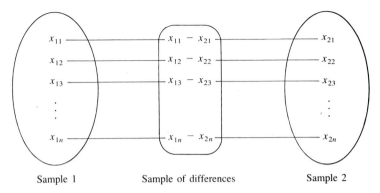

FIGURE 11-3 Reducing two samples of data to a sample of differences

The sample of differences $d = x_1 - x_2$ can be thought of as a sample from the population of the differences of paired data from two populations. The mean of the population of differences has a mean equal to the difference between the population means. If a difference score is represented by $d = x_1 - x_2$, and μ_1 and μ_2 represent the two population means, then this result can be stated as follows:

$$\mu_{\bar{x}_1 - \bar{x}_2} = \mu_d = \mu_1 - \mu_2$$

The relationship can be demonstrated by considering the following two populations whose elements have been paired:

	Population 1	Population 2	Difference d
	2	5	$2 - 5 = -3$
	4	6	$4 - 6 = -2$
	6	2	$6 - 2 = 4$
	8	4	$8 - 4 = 4$
	10	8	$10 - 8 = 2$
Sum	30	25	5
Mean	6	5	1

The difference between population means is

$$\mu_1 - \mu_2 = 6 - 5 = 1$$

and the mean of the population of differences is

$$\mu_d = \frac{\sum d}{N} = \frac{5}{5} = 1$$

Hence, we see that the mean of the population of differences is equal to the difference between the population means. By following the same line of reasoning, we can show that for two dependent samples, the mean of the sample of differences is equal to the difference between the sample means. That is, if $x_1 - x_2 = d$, then $\bar{x}_1 - \bar{x}_2 = \bar{d}$.

If we have a random sample of n data pairs and if the differences d are normally distributed, then the statistic

$$\frac{\bar{d} - \mu_d}{s_d/\sqrt{n}}$$

has a sampling distribution which is a t distribution with df $= (n-1)$, where s_d represents the standard deviation of the sample of difference scores. Hence, the t value for $\bar{d}$ is given by the following formula:

$$t = \frac{\bar{d} - \mu_d}{s_d/\sqrt{n}} \qquad (11\text{-}16)$$

with df $= n - 1$.

As a result of the above remarks, if two dependent samples are used to compare two population means, then the one-sample t statistic presented in Chapter 10 can be used. Even though two populations are involved, we can interpret the problem of comparing μ_1 and μ_2 as comparing $\mu_{\bar{x}_1 - \bar{x}_2}$ with some fixed value (which is usually 0), based on one sample of difference scores from one population of differences.

Inferences Concerning $(\mu_1 - \mu_2)$ Using Dependent Samples

When testing the null hypothesis that $\mu_1 - \mu_2 = 0$ and paired data are involved, we can write the null hypothesis as $H_0: \mu_d = 0$, where d represents the difference scores. If the population of difference scores is normal, then the test statistic is given by

$$t = \frac{\bar{d}}{s_d/\sqrt{n}}$$

where $\bar{d}$ is the mean of the sample of difference scores, s_d is the standard deviation of the sample of difference scores, and n is the size of the sample of difference scores. The t statistic has $(n-1)$ degrees of freedom associated with it.

Limits for a $(1-\alpha)100\%$ confidence interval for $(\mu_1 - \mu_2)$ are given by

$$\bar{d} \pm t_{\alpha/2}(\text{df})\left(\frac{s_d}{\sqrt{n}}\right) \qquad (11\text{-}17)$$

The following examples illustrate the methods involved in making inferences about $(\mu_1 - \mu_2)$ using paired data.

EXAMPLE 11-21 The agricultural departments of five different universities were awarded grants to study the potential yields of two new varieties of corn. Three acres of each variety were planted at each university. The following data (yields in bushels) were collected:

University	Variety 1	Variety 2
1	58	60
2	61	64
3	52	52
4	60	65
5	71	75

If it is known that the population of difference scores d is normal, determine, using $\alpha = .05$, if there is a difference in the mean yields for the two varieties.

Solution   The two samples of measurements are dependent, since the sources are universities, and each university produces two measurements. By planting both varieties at each university, we can control for the differences in soil, climate, and weather that could influence the experiment and bias the results. Let x_1 represent the yields for variety 1 and x_2 represent the yields for variety 2. We first calculate the value of s_d. The following table is used to facilitate calculation:

x_1	x_2	d	d^2
58	60	-2	4
61	64	-3	9
52	52	0	0
60	65	-5	25
71	75	-4	16
	Sums	-14	54

The mean of the differences is

$$\bar{d} = \frac{-14}{5} = -2.8$$

In order to calculate s_d, we first compute the sum of squares for the difference scores:

$$SS = \sum d^2 - \frac{(\sum d)^2}{n}$$

$$= 54 - \frac{(-14)^2}{5} = 14.8$$

Thus, the standard deviation of the difference scores is

$$s_d = \sqrt{\frac{SS}{n-1}}$$

$$= \sqrt{\frac{14.8}{4}} \simeq 1.92$$

Therefore, by (11-16), the value of the t statistic is

$$t = \frac{\bar{d} - \mu_d}{s_d/\sqrt{n}}$$

$$= \frac{-2.8 - 0}{1.92/\sqrt{5}} = -3.26$$

In addition, the value for df is

$$\text{df} = n - 1 = 5 - 1 = 4$$

The statistical hypotheses are

$$H_0: \quad \mu_1 - \mu_2 = 0$$
$$H_1: \quad \mu_1 - \mu_2 \neq 0$$

The critical values for t are $\pm t_{.025}(4) = \pm 2.776$. Since the test statistic $t = -3.26$ is less than -2.776, we reject H_0, and conclude that the samples suggest that the yields of the two varieties of corn are different.   ∎

EXAMPLE 11-22   A study was conducted to determine if jogging reduces a person's at-rest heart rate. Ten volunteers were examined before and following a 6-month jogging program. Their at-rest heart rates (beats per minute) were recorded as follows:

Before	73	77	68	62	72	80	76	64	70	72
After	68	72	64	60	71	77	74	60	64	68

By using $\alpha = .05$, determine if jogging reduces at-rest heart rates.

Solution   The difference scores are 5, 5, 4, 2, 1, 3, 2, 4, 6, and 4. The mean difference score is $\bar{d} = 3.6$ and the standard deviation of the sample of difference scores is $s_d = 1.58$.

1. The null hypothesis is $H_0: \mu_1 - \mu_2 \leq 0$.
2. The alternative hypothesis is $H_1: \mu_1 - \mu_2 > 0$ (one-tailed test).
3. The level of significance is $\alpha = .05$.
4. The test statistic is the t value for $\bar{d}$:

$$t = \frac{3.6 - 0}{1.58/\sqrt{10}} = 7.20$$

5. The critical value is $t_{.05}(9) = 1.833$.
6. The decision is to reject H_0, since $t = 7.20 > 1.833$. Thus, one can conclude that the data suggest that jogging significantly lowers at-rest heart rates.
7. With this decision, we may have committed a Type I error; $\alpha = .05$.
8. The p-value is less than .005.   ∎

EXAMPLE 11-23   Ten men were placed on a special diet; their weights were recorded before starting the diet and after 1 month on the dieting program. The results of weighings (in pounds) before and after are as follows:

Male	A	B	C	D	E	F	G	H	I	J
Before	181	172	190	186	210	202	166	173	183	184
After	178	175	185	184	207	201	160	168	180	189

Test using $\alpha = .05$ to determine whether dieting makes a difference (positive or negative).

Solution   The difference scores d are 3, -3, 5, 2, 3, 1, 6, 5, 3, and -5. The mean and standard deviation are $\bar{d} = 2$ pounds and $s_d = 3.53$ pounds, respectively. For illustrative purposes, we will determine if the diet is effective for a change in weight by using hypothesis testing and estimation procedures.

Hypothesis testing.

1. The null hypothesis is H_0: $\mu_{\text{Before}} - \mu_{\text{After}} = 0$.
2. The alternative hypothesis is H_1: $\mu_{\text{Before}} - \mu_{\text{After}} \neq 0$ (two-tailed test).
3. The level of significance is $\alpha = .05$.
4. The test statistic is the t value for $\bar{d}$:

$$t = \frac{2}{3.53/\sqrt{10}} = 1.79$$

5. The critical values are $\pm t_{.025}(9) = \pm 2.262$.
6. We cannot reject H_0, since $t = 1.79 < 2.262$.
7. With this decision, we may be committing a Type II error; β is unknown. Hence, we have no statistical evidence to support that the diet is effective in changing weight.
8. The p-value is between .05 and .10.

Estimation.   The limits for a 95% confidence interval for $(\mu_1 - \mu_2)$ are given by (11-17):

$$\bar{d} \pm t_{\alpha/2}(\text{df})\left(\frac{s_d}{\sqrt{n}}\right)$$

$$2 \pm 2.262\left(\frac{3.53}{\sqrt{10}}\right)$$

$$2 \pm 2.53$$

The 95% confidence interval is $(-.53, 4.53)$. Since this interval contains 0, we cannot conclude that the diet is effective in changing weight.   ∎

Problem Set 11.5

A

1. Given the following paired-sample data:

x	y
4	3
2	1
6	3
4	5

a. Determine the value of the t statistic.

b. How many degrees of freedom does this t have?

2. A study was undertaken to determine if undercoating late-model automobiles inhibits the development of body rust. A random sample of 20 cars of a recent make and model were used in the study. Of these, 12 were undercoated and 8 were not. At the end of a 4-year period of use, the amount of rust (in square inches) was determined for each car. The following are the results:

Undercoated	Not undercoated
$\bar{x}_1 = 30.7$	$\bar{x}_2 = 38.5$
$s_1 = 4.9$	$s_2 = 4.1$
$n_1 = 12$	$n_2 = 8$

Assume that the populations are normal and have equal variances. By using $\alpha = .05$, determine whether the average amount of rust for cars that are undercoated is less than the average amount of rust for cars that are not undercoated.

3. A new method of study was developed to increase SAT mathematics scores. The method was used on ten randomly chosen students who had already taken the SAT test once. To test the effectiveness of the method, an SAT test was again administered to the ten students. Their before and after scores were as shown in the accompanying table.

Student	A	B	C	D	E
Before	596	610	598	613	588
After	599	612	607	610	588

Student	F	G	H	I	J
Before	592	606	619	600	597
After	610	607	623	591	599

If the difference scores (After − Before) are assumed to be normally distributed with $\mu_d = 0$, calculate the value of the t statistic and determine its degrees of freedom.

4. A study was conducted on the effectiveness of an industrial safety program to reduce lost-time accidents. The results (expressed in mean man-hours lost per month over a period of 1 year) were taken at six plants before and after an industrial safety program was put into effect. Do the data in the accompanying table provide sufficient evidence (at $\alpha = .05$) to indicate that the program was effective in reducing time lost due to industrial accidents?

Plant	1	2	3	4	5	6
Before	40	66	44	72	60	32
After	33	60	45	67	54	31

5. A tire manufacturer wanted to compare the wearing qualities of two types of tires, A and B. Six cars of similar weight and design were randomly chosen for the experiment. For each car, a tire of type A and a tire of type B were randomly placed on the rear wheels. The cars were then operated for 5000 miles over similar terrain under similar driving conditions, and the amount of tread wear (in thousandths of an inch) was recorded, with the following results:

Car	1	2	3	4	5	6
Tire A	10.7	9.8	11.3	9.5	8.6	9.4
Tire B	10.3	9.4	11.5	9.0	8.2	9.0

Do the data provide sufficient evidence to indicate a difference between average tread wear for tires A and B? Use $\alpha = .05$. Assume that the difference scores are normally distributed.

6. For Problem 5 construct a 95% confidence interval for the difference between average tread wear for tires A and B.

7. Two drugs were compared for their effects on blood cholesterol levels. Two groups of patients were randomly selected to compare the two drugs, A and B. The results (serum cholesterol in milligrams per deciliter) are as follows:

Drug A	Drug B
$n_1 = 18$	$n_2 = 13$
$\bar{x}_1 = 175$	$\bar{x}_2 = 201$
$s_1 = 14.07$	$s_2 = 20.86$

a. Test the variance assumption $H_0: \sigma_1^2 = \sigma_2^2$ at $\alpha = 10\%$.

b. Do the data provide sufficient evidence to indicate that the mean cholesterol level associated with drug A is lower than the corresponding mean for drug B? Use $\alpha = .01$ and assume the blood cholesterol levels are normally distributed.

8. The following data resulted from independent random samples taken from two normal populations with equal variances:

Sample 1	Sample 2
$n_1 = 11$	$n_2 = 9$
$\bar{x}_1 = 14$	$\bar{x}_2 = 17$
$s_1 = 6$	$s_2 = 8$

a. At $\alpha = .05$, do the data indicate a difference between the population means?

b. Construct a 95% confidence interval for $(\mu_1 - \mu_2)$.

9. Two methods for teaching calculus were applied to two randomly selected groups of college students and compared by means of a standardized calculus test given at the end of the course. The results were as follows:

Method A	Method B
$n_1 = 10$	$n_2 = 13$
$\bar{x}_1 = 63$	$\bar{x}_2 = 68$
$s_1 = 7.2$	$s_2 = 8.4$

a. Do the data present sufficient evidence to indicate a difference in mean population scores? Use $\alpha = .01$.

b. Construct a 99% confidence interval for the difference in population means $(\mu_A - \mu_B)$.

10. Two different surgical procedures were being compared with respect to the length of time (in days) to recovery. The following data resulted:

Procedure A	Procedure B
$n_1 = 15$	$n_2 = 16$
$\bar{x}_1 = 6.8$	$\bar{x}_2 = 7.5$
$s_1 = 1.25$	$s_2 = 1.49$

a. At $\alpha = .05$, do the data indicate a difference in mean recovery times for the two procedures?

b. Construct a 95% confidence interval for the difference in mean recovery times.

11. Fourteen heart patients were placed on a special diet to lose weight. Their weights (in kilograms) were recorded before starting the diet and after 1 month on the diet. The following data were obtained:

Patient	1	2	3	4	5	6	7
Weight before	62	62	65	88	76	57	60
Weight after	59	60	63	78	75	58	60

Patient	8	9	10	11	12	13	14
Weight before	59	54	68	65	63	60	56
Weight after	52	52	65	66	59	58	55

Assuming that the differences in weights are normally distributed and using $\alpha = .05$, test to determine if the diet is effective.

12. Independent random samples were taken from different normal populations with equal variances. The resulting statistics are as follows:

Sample 1	Sample 2
$n_1 = 20$	$n_2 = 25$
$\bar{x}_1 = 212$	$\bar{x}_2 = 231$
$s_1 = 8.5$	$s_2 = 9.9$

a. Find 95% confidence intervals for $(\mu_1 - \mu_2)$ and $(\mu_2 - \mu_1)$.

b. At $\alpha = .05$ test $H_0: \mu_1 - \mu_2 = 0$ against $H_1: \mu_1 - \mu_2 \neq 0$.

13. The following summary information was obtained from independent random samples taken from two normal populations with equal variances:

Sample 1	Sample 2
$n_1 = 12$	$n_2 = 10$
$\bar{x}_1 = 13$	$\bar{x}_2 = 16$
$s_1 = 3.9$	$s_2 = 4.5$

Determine if the difference $(x_1 - x_2) = -3$ can be attributed to sampling error or to the possibility that $\mu_1 \neq \mu_2$ by using each of the following procedures:
a. Estimation with $1 - \alpha = .95$
b. Hypothesis testing with $\alpha = .05$

14. In an effort to determine which river, A or B, has brook trout of the greater average length, two samples of trout were compared. The following data were obtained (lengths were measured in inches):

River A	River B
$\bar{x}_1 = 11.2$	$\bar{x}_2 = 12.4$
$s_1 = 2.2$	$s_2 = 1.9$
$n_1 = 14$	$n_2 = 11$

By using $\alpha = .05$, determine if $\mu_A \neq \mu_B$ by using each of the following procedures:
a. Estimation
b. Hypothesis testing

15. Type A fertilizer was applied to six plots and type B fertilizer was applied to seven plots. The following represent the yields of corn (in bushels) per plot:

Type A	Type B
50	57
59	50
50	60
58	51
48	63
44	49
	55

a. Construct a 95% confidence interval for $(\mu_B - \mu_A)$.
b. Test the null hypothesis $H_0: \mu_B - \mu_A = 0$ against the alternative hypothesis $H_1: \mu_B - \mu_A \neq 0$ by using $\alpha = .05$.

16. A union organization conducted a study of the ages of workers in two large industries, A and B. A random sample of 35 workers from industry A had a mean age of 36.1 years and a standard deviation of 5.5 years, while a random sample of 40 workers from industry B had a mean age of 32.4 years and a standard deviation of 6.3 years. Assume that the populations of ages are normal and that the variances are equal. Construct a 95% confidence interval for the difference in mean ages for employees in industries A and B to determine if there is a difference in the mean ages.

B

17. For two dependent samples, the standard error of the difference between sample means $\sigma_{\bar{x}_1 - \bar{x}_2}$ is sometimes approximated by $\hat{\sigma}_{\bar{x}_1 - \bar{x}_2}$, where

$$\hat{\sigma}_{\bar{x}_1 - \bar{x}_2} = \sqrt{\frac{s_1^2}{n} + \frac{s_2^2}{n} - \frac{2rs_1 s_2}{n}}$$
$$= \sqrt{\frac{s_1^2 + s_2^2 - 2rs_1 s_2}{n}}$$

and r is the correlation coefficient between x_1 and x_2. For the data in Problem 3, show that

$$s_d = \sqrt{s_1^2 + s_2^2 - 2rs_1 s_2}$$

18. Consider the following computer printout for computing the t statistic for a sample of nine bivariate pairs. Using the statistics in the printout, find s_d and determine whether the population means are significantly different at $\alpha = .01$.

```
          T-TEST RESULTS
VARIABLE X:  BEFORE        VARIABLE Y:   AFTER
MEAN OF X = 7.27667        MEAN OF Y = 11.27223
S.D. OF X = 2.60702        S.D. OF Y = 5.24589
S.E. MEAN = .869007        S.E. MEAN = 1.74863

        NUMBER OF PAIRS (N) = 9
   CORRELATION OF X WITH Y (R) = 0.622
   DIFFERENCE (MEAN X - MEAN Y) = -3.99556

     DEGREES OF FREEDOM (DF) = 8
T-RATIO FOR THE DIFFERENCE = -2.71684
PROBABILITY (2 TAILED TEST) = 0.025
```

19. For small independent samples from normal populations with unknown and unequal population variances, the sampling distribution of the statistic

$$\frac{\bar{x}_1 - \bar{x}_2}{\sqrt{(s_1^2/n_1) + (s_2^2/n_2)}}$$

is approximately a t distribution with df found to be the closest integer to

$$\frac{(s_1^2/n_1) + (s_2^2/n_2)}{[(s_1^2/n_1)/(n_1 - 1)] + [(s_2^2/n_2)/(n_2 - 1)]}$$

Use this result to solve the following problem. The data resulted from independent random samples taken from two normal populations with unequal variances.

Sample 1	Sample 2
$n_1 = 11$	$n_2 = 9$
$\bar{x}_1 = 14$	$\bar{x}_2 = 17$
$s_1 = 6$	$s_2 = 8$

a. At $\alpha = .05$, do the data indicate a difference between population means?

b. Construct a 95% confidence interval for $(\mu_2 - \mu_1)$.

CHAPTER SUMMARY

In this chapter we discussed estimation and hypothesis testing involving two populations. The methods are based on the statistics from two random samples. For independent samples, the two-sample z statistic is used to compare population means when both samples are large, the two-sample t statistic is used to compare population means when sampling involves small samples from normal populations with equal variances, and the F statistic is used to compare population variances when samples are from normal populations. For dependent samples, the one-sample t statistic is used whenever the population of difference scores is normal and the variance of the population of difference scores is unknown. The statistical procedures for analyzing the differences between population proportions are similar to the procedures for analyzing the differences between population means; both involve sampling distributions that are approximately normal for large samples. Confidence intervals for $(\mu_1 - \mu_2)$ and $(p_1 - p_2)$ have the following form:

$$\text{(Test statistic)} \pm [z_{\alpha/2} \text{ or } t_{\alpha/2}(\text{df})].$$

$$\text{(Standard error of test statistic)}$$

When dealing with small independent samples, we should check the normality assumption and the assumption of equal variances whenever possible. These assumptions, along with the assumption that H_0 is true,

make it possible to reject H_0 whenever the test statistic lands in the rejection region. For if both the normality and variance assumptions are tenuous and the test statistic falls in the rejection region, we would not know whether the unlikely event of rejection should be attributed to violation of the assumptions underlying the test or to the possibility that H_0 could be false. Of course, nonrandom samples could also lead to false conclusions, and both types of error can be affected when the assumptions are violated.

The t tests are called **robust tests**, meaning that the tests work reasonably well in the face of minor violations of their theoretical assumptions. This is particularly so when two-sample tests involve samples of equal sizes. When there are major violations of the assumptions, a nonparametric test should be investigated. These will be discussed in Chapter 15.

There are many methods available to test the normality assumption underlying the χ^2 and t tests. Graphical procedures, such as the z score plot presented in Section 7.2, are easy to use. Other tests, such as the chi-square goodness-of-fit test and Lilliefors' test,[1] are also available. Both of these tests are beyond the scope of this text and will not be treated.

Table 11-1 summarizes the methods discussed in this chapter for drawing inferences concerning the comparison of two population parameters.

---

[1] Lilliefors, H. N. (1967). "On the Kolmogorov–Smirnov Test for Normality with Mean and Variance Unknown," *Journal of the American Statistical Association, 62:* 399–402.

Table 11-1 Inferences Concerning the Comparison of Two Parameters

Samples	Parameter comparison	Assumptions	Test statistic and sampling distribution	t or z value for test statistic	Confidence interval limits
Small, dependent	$\mu_1 - \mu_2$	1. Samples are random. 2. Sampling distribution of d is normal.	$\bar{x}_1 - \bar{x}_2 = \bar{d}$	$t = \dfrac{\bar{d}}{s_d/\sqrt{n}}$, df $= n - 1$	$\bar{d} \pm t_{\alpha/2}(\text{df})\left(\dfrac{s_d}{\sqrt{n}}\right)$
Small, independent	$\mu_1 - \mu_2$	1. Samples are random. 2. Both populations are normal. 3. Samples are independent. 4. $\sigma_1^2 = \sigma_2^2$	$\bar{x}_1 - \bar{x}_2$	$t = \dfrac{\bar{x}_1 - \bar{x}_2}{s_p\sqrt{(1/n_1)+(1/n_2)}}$ $s_p = \sqrt{\dfrac{(n_1-1)s_1^2 + (n_2-1)s_2^2}{n_1 + n_2 - 2}}$ df $= n_1 + n_2 - 2$.	$(\bar{x}_1 - \bar{x}_2) \pm t_{\alpha/2}(\text{df}) \cdot s_p\sqrt{\dfrac{1}{n_1} + \dfrac{1}{n_2}}$
Large, independent	$\mu_1 - \mu_2$	1. Samples are random. 2. Samples are independent.	$\bar{x}_1 - \bar{x}_2$	$z = \dfrac{\bar{x}_1 - \bar{x}_2}{\sqrt{(\sigma_1^2/n_1) + (\sigma_2^2/n_2)}}$	$(\bar{x}_1 - \bar{x}_2) \pm z\sqrt{\dfrac{\sigma_1^2}{n_1} + \dfrac{\sigma_2^2}{n_2}}$
Large, independent	$p_1 - p_2$	1. Samples are random. 2. Samples are independent. 3. $np \geq 5$, $n(1-p) \geq 5$	$\hat{p}_1 - \hat{p}_2$	$z = \dfrac{\hat{p}_1 - \hat{p}_2}{\sqrt{\hat{p}(1-\hat{p})(1/n_1 + 1/n_2)}}$ $\hat{p} = \dfrac{x_1 + x_2}{n_1 + n_2}$	$(\hat{p}_1 - \hat{p}_2) \pm z\sqrt{\dfrac{\hat{p}_1(1 - \hat{p}_1)}{n_1} + \dfrac{\hat{p}_2(1 - \hat{p}_2)}{n_2}}$
Any size, independent	$\dfrac{\sigma_1^2}{\sigma_2^2}$ or $\dfrac{\sigma_1}{\sigma_2}$	1. Samples are random. 2. Samples are independent. 3. Populations are normal.	$\dfrac{s_1^2}{s_2^2}$	$F = \dfrac{s_1^2}{s_2^2}$	$\dfrac{s_1^2}{s_2^2} \cdot F_{1-\alpha/2}(n_2 - 1, n_1 - 1) = L_1$ $\dfrac{s_1^2}{s_2^2} F_{\alpha/2}(n_2 - 1, n_1 - 1) = L_2$

C H A P T E R   R E V I E W

IMPORTANT TERMS

For each of the following terms, provide a definition in your own words. Then check your responses against the definitions given in the chapter.

dependent sample

F distribution

F statistic

independent sample

pooled estimate for p

robust tests

sampling distribution of the statistic

source

$(1 - \alpha)100\%$ confidence interval for σ_1^2/σ_2^2

$(1 - \alpha)100\%$ confidence interval for σ_1/σ_2

IMPORTANT SYMBOLS

$\mu_{\bar{x}_1 - \bar{x}_2}$, the mean of the sampling distribution of $(\bar{x}_1 - \bar{x}_2)$

$\sigma_{\bar{x}_1 - \bar{x}_2}$, the standard error of $(\bar{x}_1 - \bar{x}_2)$

$\mu_{\hat{p}_1 - \hat{p}_2}$, the mean of the sampling distribution of $(\hat{p}_1 - \hat{p}_2)$

$\hat{p}$, pooled estimate for p

$F(n_1 - 1, n_2 - 1)$, F distribution with df $= (n_1 - 1, n_2 - 1)$

$F_{1 - \alpha/2}(n_2 - 1, n_1 - 1)$, left-hand critical value for F distribution

$F_{\alpha/2}(n_1 - 1, n_2 - 1)$, right-hand critical value for F distribution

s_p, pooled estimate for σ

d, difference score

$\bar{d}$, the mean of the sample of difference scores

s_d, the standard deviation of the sample of difference scores

IMPORTANT FACTS AND FORMULAS

1. $\mu_{\bar{x}_1 - \bar{x}_2} = \mu_1 - \mu_2$, mean of the sampling distribution of $(\bar{x}_1 - \bar{x}_2)$

2. $\sigma_{\bar{x}_1 - \bar{x}_2} = \sqrt{\dfrac{\sigma_1^2}{n_1} + \dfrac{\sigma_2^2}{n_2}}$, standard error of $(\bar{x}_1 - \bar{x}_2)$

3. $\mu_{\hat{p}_1 - \hat{p}_2} = p_1 - p_2$, mean of the sampling distribution of $(\hat{p}_1 - \hat{p}_2)$

4. $\sigma_{\hat{p}_1 - \hat{p}_2} = \sqrt{\dfrac{p_1(1 - p_1)}{n_1} + \dfrac{p_2(1 - p_2)}{n_2}}$, standard error of $(\hat{p}_1 - \hat{p}_2)$

5. $\hat{p} = \dfrac{x_1 + x_2}{n_1 + n_2}$, pooled estimate for p

6. $z = \dfrac{\hat{p}_1 - \hat{p}_2}{\sqrt{p(1 - p)\left[(1/n_1) + (1/n_2)\right]}}$, z value for $(\hat{p}_1 - \hat{p}_2)$ when $p_1 = p_2$

7. $F = \dfrac{s_1^2}{s_2^2}$, df $= (n_1 - 1, n_2 - 1)$, used to test population variances

8. $F_{1 - \alpha/2}(n_2 - 1, n_1 - 1) = \dfrac{1}{F_{\alpha/2}(n_1 - 1, n_2 - 1)}$

9. $t = \dfrac{\bar{d}}{s_d/\sqrt{n}}$, t value for $\bar{d}$

10. $s_p = \sqrt{\dfrac{(n_1 - 1)s_1^2 + (n_2 - 1)s_2^2}{n_1 + n_2 - 2}}$, pooled estimate for σ

REVIEW PROBLEMS

1. Two different types of brake lining were tested for differences in wear. Twenty-four cars were used. A sample of each brand was tested with the results (listed in hundreds of miles) given in the table.

Brand A	42	58	64	40	47	50
Brand B	48	40	30	44	54	38
Brand A	62	54	42	38	66	52
Brand B	32	42	40	62	50	34

a. Assuming that the populations are normal and the population standard deviations are equal to 8, test for a difference in average brake-lining wear using $\alpha = .05$. Find the p-value.

b. Construct a 95% confidence interval for $(\mu_A - \mu_B)$.

2. Two groups of males are polled concerning their interest in a new electric razor that has four cutting edges. A sample of 64 males under age 40 indicated that only 12 were interested, while in a sample of 36 males over age 40, only 8 indicated an interest. Test whether there is a difference between the proportions of the two age groups who indicate an interest in the new razor. Use $\alpha = .05$ and determine the p-value.

3. For Problem 2, construct a 95% confidence interval for the difference between age-group proportions and determine if there is a difference between population proportions.

4. Eleven workers performed a task using two different methods. The completion times (in minutes) for each task are given below:

Worker	Method A	Method B
1	15.2	14.5
2	14.6	14.8
3	14.2	13.8
4	15.6	15.6
5	14.9	15.3
6	15.2	14.3
7	15.6	15.5
8	15.0	15.0
9	16.2	15.6
10	15.7	15.2
11	15.6	14.8

If the completion times are normally distributed for each method, test to determine whether $\mu_A \neq \mu_B$ using $\alpha = .05$.

5. Before the primary elections, candidate A showed the following voter support for president in polls taken in Maryland and Pennsylvania:

Maryland	Pennsylvania
$n_1 = 1000$	$n_2 = 540$
$x_1 = 720$	$x_1 = 324$

Let p_1 represent the proportion of Maryland voters who support candidate A and let p_2 represent the proportion of Pennsylvania voters who support A. Test at the .05 level of significance the null hypothesis $H_0: p_1 - p_2 = 0$ against the alternative hypothesis $H_1: p_1 - p_2 \neq 0$.

6. For Problem 5, construct a 95% confidence interval for $(p_1 - p_2)$, the difference between voter proportions favoring candidate A in the two states.

7. Independent samples of male and female employees selected from a large industrial plant yielded the following hourly wage results:

Males	Females
$n_1 = 45$	$n_2 = 32$
$\bar{x}_1 = \$6.00$	$\bar{x}_2 = \$5.75$
$s_1 = \$.95$	$s_2 = \$.75$

Test the statistical hypothesis that the mean hourly wage for male employees exceeds that for female employees. Use $\alpha = .01$ and determine the p-value.

8. A certain precision part is manufactured by two different companies and its length is critical. Random samples of parts were obtained from the two manufacturers and measured for length (in millimeters), with the following results:

Company A	Company B
$n_1 = 15$	$n_2 = 31$
$s_1 = .008$	$s_2 = .012$

Test using the .02 level of significance to determine whether there is a difference between the variances of the lengths of the parts manufactured by the two companies. Assume that the lengths of the parts are normally distributed.

9. For Problem 8, construct a 98% confidence interval for σ_2/σ_1 to determine if there is a difference between the variances of the lengths of the parts made by the two companies.

10. In order to compare the nicotine contents (in milligrams) of two brands of cigars, A and B, the following results were obtained:

Brand A	Brand B
$\bar{x}_1 = 15.3$	$\bar{x}_2 = 16.7$
$s_1 = 1.7$	$s_2 = 2$
$n_1 = 60$	$n_2 = 40$

a. Test using $\alpha = .05$ to determine if the mean nicotine contents for the two brands of cigars are unequal.

b. Construct a 95% confidence interval for $(\mu_A - \mu_B)$. Does it contain 0?

11. Two varieties of apples, A and B, were analyzed for their potassium content (in milligrams). The following information was obtained:

Variety A	Variety B
$n_1 = 100$	$n_2 = 150$
$\bar{x}_1 = .30$	$\bar{x}_2 = .27$
$s_1 = .07$	$s_2 = .05$

Do the data indicate that the average potassium contents differ for the two varieties? Use $\alpha = .01$. Can we conclude that $\mu_B < \mu_A$?

12. Answer Problem 11 by constructing a 99% confidence interval for the difference between the average potassium contents for the two varieties of apples.

13. Two different types of brake lining were tested for differences in wear. Twelve cars were used; each car had both brands randomly placed on the rear wheels. A sample of each brand was tested, with the results (listed in hundreds of miles) given in the table.

Car	1	2	3	4	5	6
Brand A	42	58	64	40	47	50
Brand B	48	40	30	44	54	38

Car	7	8	9	10	11	12
Brand A	62	54	42	38	66	52
Brand B	32	42	40	62	50	34

a. Assuming the population of difference scores is normal, test for a difference in average brake-lining wear using $\alpha = .05$. Find the p-value.

b. Construct a 95% confidence interval for $(\mu_A - \mu_B)$.

14. In order to evaluate two methods of teaching German, students were randomly assigned to two groups. Students in one group were taught by method A and students in the other group were taught by method B. At the end of the instructional units, both groups were given the same achievement test. The scores for the two groups are given in the accompanying table. At the .05 level of significance, determine whether method B produces higher average results than method A using the t test.

Group A	84	86	91	93	84	88	
Group B	90	88	92	94	84	85	92

15. Eleven workers performed a task using two different methods. The completion times (in minutes) for each task are given below:

Worker	Method A	Method B
1	15.2	24.5
2	14.6	14.8
3	14.2	13.8
4	15.6	15.6
5	14.9	15.3
6	15.2	14.3
7	15.6	15.5
8	15.0	15.0
9	16.2	15.6
10	15.7	15.2
11	15.6	14.8

a. Use the *t* test to determine whether there is a difference in average completion times for the two methods. Use $\alpha = .01$.

b. Construct a 99% confidence interval for the difference in average completion times.

16. An auto manufacturing plant plans to institute a new employee incentive plan. To evaluate the new plan, five employees use the incentive plan for an experimental period. Their work outputs before and after implementation of the new plan are as follows:

Employee	Output before	Output after
A	20	23
B	17	19
C	23	24
D	20	23
E	21	23

At $\alpha = .05$, use the *t* test to determine whether the new incentive plan results in greater average output.

17. For Problem 16, construct a 95% confidence interval for $(\mu_1 - \mu_2)$, where μ_1 denotes the population mean output before the new plan was implemented and μ_2 denotes the population mean output after the plan was implemented. Does this result contradict the result found in Problem 17? Explain.

18. The military conducted a study to determine the accuracy that can be obtained with two types of rifles. A random sample of equally proficient soldiers was chosen and divided into two groups. Each group shot only one type of rifle. Accuracy ratings are given in the table, where higher values signify greater accuracy. Use the *t* test at the .01 level of significance to determine whether rifle B is more accurate, on the average, than rifle A.

Rifle A	88	84	88	90	86	92	88
Rifle B	90	94	92	90	88	86	

19. Two groups of overweight teachers are randomly selected to test two diets. One group followed diet A for 1 month and the other group followed diet B for 1 month. Their weight losses (in pounds) were recorded at the end of the 1-month period, with the following results:

Diet A	Diet B
$n_1 = 50$	$n_2 = 40$
$\bar{x}_1 = 10.2$	$\bar{x}_2 = 8.4$
$s_1 = 2.6$	$s_2 = 1.8$

Test at $\alpha = .05$ to determine if there is a difference in the average weight losses for the two diets.

20. For Problem 19, determine if there is a difference in the average weight losses for the two groups of teachers by constructing a 90% confidence interval.

21. For Problem 18, test at $\alpha = .10$ to determine whether the population variances are different. Does this result affect the use of the *t* test? Explain.

22. Two brands of feed are being compared to determine if chickens fed one brand will weigh more than chickens fed the other brand. A random sample of 10,000 chickens was chosen and fed with brand A feed and another random sample of 10,000 chickens was chosen and fed with brand B. The following results were obtained (weights were measured in pounds):

Brand A	Brand B
$\bar{x}_1 = 9.072$	$\bar{x}_2 = 9.023$
$s_1 = 1.05$	$s_2 = 1.14$

Determine if there is a difference in the average weights of chickens fed with feeds A and B. Use $\alpha = .05$.

CHAPTER ACHIEVEMENT TEST

(15 points) **1.** The following information is based on random samples selected from two normal populations:

Sample 1	Sample 2
$n_1 = 31$	$n_2 = 31$
$\bar{x}_1 = 82.4$	$\bar{x}_2 = 85.5$
$s_1^2 = 21.2$	$s_2^2 = 16$

a. Test $H_0: \sigma_1^2 \leq \sigma_2^2$ against $H_1: \sigma_1^2 > \sigma_2^2$ using $\alpha = .05$.
b. Test $H_0: \mu_2 - \mu_1 \leq 0$ against $H_1: \mu_2 - \mu_1 > 0$ using $\alpha = .05$.
c. Construct a 90% confidence interval for $(\mu_2 - \mu_1)$.

(20 points) **2.** In order to test $H_0: p_1 - p_2 = 0$, the following information was obtained:

Sample 1	Sample 2
$n_1 = 50$	$n_2 = 60$
$x_1 = 29$	$x_2 = 33$

a. Find $\hat{p}$, the pooled estimate for p.
b. Find the z value for $(\hat{p}_1 - \hat{p}_2)$ under the assumption that H_0 is true.
c. Test $H_0: p_1 - p_2 \leq 0$ versus $H_1: p_1 - p_2 > 0$ using $\alpha = .05$.
d. Construct a 95% confidence interval for $(p_1 - p_2)$.

(10 points) **3.** Refer to Problem 1.
a. Construct a 90% confidence interval for σ_1^2/σ_2^2.
b. Construct a 90% confidence interval for σ_2/σ_1.

(25 points) **4.** The following information was obtained from two random samples selected from large populations:

Sample 1	Sample 2
$n_1 = 49$	$n_2 = 64$
$\bar{x}_1 = 61$	$\bar{x}_2 = 54$
$s_1 = 14$	$s_2 = 16$

a. Find $\sigma_{\bar{x}_1 - \bar{x}_2}$.
b. Find the z value for $(\bar{x}_1 - \bar{x}_2)$ if $\mu_1 - \mu_2 = 1$.
c. Test $H_0: \mu_1 - \mu_2 \leq 0$ against $H_1: \mu_1 - \mu_2 > 0$ at $\alpha = .05$.
d. Find the p-value for the test conducted in part (c).
e. Construct a 90% confidence interval for $(\mu_1 - \mu_2)$.

(30 points)   **5.** The following information was obtained from random samples selected from two normal populations:

Sample 1	Sample 2
$n_1 = 15$	$n_2 = 15$
$\bar{x}_1 = 82.4$	$\bar{x}_2 = 85.5$
$s_1^2 = 21.2$	$s_2^2 = 16$

a. Find s_p, the pooled estimate of σ.
b. Test $H_0: \sigma_1^2 \leq \sigma_2^2$ against $H_1: \sigma_1^2 > \sigma_2^2$ using $\alpha = .05$.
c. Test $H_0: \mu_2 - \mu_1 \leq 0$ against $H_1: \mu_2 - \mu_1 > 0$ using $\alpha = .05$.
d. Construct a 90% confidence interval for $(\mu_2 - \mu_1)$.
e. Construct a 90% confidence interval for $(\mu_1 - \mu_2)$.

12

ANALYSES OF COUNT DATA

Chapter Objectives

In this chapter you will learn:

- *what a contingency table is*
- *how to test the equality of two or more population proportions using the chi-square test statistic*
- *how to conduct multinomial tests using the chi-square test statistic*
- *how to conduct chi-square goodness-of-fit tests*

- *how to test for independence using the chi-square test statistic*
- *how to test for homogeneity using the chi-square test statistic*

Chapter Contents

In this chapter we will examine and illustrate additional statistical methods for dealing with data in the form of frequencies. All these methods are based on the chi-square distributions.

Introduction

Frequencies or count data often arise when variables of classification, each with several levels, are used to analyze relationships. Such would be the case when each observation is measured by two qualitative variables. For example, suppose a researcher is interested in the distribution of hypertension (high blood pressure) in a certain population, and a sample of 200 individuals is studied. The two variables of classification might be degree of hypertension and amount of smoking. Let's suppose the hypertension variable has three levels: none, mild, and severe. And suppose the smoking variable also has three levels: none, moderate, and heavy. The two variables, each with three levels, give rise to Table 12-1, called a **three-by-three (3 × 3) contingency table**.

Table 12-1   3 × 3 Contingency Table

| | | Amount of smoking | | |
		None	Moderate	Heavy
Degree of hypertension	Severe	10	14	20
	Mild	20	18	31
	None	40	22	25

If a contingency table has r rows and c columns, then it is called an **$r \times c$ contingency table**. The nine nonoverlapping groups that make up the contingency table in Table 12-1 are called **cells**. An $r \times c$ contingency table has $r \cdot c$ cells. The group of individuals who have mild hypertension and are heavy smokers consists of 31 people. Note also that these 31 people are counted only in this one cell. The entries in each cell consist of count data, called **observed frequencies**, and are denoted by O_{ij}. Note that O_{ij} represents the observed frequency for the cell located at the ith row and the jth column. Also note that the sum of the cell frequencies is the size of the sample, n:

$$\sum O_{ij} = 200 = n$$

where n represents the total frequency count or the size of the sample. Each cell in a contingency table will have two kinds of frequencies associated with it, observed and expected. The **expected frequencies** are denoted by E_{ij} and must be calculated. We will see how this is done shortly. As with the observed frequencies,

$$\sum E_{ij} = n$$

For example, suppose we are interested in determining whether a certain coin is fair. To decide, we toss it 200 times and record the number of heads and tails. Assume that we observed 90 heads and 110 tails. If the coin is fair, we would *expect* 100 heads and 100 tails; that is, the expected frequency for each of the two outcomes is 100. This information can be summarized in the following 1×2 contingency table:

Position of coin	
Head	**Tail**
$O_{11} = 90$	$O_{12} = 110$
$E_{11} = 100$	$E_{12} = 100$

We need to determine whether the difference between O_{ij} and E_{ij} for each of the two cells is due to sampling error or to the coin being biased. For, if we were to toss the coin another 200 times, we would not expect to get 90 heads and 110 tails again. Thus, we want to determine how well the observed values fit the expected values in the above table. Toward this end, we consider the difference $(O_{ij} - E_{ij})$ for each cell. The closer O_{ij} is to E_{ij}, the smaller the value of the difference $(O_{ij} - E_{ij})$ and, the further apart they are, the larger the value of the difference $(O_{ij} - E_{ij})$. For cell (1, 1), we have $(O_{11} - E_{11}) = -10$, and for cell (1, 2), $(O_{12} - E_{12}) = 10$. Notice that the differences sum to 0:

$$\sum (O_{ij} - E_{ij}) = 0$$

Since the total of the differences is 0, we square them first before we add, just as we did for the deviation scores when we studied the concept of variance in Chapter 3. Since large E_{ij} values generally correspond to large squared differences $(O_{ij} - E_{ij})^2$, we write each cell's squared difference value $(O_{ij} - E_{ij})^2$ as a percentage of its expected frequency E_{ij}; as a result, we get a more accurate measure of the difference for each cell.

If the coin is fair, an individual cell's contribution to lack of fit is determined by the value of $(O_{ij} - E_{ij})^2/E_{ij}$, and the contribution of all cells is measured by the statistic

$$\sum \frac{(O_{ij} - E_{ij})^2}{E_{ij}}$$

For our example, the value of this statistic is found to be 2, since

$$\frac{(90 - 100)^2}{100} + \frac{(110 - 100)^2}{100} = 2$$

It can be shown that the statistic $\sum (O_{ij} - E_{ij})^2/E_{ij}$ has a sampling distribution which is approximately a χ^2 distribution. Thus, we have

$$\chi^2 = \sum \frac{(O_{ij} - E_{ij})^2}{E_{ij}} \tag{12-1}$$

For our example, the value of the chi-square statistic is 2. To determine if the coin is biased, we ask whether $\chi^2 = 2$ has a larger value than can reasonably be attributed to sampling error. The maximum value that χ^2 can assume and still be attributed to chance is found in the χ^2 table. As we shall learn later, the critical value for our illustration is 3.841 for $\alpha = .05$. Thus, our value of $\chi^2 = 2$ can reasonably be accounted for by sampling error; as a result, we have no evidence to suggest that the coin is biased.

12.1 TESTING TWO OR MORE POPULATION PROPORTIONS

In Section 11.3 we learned how to test to determine whether two population proportions are different by using two random samples drawn from dichotomous or binomial populations. Each population consists of two categories, called successes and failures. The null hypothesis has the following form: H_0: $p_1 = p_2$, where p_1 is the proportion of successes in one population and p_2 is the proportion of successes in the other population. The probability of obtaining a success remains constant (property of a binomial experiment). We now want to extend the z test learned in Section 11.3 to more than two binomial populations. This extension, when restricted to two populations, will be consistent with the z test covered in Section 11.3.

In this section we are interested in testing null hypotheses of the form

$$H_0: \quad p_1 = p_2 = p_3 = \cdots = p_k$$

where $k \geq 2$ and p_i represents the proportion of successes in the ith binomial population. For each such null hypothesis, the alternative hypothesis is

H_1: At least two population proportions are unequal.

The two variables of classification for problems of this type are the outcome category (success or failure) and sample number (with k levels). The $2 \times k$ contingency table takes on the following form:

	Sample 1	Sample 2	Sample 3	$\cdots$	Sample k
Success					
Failure					

For example, suppose a certain large university is proposing a 3-hour statistics requirement for graduation. To determine student opinion, random samples of sophomores, juniors, and seniors are polled. The two outcomes in the dichotomous population are favoring the requirement and opposing the requirement. The results of the poll are contained in the 2×3 contingency table shown in Table 12-2.

Table 12-2 Results of Student Opinion Poll

		Category		
		Sophomores	**Juniors**	**Seniors**
Outcome	Favor	12	5	13
	Oppose	13	15	17

If we let $p_1, p_2,$ and p_3 represent the proportions of sophomores, juniors, and seniors, respectively, who favor the new graduation requirement, we are interested in testing the following statistical hypotheses at $\alpha = .05$:

H_0: $p_1 = p_2 = p_3$

H_1:   At least two proportions are unequal.

Continuing with our example, we first find the row and column totals (called **marginal frequencies**), which are shown in Table 12-3. The column totals give us the size of each random sample, and the row totals give us the total number of students who favor the new requirement and the total number of students who oppose the new requirement. Note that the sum of the column totals equals the sum of the row totals:

$$25 + 20 + 30 = 30 + 45 = 75$$

The total number of responses is $n = 75$.

Table 12-3   Marginal Frequencies for Student Opinion Poll

		Sample			
		Sophomores	**Juniors**	**Seniors**	**TOTAL**
	Favor	12	5	13	30
Outcome					
	Oppose	13	15	17	45
	TOTAL	25	20	30	75 = n

Before we can calculate the value of the test statistic, we must calculate the expected frequency for each cell. The expected frequencies are found by assuming the null hypothesis H_0: $p_1 = p_2 = p_3$ is true. Since each cell will have two frequencies (observed and expected) associated with it, we will place parentheses around the expected frequencies in each cell for identification purposes, as depicted here:

		Category			
		Sophomores	**Juniors**	**Seniors**	**TOTAL**
	Favor	12	5	13	30
		( )	( )	( )	
Outcome					
	Oppose	13	15	17	45
		( )	( )	( )	
	TOTAL	25	20	30	75 = n

Since H_0: $p_1 = p_2 = p_3$ is assumed to be true, we shall obtain a pooled estimate for $p_1 = p_2 = p_3$ ($= p$). We obtain the pooled estimate by combining the three samples and determining the proportion $\hat{p}$ of this combined group who are in favor of the new graduation requirement. Thus, we have

$$\hat{p} = \frac{12 + 5 + 13}{25 + 20 + 30} = \frac{30}{75} = \frac{2}{5}$$

To obtain E_{11}, the expected number of sophomores in favor of the new requirement, we multiply $\hat{p} = \frac{2}{5}$ by 25, the number of sophomores, to get $E_{11} = (\frac{2}{5})(25) = 10$. Thus, we would expect $E_{11} = 10$ sophomores to be in favor of the new requirement and $E_{21} = 15$ sophomores to be opposed ($E_{21} = 25 - 10$). Similarly, to get E_{12}, the expected number of juniors in favor of the requirement, we multiply $\hat{p}$ by 20, the number of juniors, to get $E_{12} = (\frac{2}{5})(20) = 8$. The expected number of juniors opposed to the requirement is $E_{22} = 20 - 8 = 12$. Since $E_{11} + E_{12} + E_{13} = 30$, it follows that $E_{13} = 12$. And since $E_{13} + E_{23} = 30$, we have $E_{23} = 30 - 12 = 18$. The complete 2×3 contingency table showing observed and expected frequencies is given in Table 12-4.

Table 12-4   2 × 3 Contingency Table for Student Opinion Poll

		\multicolumn{3}{c}{Sample}		
		Sophomores	**Juniors**	**Seniors**	**TOTAL**
	Favor	12 (10)	5 (8)	13 (12)	30
Outcome	Oppose	13 (15)	15 (12)	17 (18)	45
	TOTAL	25	20	30	$75 = n$

There is a more general procedure for obtaining the expected frequencies in a contingency table. For the student opinion poll example, we would first calculate the probability that one of the 75 responses is favorable and made by a sophomore under the assumption that the null hypothesis $H_0: p_1 = p_2 = p_3$ is true. Note that if the population proportions are equal, then the proportion of favorable responses is independent of the class category. We learned in Chapter 5 that E and F are independent events if $P(E \text{ and } F) = P(E)P(F)$. Thus, the probability that a response is favorable and made by a sophomore is

$$P(\text{Favor and Sophomore}) = P(\text{Favor}) \cdot P(\text{Sophomore})$$
$$= \left(\frac{30}{75}\right)\left(\frac{25}{75}\right)$$

We multiply this probability by the total number of students, n, to obtain E_{11}, the expected number of students who are sophomores and favor the new graduation requirement. Hence, we have

$$E_{11} = P(\text{Favor and Sophomore}) \cdot n$$
$$= \left(\frac{30}{75}\right)\left(\frac{25}{75}\right)(75)$$
$$= \frac{30 \times 25}{75} = 10$$

Since $E_{11} + E_{21} = 25$, by subtraction we have

$$E_{21} = 25 - E_{11} = 25 - 10 = 15$$

Note that this agrees with the results found previously. Note also that

$$E_{11} = \frac{(25)(30)}{75}$$

$$= \frac{(\text{Column total})(\text{Row total})}{n}$$

It can be shown in general that any expected frequency E_{ij} can be found by using the following relation:

$$E_{ij} = \frac{(\text{Column total}) \cdot (\text{Row total})}{n} \tag{12-2}$$

We next compute the value of the χ^2 test statistic using (12-1):

$$\chi^2 = \sum \frac{(O_{ij} - E_{ij})^2}{E_{ij}}$$

Table 12-5 helps to organize our computations for the χ^2 statistic.

Table 12-5   Computations Required for χ^2 Statistic

Cell	O_{ij}	E_{ij}	$(O_{ij} - E_{ij})^2$	$\dfrac{(O_{ij} - E_{ij})^2}{E_{ij}}$
1, 1	12	10	4	.4000
1, 2	5	8	9	1.1250
1, 3	13	12	1	.0833
2, 1	13	15	4	.2667
2, 2	15	12	9	.7500
2, 3	17	18	1	.0556
				2.6806

From Table 12-5 we see that the value of the χ^2 test statistic is $\chi^2 = 2.6806$. Note that if the null hypothesis were true and there were no sampling error involved, then the expected frequency and observed frequency for each cell would be equal. As a result, the value of the χ^2 statistic would equal zero. The question we need to answer is: How large must the value of χ^2 be in order to conclude that it no longer reflects sampling or random error?

The critical value of χ^2 that is tabulated in the χ^2 table (Table 2 of Appendix B) determines the rejection region and the maximum value of the χ^2 test statistic that can be tolerated and yet reflect random or sampling error; beyond this value, the test statistic reflects that H_0 is false and that at least two population proportions are unequal. As a consequence, all χ^2 tests for equality of population proportions are right-tailed tests.

Before we can find the critical value from the chi-square table, we must know the degrees of freedom for χ^2. The number of degrees of freedom for an

$r \times c$ contingency table is given by

$$df = (r - 1)(c - 1) \tag{12-3}$$

Consequently, the number of degrees of freedom associated with a 2×3 contingency table is 2.

Recall that $\chi^2(df)$ is used to indicate the chi-square distribution with df degrees of freedom. For our example, the chi-square critical value for $\alpha = .05$ and $df = 2$ is denoted by $\chi^2_{.05}(2)$. It is found in Table 2 (Appendix B) to be $\chi^2_{.05}(2) = 5.991$, as indicated in Figure 12-1. The value of the test statistic was found to be $\chi^2 = 2.68$. Since this statistic is not located in the rejection region, we cannot reject H_0, and we can conclude that $\chi^2 = 2.68$ reflects only sampling error. As a result, we have no statistical evidence to indicate that the population proportions are unequal. Of course, with this decision, we may have committed a Type II error. But the probability of a Type II error, β, is unknown.

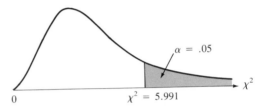

FIGURE 12-1 Rejection region for chi-square test

Let's examine the concept of degrees of freedom more closely. Consider the following example.

EXAMPLE 12-1   How many degrees of freedom are there in choosing the measures of the angles in a triangle?

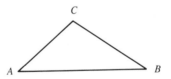

Solution   Consider triangle ABC shown in the accompanying figure. Since $A + B + C = 180°$, we are free to choose the measures of any two angles, since the third angle can be found by subtraction. Hence, there are 2 degrees of freedom associated with choosing the measures of the three angles in a triangle.   ■

For the student opinion poll example involving a 2×3 contingency table (Table 12-4), we can determine the number of degrees of freedom we have when choosing the cell frequencies once the column and row totals are known. Initially, we are free to choose any cell frequency. Suppose we choose cell (2, 2). So far we have enjoyed 1 degree of freedom. We have no freedom to choose cell (1, 2), since $O_{22} + O_{12} = 20$. Thus, we are free to choose any of the four remaining cells $(1, 1)$, $(1, 3)$, $(2, 1)$, or $(2, 3)$. Let's choose cell $(2, 1)$. So far, we have enjoyed 2 degrees of freedom. We have no freedom with cell $(2, 3)$, since $O_{21} + O_{22} + O_{23} = 45$. Similarly, we have no freedom with cell $(1, 3)$, or with cell $(1, 1)$. We have indicated freedom in our choice of cells by using a " ↓ " in the accompanying diagram. Cells marked " × " were not free to be chosen. Thus, for our example, df = 2. Note that this result is the same as that obtained by using formula (12-3).

×	×	×
↓	↓	×

Computational Formula for χ^2

There is a computational formula for computing the value of the chi-square test statistic given by (12-1). It is as follows:

$$\chi^2 = \sum \left(\frac{O_{ij}^2}{E_{ij}} \right) - n \tag{12-4}$$

Let's use (12-4) to compute the value of the chi-square test statistic for the student poll illustration used in this section. The calculations are organized in Table 12-6. The value of the chi-square test statistic is

$$\chi^2 = \sum \left(\frac{O_{ij}^2}{E_{ij}} \right) - n$$
$$= 77.68 - 75 = 2.68$$

which agrees with the result found previously by using (12-1).

Table 12-6   Calculation of χ^2 Using Computational Formula

Cell	O_{ij}	E_{ij}	$\dfrac{O_{ij}^2}{E_{ij}}$
1, 1	12	10	14.4000
1, 2	5	8	3.1250
1, 3	13	12	14.0833
2, 1	13	15	11.2667
2, 2	15	12	18.7500
2, 3	17	18	16.0556
			77.6806

In order for the sampling distribution of the statistic $\sum (O_{ij} - E_{ij})^2 / E_{ij}$ to be approximately a χ^2 distribution with df = $(r - 1)(c - 1)$, the following condition must be met:

At least 80% of the expected frequencies should be at least 5 and no expected frequency should be less than 1.

A number less than 1 in the denominator of $(O_{ij} - E_{ij})^2/E_{ij}$ tends to inflate the value of the χ^2 statistic for that cell and, as a result, leads to erroneous decisions. Let's consider another example.

EXAMPLE 12-2   A candidate for public office in a three-county area is interested in knowing if the proportion of voters who favor her is the same for each county. To determine this she took a random sample of voter opinions in each county and obtained the following results:

County 1	County 2	County 3
$x_1 = 46$	$x_2 = 48$	$x_3 = 42$
$n_1 = 120$	$n_2 = 125$	$n_3 = 110$

Do the data indicate that the true proportions of voters who favor her differ among the counties? Use $\alpha = .05$.

Solution   **1.** The null hypothesis is $H_0: p_1 = p_2 = p_3$.
2. The alternative hypothesis is H_1: at least two proportions are unequal.
3. The level of significance is $\alpha = .05$.
4. The sampling distribution is the distribution of $\sum(O_{ij} - E_{ij})^2/E_{ij}$, which is approximately $\chi^2(2)$. In order to calculate the value of the test statistic, we first calculate the expected frequencies under the assumption that H_0 is true. The expected and observed frequencies are listed in the 2×3 contingency table (Table 12-7). The first two expected frequencies in the first row were found by using (12-2):

$$E_{11} = \frac{(120)(136)}{355} = 45.97$$

$$E_{12} = \frac{(125)(136)}{355} = 47.89$$

Table 12-7   **2 × 3 Contingency Table for Example 12-2**

		County 1	County 2	County 3	TOTAL
	Yes	46 (45.97)	48 (47.89)	42 (42.14)	136
Favor					
	No	74 (74.03)	77 (77.11)	68 (67.86)	219
	TOTAL	120	125	110	$355 = n$

The remaining four expected frequencies were obtained through subtraction. In fact, for any $r \times c$ contingency table, formula (12-2) need be used to find only $df = (r - 1)(c - 1)$ expected frequencies; the remaining $(r + c - 1)$ expected frequencies can then be found by subtraction. Note that each expected frequency in this example is at least 5.

Next, we compute the value of the chi-square test statistic by using (12-4). Computations are organized in the following table:

Cell	O_{ij}	E_{ij}	$\dfrac{O_{ij}^2}{E_{ij}}$
1, 1	46	45.97	46.0300
1, 2	48	47.89	48.1103
1, 3	42	42.14	41.8605
2, 1	74	74.03	73.9700
2, 2	77	77.11	76.8902
2, 3	68	67.86	68.1403
			355.0013

Thus,

$$\chi^2 = \Sigma\left(\frac{O_{ij}^2}{E_{ij}}\right) - n$$
$$= 355.0013 - 355 = .0013$$

The degrees of freedom are found by using (12-3):

$$df = (r - 1)(c - 1)$$
$$= (2 - 1)(3 - 1) = 2$$

5. The critical value is $\chi^2_{.05}(2) = 5.991$.

6. The decision is that we cannot reject H_0, since $\chi^2 = .0013 < 5.991$.

7. With this decision, we may have committed a Type II error, but β is unknown. Hence, there is no statistical evidence to suggest that the true proportions of the voters who favor the candidate differ among the counties. ■

In the case of testing the null hypothesis H_0: $p_1 = p_2$ against the alternative hypothesis H_1: $p_1 \neq p_2$, the χ^2 test is equivalent to the z test presented in Chapter 11. This is because z^2 has a sampling distribution which is a χ^2 distribution with 1 degree of freedom. That is,

$$z^2 = \chi^2(1)$$

Since the z test for this application is nondirectional and the χ^2 test is directional, it can be shown that the two critical values have the following relationship:

$$z^2_{\alpha/2} = \chi^2_\alpha(1)$$

If $\alpha = .05$, we have $z_{.025} = 1.96$ and $z^2 = (1.96)^2 = 3.8416$. Notice that this value agrees to two decimal places with the value $\chi^2_{.05}(1)$ found in Table 2 of Appendix B. The difference in the two values is due to rounding error.

EXAMPLE 12-3   A study was conducted to determine the extent to which adults drink alcoholic beverages. Two random samples of data were gathered, one from males and the other from females. Of 150 females polled, 72 indicated they drink alcohol, while of 200 males polled, 104 indicated they drink alcohol. At $\alpha = .05$, test to determine if the proportions of adult males and adult females who drink alcohol are different.

Solution   Let p_1 denote the proportion of adult males who drink alcohol and p_2 denote the proportion of adult females who drink alcohol.

1. The null hypothesis is H_0: $p_1 = p_2$.
2. The alternative hypothesis is H_1: $p_1 \neq p_2$.
3. The level of significance is $\alpha = .05$.

For illustrative purposes, we shall test the null hypothesis by using two procedures, the z test and the χ^2 test.

Procedure 1: Using the z test.
4. The sampling distribution is the distribution of the differences between sample proportions. The test statistic is the z value for $(\hat{p}_1 - \hat{p}_2)$. The sample proportions are

$$\hat{p}_1 = \frac{104}{200} = .52$$

$$\hat{p}_2 = \frac{72}{150} = .48$$

and the pooled estimate for p is

$$\hat{p} = \frac{104 + 72}{200 + 150} = .50$$

The standard error of $(\hat{p}_1 - \hat{p}_2)$ is approximately equal to

$$\hat{\sigma}_{\hat{p}_1 - \hat{p}_2} = \sqrt{(.50)(.50)\left(\frac{1}{200} + \frac{1}{150}\right)}$$

$$= .054$$

The z value is

$$z = \frac{(\hat{p}_1 - \hat{p}_2)}{\hat{\sigma}_{\hat{p}_1 - \hat{p}_2}}$$

$$= \frac{.52 - .48}{.054} = .74$$

5. The critical values are $\pm z_{.025} = \pm 1.96$.
6. We cannot reject H_0, since $-1.96 < z = .74 < 1.96$.
7. With this decision we may have committed a Type II error; the probability of doing this is unknown. Hence, there is no statistical evidence to suggest that the proportions of adult males and adult females who drink alcohol are different.

Procedure 2: Using a χ^2 test.   Assuming H_0: $p_1 = p_2$ is true, we compute the expected frequencies. They are shown with the observed frequencies in the 2×2 contingency table (Table 12-8).

Table 12-8 2×2 Contingency Table for Example 12-3

	Males	Females	TOTAL
Alcohol	104 (100.57)	72 (75.43)	176
No alcohol	96 (99.43)	78 (74.57)	174
TOTAL	200	150	$350 = n$

The expected frequency for cell $(1, 1)$ is

$$E_{11} = \frac{(200)(176)}{350} = 100.57$$

E_{12}, E_{21}, and E_{22} are found by subtraction:

$$E_{12} = 176 - E_{11} = 176 - 100.57 = 75.43$$
$$E_{21} = 200 - E_{11} = 200 - 100.57 = 99.43$$
$$E_{22} = 174 - E_{21} = 174 - 99.43 = 74.57$$

The following table organizes the calculations for finding the value of the χ^2 test statistic using (12-4):

Cell	O_{ij}	E_{ij}	$\dfrac{O_{ij}^2}{E_{ij}}$
1, 1	104	100.57	107.5470
1, 2	72	75.43	68.7260
2, 1	96	99.43	92.6883
2, 2	78	74.57	81.5878
			350.5491

Hence, the value of the chi-square test statistic is

$$\chi^2 = \Sigma \left(\frac{O_{ij}^2}{E_{ij}} \right) - n$$
$$= 350.55 - 350 = .55$$

The critical value obtained from Table 2 (Appendix B) is $\chi^2_{.05}(1) = 3.841$. Since $\chi^2 = .55 < \chi^2_{.05}(1) = 3.841$, we cannot reject H_0. We thus conclude that there is no real evidence to suggest that the proportions of adult males and adult females who drink alcohol are different. Also note the following:

1. $z^2 = (.74)^2 \simeq .55 = \chi^2$
2. $\chi^2_{.05}(1) = 3.841$ and $(z_{.025})^2 = 1.96^2 = 3.8416 \simeq \chi^2_{.05}(1)$ ∎

The interested student might wonder why we are using a χ^2 test to test the null hypothesis $H_0: p_1 = p_2 = p_3$ instead of the three pairwise z tests associated with the following null hypotheses:

1. H_0: $p_1 = p_2$
2. H_0: $p_2 = p_3$
3. H_0: $p_1 = p_3$

There are several good reasons. The method involving three separate tests is too time-consuming and not very efficient. However, the most important reason is that sizable and unknown error rates result. For example, if each of the above three tests uses $\alpha = .05$ as the significance level and if the tests are independent of each other (in fact, they are *not*), then *the probability of committing at least one Type I error in three independent tests* is found to be:

$$1 - \text{Probability of making no Type I error in three tests}$$

$$= 1 - (.95)^3$$

$$\simeq .14$$

Thus, the error rate for the experiment is somewhere between .05 and .14, instead of .05, the error rate for one χ^2 test. As the number of tests increases, the error rate for the experiment increases.

Problem Set 12.1

A

1. For the following contingency tables, find the degrees of freedom and the appropriate critical value of χ^2 at the indicated significance level.
   a. 6×2, $\alpha = .05$
   b. 5×6, $\alpha = .01$
   c. 4×7, $\alpha = .05$
   d. 4×4, $\alpha = .01$

2. The accompanying data represent the results of a college's survey to study the graduation rate for male and female students admitted as first-time freshman students and graduating within 5 years.

	Graduated	Did not graduate
Male	16	28
Female	18	19

   a. By using the .05 level of significance and the z test, determine if the true proportions of males and females who graduate within 5 years differ.
   b. Determine by using the χ^2 test whether there is a difference between the proportions of males and females who graduate.

   c. Show that the value of the χ^2 test statistic is equal to the square of the value of the z statistic; i.e., $\chi^2(1) = z^2$.

3. At the end of a semester, the grades for Mathematics 101 were tabulated in the following 3×2 contingency table to study the relationship between class attendance and grade received:

		Grade received	
		Pass	Fail
	0–3	135	110
Number of days absent	4–6	36	4
	7–45	9	6

   At $\alpha = .05$, do the data indicate that the proportions of students who pass differ among the three absence categories?

4. A survey was undertaken to determine if the proportions of voters in Frostburg, LaVale, and Cumberland who favor an elected county school board are different. The results of the survey are given in the

accompanying table. Determine if the true proportions for the three areas differ by using $\alpha = .05$.

	Frostburg	LaVale	Cumberland
Favor	125	150	133
Oppose	130	160	102

5. An electronic supply center wants to determine if there are any differences in the proportions of service calls for four major brands of television sold by them in a certain city. The following data were collected during a 2-year period:

	Brand			
	A	**B**	**C**	**D**
Service	20	30	55	45
No service	280	289	350	89

Can the supply center conclude using $\alpha = .01$ that the proportions of defective televisions differ among brands?

6. A study was undertaken to determine if the proportions of male and female babies are the same for mothers in three different areas: the United States, Europe, and Africa. The results shown in the table were obtained. By using $\alpha = .05$, can we conclude that the proportions of male babies differ by country?

	U.S.	Europe	Africa
Male	261	207	50
Female	174	169	139

7. Women in four sections of the United States were polled and asked if they watch Monday night football on television. The results are summarized in the accompanying 2×4 contingency table. Test using $\alpha = .05$ to determine whether the proportions of women who watch Monday night football on TV differ by section of the United States.

	East	West	North	South
Yes	90	57	70	97
No	110	93	105	128

8. A certain fraternity on a large college campus conducted a study to determine the proportion of students failed by three statistics instructors in Math 209. The results are summarized in the following 2×3 contingency table:

	Instructor		
	A	**B**	**C**
Failed	8	12	10
Passed	67	88	115

By using the .01 level of significance, test to determine whether there is a difference among the proportions of students failed by each instructor.

9. A study was conducted to determine if nest location has any effect on the proportion of young birds of a certain species that survive during the nesting period. Four nest-site locations identified for study were bridges, buildings, natural rock formations, and road cuts. At each site, the nest histories were recorded for one season, with the results shown in the table. By using $\alpha = .05$ test the null hypothesis of equal survival proportions for the four nest locations.

	Nest location			
	Bridges	**Buildings**	**Rock formations**	**Road cuts**
Number of eggs hatched	15	40	32	28
Number of eggs in nest	20	42	44	35

10. As part of a health inventory questionnaire, samples of third-, sixth-, and ninth-grade students in a Missouri school district were asked the following question: "Do you eat breakfast?" Their responses are as follows:

		Grade		
		Third	**Sixth**	**Ninth**
Response	Yes	132	168	126
	No	52	33	46

At significance level $\alpha = .05$, do the data provide sufficient evidence to reject the null hypothesis that the proportions of positive responses for the three grades are equal?

B

11. Prove that $\sum(O_{ij} - E_{ij}) = 0$.

12. The value of χ^2 for the following 2×2 contingency table is given by $\chi^2 = (ad - bc)^2 m/(efgh)$. (Assume that none of the totals is equal to zero.)

	Category 2		TOTAL
Category 1	a	b	g
	c	d	h
TOTAL	e	f	m

a. Use this result to find the value of χ^2 for the following 2×2 contingency table:

18	16
22	44

b. Determine the value of χ^2 by using the computational formula.

13. Prove that $\sum(O_{ij} - E_{ij})^2/E_{ij} = \sum(O_{ij}^2/E_{ij}) - n$.

14. Calculate the χ^2 statistics for the following 2×2 contingency tables and compare your results (assume that the observed frequencies and x are different from zero):

Table 1

a	b
c	d

Table 2

ax	bx
cx	dx

15. Prove the formula for χ^2 given in Problem 12.

12.2 MULTINOMIAL TESTS

In Section 12.1 we discussed testing hypotheses of the form $H_0: p_1 = p_2 = p_3 = \cdots = p_k$ for k dichotomous (binomial) populations. For these applications, qualitative data were classified into one of two distinct outcome categories, which we labeled as success or failure. The purpose of this section is to generalize these tests to **multinomial experiments** involving populations of qualitative data having more than two distinct outcome categories. For such applications, we draw one sample from a multinomial population having m distinct categories to aid us in determining whether the population categories have proportions equal to specified values.

A multinomial experiment has properties closely resembling those of a binomial experiment. The properties of a multinomial experiment include the following:

1. The experiment consists of n trials.
2. Each trial results in exactly one of m outcomes.
3. The trials are independent.
4. The probability of any outcome remains constant from trial to trial, and the sum of the m probabilities is 1.

By examining these four properties, we can see that a binomial experiment is a special case of a multinomial experiment with $m = 2$ outcomes. For all of our applications, we will assume that sampling occurs from large (or infinite) populations.

Consider the following examples of multinomial populations:

1. The outcomes from tossing a six-sided die
2. The responses to answering a survey question by "yes," "no," or "undecided"
3. The results of achievement in terms of letter grades, A, B, C, D, or F
4. The classifications of religious affiliation as Catholic, Protestant, Jewish, or other

Possible null hypotheses for each might be

1. H_0: $p_1 = p_2 = p_3 = p_4 = p_5 = p_6 = \dfrac{1}{6}$
2. H_0: $p_y = .2, p_n = .7, p_u = .1$
3. H_0: $p_A = .1, p_B = .2, p_C = .3, p_D = .2, p_F = .2$
4. H_0: $p_C = .3, p_P = .4, p_J = .2, p_0 = .1$

Note that in each example, $\sum p_i = 1$.

The statistical calculations for testing multinomial parameters are similar to those involved in Section 12.1. The only difference is in the calculation of the degrees of freedom for the χ^2 test statistic. For testing multinomial parameters, df is equal to 1 less than the number of possible categories of distinct outcomes in a multinomial experiment. That is,

$$\text{df} = (\text{Number of outcome categories}) - 1 \qquad (12\text{-}5)$$

The following examples illustrate testing parameters for multinomial experiments.

EXAMPLE 12-4   A sportsmen's association stocked its lake 10 years ago with trout, bass, bluegill, and catfish in the following respective percentages: 20, 15, 40, and 25. Has the original distribution of fish changed over 10 years if a recent sample of fish provided the following numbers? Use $\alpha = .05$.

Trout	Bass	Bluegill	Catfish
132	100	200	168

Solution   Let $p_1 =$ percentage of trout, $p_2 =$ percentage of bass, $p_3 =$ percentage of bluegill, and $p_4 =$ percentage of catfish.

1. H_0: $p_1 = .20, p_2 = .15, p_3 = .40, p_4 = .25$
2. H_1:   At least one percentage is incorrect.
3. The level of significance is $\alpha = .05$.

The total number of fish caught is $n = 132 + 100 + 200 + 168 = 600$. For this application, we have the 1×4 contingency table shown in Table 12-9. To find the expected frequencies, we multiply the hypothesized percentages by the

Table 12-9 **Contingency Table for Example 12-4**

Trout	Bass	Bluegill	Catfish
132	100	200	168
(120)	(90)	(240)	(150)

total, $n = 600$. Thus, we have

$$E_{11} = (600)(.20) = 120$$
$$E_{12} = (600)(.15) = 90$$
$$E_{13} = (600)(.40) = 240$$
$$E_{14} = (600)(.25) = 150$$

Note that E_{14} could also have been found by subtraction:

$$E_{14} = 600 - 120 - 90 - 240 = 150$$

The value of the test statistic is computed using the computational formula:

$$\chi^2 = \sum \left(\frac{O_{ij}^2}{E_{ij}} \right) - n$$

The following table organizes the computations:

Cell	O_{ij}	E_{ij}	$\dfrac{O_{ij}^2}{E_{ij}}$
1	132	120	145.2000
2	100	90	111.1111
3	200	240	166.6667
4	168	150	188.1600
			611.1378

Hence, the value of the chi-square test statistic is

$$\chi^2 = 611.1378 - 600 \simeq 11.14$$

The degrees of freedom are found by using (12-5):

$$df = (\text{Number of outcome categories}) - 1$$
$$= 4 - 1 = 3$$

4. The critical value is $\chi_{.05}^2(3) = 7.815$.
5. Since $\chi^2 = 11.14 > \chi_{.05}^2(3) = 7.815$, we reject H_0 and conclude that the original distribution of fish has changed over the 10-year period.
6. With this decision we may be committing a Type I error; the probability of this is $\alpha = .05$. ■

EXAMPLE 12-5   A six-sided die is tossed with the results shown in Table 12-10. At $\alpha = .05$, do these results indicate that the die is biased?

Table 12-10   Observed Frequencies for Example 12-5

Face	1	2	3	4	5	6
Frequency	45	38	37	40	37	43

Solution   **1.** The null hypothesis is that the die is fair; that is,

$$H_0: \quad P(1) = P(2) = P(3) = P(4) = P(5) = P(6) = \frac{1}{6}$$

2. The alternative hypothesis is that the die is not fair; that is,

$$H_1: \quad \text{For some } i, P(i) \neq \frac{1}{6}.$$

3. The level of significance is $\alpha = .05$.

The die was tossed $n = 45 + 38 + 37 + 40 + 37 + 43 = 240$ times. The resulting 1×6 contingency table is as follows:

	Face showing					
	1	**2**	**3**	**4**	**5**	**6**
Frequency	45 (40)	38 (40)	37 (40)	40 (40)	37 (40)	43 (40)

The expected frequencies are all found to be $(\frac{1}{6})(240) = 40$.

The value of the test statistic is found by using

$$\chi^2 = \Sigma \left(\frac{O_{ij}^2}{E_{ij}} \right) - n$$

The computations are organized in the following table:

Cell	O_{ij}	E_{ij}	$\dfrac{O_{ij}^2}{E_{ij}}$
1	45	40	50.625
2	38	40	36.100
3	37	40	34.225
4	40	40	40.000
5	37	40	34.225
6	43	40	46.225
			241.400

Thus, the value of χ^2 is

$$\chi^2 = 241.4 - 240 = 1.4$$

The number of degrees of freedom is

$$df = (\text{Number of outcome categories}) - 1$$
$$= 6 - 1 = 5$$

4. The critical value is $\chi^2_{.05}(5) = 11.070$.

5. The decision is that we cannot reject H_0, since $\chi^2 = 1.4 < \chi^2_{.05}(5) = 11.070$. Thus, we have no statistical evidence that the die is biased.

6. With this decision we may have committed a Type II error; β is unknown. ∎

Goodness-of-Fit Tests

The multinomial chi-square goodness-of-fit test can be used to test whether a given sample belongs to a specified population. Consider the following examples.

EXAMPLE 12-6   Table 12-11 contains the observed frequencies for the first 100 digits in a random number table. Test to determine whether the digits in the random number table occur with different probabilities using $\alpha = .05$.

Table 12-11   Observed Frequencies for Example 12-6

Digit	0	1	2	3	4	5	6	7	8	9
f	11	9	15	11	9	11	6	6	13	9

Solution   Let p_i be the percentage of occurrences of digit i.

1. H_0: $p_i = .1$ for $i = 0, 1, 2, \ldots, 9$
2. H_1: At least one percentage differs from .1
3. $\alpha = .05$

We compute the expected frequencies under the assumption that H_0 is true. If H_0 is true, we would expect $(.1)(100) = 10$ occurrences of each digit.

We next calculate the value of the χ^2 statistic. The following table organizes the computations:

O_{ij}	E_{ij}	$\dfrac{O_{ij}^2}{E_{ij}}$
11	10	12.1
9	10	8.1
15	10	22.5
11	10	12.1
9	10	8.1
11	10	12.1
6	10	3.6
6	10	3.6
13	10	16.9
9	10	8.1
		107.2

By applying (12-4), we obtain the value of the χ^2 statistic:

$$\chi^2 = \sum \left(\frac{O_{ij}^2}{E_{ij}}\right) - n$$
$$= 107.2 - 100 = 7.2$$

4. Critical value: $\chi_{.05}^2(9) = 16.919$
5. Decision: We cannot reject H_0, since $\chi^2 = 7.2 < \chi_{.05}^2(9) = 16.919$. Thus, we have no statistical evidence to suggest that the frequencies are different.
6. Type of error possible: Type II; β is unknown. ∎

EXAMPLE 12-7 Three coins were tossed 80 times and the number of heads, X, occurring each time was recorded. The resulting data are provided in Table 12-12. Test the null hypothesis that X is binomial with $n = 3$ and $p = .5$. Use $\alpha = .05$.

Table 12-12 Frequency Table for Example 12-7

x	0	1	2	3
f	20	38	18	4

Solution Let p be the probability of getting a head.

1. H_0: X is binomial with $n = 3$ and $p = .5$.
2. H_1: X is not binomial with $n = 3$ and $p = .5$.
3. $\alpha = .05$

We assume H_0 is true and determine the expected frequencies. The binomial probability table (Table 1 of Appendix B) is used to find the binomial probabilities $p(x)$ for $n = 3$ and $p = .5$. The expected frequencies are found by multiplying $p(x)$ by 80, as indicated in the table:

x	$p(x)$	$80p(x)$
0	.125	10
1	.375	30
2	.375	30
3	.125	10

We next calculate the value of the χ^2 statistic. The following table organizes the results:

O_{ij}	E_{ij}	$\dfrac{O_{ij}^2}{E_{ij}}$
20	10	40.0000
23	30	48.1333
18	30	10.8000
4	10	1.6000
		100.5333

From (12-4), the value of the χ^2 statistic is

$$\chi^2 = \sum \left(\frac{O_{ij}^2}{E_{ij}} \right) - n$$

$$= 100.5333 - 80 \simeq 20.53$$

4. Critical value: $\chi_{.05}^2(3) = 7.815$

5. Decision: Since $20.53 > \chi_{.05}^2(3) = 7.815$, we reject H_0. Hence, we conclude that X is not binomial with $n = 3$ and $p = .5$.

6. Type of error possible: Type I; $\alpha = .05$ ■

EXAMPLE 12-8   In a study to determine the percentage of television viewers who watch the 11:00 P.M. news, a random sample of viewers was obtained. It was found that out of 500 viewers, 190 watch the late news on TV. Use $\alpha = .05$ to determine whether p, the true percentage of viewers who watch the late news on TV, differs from 40%.

Solution   **1.** H_0: $p = .4$

2. H_1: $p \neq .4$

3. $\alpha = .05$

If the null hypothesis H_0: $p = .4$ is true, then $(.4)(500) = 200$ of the sampled viewers would be expected to watch the late news, and 300 viewers would not be expected to watch the late news. The value of the χ^2 test statistic is then

$$\chi^2 = \sum \frac{(O_{ij} - E_{ij})^2}{E_{ij}}$$

$$= \frac{(190 - 200)^2}{200} + \frac{(310 - 300)^2}{300} = .83$$

4. Critical value: $\chi_{.05}^2(1) = 3.841$.

5. Decision: We cannot reject H_0, since $\chi^2 = .83 < \chi_{.05}^2(1)$. Hence, we have no statistical evidence to suggest that the percentage of viewers who watch the late news differs from .40.

6. Type of error possible: Type II; β is unknown.

Note that a z test could also have been used to test H_0: $p = .4$. The value of the z statistic for $\hat{p} = \dfrac{190}{500} = .38$ is

$$z = \frac{\hat{p} - p}{\sqrt{p(1 - p)/n}}$$

$$= \frac{.38 - .40}{\sqrt{(.4)(.6)/500}} = -.91$$

The critical values are ± 1.96. Since $-1.96 < -.91 < 1.96$, the null hypothesis cannot be rejected, which agrees with the χ^2 test. Note that $z^2 = (-.91)^2 \simeq .83 = \chi^2$.   ■

Problem Set 12.2

A

1. For the past 5 years, the percentages of A's, B's, C's, D's, and F's in a statistics class taught by Dr. X have been 11%, 18%, 35%, 25%, and 11%, respectively. Dr. Y taught the course last year and he gave 7 A's, 13 B's, 30 C's, 16 D's, and 9 F's. Can we conclude that Dr. Y's grades differ from Dr. X's grades? Use $\alpha = .01$.

2. An elementary school teacher is supposed to spend an equal amount of time on each subject taught. Do the accompanying data collected for a 4-week period indicate that the teacher does not spend the same amount of time on each subject? Use $\alpha = .01$.

Subject	Time Spent (hours)
Science	15
Reading	30
Writing	10
Spelling	27
Arithmetic	13
History	25

3. In order to plan his staffing assignments, a city police chief assumes that automobile accidents occur daily with equal frequencies during the summer months. To test the assumption, the number of automobile accidents was recorded for a random sample of 10 summer days. The results were 4, 9, 6, 15, 10, 2, 20, 8, 14, 12. Do the data indicate that the frequencies of accidents differ during the summer months? Use $\alpha = .05$.

4. Families with four children each were involved in a study to determine the sex distribution. The results listed in the table were obtained. Test at the .01 level of significance to determine if sex distribution follows a binomial distribution with $p = .5$.

Number of boys	0	1	2	3	4
Number of families	3	12	11	9	5

5. Color preferences for new cars are indicated by a random sample of potential customers. The information in the accompanying table was obtained. Test using $\alpha = .05$ to determine if color preferences are different.

Color	Red	Yellow	Blue	Green	Brown
Frequency	40	64	46	36	14

6. Grades are normally distributed if the following percentages are followed: 2% A's, 14% B's, 68% C's, 14% D's, and 2% F's. A sample of 120 statistics grades yielded the following results: 5 A's, 24 B's, 60 C's, 10 D's, and 21 F's. Using $\alpha = .05$, test to determine if the statistics grades are normally distributed.

7. In an experiment of tossing three coins 200 times, the observed frequencies were recorded as shown in the table. At the .05 level of significance, determine whether the results fit a binomial distribution with $n = 3$ and $p = .5$.

Number of heads	0	1	2	3
f	17	63	82	38

8. In an experiment, five balls were drawn, one at a time with replacement, from a box containing an equal number of red and white balls. The experiment was repeated 800 times and the following distribution of the number of white balls resulted:

Number of white balls	0	1	2	3	4	5
f	30	121	270	220	132	27

Test at $\alpha = .05$ to determine if the observed frequencies fit a binomial distribution.

9. The following is a calculation of π that has been carried out to 200 decimal places:

$\pi \simeq 3.14159$ 26535 89793 23846 26433 83279
50288 41971 69399 37510 58209 74944
59230 78164 06286 20899 86280 34825
34211 70679 82148 08651 32823 06647
09384 46095 50582 23172 53594 08128
48111 74502 84102 70193 85211 05559
64462 29489 54930 38196

Test to determine whether the digits of π to the right of the decimal point occur with unequal frequencies. Use $\alpha = .05$.

10. A biologist has collected information regarding the age and sex of 646 birds. He wants to determine if the sex ratios among adult and juvenile birds are the same this year as they were in the previous year. Last year there were 3 adult males to 7 adult females. In this year's sample there are 121 adult males, 166 adult females, 179 juvenile males, and 180

juvenile females. At the .05 significance level, does the sex ratio for adult birds differ from 3:7 this year? Does the sex ratio for juvenile birds differ from 3:7?

11. By using $\alpha = .05$ and the digits in rows 6 to 9, inclusive, of the random number table (Table 8-1 on page 246) determine if the randomization process produces digits of unequal frequencies.

B

12. A machine was found to be producing 10% defective parts. After the machine was repaired, it produced 7 defective parts in the first batch of 100 parts produced.

a. By using the z test at $\alpha = .025$, determine if the proportion of defective parts has been reduced.

b. Can we arrive at the same result by using the χ^2 statistic? Explain.

13. Both the χ^2 test and the z test can be used to test the null hypothesis $H_0: p = p_0$, provided $np \geq 5$ and $n(1 - p) \geq 5$. Prove that $\chi^2(1) = z^2$.

12.3 CHI-SQUARE TESTS FOR INDEPENDENCE

Frequently in statistical applications we are interested in determining whether two variables of classification (either quantitative or qualitative) are independent or associated. Consider the following examples:

- Are reading habits associated with sex of reader?
- Are grades received in a course related to the number of days absent?
- Is opinion on foreign policy independent of political party?
- Is sex (male, female) associated with having a college education?
- Are heart disease and smoking associated?
- Are family size and the level of education of parents independent?
- Is grade distribution independent of major subject for college students?
- Is unemployment associated with an increase in crime?
- Are faculty rank and the amount of time spent working with college students related?
- Are class size and the number of students who drop a college course associated?

Another way to express the fact that two variables are independent is to say that the two variables have nothing to do with one another; they are not related or associated.

Care should be taken *not* to conclude that two variables are correlated if they are not independent. Two variables can be uncorrelated without being independent, but all independent variables are uncorrelated.

If we know that two quantitative variables are related, we can estimate the strength of the linear relationship using the correlation coefficient, r; and using regression analysis, we can estimate the values of one variable knowing the values of the other. Of course, regression analysis can be used only when dealing with quantitative data.

For all tests of independence, the hypotheses are:

H_0: The two variables of classification are independent.
H_1: The two variables of classification are dependent.

The methods for testing H_0 against H_1 are identical to those used for testing the differences in population proportions based on the χ^2 test presented in

Section 12.1. Again, we will compare the observed and expected frequencies (derived under the assumption that H_0 is true) to determine how large a departure can be tolerated before the hypothesis of independence can be rejected. If the value of the χ^2 test statistic is greater than the tabulated critical value, then we no longer can assume that this value could have resulted from two independent variables of classification. This is why all χ^2 tests for independence are right-tailed tests. Recall that the value of the test statistic is found by

$$\chi^2 = \sum \frac{(O_{ij} - E_{ij})^2}{E_{ij}}$$

$$= \sum \left(\frac{O_{ij}^2}{E_{ij}} \right) - n$$

The corresponding degrees of freedom are found by

$$df = (r - 1)(c - 1)$$

The following example illustrates the χ^2 test for independence.

EXAMPLE 12-9   An association of university professors wants to determine whether faculty satisfaction is independent of faculty rank. The association conducted a national survey of university faculty and obtained the results shown in Table 12-13. At $\alpha = .05$ test to determine whether job satisfaction and rank are dependent based on the 3×4 contingency table.

Table 12-13   Contingency Table for Example 12-9

			Rank		
		Instructor	Assistant professor	Associate professor	Professor
Job satisfaction	High	40	60	52	63
	Medium	78	87	82	88
	Low	57	63	66	64

Solution   **1.** H_0:   Job satisfaction and rank are independent.
        H_1:   Job satisfaction and rank are dependent.
2. We next find the marginal frequencies (totals), as shown in the following table:

			Rank			
		Instructor	Assistant professor	Associate professor	Professor	TOTAL
Job satisfaction	High	40	60	52	63	215
	Medium	78	87	82	88	335
	Low	57	63	66	64	250
	TOTAL	175	210	200	215	800 = n

3. The df is found to be $df = (3 - 1)(4 - 1) = 6$.

4. The next step is to find the expected frequencies. Since df = 6, we need to compute only six expected frequencies, say, E_{11}, E_{12}, E_{13}, E_{21}, E_{22}, and E_{23}; the remaining expected frequencies are found by subtraction. The six expected frequencies are found by using (12-2):

$$E_{ij} = \frac{(\text{Column total}) \cdot (\text{Row total})}{n}$$

The observed and expected frequencies are shown in the accompanying table.

		Rank				
		Instructor	Assistant professor	Associate professor	Professor	TOTAL
	High	40 (47.03)	60 (56.44)	52 (53.75)	63 (57.78)	215
Job satisfaction	Medium	78 (73.28)	87 (87.94)	82 (83.75)	88 (90.03)	335
	Low	57 (54.69)	63 (65.62)	66 (62.50)	64 (67.19)	250
	TOTAL	175	210	200	215	800

5. The value of the χ^2 statistic is found next. The calculations are summarized in the following table:

Cell	O_{ij}	E_{ij}	$\dfrac{O_{ij}^2}{E_{ij}}$
(1, 1)	40	47.03	34.0208
(1, 2)	60	56.44	63.7846
(1, 3)	52	53.75	50.3070
(1, 4)	63	57.78	68.6916
(2, 1)	78	73.28	83.0240
(2, 2)	87	87.94	86.0700
(2, 3)	82	83.75	80.2866
(2, 4)	88	90.03	86.0158
(3, 1)	57	54.69	59.4076
(3, 2)	63	65.62	60.4846
(3, 3)	66	62.50	69.6960
(3, 4)	64	67.19	60.9615
			802.7501

Thus, we have

$$\chi^2 = \Sigma \left(\frac{O_{ij}^2}{E_{ij}} \right) - n$$

$$= 802.7501 - 800 \simeq 2.75$$

6. The critical value is $\chi^2_{.05}(6) = 12.592$.

7. Decision: Since $\chi^2 = 2.75 < \chi^2_{.05}(6) = 12.592$, we cannot reject H_0.

8. Type of error possible: Type II, but β is unknown. Thus, we have no statistical evidence to conclude that job satisfaction and faculty rank are related (or associated). ∎

EXAMPLE 12-10   Consider Table 12-14, a hypothetical 3×4 contingency table representing four populations, each classified into three categories. By using the chi-square test for independence, determine if the two variables are dependent.

Table 12-14 3×4 Contingency Table for Example 12-10

		Population				
		I	II	III	IV	TOTAL
	A	100	150	200	50	500
Category	B	40	60	80	20	200
	C	60	90	120	30	300
	TOTAL	200	300	400	100	$n = 1000$

Solution   Note that for each cell, the expected cell frequency equals the observed cell frequency; hence, the value of the χ^2 statistic is

$$\chi^2 = \sum \frac{(O_{ij} - E_{ij})^2}{E_{ij}}$$
$$= \sum \left(\frac{0}{E_{ij}}\right)$$
$$= 0$$

The two variables of classification are purely independent. Note the following relationships between rows and columns:

1. For any two rows, the column entries are proportional; for example,

$$\frac{\text{Row A}}{\text{Row B}}: \quad \frac{100}{40} = \frac{150}{60} = \frac{200}{80} = \frac{50}{20}$$

$$\frac{\text{Row B}}{\text{Row C}}: \quad \frac{40}{60} = \frac{60}{90} = \frac{80}{120} = \frac{20}{30}$$

2. For any two columns, the row entries are proportional; for example,

$$\frac{\text{Column I}}{\text{Column II}}: \quad \frac{100}{150} = \frac{40}{60} = \frac{60}{90}$$

$$\frac{\text{Column II}}{\text{Column IV}}: \quad \frac{150}{50} = \frac{60}{20} = \frac{90}{30}$$

Although this example is hypothetical, it does illustrate the relationships that the cell frequencies must exhibit in order for the two variables of classification to be independent. For a contingency table involving sample data, minor deviations due to sampling error can be tolerated. Of course, any deviation from the contingency table (Table 12-14) would indicate dependency. For our hypothetical example, there would be no need to use a χ^2 test for independence, since we already have all the population frequencies. ∎

EXAMPLE 12-11   In a study to determine if opinions on school closings are related to profession, the 3×3 contingency table of data shown in Table 12-15 was obtained. Test using $\alpha = .05$ to determine if opinion is related to profession.

Table 12-15   Contingency Table for Example 12-11

		Profession			
		Teachers	Lawyers	Doctors	TOTAL
Opinion	Favor	3	67	15	85
	Oppose	105	50	75	230
	No opinion	4	3	8	15
	TOTAL	112	120	98	330

Solution

1. H_0: Opinion and profession are independent.

H_1: Opinion and profession are dependent.

2. We find the expected frequencies to two decimal places under the assumption that H_0 is true. The expected frequencies are as follows:

$$E_{11} = \frac{(112)(85)}{330} = 28.85$$

$$E_{12} = \frac{(120)(85)}{330} = 30.91$$

$$E_{13} = 85 - 28.85 - 30.91 = 25.24$$

$$E_{21} = \frac{(112)(230)}{330} = 78.06$$

$$E_{22} = \frac{(120)(230)}{330} = 83.64$$

$$E_{23} = 230 - 78.06 - 83.64 = 68.30$$

$$E_{31} = 112 - 28.85 - 78.06 = 5.09$$

$$E_{32} = 120 - 30.91 - 83.64 = 5.45$$

$$E_{33} = 15 - 5.09 - 5.45 = 4.46$$

The observed and expected frequencies are shown in the accompanying table.

	Teachers	Lawyers	Doctors	TOTAL
Favor	3 (28.85)	67 (30.91)	15 (25.24)	85
Oppose	105 (78.06)	50 (83.64)	75 (68.30)	230
No opinion	4 (5.09)	3 (5.45)	8 (4.46)	15
TOTAL	112	120	98	$330 = n$

3. We next compute the value of the χ^2 statistic. We shall use

$$\chi^2 = \sum \frac{(O_{ij} - E_{ij})^2}{E_{ij}}$$

The following table is used to organize the computations:

Cell	O_{ij}	E_{ij}	$\dfrac{(O_{ij} - E_{ij})^2}{E_{ij}}$
(1, 1)	3	28.85	23.1620
(1, 2)	67	30.91	42.1381
(1, 3)	15	25.24	4.1544
(2, 1)	105	78.06	9.2975
(2, 2)	50	83.64	13.5300
(2, 3)	75	68.30	.6572
(3, 1)	4	5.09	.2334
(3, 2)	3	5.45	1.1014
(3, 3)	8	4.46	2.8098
			97.0838

Thus, the value of the chi-square statistic is $\chi^2 = 97.08$.

4. df $= (r - 1)(c - 1) = (3 - 1)(3 - 1) = 4$

5. Critical value: $\chi^2_{.05}(4) = 9.488$

6. Decision: Since $\chi^2 = 97.08 > 9.488$, we reject H_0. Hence, we have statistical evidence that opinion on school closings is related to profession.

7. Type of error possible: Type I; $\alpha = .05$.

8. p-value: p-value $< .005$. ∎

Test for Homogeneity

When one of the two variables of classification in a contingency table is controlled by the researcher so that either the row totals or column totals are predetermined or fixed before the data are collected, the χ^2 test is called a **test for homogeneity**, instead of a test for independence. Suppose there are four fixed column totals and three row totals in a 3×4 contingency table. The four columns of data correspond to samples from four populations. Each sample of data is then classified into three categories or cells. The objective of a test for homogeneity is to determine whether the populations that correspond to the samples of predetermined sizes are homogeneous or alike with respect to the cell probabilities. For example, suppose 200 teachers, 300 lawyers, and 400 doctors are involved in a study to determine the extent of alcohol consumption within the three professions. The frequency counts would be listed in a contingency table similar to the following:

		Profession			
		Teachers	Lawyers	Doctors	TOTAL
Alcohol consumption	Light				
	Moderate				
	Heavy				
	TOTAL	200	300	400	$900 = n$

The null hypothesis is that the three populations are homogeneous with respect to the three categories of alcohol consumption. That is,

1. The population percentage of light drinking is the same for all three professions.
2. The population percentage of moderate drinking is the same for all three professions.
3. The population percentage of heavy drinking is the same for all three professions.

The alternative hypothesis is that the populations are not homogeneous with respect to the three categories of drinking.

Beyond the design of the experiment, the computations for a test of homogeneity are identical to those required for a test of independence. The test statistic is $\chi^2 = \sum (O_{ij} - E_{ij})^2 / E_{ij}$ with df $= (r - 1)(c - 1)$.

Summary

The chi-square applications presented in this chapter are often classified as **nonparametric methods**. The only restrictions in their use are that at least 80% of the expected cell frequencies should be at least 5 and no expected frequencies less than 1 should occur. In cases where these conditions are not satisfied, it may be possible to combine rows (or columns) by adding expected frequencies so that the limitations can be overcome, or the size of the sample can be increased. Of course, the nature of the experimental study will determine if it is reasonable to combine rows or columns.

The chi-square goodness-of-fit test can be used to test the normality assumption underlying the t and F tests. This topic is left to a more advanced treatment of the subject and will be omitted here.

Problem Set 12.3

A

1. In a study to determine if sex of student and interest in mathematics are related, a sample of sixth-grade students yielded the information given in the table. Using $\alpha = .05$, test to determine if sex and interest in mathematics are related.

		Interest		
		Low	Average	High
Sex	Male	15	50	25
	Female	10	35	15

2. In a study to determine the relationship between English grades and favorite type of book for sixth-grade students, the results were categorized in the accompanying 2×5 contingency table. Determine at $\alpha = .05$ if type of book and grade in English are dependent.

	English grade				
	A	B	C	D	F
Fiction	30	20	15	10	5
Nonfiction	20	15	8	5	4

3. In a study to determine the relationship between ability in science and interest in science, the results given in the table were obtained from a random sample of high school students. Test to determine whether interest and ability in science are dependent. Use $\alpha = .05$.

		Interest		
		Low	Average	High
Ability	High	10	15	20
	Average	5	20	15
	Low	25	30	10

4. The accompanying contingency table contains the results of a random sample taken to determine if IQ scores are independent of salaries for high school graduates. Test using $\alpha = .05$ to determine if IQ score and salary are dependent variables.

IQ	Salary		
score	< $25,000	$25,000– $50,000	> $50,000
High	5	50	30
Average	20	90	15
Low	15	70	5

5. In an effort to determine if family status and high school study program are related, the data in the contingency table shown here were gathered from a random sample of high school students in a large school district. Test using $\alpha = .01$ to determine if social status and high school program are dependent.

		Social status		
		Lower	Middle	Upper
High	Academic	25	30	60
school	Commercial	40	60	70
program	General	35	70	40
	Vocational	10	15	20

6. A popular claim among teachers is that teacher ratings by students are related to grades received by students. To test this claim, an administrator collected the accompanying data. Test the claim using $\alpha = .01$.

	Grades				
Rating	A	B	C	D	F
Truly outstanding	13	17	15	3	12
Effective and competent	20	38	60	16	10
Needs improvement	20	30	45	12	5

7. Use the data in the table to test the claim that the number of cigarettes smoked per day and blood pressure level are dependent. Use $\alpha = .05$.

Systolic blood pressure	Number of cigarettes smoked					
	Under 5	6–10	11–15	16–20	21–25	Over 25
High	12	6	3	15	15	14
Slightly elevated	13	4	7	8	6	7
Normal to low	15	11	10	10	2	0

8. A sociology class studied the types of crime occurring in the four quadrants of a large city. The accompanying table displays the frequencies for various types of crimes committed during random periods in each quadrant for the past year. At $\alpha = .05$, can we conclude that type of crime is related to city quadrant?

Quadrant	Type of crime			
	Homicide	Larceny	Assault	Burglary
1	30	120	450	150
2	25	200	1000	300
3	15	190	450	260
4	10	90	100	90

9. As part of a health inventory questionnaire, random samples of sixth- and ninth-grade students in a Missouri school district were asked the following question: "Do you use seat belts?" The results are given in the table. Do the results indicate that student responses are related to grade level? Use $\alpha = .05$.

Response		Grade	
		Sixth	Ninth
	Always	5	4
	Sometimes	70	45
	Never	121	169

10. In a large school district, a study was conducted to determine whether the grades for four academic subjects are related. The results are summarized in the 5 × 4 contingency table. Test for dependence of academic subject and grade at the .01 level of significance.

Grade	Subject			
	Math	Science	English	History
A	15	8	7	11
B	18	17	10	20
C	15	17	10	20
D	10	7	7	7
F	11	8	11	8

B

11. Find the values of the χ^2 statistic for the following two 2 × 3 contingency tables. How do the values compare? Generalize the results.

Table 1

5	10	15
7	8	5

Table 2

$3 \cdot 5$	$3 \cdot 10$	$3 \cdot 15$
$3 \cdot 7$	$3 \cdot 8$	$3 \cdot 5$

12. Prove that if the value of the chi-square statistic for a contingency table is χ_0^2, and each entry in the table is multiplied by a constant c, then the new value of the chi-square statistic is $c \cdot \chi_0^2$.

CHAPTER SUMMARY

In this chapter we presented various chi-square tests. These tests measure the differences between observed and expected frequencies under the assumption that the null hypothesis is true. The larger the chi-square statistic, the stronger the evidence that the null hypothesis should be rejected. As a result, all tests studied in this chapter are right-tailed tests. Tests were presented for the equality of two or more population proportions, multinomial parameters, goodness of fit, independence, and homogeneity.

The purpose of goodness-of-fit tests is to determine whether a hypothesized population fits a particular population of interest. A test for independence is used to determine if the two qualitative variables of classification used in the contingency table are related or dependent. Tests for homogeneity of cell probabilities in the population are used whenever the row totals or column totals (but not both) in a contingency table are determined before the sample data are gathered. The test is similar to a test for independence; only the experimental design is different.

In order to use the chi-square tests in this chapter, 80% of the expected cell frequencies should be at least 5 and no cell frequency should be less than 1.

C   H   A   P   T   E   R   R   E   V   I   E   W

IMPORTANT TERMS

For each of the following terms, provide a definition in your own words. Then check your responses against the definitions given in the chapter.

cell	goodness-of-fit test	multinomial experiments	observed frequency
contingency table	marginal frequencies	nonparametric measures	test for homogeneity
expected frequency			

IMPORTANT SYMBOLS

O_{ij}, observed frequency n, total number of counts r, number of rows χ^2, chi-square statistic

E_{ij}, expected frequency c, number of columns

IMPORTANT FACTS AND FORMULAS

1. $\chi^2 = \sum \dfrac{(O_{ij} - E_{ij})^2}{E_{ij}}$, formula for computing the value of the chi-square test statistic

2. $\chi^2 = \sum \left(\dfrac{O_{ij}^2}{E_{ij}}\right) - n$, computational formula for chi-square

3. df $= (r - 1)(c - 1)$ or, for multinomial tests, df $=$ the number of outcome categories $- 1$

4. $E_{ij} = \dfrac{(\text{Column total}) \cdot (\text{Row total})}{n}$, formula for computing expected cell frequencies

5. At least 80% of the expected cell frequencies should be at least 5 and no expected frequency should be less than 1.

REVIEW PROBLEMS

1. In a study to determine the preferred brand of coffee for coffee drinkers, the following results were obtained:

Brand	A	B	C	D	E
Numbered preferred	185	205	235	195	180

Test at $\alpha = .05$ to determine if preferences for the five brands differ.

2. A sports preference study yielded the results given in the table. Test using $\alpha = .01$ to determine if sex and sport preference are dependent.

	Sport preference			
	Baseball	**Tennis**	**Basketball**	**Football**
Male	38	25	30	48
Female	32	15	36	38

3. In a sample of 200 people with colds, a new drug was used to relieve the symptoms. The results indicated that 155 people obtained relief from their symptoms as a result of the drug. Use a chi-square test and $\alpha = .05$ to determine if the drug was 85% effective.

4. The numbers of cars sold by three salesmen at a large GM dealership over a period of 6 months are shown in the table. At $\alpha = .01$, test for dependence of salesman and type of GM car sold.

		Cars		
		Pontiacs	**Buicks**	**Chevrolets**
	A	28	24	8
Salesman	B	42	32	16
	C	30	10	20

5. A homemade six-sided die was tossed 240 times. The following results were obtained:

Face	1	2	3	4	5	6
Frequency	54	43	35	37	39	32

Using $\alpha = .05$, determine if the die is biased.

6. Voter support for a candidate for mayor in a large city was determined by sampling the four city quadrants. The results shown in the accompanying table were obtained. At $\alpha = .05$, test the null hypothesis $H_0: p_{\text{I}} = p_{\text{II}} = p_{\text{III}} = p_{\text{IV}}$.

	Quadrant			
	I	**II**	**III**	**IV**
Favor	110	90	96	38
Oppose	90	60	79	37

7. The number of orders filled by two major suppliers over a 1-month period are contained in the accompanying 2 × 2 contingency table. Determine if the two suppliers have different percentages of orders that are back-ordered. Use $\alpha = .05$.

	Supplier	
	ABC	**XYZ**
Back-ordered	35	28
Not back-ordered	165	172

8. Do Problem 7 by using a z test. Then show that $z^2 = \chi^2(1)$.

9. In a certain city last year, 75% of the drivers had no accidents, 15% had one accident, and 10% had more than one accident. This year a random sample of drivers produced the following information:

Number of accidents	0	1	more than 1
Frequency	280	72	48

At $\alpha = .05$, does this sample indicate that the accident percentages have changed?

10. A study was conducted by an independent consulting firm to investigate the reliability of automobile brakes. By studying a sample of automobile accident reports involving brake failure, the firm obtained the data given in the table. Test to determine if type of brake and cause of failure are dependent. Use $\alpha = .05$.

	Cause of failure			
Type of brake	**Wheel cylinder**	**Master cylinder**	**Scoring**	**Worn lining**
Disk	70	30	5	20
Drum	80	20	20	30

11. In an effort to determine if time of day and the number of incorrectly assembled automobiles are related, an industrial engineer collected the data given here. At $\alpha = .01$, test the null hypothesis that the proportions of incorrectly assembled units differ among the four time periods.

	Time of day			
Assembled	**8–10 A.M.**	**10–12 A.M.**	**12–2 P.M.**	**2–4 P.M.**
Correctly	44	40	52	40
Incorrectly	4	5	9	7

12. The accompanying data illustrate employment in various occupations by race. At $\alpha = .05$, determine if race and occupation are dependent.

Race	Occupation			
	White-collar	Blue-collar	Farm worker	Household services
White	560	360	76	80
Black	30	50	24	15

13. A small-town newspaper publisher conducted a study to determine whether social class and newspaper subscription are related. The data listed in the table were collected. Test at $\alpha = .05$ to determine if subscriptions and social class are dependent.

	Social class		
	Poor	Middle class	Rich
Subscribe	20	48	16
Do not subscribe	85	90	2

14. According to Mendel's law, a certain pea plant should produce offspring that have white, red, and pink flowers in the ratios $1:1:2$. A sample of 500 offspring of the pea plant were colored in the ratios $23:25:52$. At $\alpha = .01$, can Mendel's law be rejected?

15. A study was made of the opinions of undergraduate students at a large university concerning a new proposed alcoholic drinking policy for the university. Five hundred males and 500 females were polled concerning their opinions regarding the proposed

policy. The results are provided in the table. Do the data indicate that for the three response categories, the probabilities differ for the populations of males and females? Use $\alpha = .05$.

	Favor	Indifferent	Oppose
Male	55	225	220
Female	95	360	45

16. In a study to investigate the relationship between price and perceived quality, 180 adults were asked to purchase beer from three presumably different brands of beer that were in fact identical. Subjects received a 5¢ discount per bottle for choosing the low-priced beer, a 2¢ discount per bottle for choosing the medium-priced beer, and no discount for the highest-priced beer. After purchasing and consuming the beer, the subjects were asked to rate the beers. The results are given in the accompanying table. Use $\alpha = .05$ to determine if there is a relationship between price and perceived product quality.

Rating	Price level		
	High	Medium	Low
Undrinkable	4	1	4
Poor	8	21	20
Fair	26	22	23
Good	15	12	9
Very pleasant	7	4	4

CHAPTER ACHIEVEMENT TEST

(10 points) **1.** If $\chi^2(1) = 15$, find the positive value of the z statistic.

(40 points) **2.** A random sample of voters were polled and asked two questions: (1) Do you have children attending public schools? and (2) Are you in favor of increased tax support for public education? The results are classified in the following 2×2 contingency table:

	Children attending	No children attending
Favor	125	35
Opposed	25	140

Let p_A represent the percentage of voters in favor who have children in school and let p_N represent the percentage of voters in favor who do not have children in school. Use

$\alpha = .01$ and test H_0: $p_A = .90$, $p_N = .10$ and provide the following information:

a. H_1 **b.** Chi-square test statistic

c. Critical value   **d.** Decision

(30 points)   **3.** Four coins were tossed 200 times. Do the following observed frequencies follow a binomial distribution with $p = .5$? Test at $\alpha = .01$ and provide the following information.

Number of heads	0	1	2	3	4
Frequency	10	43	82	60	5

a. Test statistic   **b.** Critical value   **c.** Decision

(20 points)   **4.** If two variables of classification yield the contingency table given here, are the two variables dependent? Use $\alpha = .05$ and provide the following information

	I	II	III
A	26	24	50
B	174	176	550

a. Test statistic   **b.** Critical value   **c.** Decision

13

SINGLE-FACTOR
ANALYSIS OF VARIANCE

Chapter Objectives

In this chapter you will learn:

- *the methodology involved with analysis of variance*

- *two different approaches to viewing analysis of variance*

- *the procedures for using analysis of variance to test the null hypothesis of equality of two or more population means*

- *that analysis of variance is a generalization of the two-sample t test for comparing two population means*

- *how to test for differences in population means following a significant F test*
- *how to estimate the strength of association following a significant F test*

Chapter Contents

Analysis of variance (ANOVA) is a statistical method used to test the equality of two or more population means; it is an extension or generalization of the two-sample t test to more than two independent samples.

13.1 INTRODUCTION TO ANOVA

Suppose we are interested in determining which of three new midsized, four-cylinder automobiles (Citation, LeBaron, or Reliant) is most fuel efficient in terms of miles traveled per gallon of gasoline. All three cars have approximately the same size engine and total weight. To determine the most fuel-efficient car, we will conduct an experiment involving fifteen cars, five of each make. Each car will be tested once. All cars will be supplied with a special gas tank containing 1 gallon of gasoline and driven over a test course until it runs out of gasoline. The mileage driven will then be recorded. Fifteen professional test-car drivers will be randomly assigned to drive the fifteen cars. Suppose the following results were obtained. We will consider three possibilities, A, B, and C.

Situation A

	Mileages		
	Citation	LeBaron	Reliant
	23	30	27
	24	31	27
	24	32	26
	24	32	26
	25	30	24
$\bar{x}$	24	31	26

Situation B

	Mileages		
	Citation	LeBaron	Reliant
	22	30	29
	28	29	25
	24	33	26
	25	36	26
	21	27	24
$\bar{x}$	24	31	26

Situation C

	Mileages		
	Citation	**LeBaron**	**Reliant**
	25	23	32
	17	32	34
	28	40	23
	20	34	21
	30	26	20
$\bar{x}$	24	31	26

Our task at hand is to determine if the cars differ in average gas consumption. That is, are the sample averages (24, 31, and 26) different because the underlying population means are unequal (the cars differ in fuel efficiency)? Or can the differences be attributed to random error or chance variation? Situation A suggests that the cars may differ in average miles-per-gallon ratings. Situation B is not as suggestive of a difference in population means as situation A, while situation C depicts an erratic situation. Note that for all three situations, the sample means are the same. The three situations differ in the extent to which the mileage ratings vary within each sample. As a result, it is appropriate to take into account the extent to which the sample means vary from one another (between-sample variance) and the variability of the measurements within each sample (within-sample variance). If our experiment is well designed and uses random assignments to control for any factors that could influence the outcome, then we would not expect the within-sample variation of ratings to be sizable, and would attribute this variability to sampling error. Consider the following diagrams illustrating the between-sample variance and within-sample variance for each of the situations A, B, and C.

Situation A

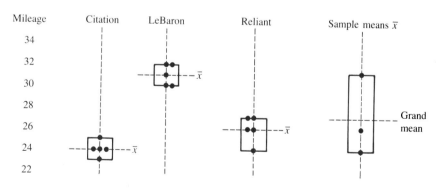

The variation within each sample appears to be small compared to the variation between the samples. By using the range as a measure of variation we have the following:

Car	Range
Citation	$25 - 23 = 2$
LeBaron	$32 - 30 = 2$
Reliant	$27 - 24 = 3$

Average range $= 2\frac{1}{3}$

The range of the sample means is $R = 31 - 24 = 7$. Thus, by using the range as a measure of variation, we see the between-sample variation is 3 times the within-sample variation. That is,

$$\frac{\text{Between-sample variation}}{\text{Within-sample variation}} = \frac{7}{2\frac{1}{3}} = 3$$

Thus, the data for situation A suggest that the average gas mileage ratings for the three cars are different.

Situation B

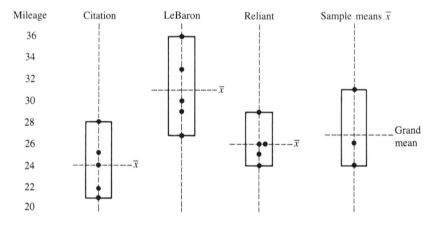

The variation within each sample appears to be approximately equal to the variation between the samples. By using the range as a measure of variation, we have the following:

Car	Range
Citation	$28 - 21 = 7$
LeBaron	$36 - 27 = 9$
Reliant	$29 - 24 = 5$

Average range $= 7$

The range of the sample means is $R = 31 - 24 = 7$. Thus, we see that

$$\frac{\text{Between-sample variation}}{\text{Within-sample variation}} = \frac{7}{7} = 1.0$$

The data for situation B do not suggest that the average miles-per-gallon fuel ratings are different.

Situation C

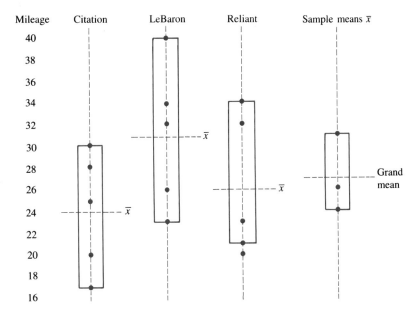

Compared to the between-sample variation, the within-sample variation appears to be greater. Again, using the range as a measure of variation, we have the following:

Car	Range
Citation	$30 - 17 = 13$
LeBaron	$40 - 23 = 17$
Reliant	$34 - 20 = 14$

Average range $= 14\frac{2}{3}$

The range of the sample means is $R = 31 - 24 = 7$; thus, we have

$$\frac{\text{Between-sample variation}}{\text{Within-sample variation}} = \frac{7}{14\frac{2}{3}} = \frac{21}{44} \simeq .5$$

The within-sample variation is greater than the between-sample variation for each make of car. This erratic behavior seemingly could have occurred by chance.

In summary, if the between-sample variation is large relative to the within-sample variation, we can conclude that the average miles-per-gallon ratings for the three cars are not equal.

Analysis of variance is a methodology for analyzing the variation between samples and the variation within samples using variances, rather than ranges. As such, it is a useful statistical method for comparing two or more population means. ANOVA enables us to test hypotheses such as

H_0: $\mu_1 = \mu_2 = \mu_3 = \cdots = \mu_k$

H_1: At least two population means are unequal.

Recall the assumptions underlying the two-sample t test involving independent samples:

1. Both populations are normal.
2. The population variances are equal; that is, $\sigma_1^2 = \sigma_2^2$.

Since ANOVA is a generalization of the two-sample t test, the assumptions for ANOVA are as follows:

1. All k populations are normal.
2. $\sigma_1^2 = \sigma_2^2 = \sigma_3^2 = \cdots = \sigma_k^2 \, (= \sigma^2)$

The ANOVA method involves calculating two independent estimates of σ^2, the common population variance. These two estimates are denoted by s_B^2 and s_W^2; s_B^2 is called the **between-samples variance estimate** and s_W^2 is called the **within-samples variance estimate**. The test statistic then becomes s_B^2/s_W^2 and has a sampling distribution which is an F distribution. Thus,

$$F = \frac{s_B^2}{s_W^2} \tag{13-1}$$

For the two-sample t test, we had two samples of data from which to compute the t statistic. For ANOVA, we have k samples of data, as shown here:

	Sample 1	Sample 2	Sample 3 ... Sample k
Sample mean	$\bar{x}_1$	$\bar{x}_2$	$\bar{x}_3$ $\bar{x}_k$
Sample standard deviation	s_1	s_2	s_3 s_k
Sample size	n_1	n_2	n_3 n_k

To simplify computations, we will assume that all samples are of the same size n. That is,

$$n_1 = n_2 = n_3 = \cdots = n_k (= n)$$

Recall that for the two-sample t test, s_p^2 is called the **pooled estimate** for σ^2 and is found by using

$$s_p^2 = \frac{(n_1 - 1)s_1^2 + (n_2 - 1)s_2^2}{n_1 + n_2 - 2}$$

If $n_1 = n_2 \ (= n)$, then s_p^2 can be written as

$$s_p^2 = \frac{(n-1)s_1^2 + (n-1)s_2^2}{2n - 2}$$

$$= \frac{(n-1)(s_1^2 + s_2^2)}{2(n-1)}$$

$$= \frac{s_1^2 + s_2^2}{2}$$

Thus, the average variance can be used as an estimate for σ^2 when samples of the same size are used. Generalizing this result to k samples, we see that the average variance can be used to estimate σ^2. We denote this estimate by s_W^2, and call it the **within-samples variance estimate**. Thus,

$$s_W^2 = \frac{s_1^2 + s_2^2 + s_3^2 + \cdots + s_k^2}{k} \tag{13-2}$$

Since this estimate is based on k samples, each with size n, and each sample has $(n-1)$ degrees of freedom associated with it (see Figure 13-1), there are $k(n-1)$ total degrees of freedom associated with the variance estimate s_W^2. This can be shown as follows:

$$\begin{aligned} df_W &= (n-1) + (n-1) + (n-1) + \cdots + (n-1) \\ &= k(n-1) \end{aligned} \tag{13-3}$$

where df_W denotes the degrees of freedom associated with s_W^2.

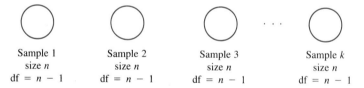

FIGURE 13-1 Degrees of freedom associated with s_W^2

Recall that the sampling distribution of the mean has a variance $\sigma_{\bar{x}}^2$ defined by

$$\sigma_{\bar{x}}^2 = \frac{\sigma^2}{n}$$

By multiplying both sides of the above equation by n, we have

$$\sigma^2 = n\sigma_{\bar{x}}^2$$

If we had an estimate for $\sigma_{\bar{x}}^2$, we could obtain an estimate for σ^2 by multiplying this estimate by n. Since we have k samples, each with a sample mean $\bar{x}$ (see Figure 13-2), the variance of the sample of k sample means can be used to estimate the variance of the sampling distribution of the mean, $\sigma_{\bar{x}}^2$. We will then multiply this estimate for $\sigma_{\bar{x}}^2$ by n, thereby obtaining our second estimate for σ^2.

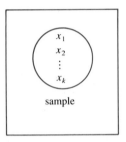

FIGURE 13-2 Sampling distribution of $\bar{x}$

The mean of the sample of k sample means is called the **grand mean** and is denoted by $\bar{x}$. Thus, we have

$$\bar{x} = \frac{\bar{x}_1 + \bar{x}_2 + \bar{x}_3 + \cdots + \bar{x}_k}{k} \qquad (13\text{-}4)$$

Consider the following table to organize the computations needed to help find the variance of the sample of k sample means:

Sample mean	(Sample mean $- \bar{x}$)	(Sample mean $- \bar{x}$)^2
$\bar{x}_1$	$\bar{x}_1 - \bar{x}$	$(\bar{x}_1 - \bar{x})^2$
$\bar{x}_2$	$\bar{x}_2 - \bar{x}$	$(\bar{x}_2 - \bar{x})^2$
$\bar{x}_3$	$\bar{x}_3 - \bar{x}$	$(\bar{x}_3 - \bar{x})^2$
$\vdots$	$\vdots$	$\vdots$
$\bar{x}_k$	$\bar{x}_k - \bar{x}$	$(\bar{x}_k - \bar{x})^2$

The variance of the sample of k sample means is the sum of the last column in the table, divided by $(k - 1)$. Hence, the estimator s_B^2 is given by

$$s_B^2 = \frac{n[(\bar{x}_1 - \bar{x})^2 + (\bar{x}_2 - \bar{x})^2 + (\bar{x}_3 - \bar{x})^2 + \cdots + (\bar{x}_k - \bar{x})^2]}{k - 1}$$

or

$$s_B^2 = \frac{n\sum(\bar{x}_i - \bar{x})^2}{k - 1} \qquad (13\text{-}5)$$

Since the sample of sample means has k elements, the between-samples variance estimate s_B^2 has $(k - 1)$ degrees of freedom associated with it; that is, the degrees of freedom associated with s_B^2 are

$$df_B = k - 1 \qquad (13\text{-}6)$$

where df_B denotes the number of degrees of freedom associated with s_B^2.

Recall from Section 11.4 that to determine whether two sample variances estimate the same population variance, we used the F statistic. Since both s_B^2 and s_W^2 are estimates for σ^2, the test statistic $F = s_B^2/s_W^2$ has a sampling distribution which is an F distribution with $df_B = (k - 1)$ and $df_W = k(n - 1)$. The critical value, obtained from the F tables (Tables 3a and 3b of Appendix B), is denoted by

$$F_\alpha(k - 1, k(n - 1))$$

where $(k - 1)$ is the degrees of freedom associated with the numerator of the F statistic and $k(n - 1)$ is the degrees of freedom associated with the denominator of the F statistic. The following example illustrates the procedures involved with ANOVA.

EXAMPLE 13-1   For the data from situation A of the gasoline mileage illustration involving the Citation, the LeBaron, and the Reliant (repeated in Table 13-1), determine using $\alpha = .05$ if the mean gasoline mileage ratings are different.

Table 13-1 Data for Example 13-1

Citation	LeBaron	Reliant
23	30	27
24	31	27
24	32	26
24	32	26
25	30	24

Solution   If we let μ_1, μ_2, and μ_3 denote the population mean mileage ratings for the Citation, LeBaron, and Reliant, respectively, then we can write the statistical hypotheses as

$$H_0: \quad \mu_1 = \mu_2 = \mu_3$$
$$H_1: \quad \text{At least two population means are unequal.}$$

The sample statistics are as follows:

Citation	LeBaron	Reliant
$n = 5$	$n = 31$	$n = 5$
$\bar{x} = 24$	$\bar{x} = 31$	$\bar{x} = 26$
$s^2 = .5$	$s^2 = 1$	$s^2 = 1.5$

1. Calculate s_W^2 using (13-2):

$$s_W^2 = \frac{s_1^2 + s_2^2 + s_3^2}{3}$$

$$= \frac{.5 + 1 + 1.5}{3} = \frac{3}{3} = 1$$

2. Calculate df_W using (13-3):

$$df_W = k(n - 1) = 3(5 - 1) = 12$$

3. Calculate s_B^2 using (13-5). The grand mean is first computed to be

$$\bar{x} = \frac{\bar{x}_1 + \bar{x}_2 + \bar{x}_3}{3}$$

$$= \frac{24 + 31 + 26}{3} = 27$$

Then

$$s_B^2 = \frac{n\sum(\bar{x}_i - \bar{x})^2}{k - 1}$$

$$= \frac{5[(24 - 27)^2 + (31 - 27)^2 + (26 - 27)^2]}{2} = 65$$

4. Calculate $df_B = k - 1 = 3 - 2 = 2$.
5. Calculate the value of F using (13-1):

$$F = \frac{s_B^2}{s_W^2} = \frac{65}{1} = 65$$

6. Find the critical value $F_{.05}(2, 12)$ in the F tables:

$$F_{.05}(2, 12) = 3.89$$

Since our test statistic F exceeds the tabulated critical value, we reject the null hypothesis $H_0: \mu_1 = \mu_2 = \mu_3$ and conclude that the population mean gas mileage ratings are unequal. ∎

If the null hypothesis is true and the two ANOVA assumptions hold, then the k samples can be thought of as being selected from the same normal population. If H_0 is true, the between-samples variance estimate s_B^2 indicates the variance of the sample means and reflects sampling error (or error variance). The within-samples variance estimate, being an average variance, reflects sampling error (or error variance) and does not depend on the null hypothesis being true or false. The measurements within each group or sample come from a single population and are not affected by a difference between population means. As a result, the values of s_W^2 and s_B^2 in this case should be close. If $H_0: \mu_1 = \mu_2 = \mu_3 = \cdots = \mu_k$ is false, then the between-samples variance estimate s_B^2 should be greater than the within-samples variance estimate s_W^2, because s_W^2 reflects only error variance and does not depend on whether the population means are equal. But if H_0 is false, the variance of the sample of sample means will reflect more than sampling error, since the population means are not equal. The tabulated critical value for F determines how much larger than 1 the test statistic can be and still reflect only sampling error. If the test statistic F is greater than the critical value (and thus falls in the rejection region), then the null hypothesis H_0 is rejected and we conclude that the test statistic reflects more than sampling error. This explains why ANOVA is a right-tailed test procedure. Let's examine another problem.

EXAMPLE 13-2   Three different makes of diesel engines were tested to determine the useful lifetimes before an overhaul was needed. If the engine lives for each make are normally distributed and have the same variance, test using $\alpha = .05$ to determine if the mean useful lives before an overhaul differ. The useful engine lives (in tens of thousands of miles) for each engine are given in Table 13-2.

Table 13-2 Data for Example 13-2

A	B	C
6	8	3
2	7	2
4	7	5
1	2	4
7	6	1

Solution   If we let μ_1, μ_2, and μ_3 denote the population mean useful lifetimes for makes A, B, and C, respectively, then we can write the statistical hypotheses as

H_0: $\mu_1 = \mu_2 = \mu_3$

H_1: At least two population means are unequal.

We shall proceed according to the following steps: (1) Find the sample means and variances; (2) find the within-samples variance estimate s_W^2 and its associated degrees of freedom df_W; (3) find the grand mean $\bar{x}$ for the sample of sample means; (4) find the between-samples variance estimate s_B^2 and its associated degrees of freedom df_B; (5) find the value of the F test statistic; (6) find the critical value for F based on df_B and df_W; (7) decide whether to reject H_0.

1. The following table displays $\sum x$ and $\sum x^2$, which are used to find the sample means and variances for the three groups A, B, and C:

	A	B	C
	6	8	3
	2	7	2
	4	7	5
	1	2	4
	7	6	1
$\sum x$	20	30	15
$\sum x^2$	106	202	55

By using a hand-held calculator, we compute the sample means to be $\bar{x}_1 = 4$, $\bar{x}_2 = 6$, and $\bar{x}_3 = 3$. We then apply the formulas

$$s^2 = \frac{SS}{n-1}$$

$$SS = \sum x^2 - \frac{(\sum x)^2}{n}$$

with the help of a calculator to compute the sample variances, $s_1^2 = 6.5$, $s_2^2 = 5.5$, and $s_3^2 = 2.5$.

2. By using (13-2), we find s_W^2:

$$s_W^2 = \frac{s_1^2 + s_2^2 + s_3^2}{k}$$

$$= \frac{6.5 + 5.5 + 2.5}{3} = 4.83$$

By using (13-3), we find df_W:

$$df_W = k(n - 1) = 3(5 - 1) = 12$$

Note that each sample has 4 degrees of freedom and that there are 3 samples; thus, $df = (4)(3) = 12$.

3. By using (13-4), we find $\bar{x}$:

$$\bar{x} = \frac{\bar{x}_1 + \bar{x}_2 + \bar{x}_3}{3}$$

$$= \frac{4 + 6 + 3}{3} = 4.33$$

4. By using (13-5), we find s_B^2:

$$s_B^2 = \frac{n[(\bar{x}_1 - \bar{x})^2 + (\bar{x}_2 - \bar{x})^2 + (\bar{x}_3 - \bar{x})^2]}{k - 1}$$

$$= \frac{5[(4 - 4.33)^2 + (6 - 4.33)^2 + (3 - 4.33)^2]}{2} = \frac{5(4.667)}{2} = 11.67$$

The degrees of freedom for s_B^2 is found by using (13-6):

$$df_B = k - 1 = 3 - 1 = 2$$

5. The value for the F test statistic is given by (13-1):

$$F = \frac{s_B^2}{s_W^2} = \frac{11.67}{4.83} = 2.42$$

6. The critical value for the test statistic is $F_{.05}(2, 12)$ and is found in the F table:

$$F_{.05}(2, 12) = 3.89$$

7. Since $F = 2.42 < F_{.05}(2, 12) = 3.89$, we cannot reject H_0.

We have no statistical evidence to support that the population means differ. The differences in sample means can be attributed to sampling error.   ■

By examining the entries in the F table (Table 3 of Appendix B), we note that all values are greater than 1. As a result, there is no need to find the critical value for the F statistic when the value for the F test statistic is less than 1. That is, there is no chance for rejecting the null hypothesis H_0 when $F \leq 1$.

We noted earlier that the ANOVA method generalizes the nondirectional two-sample t test for independent samples having the same size. This is because when only $k = 2$ samples are involved, the square of the t statistic is identical to the F statistic. That is, we have

$$t^2 = F$$

where the t distribution has $\mathrm{df} = k(n - 1) = 2(n - 1)$ and the F distribution has $\mathrm{df}_B = (k - 1) = 1$ and $\mathrm{df}_W = k(n - 1) = 2(n - 1)$. Consider the following example.

EXAMPLE 13-3 As part of a readability study conducted to determine the clarity of two different texts, 20 random pages from each text were selected. A readability score for each page was determined. Assume that the readability scores are normal for each text and have equal variances. The following data were obtained:

Text 1	Text 2
$n_1 = 20$	$n_2 = 20$
$\bar{x}_1 = 76$	$\bar{x}_2 = 70$
$s_1^2 = 60$	$s_2^2 = 100$

Test at $\alpha = .05$ to determine if the population average readability scores are different.

Solution **Method 1: t test**

1. H_0: $\mu_1 = \mu_2$
   H_1: $\mu_1 \neq \mu_2$
2. $\alpha = .05$
3. The difference between sample means is

$$\bar{x}_1 - \bar{x}_2 = 76 - 70 = 6$$

The pooled estimate for σ is

$$s_p = \sqrt{\frac{(n_1 - 1)s_1^2 + (n_2 - 1)s_2^2}{n_1 + n_2 - 2}}$$

$$= \sqrt{\frac{(19)(60) + (19)(100)}{38}} = \sqrt{80} = 8.94$$

The value of the test statistic is

$$t = \frac{6}{(8.94)\sqrt{(\frac{1}{20}) + (\frac{1}{20})}} = 2.122$$

The degrees of freedom for the test statistic are

$$\mathrm{df} = n_1 + n_2 - 2 = 20 + 20 - 2 = 38$$

4. The critical values are $\pm t_{.025}(38) = \pm 2.021$.

5. Decision: Reject H_0, since $t = 2.122 > t_{.025}(38) = 2.021$. Hence, we have statistical evidence to suggest that the average clarity scores for the texts are unequal.

6. Type of error possible: Type I; $\alpha = .05$.

7. p-value: $.025 < p$-value $< .05$.

Method 2: ANOVA

1. The within-groups variance estimate is

$$s_W^2 = \frac{s_1^2 + s_2^2}{2} = \frac{60 + 100}{2} = 80$$

2. The df for s_W^2 is

$$df_W = k(n - 1) = 2(20 - 1) = 38$$

3. The grand mean is

$$\bar{x} = \frac{\bar{x}_1 + \bar{x}_2}{2} = \frac{76 + 70}{2} = 73$$

4. The between-groups variance estimate is

$$s_B^2 = \frac{n[(\bar{x}_1 - \bar{x})^2 + (\bar{x}_2 - \bar{x})^2]}{k - 1}$$

$$= \frac{20[(76 - 73)^2 + (70 - 73)^2]}{2 - 1}$$

$$= 20(9 + 9) = 360$$

5. The degrees of freedom for s_B^2 are

$$df_B = k - 1 = 2 - 1 = 1$$

6. The value of the F statistic is

$$F = \frac{s_B^2}{s_W^2} = \frac{360}{80} = 4.50$$

7. $F_{.05}(1, 38) \simeq 4.17 = F_{.05}(1, 30)$.

8. Since $F = 4.5 > F_{.05}(1, 38)$, we reject H_0. As with the t test, we conclude that the average clarity scores are different.

Note that $t^2 = (2.122)^2 = 4.50 = F$. ∎

The F test has been determined by statisticians to be a robust test, particularly when equal sample sizes are involved. A test is called *robust* if it is relatively insensitive to minor violations of the normality and equal-variance assumptions. Determining which violations are minor is open to speculation and depends on the judgment of the individual using the test. Knowing that the F test is robust should not lead one to believe that the assumptions do not have to be checked. They should always be checked, even for equal sample sizes.

Analysis of variance is a mathematical model used to solve practical problems. As a result, it never fits a given situation exactly, and there are times when ANOVA should not be used to model a given situation. These times are when the assumptions are very tenuous, at best. In such cases, a nonparametric method may be appropriate. Nonparametric methods are presented in Chapter 15.

Problem Set 13.1

A

For Problems 1–9, assume the assumptions for ANOVA hold.

1. A readability study was done on two textbooks to investigate their clarity. Five independent determinations of readability were made for each textbook. The following data represent the readability scores for the two books:

Text A	50	49	54	57	58
Text B	48	57	61	55	60

   a. At $\alpha = .05$ use the t test to determine if the two texts have different average readability scores.
   b. At $\alpha = .05$, use ANOVA to answer the same question.
   c. Show that $t^2 = F$.

2. Five different methods are used to teach a basic unit on ANOVA to college business majors. A different teaching method was used in each of five classes and thirteen students were randomly assigned to each class. The results, based on an examination given at the end of the unit, yielded the following data:

Method A	Method B	Method C
$n_1 = 13$	$n_2 = 13$	$n_3 = 13$
$\bar{x}_1 = 75$	$\bar{x}_2 = 76$	$\bar{x}_3 = 70$
$s_1 = 7.7$	$s_2 = 7.1$	$s_3 = 9.9$

Method D	Method E
$n_4 = 13$	$n_5 = 13$
$\bar{x}_4 = 74$	$\bar{x}_5 = 76$
$s_4 = 6$	$s_5 = 6.3$

By using $\alpha = .05$, test to determine if there is a difference among the five population means.

3. Three laboratories are used by a chemical firm for performing analyses. For quality control purposes, it is decided to submit five samples of the same material to each lab and to compare their analyses to determine whether they give, on the average, the same results. The analytical results from the three labs are as follows:

Lab A	58.6	60.7	59.4	59.6	60.5
Lab B	61.6	64.8	62.8	59.2	60.4
Lab C	60.7	55.6	57.3	55.2	60.2

By using $\alpha = .05$, determine whether the labs produce, on the average, the same results.

4. The accompanying table contains the number of words typed per minute by four college secretaries at five different times using the same typewriter. By using $\alpha = .05$, determine if the typing speeds of the four secretaries differ.

A	B	C	D
82	55	69	87
79	67	72	61
75	84	78	82
68	77	83	61
65	71	74	72

5. Basic skills tests are given to ninth-grade students in four high schools in a certain county. Random samples of their scores are shown in the table. By using $\alpha = .05$, test to determine if the mean scores for each school differ.

School A	School B	School C	School D
20	24	16	19
21	21	21	20
22	22	18	21
24	25	13	20

6. A study was conducted to study the length of time (in seconds) required for students majoring in art, music, and physical education to complete a particular task involving certain motor skills. Seven people were randomly chosen from each of the three disciplines. The results are listed in the table. By using the .01 level of significance, test to determine if there is a difference in the population mean completion times for the three disciplines.

Art	Music	Physical education
17	24	25
21	18	24
25	19	25
16	22	21
19	23	24
22	20	28
18	21	19

7. Tourists were polled and asked which of three activities was most important in influencing their choice of resort and the number of hours per day they spent participating in the activity. The results of the survey for activity participation times are as follows:

Tennis	Swimming	Golf
$n = 20$	$n = 20$	$n = 20$
$\bar{x} = 3.05$	$\bar{x} = 2.15$	$\bar{x} = 3.82$
$s = 2.51$	$s = 3.26$	$s = 1.17$

At $\alpha = .05$ do the data indicate different population mean participation times for the three activities?

8. A study was conducted to investigate the contaminant level of coal gases in the atmosphere from four different utility sources that use coal to generate energy. The results given in the accompanying table were produced by taking five readings from each utility at different times. By using $\alpha = .05$, determine if there are differences among the average contaminant levels for the four utility plants.

Utility A	Utility B	Utility C	Utility D
.047	.037	.019	.041
.039	.041	.021	.037
.051	.036	.018	.038
.048	.035	.022	.047
.046	.040	.017	.048

9. In a study of the effect of diet on blood pressure, 18 adults aged 25–30 were randomly assigned to one of three diets. After subjects had been on the diets for 2 months, blood-pressure measurements were recorded for each individual, with the results shown in the table. At $\alpha = .01$, is there a difference in mean blood-pressure measurements for subjects on the three diets?

A	B	C
122	117	128
130	123	124
118	116	119
115	112	132
128	119	135
118	115	118

B

10. Suppose two random samples of size n are selected from normal populations with equal means and variances and yield the following information:

Sample 1	Sample 2
Size $= n$	Size $= n$
Mean $= \bar{x}_1$	Mean $= \bar{x}_2$
Variance $= s_1^2$	Variance $= s_2^2$

Prove that $t^2 = F$.

13.2 COMPUTATIONAL FORMULAS FOR SINGLE-FACTOR ANOVA

While the approach to single-factor ANOVA used in Section 13.1 is reasonable from a developmental standpoint, it does not lend itself to shortcut computations that minimize the amount of work involved. Nor does the approach generalize easily to ANOVA techniques involving more than one factor or to applications requiring samples with unequal sizes. In this section we will approach single-factor ANOVA from a different point of view and will present computational formulas that will facilitate the calculation of the F statistic. The methods presented can be extended or generalized to the more advanced techniques of ANOVA.

Notation

Let's assume k samples of data are used to test the statistical hypotheses

H_0: $\mu_1 = \mu_2 = \mu_3 = \cdots = \mu_k$

H_1:   At least two population means are unequal.

The following notation will be used. The jth measurement in the ith sample will be represented by x_{ij}, as depicted here:

$$x_{ij}$$

Sample   Measurement

The experimental procedure may be summarized as in Table 13-3. There are k samples and the size of each sample is n_i. The sum of all the column totals C_i is called the **grand total** and is represented by T. Thus, $T = \sum x_{ij} = \sum C_i$. We let the total number of measurements be represented by $N = \sum n_i$. If the grand total T is divided by the total number of measurements, then the quotient T/N is called the **grand mean** and is denoted by $\bar{T}$. Thus, the grand mean is represented by

$$\bar{T} = \frac{T}{N}$$

Table 13-3 Experimental Procedure for Single-Factor ANOVA

	Sample 1	Sample 2	Sample 3	$\cdots$	Sample k
	x_{11}	x_{21}	x_{31}		x_{k1}
	x_{12}	x_{22}	x_{32}		x_{k2}
	x_{13}	x_{23}	x_{33}		x_{k3}
	$\vdots$	$\vdots$	$\vdots$		$\vdots$
	x_{1n_1}	x_{2n_2}	x_{3n_3}		x_{kn_k}
Totals	C_1	C_2	C_3		C_k T

Formulas

Recall from Chapter 3 that the sum of squares SS represents the sum of the squared deviations from the mean. That is,

$$SS = \sum(x - \bar{x})^2$$

Also recall the following computational formula for SS:

$$SS = \sum x^2 - \frac{(\sum x)^2}{n}$$

The sum of squared deviations about the grand mean is called the **sum of squares for total** and is denoted by SST. That is,

$$SST = \sum(x_{ij} - \bar{T})^2 \qquad (13\text{-}7)$$

To find SST, we first find the deviation of each measurement from the grand mean. These deviations are then squared and totaled. For example, consider the three samples of data shown in Table 13-4.

Table 13-4 Data to Illustrate Calculation of SST

	Sample 1	Sample 2	Sample 3	
	3	2	4	
	5	2	3	
		4	3	
		1		
Column totals	$C_1 = 8$	$C_2 = 9$	$C_3 = 10$	$T = 27$

The grand mean $\bar{T}$ is

$$\bar{T} = \frac{T}{\sum n_i} = \frac{27}{9} = 3$$

The total sum of squares SST is

$$\begin{aligned}
SST &= (3 - 3)^2 + (5 - 3)^2 + (2 - 3)^2 + (2 - 3)^2 + (4 - 3)^2 + (1 - 3)^2 \\
&\quad + (4 - 3)^2 + (3 - 3)^2 + (3 - 3)^2 \\
&= 0 + 4 + 1 + 1 + 1 + 4 + 1 + 0 + 0 = 12
\end{aligned}$$

A computational formula for SST is given by

$$SST = \sum x_{ij}^2 - \frac{T^2}{N} \qquad (13\text{-}8)$$

where $N = \sum n_i$, the total number of measurements. For the data in Table 13-4, we have

$$\begin{aligned}
T &= C_1 + C_2 + C_3 = 8 + 9 + 10 = 27 \\
N &= \sum n_i = 2 + 4 + 3 = 9 \\
\sum x_{ij}^2 &= 3^2 + 5^2 + 2^2 + 2^2 + 4^2 + 1^2 + 4^2 + 3^2 + 3^2 = 93
\end{aligned}$$

By using (13-8), we have

$$SST = \sum x_{ij}^2 - \frac{T^2}{N} = 93 - \frac{(27)^2}{9} = 12$$

This result agrees with the answer found by using (13-7).

The sum of squares for total, SST, can be partitioned (or separated) into two component parts, called **sum of squares between groups** (SSB) and **sum of squares within groups** (SSW). That is,

$$SST = SSB + SSW \tag{13-9}$$

SSB is defined by

$$SSB = \sum n_i (\bar{C}_i - \bar{T})^2$$

and SSW is defined by

$$SSW = \sum (x_{ij} - \bar{C}_i)^2$$

Hence, the sum of the squares for total can be written as

$$\sum (x_{ij} - \bar{T})^2 = \sum n_i (\bar{C}_i - \bar{T})^2 + \sum (x_{ij} - \bar{C}_i)^2$$

A computational formula for SSB is given by

$$SSB = \sum \left(\frac{C_i^2}{n_i} \right) - \frac{T^2}{N} \tag{13-10}$$

For the data in Table 13-4,

$$SSB = \frac{C_1^2}{n_1} + \frac{C_2^2}{n_2} + \frac{C_3^2}{n_3} - \frac{T^2}{N}$$

$$= \frac{(8)^2}{2} + \frac{(9)^2}{4} + \frac{(10)^2}{3} - \frac{(27)^2}{9} = 4.58$$

As a result of (13-9), we have

$$SSW = SST - SSB = 12 - 4.58 = 7.42$$

The normal procedure for finding SSW is first to find SST and SSB and then to subtract SSB from SST. This involves less work than finding SSW by using the formula $SSW = \sum (x_{ij} - \bar{C}_i)^2$.

Each of the three sums of squares, SST, SSB, and SSW, has associated degrees of freedom. These are given by

$$df_T = N - 1 \qquad df_B = k - 1 \qquad df_W = N - k \tag{13-11}$$

The degrees of freedom for SST, df_T, are 1 less than the total number of measurements. The degrees of freedom for SSB are 1 less than the number of samples. Since the ith sample has $(n_i - 1)$ degrees of freedom associated with it, the k samples have a total number of degrees of freedom given by

$$df_W = (n_1 - 1) + (n_2 - 1) + (n_3 - 1) + \cdots + (n_k - 1)$$

$$= (n_1 + n_2 + n_3 + \cdots + n_k) - k$$

$$= N - k$$

Note that since $N - 1 = (k - 1) + (N - k)$, we have

$$df_T = df_B + df_W$$

Hence, the following two relations must hold:

1. $SST = SSB + SSW$
2. $df_T = df_B + df_W$

For the example above, we have

$$df_T = 9 - 1 = 8$$
$$df_B = 3 - 1 = 2$$
$$df_W = 9 - 3 = 6$$

As a computational check, we note that $8 = 2 + 6$.

In statistics, a sum of squares divided by its associated degrees of freedom is called a **mean square** and represents a variance estimate. Recall from Chapter 3 that

$$s^2 = \frac{SS}{n - 1} = \frac{SS}{df}$$

Since a sample of size n has $(n - 1)$ degrees of freedom associated with it, the **mean square for between groups**, denoted by MSB, is defined as

$$MSB = \frac{SSB}{df_B} = \frac{SSB}{k - 1}$$

Thus, the mean square for between groups is given by

$$MSB = \frac{SSB}{k - 1} \tag{13-12}$$

The **mean square for within groups** is denoted by MSW and is defined as

$$MSW = \frac{SSW}{df_W} = \frac{SSW}{N - k}$$

Thus, the mean square for within groups is given by

$$MSW = \frac{SSW}{N - k} \tag{13-13}$$

For the data in Table 13-4, we have

$$MSB = \frac{SSB}{k - 1} = \frac{4.58}{2} = 2.29$$

$$MSW = \frac{SSW}{N - k} = \frac{7.42}{6} = 1.24$$

As we have already pointed out, MSB and MSW are both variance estimates. MSB reflects the variation between samples and MSW reflects the variation within samples. Since each observation within a given sample comes from a population with a given mean, the differences between observations within a random sample are explained by sampling error. As a result, MSW is typically referred to as **mean square for error**.

The value of the F test statistic is given by

$$F = \frac{MSB}{MSW} \tag{13-14}$$

For our example, the value of F is

$$F = \frac{MSB}{MSW} = \frac{2.29}{1.24} = 1.85$$

The computations we have just presented are usually displayed in a special summary table, called an **ANOVA summary table**, the general form of which is shown in Table 13-5. For the data of Table 13-4, the ANOVA summary table is given in Table 13-6.

Table 13-5 General Form of ANOVA Summary Table

Source of variation	SS	df	MS	F
Between groups	SSB	$k - 1$	$\dfrac{SSB}{k-1}$	$\dfrac{MSB}{MSW}$
Within groups	SSW	$N - k$	$\dfrac{SSW}{N-k}$	
Total	SST	$N - 1$		

Table 13-6 ANOVA Summary Table for Data of Table 13-4

Source of variation	SS	df	MS	F
Between groups	4.58	2	2.29	1.85
Within groups	7.42	6	1.24	
Total	12.00	8		

Summarizing, we use the following steps to find the value for the F test statistic:

1. Find the column totals and the grand total.
2. Find SST by using (13-8).
3. Find SSB by using (13-10).
4. Find SSW by subtracting SSB from SST.
5. Find df_B and df_W by using (13-11).
6. Find MSB and MSW by using (13-12) and (13-13).
7. Find F by using (13-14).

The computations are usually summarized in an ANOVA summary table. Consider the following example.

EXAMPLE 13-4   For Example 13-2, find the value of F by using the methods presented in this section. The data are repeated for convenience in Table 13-7.

Table 13-7 Data for Example 13-4

A	B	C
6	8	3
2	7	2
4	7	5
1	2	4
7	6	1

Solution   **1.** We first find the column totals, C_i, and the grand total, T, as shown in the following table:

	A	B	C
	6	8	3
	2	7	2
	4	7	5
	1	2	4
	7	6	1
Totals	$C_1 = 20$	$C_2 = 30$	$C_3 = 15$ $T = 65$

2. Find SST using (13-8):

$$\text{SST} = \sum x_{ij}^2 - \frac{T^2}{N}$$

$$= 6^2 + 2^2 + 4^2 + \cdots + 5^2 + 4^2 + 1^2 - \frac{65^2}{15} = 81.33$$

3. Find SSB using (13-10):

$$\text{SSB} = \sum \left(\frac{C_i^2}{n_i} \right) - \frac{T^2}{N}$$

$$= \frac{(20)^2}{5} + \frac{(30)^2}{5} + \frac{(15)^2}{5} - \frac{(65)^2}{15} = 23.33$$

4. Find SSW by subtracting SSB from SST:

$$\text{SSW} = \text{SST} - \text{SSB} = 81.33 - 23.33 = 58$$

5. Find df_B and df_W by using (13-11):

$$df_B = k - 1 = 3 - 1 = 2$$
$$df_W = N - k = 15 - 3 = 12$$

6. Find MSB and MSW using (13-12) and (13-13):

$$MSB = \frac{SSB}{df_B} = \frac{23.33}{2} = 11.67$$

$$MSW = \frac{SSW}{df_W} = \frac{58}{12} = 4.83$$

Note that for Example 13-2, $s_B^2 = 11.67$ and $s_W^2 = 4.83$.

7. Find F by using (13-14):

$$F = \frac{MSB}{MSW} = \frac{11.67}{4.83} = 2.42$$

Note that this value of F agrees with the value found in Example 13-2.

The summary table is given in Table 13-8.

Table 13-8 ANOVA Summary Table for Example 13-4

Source of variation	SS	df	MS	F
Between groups	23.33	2	11.67	2.42
Within groups	58.00	12	4.83	
Total	81.33	14		

At $\alpha = .05$, the critical value is found to be $F_{.05}(2, 12) = 3.89$. Since the value of our test statistic F is less than 3.89, we cannot reject H_0. ∎

The following computer printout was obtained for the problem in Example 13-4 by using SPSS. Note that the p-value is .1315.

```
    VARIABLE LIFE
BY VARIABLE MAKE
                        ANALYSIS OF VARIANCE
                          SUM OF        MEAN          F         F
      SOURCE        DF    SQUARES      SQUARES       RATIO      PROB.

BETWEEN GROUPS       2    23.3333      11.6667       2.4138     .1315
WITHIN GROUPS       12    58.0000       4.8333
TOTAL               14    81.3333

                     STANDARD    STANDARD                    95 PCT CONF
GROUP  COUNT  MEAN   DEVIATION    ERROR    MINIMUM  MAXIMUM   INT FOR MEAN

GRP 1    5   4.0000   2.5495     1.1402    1.0000   7.0000    .8344 TO 7.1656
GRP 2    5   6.0000   2.3452     1.0488    2.0000   8.0000   3.0881 TO 8.9119
GRP 3    5   3.0000   1.5811      .7071    1.0000   5.0000   1.0368 TO 4.9632

TOTAL   15   4.3333   2.4103      .6223    1.0000   8.0000   2.9986 TO 5.6681
```

EXAMPLE 13-5   In a study to determine if the rates of unemployment are different for eastern, central, and western cities, random samples of cities were selected from the three areas of the United States for study. The data in Table 13-9 represent the extent of unemployment (in percentages). By assuming the assumptions for ANOVA hold and by using $\alpha = .05$, test the null hypothesis H_0: $\mu_E = \mu_C = \mu_W$.

Table 13-9 Data for Example 13-5

East	Central	West
5.2	11.4	7.2
11.5	9.1	15.9
6.3	6.6	10.3
6.6	10.5	9.5
7.7	3.6	
3.8		
7.6		

Solution   **1.** The following table is used to find $\sum x$ and $\sum x^2$ for the measurements of each area:

East		Central		West	
x	x^2	x	x^2	x	x^2
5.2	27.04	11.4	129.96	7.2	51.84
11.5	132.25	9.1	82.81	15.9	252.81
6.3	39.69	6.6	43.56	10.3	106.09
6.6	43.56	10.5	110.25	9.5	90.25
7.7	59.29	3.6	12.96	42.9	500.99
3.8	14.44	41.2	379.54		
7.6	57.76				
48.7	374.03				

We perform the necessary preliminary calculations:

$$T = 48.7 + 41.2 + 42.9 = 132.8$$
$$\sum x_{ij}^2 = 374.03 + 379.54 + 500.99 = 1254.56$$
$$N = 7 + 5 + 4 = 16$$

2. Find SST by using (13-8):

$$\text{SST} = \sum x_{ij}^2 - \frac{T^2}{N} = 1254.56 - \frac{(132.8)^2}{16} = 152.32$$

3. Find SSB using (13-10):

$$SSB = \sum \left(\frac{C_i^2}{n_i} \right) - \frac{T^2}{N}$$

$$= \frac{(48.7)^2}{7} + \frac{(41.2)^2}{5} + \frac{(42.9)^2}{4} - \frac{(132.8)^2}{16} = 36.16$$

4. Find SSW by subtracting SSB from SST:

$$SSW = SST - SSB = 152.32 - 36.16 = 116.16$$

5. Find df_B and df_W:

$$df_B = k - 1 = 3 - 1 = 2$$
$$df_W = N - k = 16 - 3 = 13$$

6. Find MSB and MSW using (13-12) and (13-13):

$$MSB = \frac{SSB}{df_B} = \frac{36.16}{2} = 18.08$$

$$MSW = \frac{SSW}{df_W} = \frac{116.16}{13} = 8.94$$

7. Find the value of the test statistic F:

$$F = \frac{MSB}{MSW} = \frac{18.08}{8.94} = 2.02$$

8. Find the critical value for F:

$$F_{.05}(2, 13) = 3.81$$

Since $F = 2.02 < 3.81$, we cannot reject H_0. Thus, we have no statistical evidence to suggest that the unemployment rates are different for the three sections of the United States. An ANOVA summary table for our calculations is given in Table 13-10.

Table 13-10 ANOVA Summary Table for Example 13-5

Source of variation	SS	df	MS	F
Between groups	36.16	2	18.08	2.02
Within groups	116.16	13	8.94	
Total	152.32	15		

As a final remark, we note that the F test for ANOVA tells us whether we can reject the null hypothesis that the population means are equal. If the decision is to reject H_0: $\mu_1 = \mu_2 = \mu_3 = \cdots = \mu_k$, ANOVA does not tell us where the differences are. To find which population means are different, further tests, called **post hoc tests**, are used. We will discuss one popular post hoc test in Section 13.3.

Problem Set 13.2

A

In Problems 1–5, use the computational formulas to find the value for F and summarize your computations using an ANOVA summary table.

1. Readability scores from Problem 1 of Problem Set 13.1:

Text A	50	49	54	57	58
Text B	48	57	61	55	60

2. Laboratory analysis results from Problem 3 of Problem Set 13.1:

Lab A	58.6	60.7	59.4	59.6	60.5
Lab B	61.6	64.8	62.8	59.2	60.4
Lab C	60.7	55.6	57.3	55.2	60.2

3. Typing scores from Problem 4 of Problem Set 13.1:

A	B	C	D
82	55	69	87
79	67	72	61
75	84	78	82
68	77	83	61
65	71	74	72

4. Basic skills test scores from Problem 5 of Problem Set 13.1:

School A	School B	School C	School D
20	24	16	19
21	21	21	20
22	22	18	21
24	25	13	20

5. Three sections of elementary statistics were taught by different teachers. A common final examination was given. The test scores are given in the table. Assume that the populations of test scores are normally distributed with equal variances. By using the .05 level of significance, test to determine if there is a difference in the average grades given by the three instructors. Summarize your calculations using an ANOVA summary table.

Teacher A	Teacher B	Teacher C
75	90	17
91	80	81
83	50	55
45	93	70
82	53	61
75	87	43
68	76	89
62	58	73
47	82	73
95	98	93
38	78	58
79	64	81
	80	70
	33	
	79	

6. Test scores from Problem 2 of Problem Set 13.1:

Method A	Method B	Method C	Method D	Method E
$n_1 = 13$	$n_2 = 13$	$n_3 = 13$	$n_4 = 13$	$n_5 = 13$
$\bar{x}_1 = 75$	$\bar{x}_2 = 76$	$\bar{x}_3 = 70$	$\bar{x} = 74$	$\bar{x}_5 = 76$
$s_1 = 7.7$	$s_2 = 7.1$	$s_3 = 9.9$	$s_4 = 6$	$s_5 = 6.3$

B

7. Prove that SST $= \sum (x_{ij} - \bar{T})^2 = \sum x_{ij}^2 - T^2/N$.

8. Prove that
$$\text{SSB} = \sum n_i(\bar{C}_i - \bar{T})^2 = \sum (C_i^2/n_i) - T^2/N.$$

9. Prove that SST = SSB + SSW.

10. Show that $s_B^2 = $ MSB for equal sample sizes.

11. Show that $s_W^2 = $ MSW for equal sample sizes.

12. Consider the following data:

Class 1	Class 2	Class 3
3	2	4
5	3	3
1	1	5

a. Find the value of the F statistic.
b. If each of the above observations is multiplied by 5, find the value of the F statistic for the transformed data.

c. If 4 is added to each observation above, find the value of the F statistic for the transformed data.

d. Transform each of the above observations by using the formula $Y = 5x + 4$. Then find the value of the F statistic for the transformed data.

e. Generalize the results of parts (a)–(d).

13. Each value x_{ij} satisfies the following relation:

$$x_{ij} = \bar{T} + (\bar{C}_i - \bar{T}) + (x_{ij} - \bar{C}_i)$$

Use the data in Problem 12 and the given relation to do the following:

a. For each x_{ij}, express $(x_{ij} - \bar{T})$ as a sum of two quantities.

b. Show that

$$\sum(x_{ij} - \bar{T}) = \sum(\bar{C}_i - \bar{T}) + \sum(x_{ij} - \bar{C}_i).$$

c. Show that

$$\sum(x_{ij} - \bar{T})^2 = \sum(\bar{C}_i - \bar{T})^2 + \sum(x_{ij} - \bar{C}_i)^2.$$

14. In an ANOVA problem involving samples of the same size, a $(1 - \alpha)100\%$ confidence interval for the difference between two population means μ_i and μ_j can be formed using MSW to estimate σ^2. In this case,

$$\sigma_{\bar{x}_i - \bar{x}_j} \simeq \sqrt{\frac{\text{MSW}}{n} + \frac{\text{MSW}}{n}}$$

The limits for the confidence interval become

$$(\bar{x}_i - \bar{x}_j) \pm t_{\alpha/2}(2n - 2)\sqrt{\frac{2\text{MSW}}{n}}$$

Use this result on the data in Problem 2 to find a 95% confidence interval for:

a. $\mu_A - \mu_C$

b. $\mu_B - \mu_C$

c. Can you be 95% confident of both results simultaneously? Explain.

13.3 FOLLOW-UP PROCEDURES FOR A SIGNIFICANT F ⎯⎯⎯⎯⎯

For the gasoline mileage illustration of Example 13-1, we concluded that the Citation, the LeBaron, and the Reliant have different mean gas mileage ratings. We did not determine where the differences exist among the three automobile styles. To investigate further we might use three t tests, each at the .05 significance level. The trouble with doing this is that we have three opportunities for making a Type I error. If the three tests were independent (in fact, they are not), the probability of making at least one Type I error would be

$$1 - P(\text{No Type I error}) = 1 - (.95)^3$$
$$\simeq 1 - .86$$
$$= .14$$

That is, the actual error rate for the experiment is somewhere between .05 and .14, not the chosen error rate of .05.

There are many procedures designed to test for differences among population means following the rejection of a null hypothesis of equal population means that attempt to control the error rate for the experiment. These procedures are commonly known as **multiple-comparison procedures** or **post hoc procedures**. One such procedure is **Scheffé's test**. It is one of the most flexible and conservative tests available for "data-snooping" purposes. In addition, it requires only the use of the F tables and is applicable to ANOVA tests involving unequal sample sizes. Scheffé's procedure is based on the same assumptions as ANOVA. When Scheffé's test is used, the critical difference (CD) that a positive difference of sample means, $|\bar{x}_i - \bar{x}_j|$, must exceed to be declared significant is given by

$$CD = \sqrt{(k - 1)F_\alpha(df_1, df_2)} \cdot \sqrt{MSW \cdot \left[\frac{1}{n_i} + \frac{1}{n_j}\right]} \tag{13-15}$$

where MSW represents the mean square within-groups variance estimate, n_i is the size of the ith sample, and k is the number of samples. If all the samples are of the same size, the formula for the critical difference can be simplified to

$$CD = \sqrt{(k - 1)F_\alpha(df_1, df_2)} \cdot \sqrt{\frac{2MSW}{n}} \tag{13-16}$$

where n represents the common sample size.

For the gas mileage illustration of Example 13-1, the critical difference CD can be found using (13-16), since all three samples are of size 5. Since $\alpha = .05$, $df_1 = 2$, and $df_2 = 12$, we have $F_{.05}(2, 12) = 3.89$. Scheffé's critical difference is

$$CD = \sqrt{(k - 1)F_\alpha(df_1, df_2)} \cdot \sqrt{\frac{2MSW}{n}}$$

$$= \sqrt{(2)(3.89)\left(\frac{2}{5}\right)} = 1.76$$

Thus, in order for two population means to be declared different, the positive difference in the two corresponding sample means must exceed the critical difference of 1.76. The three sample means are $\bar{x}_1 = 24$, $\bar{x}_2 = 31$, and $\bar{x}_3 = 26$. There are three pairwise comparisons to be made. The number of pairwise comparisons for k sample means can be found by evaluating the binomial coefficient $\binom{k}{2}$. For our example, $\binom{3}{2} = 3$. The three positive pairwise comparisons are

LeBaron − Citation: $\bar{x}_2 - \bar{x}_1 = 31 - 24 = 7$

LeBaron − Reliant: $\bar{x}_2 - \bar{x}_3 = 31 - 26 = 5$

Reliant − Citation: $\bar{x}_3 - \bar{x}_1 = 26 - 24 = 2$

Since all three positive differences in sample means are greater than the critical difference CD = 1.76, we can conclude that any two models differ significantly in average gasoline mileages. And with these three tests, we can be confident that the probability of committing at least one Type I error for the three comparisons is at most .05.

The following example illustrates Scheffé's procedure for unequal sample sizes.

EXAMPLE 13-6   A large appliance company is considering the purchase of a large quantity of paint from one of five paint companies with nearly equal bid prices. Since drying time (in minutes) is a critical factor for the appliance company, it was decided to paint 35 appliances, seven appliances with each of the five different brands of paint. The sample sizes and mean drying times are given in Table 13-11. The sample sizes are unequal because it was discovered too late that four appliances had been inadequately prepared for painting.

Table 13-11 Data for Example 13-6

Company	Sample size	Mean (in minutes)
A	7	29.0
B	5	25.8
C	6	25.7
D	6	26.8
E	7	31.0

By using ANOVA it was determined that the mean drying times for the paint from the five paint companies were significantly different ($\alpha = .05$). The following is the ANOVA summary table for the analysis:

Source of variation	SS	df	MS	F
Between groups	134.52	4	33.63	4.03
Within groups	216.97	26	8.34	
Total	351.49	30		

Use Scheffé's procedure and $\alpha = .05$ to test for pairwise differences between the average drying times for paint from the five paint companies.

Solution   Since $k = 5$, there are $\binom{5}{2} = 10$ pairwise comparisons to be made. To organize our results, we shall order the sample means from largest to smallest and construct the following table of positive differences:

		Sample				
		E 31.0	**A** 29.0	**D** 26.8	**B** 25.8	**C** 25.7
E	31.0	—	2	4.2	5.2	5.3
A	29.0		—	2.2	3.2	3.3
D	26.8			—	1.0	1.1
B	25.8				—	.1
C	25.7					—

The critical difference for assessing the positive difference $|\bar{x}_i - \bar{x}_j|$ is

$$\text{CD} = \sqrt{(k-1)F_\alpha(\text{df}_1, \text{df}_2)} \cdot \sqrt{\text{MSW} \cdot \left[\frac{1}{n_i} + \frac{1}{n_j}\right]}$$

$$= \sqrt{(4)(8.34)(2.74)\left[\frac{1}{n_i} + \frac{1}{n_j}\right]} = 9.56\sqrt{\frac{1}{n_i} + \frac{1}{n_j}}$$

where $F_{.05}(4, 26) = 2.74$.

The critical difference for comparing A versus E is then

$$CD = 9.56 \sqrt{\frac{1}{7} + \frac{1}{7}} = 5.11$$

Similarly, the critical differences for the other comparisons are

$$C \text{ versus } D\} \quad CD = 9.56 \sqrt{\frac{1}{6} + \frac{1}{6}} = 5.52$$

$$\left.\begin{matrix} A \text{ versus } B \\ B \text{ versus } E \end{matrix}\right\} \quad CD = 9.56 \sqrt{\frac{1}{5} + \frac{1}{7}} = 5.60$$

$$\left.\begin{matrix} A \text{ versus } C \\ A \text{ versus } D \\ E \text{ versus } C \\ E \text{ versus } D \end{matrix}\right\} \quad CD = 9.56 \sqrt{\frac{1}{7} + \frac{1}{6}} = 5.32$$

$$\left.\begin{matrix} B \text{ versus } C \\ B \text{ versus } D \end{matrix}\right\} \quad CD = 9.56 \sqrt{\frac{1}{5} + \frac{1}{6}} = 5.79$$

We see in every case that the positive difference between sample means is less than the critical difference. Thus, by using Scheffé's procedure, we cannot detect any pairwise differences. This is entirely possible with Scheffé's procedure, since it is known to be conservative. Other multiple-comparison procedures are less conservative and more appropriate when testing only pairwise differences. Three such tests are **Duncan's multiple range test**, the **Newman–Keuls test**, and **Tukey's test**. These tests involve special tables and a deeper study of statistics beyond the scope of this text. We shall leave them to a more advanced course in statistics. ∎

A Measure of Association

A significant F value for ANOVA indicates there is an association between two variables, the treatment variable and the dependent variable. The treatment variable for the gasoline mileage illustration of Example 13-1 is the type of car and the dependent variable is gas mileage rating. Since the F statistic was significant, we conclude that there is a relationship between type of car and gas mileage rating. The F statistic does not indicate the **strength of association** between these two variables. Such information is important in evaluating the outcome of an experiment since it is possible to have a small or trivial association between two variables although the association is statistically significant because sufficiently large samples are involved. In Section 10.3 we saw an example involving a new study guide for improving quantitative SAT scores in which a difference of .4 on the SAT was significant because a sample of 1,000,000 was involved. In this case the association is trivial, but statistically significant.

A useful estimator of the strength of association between two variables is **Hays' omega-squared statistic**, $\hat{\omega}^2$. The $\hat{\omega}^2$ statistic is given by

$$\hat{\omega}^2 = \frac{\text{SSB} - (k-1)\text{MSW}}{\text{SST} + \text{MSW}} \qquad (13\text{-}17)$$

where SSB is the sum of squares for between groups, SST is the total sum of squares, MSW is the mean square for within groups, and k is the number of samples.

For the gas mileage illustration, the value of $\hat{\omega}^2$ is

$$\hat{\omega}^2 = \frac{\text{SSB} - (k-1)\text{MSW}}{\text{SST} + \text{MSW}}$$

$$= \frac{130 - 2(1)}{142 + 1} = .90$$

We can conclude that the treatment variable, type of car, accounts for 90% of the variance in the dependent variable, gas mileage rating. Not only is the association between car type and gas mileage rating statistically significant, but it is also very strong.

EXAMPLE 13-7 Calculate the value of Hays' omega-squared statistic for the data given in Example 13-6.

Solution The value of Hays' omega-squared statistic is

$$\hat{\omega}^2 = \frac{\text{SSB} - (k-1)\text{MSW}}{\text{SST} + \text{MSW}}$$

$$= \frac{134.52 - (5-1)(8.34)}{351.48 + 8.34}$$

$$= \frac{101.16}{359.82} = .28$$

We can conclude that type of paint accounts for 28% of the variance in the drying times of the paints. While there is a relationship between type of paint and drying time, it cannot be considered strong. ∎

Problem Set 13.3 _____

A

For each of the following problems, use Scheffé's procedure to test for differences in means whenever appropriate, use Hays' omega-squared statistic to estimate the strength of association when it exists, and interpret your results for each procedure. Assume that the assumptions for ANOVA hold for each problem.

1. Four diets were compared for controlling blood sugar in diabetic patients. Nine patients were randomly assigned to each diet. The data in the accompanying table represent the blood glucose readings of the patients following 2 weeks on the special diets. By using ANOVA and $\alpha = .05$, determine if the four diets produce different mean effects on blood glucose readings for diabetic patients.

Diet A	Diet B	Diet C	Diet D
140	160	200	200
120	110	220	180
80	110	220	180
120	130	190	170
130	150	150	130
120	100	210	210
120	140	230	150
120	100	190	170
120	130	180	170

2. Four different assessments of acidity of high-sulfur coal at a power plant produced the pH data readings listed in the table. By using ANOVA and $\alpha = .05$, determine if there are differences in the mean pH values for the different assessments.

	Assessment			
	I	II	III	IV
n	10	10	10	10
$\bar{x}$	6.60	6.90	3.00	5.30
s	.38	.50	.43	.47

3. A large oil-drilling firm developed a new high-speed drill bit in an attempt to reduce drilling costs. The new high-speed drill bit, called the DB3, is believed to be superior to the two fastest drill bits known, the DB1 and the DB2. In an experiment to test the new drill bit, four drilling sites were randomly assigned to use each bit. The accompanying data indicate the rate of penetration (in feet per hour) after drilling 2000 feet at each site. By using ANOVA and $\alpha = .05$, determine if the mean rates of penetration differ for the three drill bits.

DB3	DB1	DB2
37.4	28.0	16.9
32.3	31.9	31.1
39.8	28.8	25.5
36.5	32.3	18.4

4. A national engineering society is interested in comparing the starting hourly rates of engineering graduates of three large universities, A, B, and C. Independent random samples of six engineering graduates of each university were selected for the

study. The table shows the starting hourly salary (in dollars) for each graduate. By using ANOVA and $\alpha = .05$, determine if there is a difference among the mean starting salaries of engineering graduates of the three universities.

	University	
A	B	C
11.25	12.50	11.75
11.25	13.05	12.00
12.35	13.12	10.85
12.25	13.35	11.61
12.00	12.55	12.10
11.85	12.60	12.15

5. The accompanying data represent yields per plot for tomatoes grown with three different fertilizers. Seven plots were grown with fertilizer A, eight with fertilizer B, and six with fertilizer C. Use ANOVA and $\alpha = .05$ to determine if there are differences in tomato yields for the three fertilizers.

	Fertilizer	
A	B	C
31.0	40.0	41.4
31.8	39.6	42.5
28.3	35.3	36.0
29.7	33.0	36.4
28.0	35.7	36.8
27.1	33.7	38.1
32.6	37.4	
	38.6	

B

6. Scheffé's procedure can be used to establish $(1 - \alpha)100\%$ multiple confidence intervals or simultaneous confidence intervals for $(\mu_i - \mu_j)$. The confidence intervals for the pairwise differences between population means can be found by using the following expression:

$$(\bar{x}_i - \bar{x}_j) \pm CD$$

where CD is the critical difference given by (13-15) or (13-16). When this procedure is used, the probability that all the confidence intervals contain the differences between population means is at least $(1 - \alpha)$. For the gasoline mileage data given in Example 13-1, construct three 95% multiple confidence intervals for the pairwise differences in population mean gasoline mileages. Compare your findings with those found earlier in this section.

7. The **Bonferroni t procedure** is another multiple comparison procedure which is less conservative than Scheffé's procedure. This method consists of testing for pairwise differences in population means $(\mu_i - \mu_j)$ in much the same way as we did for an individual difference in Section 11.5, except that a different t value is used. Instead of using a t value of $t_{\alpha/2}(\text{df})$, as we did in Section 11.5, we use the larger t value of $t_{\alpha/(2m)}(\text{df})$, where $m = \binom{k}{2}$ is the total number of pairwise differences in means.
   a. Show that in order for a pairwise difference in population means to be declared significant (with an overall significance level of α), the corresponding difference in sample means must exceed the following critical difference:

   $$t_{\alpha/(2m)}(\text{df}_w)\sqrt{\left[\frac{1}{n_i} + \frac{1}{n_j}\right]\text{MSW}}$$

   where $\text{df}_w = N - k$. [*Hint:* Let MSW be the pooled estimate of σ^2.]
   b. Use the Bonferroni t procedure and an overall level of significance of $\alpha = .05$ to test for pairwise differences for the gasoline mileage data in Example 13-1. [Since the t value is not contained in the t table (see the back endpaper) use $t_{.00833}(12) = 2.779$.] Compare the critical differences (CD) for Bonferroni's t procedure and Scheffé's procedure.

8. Following Bonferroni's t procedure, we can construct a set of $(1 - \alpha)100\%$ simultaneous confidence intervals for pairwise differences $(\mu_i - \mu_j)$ by using the following expression:

   $$(\bar{x}_i - \bar{x}_j) \pm t_{\alpha/(2m)}(\text{df}_w)\sqrt{\left[\frac{1}{n_i} + \frac{1}{n_j}\right]\text{MSW}}$$

   where $\text{df}_w = N - k$. With this procedure, the probability that all $m = \binom{k}{2}$ confidence intervals contain the differences in population means is at least $(1 - \alpha)$. Use this procedure to construct three 95% simultaneous confidence intervals for the gasoline mileage data in Example 13-1. Compare the lengths of these intervals to the corresponding lengths of those found in Problem 6. Are these results consistent with the fact that Scheffé's procedure is more conservative than the Bonferroni t procedure?

9. Five brands of cigarettes were tested for tar content. The following table gives the tar contents (in milli-grams) for five packs of cigarettes for each of the five brands:

Brand A	Brand B	Brand C	Brand D	Brand E
339	357	334	357	330
322	319	317	344	315
318	339	329	329	325
329	359	319	349	316
334	337	337	339	340

The following is an ANOVA summary table for the data:

Source of variation	SS	df	MS
Between groups	1560.75	4	390.1875
Within groups	2560.75	20	128.0375
Total	4121.50	24	

   a. Use $\alpha = .10$ to determine if there are differences in the mean tar contents for the five brands of cigarettes. [Since the value for $F_{.10}(4, 20)$ is not contained in the F table in Appendix B, use $F_{.10}(4, 20) = 2.25$.]
   b. Use the Bonferroni t procedure and an overall significance level of $\alpha = .10$ to test for pairwise differences in the population mean tar contents for the five brands of cigarettes.

10. Refer to the data given in Problem 9.
   a. Test for pairwise differences using Scheffé's procedure and an overall significance level of $\alpha = .10$, and compare your results with those found in Problem 9(b).
   b. Use Hays' omega-squared statistic to estimate the strength of association between brand of cigarettes and tar content.

11. Refer to the data in Problem 9.
   a. Construct simultaneous 90% confidence intervals using Bonferroni's t procedure for the pairwise comparisons of the population mean tar contents for the five brands of cigarettes.
   b. Construct simultaneous 90% confidence intervals using Scheffé's procedure for the pairwise comparisons of the population mean tar contents for the five brands of cigarettes.
   c. Compare the lengths of the corresponding confidence intervals for the two procedures. For which procedure are the intervals shorter? Explain.

CHAPTER SUMMARY

In this chapter we presented two different approaches to using analysis of variance (ANOVA). We discussed Scheffé's procedure for determining which population means are significantly different following a significant F test, and we learned how to estimate the strength of association by using Hays' omega-squared statistic. ANOVA is used to determine if k population means are different. The assumptions underlying the test are that the k populations are normal and the k population variances are equal. For equal sample sizes, the F test is robust, which means that minor violations in the assumptions underlying the test do not seriously affect the outcome of the test. We showed that the two-sample t test for determining whether two population means differ is a special case of ANOVA for two samples. The simultaneous use of multiple t tests instead of ANOVA is not recommended to test for differences in more than two population means because the probability of making at least one Type I error for the experiment increases as the number of pairwise tests increases.

C H A P T E R   R E V I E W

IMPORTANT TERMS

For each of the following terms, provide a definition in your own words. Then check your responses against the definitions given in the chapter.

analysis of variance

ANOVA summary table

between-samples variance estimate

Bonferroni t procedure

Ducan's multiple range test

grand mean

grand total

Hay's omega-squared statistic

mean square for between groups

mean square for error

mean square for within groups

measure of association

multiple-comparison procedures

Newman–Keuls test

pairwise comparisons

pooled estimate

post hoc procedures

Scheffé's test

strength of association

sum of squares between groups

sum of squares within groups

sum of squares for total

Tukey's test

within-samples variance estimate

IMPORTANT SYMBOLS

s_B^2, between-samples variance estimate

s_W^2, within-samples variance estimate

F, F statistic

df_B, degrees of freedom for s_B^2

df_W, degrees of freedom for s_W^2

n, sample size

n_i, size of the ith sample

k, number of samples

N, total number of data

SSB, sum of squares between groups

SSW, sum of squares within groups

SST, total sum of squares

T, grand total

C_i, sum of measurements of ith sample

x_{ij}, jth measurement in ith sample

$\bar{T}$, grand mean

$\bar{C}_i$, mean of ith sample

$\hat{\omega}^2$, omega-squared statistic

IMPORTANT FACTS AND FORMULAS

1. $s_W^2 = \dfrac{s_1^2 + s_2^2 + s_3^2 + \cdots + s_k^2}{k} = \dfrac{\sum s_i^2}{k}$, within-samples variance estimate (used when sample sizes are equal)

2. $\bar{x} = \dfrac{\bar{x}_1 + \bar{x}_2 + \bar{x}_3 + \cdots + \bar{x}_k}{k} = \sum \bar{x}_i \div k$, grand mean (used when sample sizes are equal)

3. $s_B^2 = \dfrac{n \sum (\bar{x}_i - \bar{x})^2}{k - 1}$, between-samples variance estimate (used when sample sizes are equal)

4. $df_B = k - 1$, degrees of freedom for between-samples estimate

5. $df_W = k(n - 1)$, degrees of freedom for within-samples estimate (used when sample sizes are equal)

6. $F = \dfrac{s_B^2}{s_W^2}$, the F statistic

7. $SST = \sum (x_{ij} - \bar{T})^2$, sum of squares for total

8. $SST = \sum x_{ij}^2 - \dfrac{T^2}{N}$, computational formula

9. $SSB = \sum n_i (\bar{C}_i - \bar{T})^2$, sum of squares for between groups

10. $SSB = \sum \left(\dfrac{C_i^2}{n_i} \right) - \dfrac{T^2}{N}$, computational formula

11. $SST = SSB + SSW$

12. $MSB = \dfrac{SSB}{k - 1}$, mean square for between groups

13. $MSW = \dfrac{SSW}{N - k}$, mean square for within groups

14. $df_W = N - k$, degrees of freedom for within groups

15. $CD = \sqrt{(k - 1)F_\alpha(df_1, df_2)} \cdot \sqrt{MSW \cdot \left[\dfrac{1}{n_i} + \dfrac{1}{n_j} \right]}$, critical difference

16. $\hat{\omega}^2 = \dfrac{SSB - (k - 1)MSW}{SST + MSW}$, omega-squared statistic

REVIEW PROBLEMS

1. Three random groups of students were exposed to different types of training, after which each group was administered the same standardized task to perform. The completion times (in minutes) for each group are given in the table. By using ANOVA, test at $\alpha = .05$ to determine if the three population mean task completion times are different.

2. A comparison of the mean recovery times (in days to recovery) of patients suffering from a certain disease and using three different treatments was made. The data shown in the table were obtained. Use $\alpha = .05$ to determine if there is a difference in the mean recovery times for the three treatments. Assume the assumptions for ANOVA hold.

Types of training		
A	**B**	**C**
19	16	15
17	14	11
21	12	16
18	22	19
15	12	15
20	15	13
20	21	13
25	19	18
18	16	18
17	13	14

Treatment		
A	**B**	**C**
4	8	5
9	7	4
7	10	6
10	6	3
8	6	7
6		4
5		

3. Twelve patients with a particular disease were randomly assigned to four groups. Each group received a different dosage of a new experimental drug. The accompanying data indicate the percentage absorption of the new drug for the four groups of patients.

Group 1	Group 2	Group 3	Group 4
71.0	48.6	52.2	51.1
71.1	61.1	47.1	58.5
61.8	49.5	54.0	49.2

a. By using ANOVA and $\alpha = .05$, test to determine if the average absorption rates for different dosages of the new drug are different. Summarize your results in an ANOVA summary table.

b. Test for pairwise differences using Scheffé's procedure and $\alpha = .05$.

c. Use Hays' omega-squared statistic to estimate the strength of association between the amount of the experimental drug and the average absorption rate. Interpret your finding.

4. The accompanying data represent samples of examination scores for four different calculus classes. Use ANOVA and $\alpha = .01$ to test H_0: $\mu_1 = \mu_2 = \mu_3 = \mu_4$. Summarize your results in an ANOVA summary table.

Class 1	Class 2	Class 3	Class 4
85	69	90	63
82	73	88	64
80	87	85	80
79	81	84	77
77	85	80	81
69	87	76	83
67	88		85
65	90		
	84		

5. A study was conducted to compare three different methods for training typists. Students were randomly assigned to the three methods and given a standardized typing test at the conclusion of the instruction. The results (in words typed per minute) are shown in the table. Test the null hypothesis H_0: $\mu_A = \mu_B = \mu_C$ at the .05 level of significance. Assume that the assumptions for ANOVA hold.

Typing method		
A	B	C
74	71	85
94	81	76
81	84	97
72	72	93
78		

6. Three different army companies participated in a marksmanship competition. Soldiers were randomly selected from each company for the competition. For each company, the mean number of points per ten shots per soldier was obtained. The results were recorded in the table. At $\alpha = .01$, do the data indicate that the companies differ in marksmanship?

Company A	Company B	Company C
$\bar{x} = 87.7$	$\bar{x} = 92.2$	$\bar{x} = 88.4$
$s = 14.3$	$s = 17.4$	$s = 16.5$
$n = 20$	$n = 25$	$n = 30$

CHAPTER ACHIEVEMENT TEST

(60 points)   **1.** Consider the following data set:

Group 1	Group 2	Group 3
6	5	7
3	4	3
1		5
2		

Find each of the following:

a. SST **b.** SSB **c.** SSW **d.** df_B **e.** df_W
f. MSB **g.** MSW **h.** s_B^2 **i.** s_W^2 **j.** F

(20 points) **2.** For the data in Problem 1, complete the following ANOVA summary table and test $H_0: \mu_1 = \mu_2 = \mu_3$ at $\alpha = .05$.

Source of variation	SS	df	MS	F
Between groups				
Within groups				
Total				

(10 points) **3.** For the data in Problem 1, calculate the value of $\hat{\omega}^2$, Hays' omega-squared statistic.

(10 points) **4.** What do we know about the null hypothesis if the value of the F statistic is less than or equal to 1?

14

LINEAR REGRESSION ANALYSIS

Chapter Objectives

In this chapter you will learn:

- *about two different regression models*

- *the assumptions for the linear regression model*

- *how to determine if linear regression is appropriate*

- *how to test the parameters in the linear regression model*

- *how to summarize your results using an ANOVA summary table*

- *two different correlation coefficients*

- *what the coefficient of determination is and how to interpret it*

Chapter Contents

In Chapter 4 we viewed regression and correlation primarily from a descriptive point of view; scattergrams and the magnitude of the linear correlation coefficient *r* were used to determine if a linear relationship existed between two variables. Correlation coefficients and regression equations were computed from samples of bivariate data. No attempt was made in Chapter 4 to draw inferences concerning linear relationships for paired data in populations of bivariate data from which the samples were obtained. The purpose of this chapter is to study regression and correlation from an inference-making point of view using the descriptive statistics developed in Chapter 4. We begin by discussing the linear regression model and related concepts.

14.1 THE LINEAR REGRESSION MODEL

The Environmental Protection Agency released the information in Table 14-1 comparing engine size (in cubic inches of displacement) and miles-per-gallon (mpg) estimates for eight representative models of 1984 compact automobiles.

Table 14-1 EPA Mileage Ratings

Compact car	Engine size (cid)	mpg
Chevrolet Cavalier	121	30
Datsun Nissan Stanza	120	31
Dodge Omni	- 97	34
Ford Escort	98	27
Mazda 626	122	29
Plymouth Horizon	97	34
Renault Alliance/Encore	85	38
Toyota Corolla	122	32

Suppose we view the collection of eight data pairs (cid, mpg) as constituting a sample from a population of pairs where the cid measurements can be any value within the range of values extending from 85 to 122. For each possible cid measurement, there are many mpg ratings associated with it. For example, for an engine size of 97, there are a large number of mpg ratings associated with it—one for each car with engine size 97 cid. Let's assume a linear relationship exists for the population of data pairs of cid and mpg ratings. The following **probabilistic model** can then be used to explain the behavior of the mpg ratings for the eight (six distinct) cid measurements. It is called a **linear regression model** and expresses the linear relationship between cid (*x*) and mpg (*y*):

$$y = \beta_0 + \beta_1 x + \varepsilon \tag{14-1}$$

The right-hand side of equation (14-1) involves a random variable ε (lower-case Greek letter epsilon), which expresses the random error involved in measuring *y* at a fixed value of *x*. If all the points in the population scattergram fell on a straight line, there would be no need to include the **error term** in the model, nor to involve probability in our study of the relationship between x and y. In this case, for each value of x, y could be determined exactly and the resulting linear model $y = \beta_0 + \beta_1 x$ would be called a **deterministic model**. The two parameters β_0 and β_1 would represent the y-intercept and slope, respectively, of the line. The expression $\beta_0 + \beta_1 x$ is sometimes referred to as the **deterministic component** of the linear regression model. Our sample data consisting of eight data pairs will be used to estimate the parameters β_0 and β_1 of the deterministic component.

The main difference between a probabilistic model and a deterministic model is the inclusion of a random error term in the probabilistic model. For our sample data, a probabilistic model is appropriate since the Mazda 626 and Toyota Corolla have different mpg ratings (29 and 32) for the same size engine (122 cid). The different mpg ratings for the same engine size are accounted for by the error term ε in the regression model. For each engine size, the deterministic component of the regression model is equal to a constant, say C. That is, for a fixed engine size x, the miles-per-gallon rating is equal to the constant C plus a random error. That is,

$$y = C + \varepsilon$$

The values of y (mpg ratings) vary directly with the values of the random error term. The terms y and ε that occur in (14-1) are random variables.

We shall make the following assumptions concerning the error term ε in the linear regression model $y = \beta_0 + \beta_1 x + \varepsilon$:

1. For each value of x, the random variable ε is normally distributed.
2. For each value of x, the mean or expected value of ε is 0; that is, $E(\varepsilon) = \mu_\varepsilon = 0$.
3. For each value of x, the variance of ε is the constant σ^2 (called **error variance**).
4. The values of the error term ε are independent.

We can make the following observations concerning the linear regression model:

1. The values of x are fixed. For our illustration, we have six distinct values: 85, 97, 98, 120, 121, and 122.
2. The values of the parameters β_0 and β_1 are constant, but unknown. Based on the four assumptions and the sample statistics, their values can be estimated.
3. Since the value of y for a fixed value of x is determined by adding the constant $\beta_0 + \beta_1 x$ to the random variable ε, the value for y will depend on the values for ε. Therefore, y is a random variable.
4. For a fixed value of x, the sampling distribution of y is normal, since the values of y depend on the values of ε and the values of ε are normally distributed (see Figure 14-1).

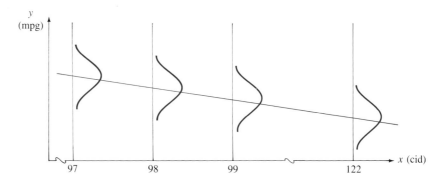

FIGURE 14-1 Sampling distribution of y for different values of x

5. The sampling distribution of y for a fixed value of x has a mean, denoted by $\mu_{y|x}$. The symbol "$y|x$" is read "y given x." That is, $E(y|x) = \beta_0 + \beta_1 x$, since $E(\varepsilon) = 0$. The equation $E(y|x) = \beta_0 + \beta_1 x$ is called the **population regression equation** (see Figure 14-2).

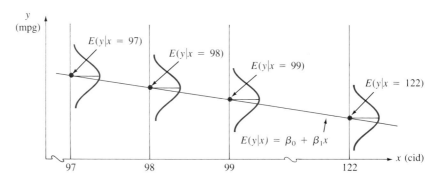

FIGURE 14-2 Population regression equation

6. The variance of the sampling distribution of y for any value of x is σ^2. This means that the y values for any two distinct x values have the same variance. As a result, all the normal distributions of the y values corresponding to the x values have the same shape, as shown in Figure 14-2.
7. For a fixed x value, the value of y can be predicted.
8. For a fixed x value, the average value of y can be estimated.
9. The unknown error variance σ^2 can be estimated.

Let's return to the illustration involving engine size and mpg ratings for the eight 1984 compact cars and calculate the estimated regression equation (or **least squares prediction equation**). For convenience, let's represent the estimated regression equation by $\hat{y} = b_0 + b_1 x$. Remember that we call $\hat{y}$ ("y-hat") the predicted value of y for a particular value of x. The constant b_0 will serve as a point estimator for β_0 and the constant b_1 will serve as a point estimator for β_1. [Note that in Chapter 4 we represented the estimated regression equation by $\hat{y} = a + bx$.] Toward computing the values for b_0 and b_1, recall the following formulas from Chapter 4:

$$SS_x = \sum x^2 - \frac{(\sum x)^2}{n}$$

$$SS_y = \sum y^2 - \frac{(\sum y)^2}{n}$$

$$SS_{xy} = \sum xy - \frac{(\sum x)(\sum y)}{n}$$

$$b_1 = \frac{SS_{xy}}{SS_x}$$

$$b_0 = \bar{y} - b_1\bar{x}$$

$$SSE = \sum(y - \hat{y})^2 = SS_y - b_1 SS_{xy}$$

We will use these formulas to calculate the least squares estimates of β_0 and β_1.

The following table organizes the computations for the gas mileage data given in Table 14-1:

x	y	x^2	y^2	xy
121	30	14,641	900	3,630
120	31	14,400	961	3,720
97	34	9,409	1,156	3,298
98	27	9,604	729	2,646
122	29	14,884	841	3,538
97	34	9,409	1,156	3,298
85	38	7,225	1,444	3,230
122	32	14,884	1,024	3,904
862	255	94,456	8,211	27,264

1. $SS_x = \sum x^2 - \dfrac{(\sum x)^2}{n}$

$$= 94,456 - \frac{(862)^2}{8} = 1575.5$$

2. $SS_y = \sum y^2 - \dfrac{(\sum y)^2}{n}$

$$= 8211 - \frac{(255)^2}{8} = 82.875$$

3. $SS_{xy} = \sum xy - \dfrac{(\sum x)(\sum y)}{n}$

$$= 27,264 - \frac{(862)(255)}{8} = -212.25$$

4. $b_1 = \dfrac{SS_{xy}}{SS_x}$

$\quad = -\dfrac{212.25}{1575.5} = -.1347$

5. $b_0 = \bar{y} - b_1\bar{x}$ $\qquad\qquad\qquad\qquad$ $\bar{y} = \dfrac{\sum y}{n} = \dfrac{255}{8} = 31.875$

$\quad = (31.875) - (-.1347)(107.75)$ $\qquad$ $\bar{x} = \dfrac{\sum x}{n} = \dfrac{862}{8} = 107.75$

$\quad = 46.3889$

The least squares prediction equation thus becomes

$\hat{y} = 46.3889 - .1347x$

The regression equation for our sample of eight pairs of data is $\hat{y} = 46.3889 - .1347x$. It is always a good idea when computing the regression constants b_0 and b_1 to carry as many significant digits as possible in order to minimize the effects of rounding errors. For the remainder of this chapter we will use at least four-decimal-place accuracy in all computations involving b_0 and b_1.

Figure 14-3 contains the graph of our regression equation. The errors $e_i = y - \hat{y}$ in using the equation $\hat{y} = 46.3889 - .1347x$ to predict the mpg ratings y for each engine size are indicated on the graph by vertical line segments. Recall from Section 4.3 that by using the least squares criterion to obtain the line that best fits the data, we can calculate the minimum value for the sum of squares for error, $SSE = \sum(y - \hat{y})^2$. The errors e_i are frequently called **residuals**.

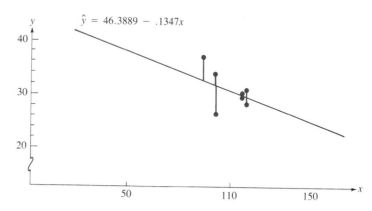

FIGURE 14-3 Scattergram and regression equation for data of Table 14-1

The variance of the errors e_i is called **residual variance** and is denoted by s_e^2. The residual variance is found by using the following formula:

$$s_e^2 = \frac{\sum(y - \hat{y})^2}{n - 2} = \frac{SSE}{n - 2} \qquad\qquad (14\text{-}2)$$

where $(n-2)$ is the number of degrees of freedom associated with the variance of the residuals. Recall from Section 3.2 that we divided SS by $(n-1)$ to obtain s^2, an estimator for σ^2. We used $(n-1)$ in the denominator because 1 degree of freedom was lost by using $\bar{x}$ to estimate μ in computing $SS = \sum(x-\mu)^2$. So to estimate σ^2 by using $SSE = \sum(y-\hat{y})^2 = \sum(y-b_0-b_1x)^2$, we lose 2 degrees of freedom because β_0 and β_1 are estimated by b_0 and b_1, respectively. To obtain the estimator s_e^2 for σ^2 we thus divide SSE by $(n-2)$ degrees of freedom. The estimator s_e^2 provides a measure of how the points are scattered about the regression line. The dispersion varies directly with the magnitude of s_e^2. The positive square root of the residual variance is called the **standard error of estimate**. We will use it in the next section to draw inferences concerning the linear regression model.

Recall from Section 4.3 that the sum of squares for error, SSE, can be found by using the computational formula (b has been replaced by b_1)

$$SSE = SS_y - b_1 SS_{xy} \qquad (14\text{-}3)$$

As a result, the sum of squares for error for our illustration is found to be

$$SSE = SS_y - b_1 SS_{xy}$$
$$= 82.875 - (-.1347)(-212.25) = 54.2849$$

Hence, the residual variance is

$$s_e^2 = \frac{SSE}{n-2} = \frac{54.2849}{6} = 9.0475$$

If we assume that our data satisfy the probabilistic model $y = \beta_0 + \beta_1 x + \varepsilon$, then we can use the following point estimators to estimate values associated with the model:

Model parameter	Estimator
β_0	b_0
β_1	b_1
σ^2	s_e^2
$E(y\mid x)$	$\hat{y}$

For our illustration, we have the following estimates:

Parameter	Estimate
β_0	$46.3889\ (=b_0)$
β_1	$-.1347\ (=b_1)$
σ^2	$9.0475\ (=s_e^2)$
$E(y\mid x = 98)$	$33.1883\ (=\hat{y}(98))$

The estimate for $E(y\mid x = 98)$, $\hat{y}(98)$, was found by substituting $x = 98$ into the regression equation $\hat{y} = 46.3889 - .1347x$ and solving for $\hat{y}$. The value $\hat{y}$ is used to predict the value of y and to estimate the value of $E(y\mid x)$. For an engine size of 98 cid, we can use $\hat{y}(98) = 33.2$ mpg to estimate the average mpg rating for all Ford Escort cars, as well as to predict the mpg rating for a particular Ford Escort car.

To Predict or to Estimate?

In statistics we estimate only parameters, not random variables. The mean value of y when $x = 98$ is greatly different from some value of y chosen at random from the collection of all y values for which $x = 98$. To make this distinction clear, we shall say that we *predict* the value of a random variable and *estimate* the value of a parameter.

Relevant Range of Prediction

To use the sample regression equation for prediction purposes, we must consider only the relevant range of the independent variable x. For our example, the relevant range is the set of six specifically selected engine sizes. Any interpolations between or beyond these six values must be done with extreme caution, because the estimated values are valid only for our six distinct values of x. Even in situations where the linear trend can be expected to continue, extrapolation should be used only with extreme caution. For example, suppose a high linear correlation was found between the yields of a certain variety of tomato plant and the amount of fertilizer used per plot, and the amount of fertilizer that was used varied from 10 to 60 pounds per plot in 10-pound increments. At 50 pounds of fertilizer per plot, suppose a bumper tomato crop is predicted. The residual corresponding to $x = 50$ will be small. If the fertilizer level is raised to 100 pounds per plot, the *predicted* tomato yield might triple. But if this much fertilizer is used, the harvest would yield no tomatoes, since all the plants would die as a result of being overfertilized. Too much fertilizer is as bad for crop yields as is too little fertilizer. The point to be made is that the regression equation is to be used for prediction purposes only for those values close to the values used to arrive at the point estimates b_0 and b_1.

Fixed and Random x Values

If the x values are fixed, it makes sense to predict the y values; the y values are random variables. But it makes no sense to predict the values of x from the y values if the values of x are fixed. That is, given fixed values of x, it makes no sense to find the prediction equation for x given the values for y. For example, for the illustration involving engine size and mpg rating, it would make no sense to find the line relating engine size to mpg rating. In particular, solving the prediction equation $\hat{y} = 46.3889 - .1347x$ for x in terms of $\hat{y}$ and then using the resulting equation to predict engine sizes given mpg ratings makes no sense and should not be attempted, since the engine sizes are fixed values.

Summary

In summary, if there is a linear relationship between x and y, the population regression model can be expressed as

$$y = \beta_0 + \beta_1 x + \varepsilon$$

This functional relationship can be reexpressed as

$$y = E(y|x) + \varepsilon$$

$\uparrow$ $\uparrow$ $\uparrow$

Observed   *Fitted-*   *Random*
value   *model*   *error*
        value

where the fitted model, (population regression equation) is given by $E(y|x) = \beta_0 + \beta_1 x$. Since we do not have access to the entire population to determine the values of the constants β_0 and β_1, we can estimate their values from a bivariate sample using the method of least squares. The sample regression equation is $\hat{y} = b_0 + b_1 x$.

Problem Set 14.1

For Problems 1–5, assume that the linear model $y = \beta_0 + \beta_1 x + \varepsilon$ holds.

1. a. Write the regression equation if $\beta_0 = 3$ and $\beta_1 = 4$.
  b. For the equation in part (a), find the value of y for $x = 2$ if $\varepsilon = .05$.
  c. For the equation in part (a), find the error ε if $x = 2$ and $y = 6$.

2. If the regression equation for a sample of pairs is $\hat{y} = 4x - 8$, find:
  a. the predicted value $\hat{y}(3)$.
  b. a point estimate of β_0.
  c. a point estimate of β_1.
  d. a point estimate of the average value of y if $x = 3$, $E(y|x = 3)$.

3. Jones Realty collected the accompanying data comparing the selling price y of new homes with the size x of the living space (in hundreds of square feet).

Living space, x (in hundreds of square feet)	Selling price, y (in thousands of dollars)
20	116
22	118
18	91
30	145
23	105
25	121

  a. Find a point estimate for the error variance σ^2.
  b. Find a point estimate for β_0.
  c. Find a point estimate for β_1.
  d. Find a point estimate for $E(y|x = 18)$.
  e. Predict a value for $y(24)$.

f. If it has been determined that $\beta_1 = 0$, what is the best point estimate for y? for $E(y)$?

4. A major discount department store chain keeps extensive records on its salespeople. A sample of six salespeople produced the data given in the table concerning months on the job, x, and monthly sales in thousands of dollars, y.

x	2	3	6	11	12	7
y	2.4	3.7	7.6	14.2	15	8

  a. Find a point estimate for β_0.
  b. Find a point estimate for β_1.
  c. Find a point estimate for the error variance.
  d. Find a point estimate for the average monthly sales for a salesperson who has been on the job 11 months.
  e. Predict a value for the monthly sales for a salesperson who has been on the job 6 months.

5. A car rental firm compiled the accompanying data concerning car maintenance costs, y, and miles driven, x, for seven of its automobiles.

Car	Miles driven, x (in thousands of miles)	Maintenance costs, y (in dollars)
A	55	299
B	27	160
C	36	215
D	42	255
E	65	350
F	48	275
G	29	207

a. Find a point estimate for β_0.
b. Find a point estimate for β_1.
c. Find a point estimate for the error variance σ^2.

d. Find a point estimate for the average cost of driving a car 36,000 miles.
e. Predict the cost of driving a car 29,000 miles.

14.2 INFERENCES CONCERNING
THE LINEAR REGRESSION MODEL

In order to use the regression equation $\hat{y} = b_0 + b_1 x$ for predictive purposes, we want to be reasonably sure that the **slope β_1 of the regression equation** $E(y|x) = \beta_0 + \beta_1 x$ is not 0. For if $\beta_1 = 0$, then for every value of x, $E(y|x)$ would be identically equal to β_0, as shown in Figure 14-4. Toward the goal of determining whether the slope of the population regression equation is different from 0, we shall separate SS_y into two components, SSE and SSR.

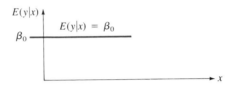

FIGURE 14-4 $E(y|x) = \beta_0$ for all x when $\beta_1 = 0$

Partitioning SS_y

The deviation score $(y - \bar{y})$ can be partitioned into the following two deviation scores:

$$y - \bar{y} = (y - \hat{y}) + (\hat{y} - \bar{y})$$

Consider the following diagram:

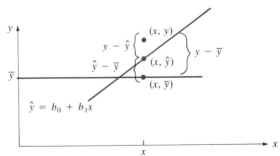

The component $(y - \hat{y})$ is due to error and the component $(\hat{y} - \bar{y})$ is due to linear regression. That is,

$$y - \bar{y} = (y - \hat{y}) + (\hat{y} - \bar{y})$$

$$\underset{\substack{\uparrow \\ \text{Deviation} \\ \text{score}}}{} \qquad \underset{\substack{\uparrow \\ \text{Due to} \\ \text{error}}}{} \qquad \underset{\substack{\uparrow \\ \text{Due to} \\ \text{regression}}}{}$$

It can be shown algebraically that SS_y is equal to the sum of squares for error (SSE) plus the sum of squares for regression (SSR), or

$$\underset{\underset{\text{SS}_y}{\uparrow}}{\sum(y - \bar{y})^2} = \underset{\underset{\text{SSE}}{\uparrow}}{\sum(y - \hat{y})^2} + \underset{\underset{\text{SSR}}{\uparrow}}{\sum(\hat{y} - \bar{y})^2}$$

where $\sum(\hat{y} - \bar{y})^2$ is called the **sum of squares for regression** and is denoted by SSR. Thus, we have the following relationship:

$$SS_y = SSR + SSE \tag{14-4}$$

Recall from Section 14.1 that $SSE = SS_y - b_1 SS_{xy}$. It thus follows that SSR must be identically equal to $b_1 SS_{xy}$. That is,

$$SSR = b_1 SS_{xy}$$

This formula is called the **computational formula for computing SSR**. It is easier to use than the formula $SSR = \sum(\hat{y} - \bar{y})^2$.

Testing the Appropriateness of the Linear Model

We saw earlier that a value of the residual variance s_e^2 provides an estimate for the error variance σ^2. If we assume that $\beta_1 = 0$ is true, then SSR provides another independent estimate for σ^2. Hence, if the slope of the population regression equation is 0, then we have two independent estimates for σ^2. Since the quotient of two independent variance estimators is an F statistic, the sampling distribution of the statistic SSR/s_e^2 is an F distribution with $df = (1, n - 2)$. That is, if $H_0: \beta_1 = 0$ is assumed to be true, the F statistic serves as our test statistic, where F is defined as

$$F = \frac{SSR}{s_e^2} \tag{14-5}$$

with $df = (1, n - 2)$, or $df = 1$ for SSR and $df = (n - 2)$ for s_e^2. The F statistic can be used to determine if β_1 is different from 0. If the slope of the population regression equation is different from 0, then the equation can be used for prediction purposes.

EXAMPLE 14-1   For the EPA gas mileage data in Section 14.1, test using $\alpha = .05$ to determine if $\beta_1 \neq 0$.

Solution   The following values were computed in Section 14.1:

$$SS_y = 82.875$$
$$SSE = 54.2849$$

1. Compute the sum of squares for regression, SSR:

$$SSR = SS_y - SSE = 82.875 - 54.2849 = 28.5901$$

2. Compute the residual variance, s_e^2:

$$s_e^2 = \frac{\text{SSE}}{n-2}$$

$$= \frac{54.2849}{8-2} = 9.0475$$

3. Compute the value of the F statistic:

$$F = \frac{\text{SSR}}{s_e^2} = \frac{28.5901}{9.0475} = 3.16$$

4. Find the critical value: $F_{.05}(1, 6) = 5.99$.
5. Since $F < F_{.05}(1, 6)$, we cannot reject H_0: $\beta_1 = 0$. Thus, we conclude that the equation $\hat{y} = 46.3889 - .1347x$ should not be used for predictive purposes, and we have no evidence to support that a linear model is correct for our data. A nonlinear model, such as $y = \beta_0 + \beta_1 x + \beta_2 x^2 + \varepsilon$, may be appropriate.

■

It can be shown that $\text{df}_y = \text{df}_E + \text{df}_R$, where df_y is the number of degrees of freedom for SS_y, df_E is the number of degrees of freedom for SSE, and df_R is the number of degrees of freedom for SSR. Since $\text{df}_y = (n-1)$ and $\text{df}_E = (n-2)$, it follows that the degrees of freedom for SSR is

$$\text{df}_R = \text{df}_y - \text{df}_E = (n-1) - (n-2) = 1$$

Hence, we see that SSR has 1 degree of freedom associated with it.

Mean Squares

Any sum of squares divided by its associated degrees of freedom provides a variance estimate, called a **mean square**, denoted by MS. That is,

$$\text{MS} = \frac{\text{SS}}{\text{df}}$$

As a result, we can define the following two mean squares:

$$\text{MSR} = \frac{\text{SSR}}{1} = \text{SSR}$$

$$\text{MSE} = \frac{\text{SSE}}{n-2} = s_e^2$$

Notice that SSR is a mean square and thus a variance estimate.

By using the concept of mean squares, we can write the F statistic as SSR/s_e^2, since SSR and s_e^2 are variance estimates. The computations involved in testing the null hypothesis H_0: $\beta_1 = 0$ against the alternative hypothesis H_1: $\beta_1 \neq 0$ can then be summarized by using an **ANOVA summary table**, the general form of which is shown in Table 14-2. For the gasoline mileage example, the results are summarized in Table 14-3.

Table 14-2 General Form of ANOVA Summary Table

Source of variation	SS	df	MS	F
Regression	SSR	1	MSR	$\dfrac{SSR}{s_e^2}$
Error	SSE	$n-2$	$s_e^2 = \dfrac{SSE}{n-2}$	
Total (y)	SS_y	$n-1$		

Table 14-3 ANOVA Summary Table for Data in Table 14-1

Source of variation	SS	df	MS	F
Regression	28.5901	1	28.5901	3.16
Error	54.2849	6	9.0475	
Total	82.8750	7		

To summarize the discussion thus far, we note that the following computational steps are typically followed when testing $H_0: \beta_1 = 0$:

1. Compute SS_y using the formula $SS_y = \sum y^2 - (\sum y)^2/n$.
2. Compute SSR using the formula $SSR = b_1 SS_{xy}$.
3. Compute SSE using the relation $SSE = SS_y - SSR$.
4. Compute s_e^2 using the formula $s_e^2 = SSE/(n-2)$.
5. Compute F using $F = SSR/s_e^2$.
6. Find the critical value $F_\alpha(1, n-2)$ in the F tables.
7. Compare the F test statistic with the critical value $F_\alpha(1, n-2)$; if $F > F_\alpha(1, n-2)$, then we reject H_0.

Computers are often used to carry out the analyses involved in regression. The following computer printout was obtained using SPSS to perform a regression analysis for the EPA gas mileage data:

```
MULTIPLE R            -.58739           ANALYSIS OF VARIANCE
R SQUARE               .34503              DF   SUM OF SQUARES    MEAN SQ
ADJUSTED R SQUARE      .23587   REGRESSION   1      28.59414      28.59414
STANDARD ERROR        3.00779   RESIDUAL     6      54.28086       9.04681
              F = 3.16069     SIGNIF F = .1258

                       VARIABLES IN THE EQUATION
VARIABLE         B           SE B        BETA          T        SIG T
CID           -.13472        .07578     -.58739     -1.778      .1258
(CONSTANT)   46.39099       8.23395                  5.634      .0013
```

The correspondence between our notation and the above printout notation is as follows:

Printout notation	**Our notation**
MULTIPLE R	r
R SQUARE	r^2
RESIDUAL	Error
SIGNIF F	p-value for F
CID	b_1
(CONSTANT)	b_0
SE B	Standard error
T	t value
SIG T	p-value for t

The information labeled ADJUSTED R SQUARE, STANDARD ERROR, and BETA is irrelevant for our purposes. The portion of the printout below the VARIABLES IN THE EQUATION line will be explained subsequently.

Testing H_0: $\beta_1 = 0$ Using the *t* Distributions

We mentioned in Section 13.1 that $t^2 = F$ for an F statistic with df $= (1, n - 2)$. Since $t = \sqrt{F}$, the t statistic can be used to test the null hypothesis H_0: $\beta_1 = 0$. The t statistic is defined as

$$t = \frac{\sqrt{MSR}}{s_e}$$

where $s_e = \sqrt{s_e^2}$ is an estimate of σ. Since $MSR = b_1 SS_{xy}$ and $b_1 = SS_{xy}/SS_x$, we have

$$MSR = b_1^2 SS_x$$

By using these facts, we can express the t statistic as

$$t = \frac{\sqrt{b_1^2 SS_x}}{s_e}$$

$$= \frac{|b_1|}{s_e/\sqrt{SS_x}}$$

Hence, if the null hypothesis H_0: $\beta_1 = 0$ is assumed to be true, then the test statistic can be expressed as

$$t = \frac{b_1}{s_e/\sqrt{SS_x}} \tag{14-6}$$

where df $= (n - 2)$. The two critical values, $t_{\alpha/2}^2(n - 2)$ and $F_\alpha(1, n - 2)$ are related as follows:

$$t_{\alpha/2}^2(n - 2) = F_\alpha(1, n - 2)$$

EXAMPLE 14-2   For the EPA gas mileage data, test to determine if $\beta_1 \neq 0$ by using the t test. Use $\alpha = .05$.

Solution   By using (14-6), we have

$$t = \frac{b_1}{s_e/\sqrt{SS_x}}$$

$$= \frac{-.1347}{\sqrt{9.0475}/\sqrt{1575.5}} = -1.7775$$

The critical values $\pm t_{.025}$ for df = 6 are $\pm t_{.025}(6) = \pm 2.447$. Since $-t_{.025}(6) < t < t_{.025}(6)$, we cannot reject H_0: $\beta_1 = 0$. Hence, we have no evidence to suggest that the linear model is appropriate for our data. Note that

$$t^2 = (-1.7775)^2 = 3.16 = F$$

and

$$[t_{.025}(6)]^2 = (2.447)^2 = 5.99 = F_{.05}(1, 6)$$ ■

Confidence Intervals for β_1

The limits for a $(1 - \alpha)100\%$ confidence interval for β_1 can be found by using

$$b_1 \pm t_{\alpha/2}(\text{df})\left(\frac{s_e}{\sqrt{SS_x}}\right) \tag{14-7}$$

By using (14-7), we can determine a 95% confidence interval for β_1, the slope of the population regression equation for the EPA gas mileage data. The confidence interval limits are given by

$$b_1 \pm t_{\alpha/2}(\text{df})\left(\frac{s_e}{\sqrt{SS_x}}\right)$$

$$-.1347 \pm (2.447)(.076)$$

$$-.1347 \pm .1854$$

Thus, a 95% confidence interval for β_1 is $(-.3201, .0507)$. Since the interval contains 0, we cannot conclude that $\beta_1 \neq 0$. We can be 95% confident that the value of β_1, the population regression slope, is between $-.3201$ and $.0507$.

Confidence Intervals for $E(y|x_0)$

Once the sample regression equation has been found and it has been determined that the model $E(y|x) = \beta_0 + \beta_1 x$ is appropriate for the data, then we can use the regression equation $\hat{y} = b_0 + b_1 x$ to make predictions. In so doing, we will want to estimate the average value for y given x, $E(y|x)$, as well as to predict the random y value for a given value x.

The limits for a $(1 - \alpha)100\%$ confidence interval for $E(y|x_0)$ are given by

$$\hat{y} \pm t_{\alpha/2}(n - 2)s_e\sqrt{\frac{1}{n} + \frac{(x_0 - \bar{x})^2}{SS_x}} \tag{14-8}$$

The confidence interval is shortest when $x_0 = \bar{x}$, and the length of the interval increases as x_0 moves away from the mean $\bar{x}$.

Consider the following example, which illustrates the construction of a $(1 - \alpha)100\%$ confidence interval for $E(y|x_0)$.

EXAMPLE 14-3 In a study of how corn yield depends on the amount of fertilizer, the following data were obtained for six different plots of land:

x (fertilizer in 100 lb/acre)	1	2	3	4	5	6
y (yield in bu/acre)	40	50	50	70	65	80

Determine if the linear model is appropriate using $\alpha = .05$ and construct a 95% confidence interval for $E(y|x = 4)$.

Solution With the aid of a hand-held calculator we determined that the regression equation is $\hat{y} = 32.6667 + 7.5714x$, the sum of squares for x is $SS_x = 17.5$, the sum of squares for y is $SS_y = 1120.8333$, the sum of cross products is $SS_{xy} = 132.5$, and the residual variance is $s_e^2 = 29.4123$. The value of $\hat{y}(4)$ is found by using the regression equation:

$$\hat{y}(4) = 32.6667 + 7.5714(4) = 62.9523$$

To determine if β_1 is different from 0, we use (14-6) to determine the value of the t statistic corresponding to b_1:

$$t = \frac{b_1}{s_e/\sqrt{SS_x}}$$

$$= \frac{7.5714}{5.4233/\sqrt{17.5}} = 5.84$$

The positive critical value is found to be $t_{.025}(4) = 2.776$. Since $t > t_{.025}(4)$, we reject $H_0: \beta_1 = 0$ and conclude that the linear model is appropriate. Therefore, we can use the equation $\hat{y} = 32.6667 + 7.5714x$ for prediction purposes.

By using (14-8), we find the limits of a 95% confidence interval for $E(y|x = 4)$ as follows:

$$\hat{y} \pm t_{.025}(4)s_e\sqrt{\frac{1}{n} + \frac{(x_0 - \bar{x})^2}{SS_x}}$$

$$62.9523 \pm (2.447)(5.4233)\sqrt{\frac{1}{6} + \frac{(4 - 3.5)^2}{17.5}}$$

$$62.9523 \pm 5.6452$$

Thus, a 95% confidence interval for the average yield of corn (in bushels) given $x = 400$ pounds of fertilizer is (57.31, 68.60). ∎

Figure 14-5 illustrates the 95% **confidence interval band** for the line representing the regression equation $\hat{y} = 32.6667 + 7.5714x$ from Example 14-3. One side of the band was formed by joining the lower limits of the 95% confidence

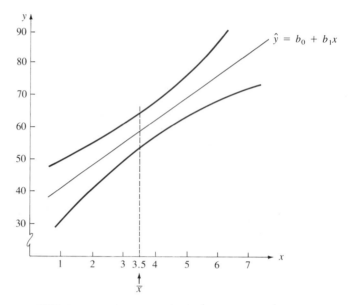

FIGURE 14-5 Confidence interval band for Example 14-3

intervals for the average values of y at each value of x, while the other side of the band was formed by joining the upper limits of the 95% confidence intervals for the average values of y at each value of x. Note that the band is narrowest at $x = \bar{x}$ and gets wider as x gets farther away from $\bar{x}$. (See also Table 14-4.)

Table 14-4 95% Confidence Intervals
for $E(y|x)$

x	L_1	L_2	Width
1	30.63	49.84	19.21
2	40.60	55.02	14.42
3	49.74	61.03	11.29
3.5	53.75	64.58	10.83
4	57.31	68.60	11.29
5	63.31	77.73	14.42
6	68.49	87.70	19.21

Prediction Intervals for y

There are occasions when we shall want to predict an individual value of y for a given value of x, instead of estimating the average value for all the y values for a particular x value. For example, with the corn/fertilizer data of Example 14-3, we may want to predict the corn yield in bushels per acre for a plot planted with $x = 400$ pounds of fertilizer per acre instead of estimating the average yield for all plots that are fertilized at a level of 400 pounds per acre.

The limits for a $(1 - \alpha)100\%$ prediction interval for the value of a single random y value for a given value of $x = x_0$ are given by

$$\hat{y} \pm t_{\alpha/2}(n - 2)s_e \sqrt{1 + \frac{1}{n} + \frac{(x_0 - \bar{x})^2}{SS_x}} \qquad (14\text{-}9)$$

Note once again that the prediction band widens as the distance between x_0 and $\bar{x}$ increases. The basic difference between the formulas for the limits of a prediction interval for y given x_0 and the limits for a confidence interval for the average value of y given x_0 is the inclusion of the number 1 in the expression under the radical sign of (14-9).

EXAMPLE 14-4   For the corn/fertilizer data in Example 14-3, construct a 95% prediction interval for the corn yield of a particular field if 400 pounds per acre of fertilizer are used.

Solution   We use (14-9) to find the prediction interval limits:

$$\hat{y} \pm t_{\alpha/2}(n - 2)s_e \sqrt{1 + \frac{1}{n} + \frac{(x_0 - \bar{x})^2}{SS_x}}$$

$$62.9523 \pm (2.447)(5.4233) \sqrt{1 + \frac{1}{6} + \frac{(4 - 3.5)^2}{17.5}}$$

$$62.9523 \pm 14.4216$$

$$(48.53, 77.37)$$

Thus, the 95% prediction interval for the yield of corn when 400 pounds of fertilizer per acre are used is $(48.53, 77.37)$. We can be 95% confident that the corn yield is at most 77.37 bushels per acre and at least 48.53 bushels per acre. Note that this prediction interval has a width $w = 77.37 - 48.53 = 28.84$, while the width for the 95% confidence interval for $E(y|4)$ found in Example 14-3 is $w = 68.60 - 57.31 = 11.29$. In general, it can be shown that for fixed values of x_0 and α, the width of the prediction interval for y given x_0 will exceed the width of the confidence interval for $E(y|x_0)$. ■

In Section 14.1 we saw that the population regression model can be expressed as $y = E(y|x) + \varepsilon$, where, for a fixed value of x, y is the observed value, $E(y|x)$ is the average value of y, and ε is random error. This model can be approximated by the least squares equation, $\hat{y} = b_0 + b_1 x$. For a particular value of x, for example, x_0, the vertical distance between a particular value of y, y_0, and the predicted value of y, $\hat{y}$, represents the error of prediction. This error is shown in Figure 14-6. We can see that the vertical distance between the two lines represents the error of estimate when $\hat{y}$ is used to estimate the average value of y, $E(y|x)$. When $\hat{y}$ is used to predict y_0, the prediction error is the sum of two errors—the error of estimate and random error. As a consequence of this relationship, for a fixed value of x, the error of predicting a particular value of y is usually larger than the error of estimating the average value of y. And for a particular value of x, the width of the prediction interval for y is usually greater than the width of the confidence interval for $E(y|x)$. We can also see in Figure 14-6 that both the error of estimate and the error of prediction decrease as x_0 approaches $\bar{x}$ (the two lines cross at the point where $x = \bar{x}$).

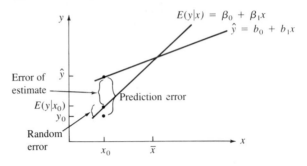

FIGURE 14-6 Errors of estimating the average value of y and predicting a future value of y for a given value of x

Problem Set 14.2

A

1. The population (in thousands) of seven countries (x) and the corresponding number of practicing physicians (y) are given in the table.

x	38	52	75	32	60	43	57
y	20	23	35	20	26	25	22

a. Use the F test and $\alpha = .05$ to determine if the data satisfy the linear regression model.
b. Construct a 95% confidence interval for β_1.

2. The numbers of murders y committed annually in the state of Maryland from 1978 to 1982 using a blunt object are listed in the accompanying table.

Year, x	1978	1979	1980	1981	1982
Number of murders, y	10	12	17	22	20

a. By coding the years using the relation $x =$ Year $-$ 1978, determine if the pairs (x, y) satisfy the linear model. Use the F test and $\alpha = .05$.
b. Construct a 95% confidence interval for β_1.

3. The data in the table indicate the length of confinement in days (x) and the cost in dollars (y) at Memorial Hospital.

x	3	1	11	4	8
y	685	270	2310	885	1700

a. By using the F test and $\alpha = .05$, test to determine if $\beta_1 \neq 0$.
b. Construct a 95% confidence interval for the average cost for 4 days of hospital confinement.
c. Construct a 95% prediction interval for the cost of a 4-day stay in the hospital.
d. Construct a 95% prediction interval for the cost of an 8-day stay in the hospital.
e. What is the additional cost for 1 more day in the hospital?

4. The data given here resulted from an experiment to study the relationship between age x in months and size of vocabulary y for young children aged 18 to 48 months.

x	18	24	30	36	42	48
y	100	250	460	890	1210	1530

a. Construct a 99% confidence interval for β_1.
b. Construct a 99% confidence interval for $E(y|x = 36)$.
c. Construct a 99% prediction interval for $y(36)$.

5. Use the t test to determine if the linear model is appropriate for the data in Problem 1.

6. Use the t test to determine if the linear model is appropriate for the data in Problem 2.

7. Use the t test to determine if the linear model is appropriate for the data in Problem 3.

8. Use the t test to determine if the linear model is appropriate for the data in Problem 4.

9. For the data in Problem 4, show that if each x value is multiplied by $\frac{1}{12}$ and each y value is multiplied by .1, then the value of the t statistic for testing $H_0: \beta_1 = 0$ versus $H_1: \beta_1 \neq 0$ for the new data pairs is equal to the value of the t statistic for the original data pairs.

B

10. Suppose the value of the t statistic for testing H_0: $\beta_1 = 0$ versus $H_1: \beta_1 \neq 0$ is t_0 and c and d are posi-

tive numbers. If each value of x is multiplied by c and each value of y is multiplied by d, show that the value of the t statistic for testing $H_0: \beta_1 = 0$ versus $H_1: \beta_1 \neq 0$ for the transformed data is t_0.

11. The regression slope b_1 can be thought of as a random variable. After examining formula (14-6) and recalling the definition of the t statistic, speculate as to the mean and standard error of b_1.

14.3 CORRELATION ANALYSIS ─────────────────────

In this section we want to reexamine the linear correlation coefficient r that was introduced in Section 4.2 and extend its use to forming inferences about the population correlation coefficient. The population correlation coefficient is commonly denoted by the Greek letter ρ (rho).

Let's recall several formulas from Section 4.2:

$$r = \frac{SS_{xy}}{\sqrt{SS_x SS_y}}$$

$$b_1 = r\left(\frac{s_y}{s_x}\right) \tag{14-10}$$

The regression slope b_1 and the correlation coefficient r for a sample of bivariate data are related by the second formula above; the same relationship holds for the population of bivariate data if the assumptions of the linear regression model hold:

$$\beta_1 = \rho\left(\frac{\sigma_y}{\sigma_x}\right) \tag{14-11}$$

A statistical test for the significance of a linear relationship between x and y can be performed by testing the following statistical hypotheses:

$$H_0: \quad \rho = 0$$
$$H_1: \quad \rho \neq 0$$

By examining (14-11) we see that β_1 and ρ agree in sign (either both positive or both negative) and β_1 is 0 if and only if $\rho = 0$. As a result, testing the above statistical hypotheses concerning ρ is equivalent to testing the following hypotheses concerning β_1:

$$H_0: \quad \beta_1 = 0$$
$$H_1: \quad \beta_1 \neq 0$$

In some situations, we are not interested in determining the linear regression equation for predictive purposes. In other situations, if both x and y are random variables (instead of just y), then the linear regression model we developed is not appropriate. Therefore, it may be desirable to test for a significant linear relationship between the two variables without performing a regression analysis. If we assume that samples of n data pairs (x, y) are drawn from two normal populations, the x values from one and the y values from the other, then we can perform a hypothesis test to determine if $\rho \neq 0$. If the null hypothesis $H_0: \rho = 0$ is assumed to be true, it can be shown that the sampling distribution of the statistic

$$\frac{r\sqrt{n-2}}{\sqrt{1-r^2}}$$

is a t distribution with $(n-2)$ degrees of freedom. Thus, we have the following relationship:

$$t = \frac{r\sqrt{n-2}}{\sqrt{1-r^2}} \tag{14-12}$$

with df $= (n-2)$.

EXAMPLE 14-5   For the EPA gas mileage data of Section 14.1, test using $\alpha = .05$ to determine if $\rho < 0$.

Solution   The statistical hypotheses are

$$H_0: \quad \rho \geq 0$$
$$H_1: \quad \rho < 0$$

The sample correlation coefficient is

$$r = \frac{SS_{xy}}{\sqrt{SS_x SS_y}}$$

$$= \frac{-212.25}{\sqrt{(1575.5)(82.875)}} = -.5874$$

From (14-12), the value of the t statistic is

$$t = \frac{r\sqrt{n-2}}{\sqrt{1-r^2}}$$

$$= \frac{-.5874\sqrt{8-2}}{\sqrt{1-(-.5874)^2}} = -1.78$$

By using the t table (see the back endpaper) we find $-t_{.05}(6) = -1.943$. Since $-t_{.05}(6) < t = -1.78$, we fail to reject the null hypothesis $H_0: \rho \geq 0$.   ∎

Coefficient of Determination

Recall from Section 14.2 that SS_y can be partitioned into two sums of squares, SSE and SSR:

$$SS_y = SSR + SSE$$

Dividing both sides of this equation by SS_y, we have

$$1 = \frac{SSR}{SS_y} + \frac{SSE}{SS_y}$$

Notice that both SSR/SS_y and SSE/SS_y are nonnegative numbers that sum to 1. The expression SSR/SS_y represents the percentage of the total sum of squares that is explained by the regression equation and is called the **coefficient of determination**. It is denoted by r^2. Thus, we have

$$r^2 = \frac{\text{Sum of squares explained by regression}}{\text{Total sum of squares (before regression)}}$$

The expression $(1 - r^2)$ represents the percentage of the total sum of squares that cannot be attributed to regression. The coefficient of determination is quite useful in estimating the strength of the linear relationship between two variables. It can be shown that the coefficient of determination r^2 is also the square of the correlation coefficient. This is the reason the coefficient of determination is denoted by r^2.

EXAMPLE 14-6   For the EPA gas mileage data in Table 14-1, determine the coefficient of determination and interpret your result.

Solution   From Example 14-5 we have $r = -.5874$. The coefficient of determination is then

$$r^2 = (-.5874)^2 = .345$$

Approximately 35% of the total sum of squares (before regression) can be attributed to the linear relationship between mileage rating and engine size, and approximately 65% of SS_y is not explained by the linear model $y = \beta_0 + \beta_1 x + \varepsilon$. A nonlinear model, such as the quadratic model $y = \beta_0 + \beta_1 x + \beta_2 x^2 + \varepsilon$, may account for a larger percentage of SS_y.   ∎

In regression analyses involving more than one independent variable, there are many kinds of correlation coefficients. One important coefficient is the multiple correlation coefficient. Like r^2, the **multiple correlation coefficient** R^2 gives the proportion of the total sum of squares SS_y that can be attributed to (or explained by) all the independent variables. We shall leave the development of R^2 to a more advanced course in statistics.

Spearman's Rank Correlation Coefficient

There are many occasions where it is desirable to measure the association between two variables measured on an ordinal scale. Spearman's rank correlation coefficient r_s can be used for this purpose. Ranks are assigned to the x measurements and to the y measurements. The correlation coefficient of the ranked data is then found by using the formula $r = SS_{xy}/\sqrt{SS_x SS_y}$. When ranks are used instead of the original measurements, the correlation coefficient is called **Spearman's rank correlation coefficient** and is denoted by r_s. If there are no values for either set of data that occur with a frequency greater than 1 (that is, there are no ties), then the formula for r_s can be simplified to

$$r_s = 1 - \frac{6\sum d^2}{n(n^2 - 1)} \qquad (14\text{-}13)$$

where d is the difference between ranks for each pair and n is the number of pairs. The following example explains the procedures.

EXAMPLE 14-7   A sample of ten students from a statistics class had the SAT math and verbal scores shown in Table 14-5. Determine Spearman's rank correlation coefficient.

Table 14-5 Data for Example 14-7

Math	Verbal
425	535
358	375
515	500
672	550
378	414
397	435
715	750
638	515
478	482
350	410

Solution   We first rank both sets of data in increasing order:

Math	Rank	Verbal	Rank
425	5	535	8
358	2	375	1
515	7	500	6
672	9	550	9
378	3	414	3
397	4	435	4
715	10	750	10
638	8	515	7
478	6	482	5
350	1	410	2

For illustrative purposes, we will find r_s using formulas (14-10) and (14-13).

1. The following table organizes the computations for r_s, using formula (14-10):

x	y	x^2	y^2	xy
5	8	25	64	40
2	1	4	1	2
7	6	49	36	42
9	9	81	81	81
3	3	9	9	9
4	4	16	16	16
10	10	100	100	100
8	7	64	49	56
6	5	36	25	30
1	2	1	4	2
Sums 55	55	385	385	378

The values of SS_x, SS_y, and SS_{xy} are found as follows:

$$SS_x = \sum x^2 - \frac{(\sum x)^2}{n} = 385 - \frac{(55)^2}{10} = 82.5$$

$$SS_y = \sum y^2 - \frac{(\sum y)^2}{n} = 385 - \frac{(55)^2}{10} = 82.5$$

$$SS_{xy} = \sum xy - \frac{(\sum x)(\sum y)}{10} = 378 - \frac{(55)(55)}{10} = 75.5$$

Spearman's correlation coefficient is computed using (14-10):

$$r_s = \frac{SS_{xy}}{\sqrt{SS_x SS_y}} = \frac{75.5}{\sqrt{(82.5)(82.5)}} = .91515$$

2. The following table organizes the computations involving the paired ranks, using formula (14-13):

x	y	d	d^2
5	8	−3	9
2	1	1	1
7	6	1	1
9	9	0	0
3	3	0	0
4	4	0	0
10	10	0	0
8	7	1	1
6	5	1	1
1	2	−1	1
			$14 = \sum d^2$

The value of Spearman's correlation coefficient is obtained from (14-13):

$$r_s = 1 - \frac{6\sum d^2}{n(n^2 - 1)}$$

$$= 1 - \frac{6(14)}{10(100 - 1)} = 1 - \frac{84}{990} = .91515$$

We can see by comparing the two methods that both formulas (14-10) and (14-13) provide identical results, but formula (14-10) involves less work; hence, it is the preferred formula to use when paired data are involved with no ties. ∎

Spearman's rank correlation coefficient is used for a variety of reasons. It provides an alternative when the normality assumption is tenuous or when it is desired to have a more general coefficient that is not restricted by a linear relationship, but is monotonic. A relationship is **monotone increasing** if increases in one variable correspond to increases in the other; a relationship is **monotone decreasing** if increases in one variable correspond to decreases in the other variable. Values of r_s near 1 indicate a monotone increasing relationship and values of r_s near -1 indicate a monotone decreasing relationship (see Figure 14-7).

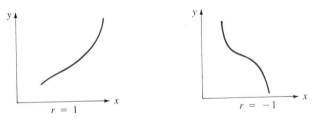

FIGURE 14-7 Monotonic relationships

In addition, r_s can be used to determine the strength of relationship or agreement when the original data consist of paired ranks. Such would be the case, for example, when two judges rank the floats in a homecoming parade and it is desired to have a measure of consistency for the two judges.

It is of interest to note that $\sum d^2$ provides a measurement of the agreement between two sets of ranks. For the data in Example 14-7, $\sum d^2 = 14$. If $\sum d^2$ is small there is high agreement between the ranks, and if $\sum d^2$ is large, there is little agreement between the ranks. Total agreement occurs when the ranks for each pair are equal. Such would be the case for the following pairing of data sets:

Student	A	B	C	D	E	F	G	H	I	J
Rank on math SAT	1	2	3	4	5	6	7	8	9	10
Rank on verbal SAT	1	2	3	4	5	6	7	8	9	10

In this case, $\sum d^2 = 0$. Total disagreement occurs when the data are ranked in reverse order as follows:

Student	A	B	C	D	E	F	G	H	I	J
Rank on math SAT	1	2	3	4	5	6	7	8	9	10
Rank on verbal SAT	10	9	8	7	6	5	4	3	2	1

In this case, $\sum d^2 = (-9)^2 + (-7)^2 + (-5)^2 + (-3)^2 + (-1)^2 + 1^2 + 3^2 + 5^2 + 7^2 + 9^2 = 330$. Thus, for ten pairs of ranks, the value of $\sum d^2$ must fall somewhere between 0 and 330, that is

$$0 \le \sum d^2 \le 330$$

Since $\sum d^2 = 14$ for our example, we have $0 \le 14 \le 330$. If this interval is standardized (transformed) so that the interval $[0, 330]$ becomes the interval $[-1, 1]$, then $\sum d^2 = 14$ is transformed to the value for r_s. The interval $[0, 330]$ can be transformed to $[-1, 1]$ by carrying out the following steps on the inequalities $0 \le 14 \le 330$:

Steps	Interval $(0 \le 14 \le 330)$
1. Multiply by -1	$0 \ge -14 \ge -330$
2. Rewrite	$-330 \le -14 \le 0$
3. Divide by 165 $(= 330/2)$	$-2 \le \dfrac{-14}{165} \le 0$
4. Add 1	$-1 \le 1 - \dfrac{14}{165} \le 1$

Note that the value $1 - \left(\frac{14}{165}\right)$ is equal to $r_s = .91515$.

In practice, Spearman's formula is also used when ties are present in the data sets. In such a case, each of the tied measurements is assigned the average of the ranks that would have resulted had there been no ties. For example, consider the following measurements x:

x	Rank	Rank if no ties present	
10.4	1	1	
11.2	3	2	
11.2	3	3	The average of the ranks is 3.
11.2	3	4	
12.3	5.5	5	
12.3	5.5	6	The average of the ranks is 5.5.
13.7	7	7	
14.8	8.5	8	
14.8	8.5	9	The average of the ranks is 8.5.

EXAMPLE 14-8   For ten automobiles made recently, weight (x, in thousands of pounds) and average miles per gallon of gasoline (y) were recorded in Table 14-6. Determine Spearman's rank correlation coefficient r_s.

Table 14-6 Data for Example 14-8

x	2.7	4.1	3.5	2.7	2.2	3.9	2.2	2.2	3.5	2.2
y	28	19	22	30	26	19	38	40	26	45

Solution   We first rank the x values. Since there are four measurements of magnitude 2.2, we assign each of the four values the rank of the average of the first four rank values:

$$\frac{1 + 2 + 3 + 4}{4} = 2.5$$

Next, since the fifth and sixth measurements are equal and occupy the fifth and sixth ranks, they are each assigned the average rank of 5.5. The seventh and eighth values are equal and are assigned the average rank of 7.5. Similarly, ranks are provided for the y measurements. After ranks have been assigned, we compute the square of the difference d for each pair. The sum of the squared differences is then used in formula (14-13) to find r_s. The computations are summarized in the following table:

x	Rank x	y	Rank y	d	d^2
2.7	5.5	28	6	$-.5$	.25
4.1	1	19	1.5	8.5	72.25
3.5	7.5	22	3	4.5	20.25
2.7	5.5	30	7	-1.5	2.25
2.2	2.5	26	4.5	-2	4
3.9	9	19	1.5	7.5	56.25
2.2	2.5	38	8	-5.5	30.25
2.2	2.5	40	9	-6.5	42.25
3.5	7.5	26	4.5	3	9
2.2	2.5	45	1	-7.5	56.25
					$\sum d^2 = 293$

The value of Spearman's correlation coefficient is

$$r_s = 1 - \frac{6\sum d^2}{n(n^2 - 1)}$$

$$= 1 - \frac{(6)(293)}{10(100 - 1)} = -.7758$$

The value $r_s = -.78$ indicates a monotone decreasing relationship for automobile weights and gasoline mileages; the heavier the car, the lower the gasoline mileage.   ∎

If the number of ties is small compared to the number of data pairs, little error results when using Spearman's formula instead of (14-10). By using formula (14-10) on the ranked pairs in Example 14-8, we find that $r = -.8549$. For this data set, the total number of tied observations is 12, which is 60% of the total number of x and y observations. These ties reflect an error of .08 when using Spearman's formula to compute the correlation coefficient r for the ranked pairs.

Testing H_0: $\rho_s = 0$

If $n \geq 10$ and the population correlation coefficient ρ_s of ranked data is 0, the sampling distribution of r_s is approximately normal with a mean of 0 and a standard deviation given by

$$\sigma_{r_s} = \frac{1}{\sqrt{n-1}}$$

As a consequence, for a population of ranked pairs we can determine if $\rho_s \neq 0$ by first finding the z value for r_s under the assumption that $\rho_s = 0$:

$$z = \frac{r_s - 0}{1/\sqrt{n-1}}$$
$$= r_s\sqrt{n-1}$$

The value of the test statistic for testing the null hypothesis H_0: $\rho_s = 0$ is given by

$$z = r_s\sqrt{n-1} \tag{14-14}$$

EXAMPLE 14-9   For the paired ranked data in Example 14-7, test the statistical hypothesis that there is no correlation between SAT math scores and SAT verbal scores. Use the .05 level of significance.

Solution   **1.** H_0: $\rho_s = 0$

2. H_1: $\rho_s \neq 0$ (two-tailed test)

3. $\alpha = .05$

4. Sampling distribution: Distribution of r_s. The test statistic is the z value for r_s:

$$z = r_s\sqrt{n-1}$$
$$= (.92)\sqrt{10-1} = 2.76$$

5. Critical values: $\pm z_{.025} = \pm 1.96$.

6. Decision: Reject H_0, since $z > z_{.025} = 1.96$. Hence, we have evidence to suggest that SAT math scores and SAT verbal scores are correlated.

7. Type of error possible: Type I; $\alpha = .05$.

8. p-value: $2P(z > 2.76) = 2(.5 - .4971) = .0058$.   ■

EXAMPLE 14-10   For the gasoline mileage data in Example 14-8 test the null hypothesis $H_0: \rho_s \geq 0$ against the alternative hypothesis $H_1: \rho_s < 0$ using the .01 level of significance.

Solution   **1.** H_0: $\rho_s \geq 0$

2. H_1: $\rho_s < 0$ (left-tailed test)

3. $\alpha = .01$

4. Sampling distribution: Distribution of r_s. The test statistic is the z value for r_s:

$$z = r_s\sqrt{n-1}$$
$$= (-.78)\sqrt{10-1} = -2.34$$

5. Critical value: $-z_{.01} = -2.33$

6. Decision: Since $z < -z_{.01}$, we can reject H_0 and conclude that the data suggest that weights of cars and gasoline mileage have a monotone relationship.

7. Type of error possible: Type I; $\alpha = .01$

8. *p*-value: $P(z < -2.34) = .0096$ ■

Problem Set 14.3

A

1. A small retail business has determined that the correlation coefficient between monthly expenses and profits for the past year is $r = .56$. Assuming that both expenses and profits are approximately normal, test at the .05 level of significance the null hypothesis $H_0: \rho = 0$ against the alternative hypothesis $H_1: \rho \neq 0$.

2. A medical pathologist has found from a sample of 12 human skeletons that body size and bone size have a correlation $r = .46$. Based on this one sample, should bone sizes from skeleton remains be used to predict body sizes? Use $\alpha = .01$ and assume that bone sizes and body weights are normally distributed.

3. A small business determined that the correlation between weekly sales (in dollars) and weekly costs (in dollars) of advertising was $r = .61$ for a 10-week period. Determine if there is a positive correlation between weekly sales and weekly advertising costs. Assume that weekly sales and weekly advertising costs follow normal distributions. Use $\alpha = .01$.

4. A bookstore owner, using a sample of 30 books, found the linear correlation between book cost and book thickness to be $r = .81$. Assuming that book cost and thickness are normally distributed, test the null hypothesis $H_0: \rho \leq 0$ against the alternative hypothesis $H_1: \rho > 0$. Use $\alpha = .05$.

5. In a study to determine the relationship between weights of 1-year-old babies and their weights 24 years later, a sample of 20 medical files for females showed that $r = .57$. Test to determine if the population correlation coefficient is positive, assuming that the population of weights of 1-year-olds and the population of weights of 25-year-olds are normally distributed. Use $\alpha = .05$.

6. Ten baseball players were ranked by a scout on running speed and hitting ability as they demonstrated their baseball-playing talents at a major-league tryout. The results are shown in the accompanying table.

Player	Speed	Hitting
A	1	7
B	2	9
C	3	3
D	4	4
E	5	1
F	6	6
G	7	8
H	8	2
I	9	10
J	10	5

a. Calculate Spearman's rank correlation coefficient r_s to measure the association between speed and hitting.

b. At the .05 level of significance, test to determine if ρ_s is different from 0.

7. Two teachers rated ten mathematics students according to ability. The ratings are given in the table.

Student	A	B	C	D	E	F	G	H	I	J
Teacher X	1	9	6	2	5	8	7	3	10	4
Teacher Y	3	10	8	1	7	5	6	2	9	4

a. Calculate Spearman's rank correlation coefficient to measure the consistency of the ratings.

b. Test the null hypothesis $H_0: \rho_s \le 0$ against the alternative hypothesis $H_1: \rho_s > 0$ at the .01 level of significance.

8. In an effort to determine the relationship between annual wages for employees and the number of days absent from work due to sickness, a large corporation studied the personnel records for a random sample of twelve employees. The paired data are provided in the table.

Employee	Annual wages (in thousands of dollars)	Days missed
1	15.7	4
2	17.2	3
3	13.8	6
4	24.2	5
5	15.0	3
6	12.7	12
7	13.8	5
8	18.7	1
9	10.8	12
10	11.8	11
11	25.4	2
12	17.2	4

a. Determine the rank correlation coefficient r_s.

b. Test $\alpha = .05$ to determine if the number of days missed is related to annual wages.

9. Refer to Problem 8.

a. Find the correlation coefficient using formula (14-10).

b. Test the null hypothesis $H_0: \rho = 0$ against the alternative hypothesis $H_1: \rho \ne 0$ by using formula (14-12) and $\alpha = .05$.

c. Compare the results with those obtained from Problem 8.

10. For Problem 6, find and interpret the coefficient of determination.

11. For Problem 8, find and interpret the coefficient of determination.

B

12. For Problem 1, construct a 95% confidence interval for ρ.

13. For Problem 2, construct a 99% confidence interval for ρ.

14. For Problem 6, construct a 95% confidence interval for ρ_s.

15. For Problem 8, construct a 95% confidence interval for ρ_s.

16. Prove that $b_1 = r s_y / s_x$.

17. Explain how to define the linear correlation coefficient r in terms of the coefficient of determination.

18. One interpretation of r^2 is that it indicates the percentage of the variation in y accounted for by the variation in x. Explain.

19. Show that the maximum value of $\sum d^2$ is $(n^3 - n)/3$ for any set of n paired ranks.

20. By using the result of Problem 19, show that

$$r_s = \frac{1 - 6\sum d^2}{n(n^2 - 1)}$$

CHAPTER SUMMARY

In Chapter 4 we learned that correlation analysis is used to determine whether a linear relationship exists for a population of bivariate data. If a linear relationship exists, regression analysis can be used to describe the relationship. In this chapter we have examined correlation and regression from an inference-making point of view. We saw how inferences about the population regression equation can be formed by using a sample regression equation. To judge the accuracy of prediction, we learned how to draw inferences concerning the parameters in the linear regression model by using hypothesis testing and interval estimation. We also learned how to construct prediction intervals for a single value of y for a given value of x and confidence intervals for the average value of y for a given value of x. The coefficient of determination is useful for interpreting

the correlation coefficient. It is used to determine the strength of the linear relationship between the two variables. And it suggests the percentage of the variance in the dependent variable that is accounted for by the variance in the independent variable. Finally, we learned that Spearman's correlation coefficient can be used to determine if ordinal data form a monotone relationship.

C H A P T E R   R E V I E W

IMPORTANT TERMS

For each of the following terms, provide a definition in your own words. Then check your responses against the definitions given in the chapter.

coefficient of
  determination

computational formula
  for computing SSR

confidence interval bond

correlation analysis

deterministic component

deterministic model

error term

error variance

least squares prediction
  equation

linear regression model

mean square

monotone decreasing

monotone increasing

multiple correlation
  coefficient R^2

population regression
  equation

probabilistic model

residual

residual variance

slope β_1 of regression
  line equation

Spearman's rank
  correlation coefficient

standard error of
  estimate

sum of squares for
  regression

IMPORTANT SYMBOLS

σ^2, error variance

β_0, y-intercept of linear
  model

β_1, slope of linear
  model

ε, random error

$E(y|x)$, average value
  of y given x for the
  linear model

SSR, sum of squares
  for regression

SSE, sum of squares for
  error

MSR, mean square for
  regression

MSE, mean square for
  error

$y(x)$, value for y given x
  for the linear model

$\hat{y}$, predicted value for y

$\hat{y}(x)$, predicted value for
  y at x

$y - \bar{y}$, deviation due
  to the mean

$\hat{y} - \bar{y}$, component of
  deviation due to
  regression

$y - \hat{y}$, component of
  deviation due to
  error

s_e^2, residual variance

r^2, coefficient of
  determination

r, linear correlation
  coefficient

r_s, Spearman's
  correlation coefficient

ρ, population correlation
  coefficient

ρ_s, Spearman's rank
  population correlation
  coefficient

b_0, y-intercept of sample
  regression line

b_1, slope of sample
  regression line

IMPORTANT FACTS AND FORMULAS

1. $SS_y = SSR + SSE$

2. $SSE = \sum(y - \hat{y})^2$, sum of squares for error

3. $r = \dfrac{SS_{xy}}{\sqrt{SS_x SS_y}}$, correlation coefficient

4. $b_1 = r\left(\dfrac{s_y}{s_x}\right)$, slope of the sample regression equation

5. $y = \beta_0 + \beta_1 x + \varepsilon$, linear regression model

6. $E(y|x) = \beta_0 + \beta_1 x$, population regression equation

7. $E(\varepsilon) = 0$

8. $\sigma_\varepsilon^2 = \sigma^2$

9. $\hat{y} = b_0 + b_1 x$, sample regression equation

10. $SSR = bSS_{xy}$, sum of squares for regression
$SSR = \sum(\hat{y} - \bar{y})^2$

11. $r^2 = \dfrac{SSR}{SS_y}$, coefficient of determination

12. $MSR = SSR$

13. $MSE = \dfrac{SSE}{n - 2}$

$MSE = s_e^2$

14. $s_e^2 = \dfrac{SSE}{n - 2}$, estimate of σ^2, the error variance

15. Limits for confidence interval estimate of $E(y|x)$:

$$\hat{y}(x) \pm t_{\alpha/2}(n - 2)s_e \sqrt{\frac{1}{n} + \frac{(x_0 - \bar{x})^2}{SS_x}}$$

16. $e = y - \hat{y}$, prediction error or residual

17. Limits for prediction interval for y:

$$\hat{y} \pm t_{\alpha/2}(n - 2)s_e \sqrt{1 + \frac{1}{n} + \frac{(x_0 - \bar{x})^2}{SS_x}}$$

18. Limits for interval estimate of β_1:

$$b_1 \pm t_{\alpha/2}(n - 2)\frac{s_e}{\sqrt{SS_x}}$$

19. $r_s = \dfrac{1 - 6\sum d^2}{n(n^2 - 1)}$, Spearman's correlation coefficient

20. $F = \dfrac{SSR}{s_e^2}$, used to test whether the linear model is appropriate

21. $t = \dfrac{b_1}{s_e/\sqrt{SS_x}}$, used to test whether the linear model is appropriate

22. $t = \dfrac{r\sqrt{n - 2}}{\sqrt{1 - r^2}}$, t value for r

23. $z = r_s\sqrt{n - 1}$, z-value for r_s

REVIEW PROBLEMS

1. Over a 6-month period, a large department store compared the number of salespeople available with gross sales receipts. A sample of six pairs of data yielded $r = .47$. Assuming that gross sales and number of salespeople are approximately normal, test at $\alpha = .05$ to determine if $\rho \neq 0$.

2. Two judges ranked each of ten students who had entered a project in a mathematics fair. The rankings are given in the accompanying table.

Student	1	2	3	4	5	6	7	8	9	10
Judge A	6	3	8	1	4	7	2	10	5	9
Judge B	4	1	10	8	3	5	2	9	6	7

a. Compute r_s.

b. At $\alpha = .05$ test the null hypothesis $H_0: \rho_s = 0$ against the alternative hypothesis $H_1: \rho_s \neq 0$.

3. The data listed in the table represent SAT math scores and scores made in an introductory statistics course for a sample of ten college students.

SAT math score	476	525	619	515	475
Statistics score	77	90	69	88	72

SAT math score	379	517	415	405	616
Statistics score	62	92	68	73	95

a. Compute the rank correlation coefficient r_s.

b. Test to determine if $\rho_s > 0$ using $\alpha = .05$.

4. A study of 15 cars revealed a correlation of .48 between engine size and average gasoline mileage. Assuming engine sizes and average gasoline mileages are normally distributed, use the .05 level of significance to test whether ρ is different from 0.

5. In an attempt to determine the relationship between attendance and concession sales at football games, a high school principal obtained the data given here for five randomly chosen football games during the previous three seasons.

Attendance, x (in hundreds of people)	Sales, y (in hundreds of dollars)
5	15
13	36
10	20
21	64
11	32

a. Using $\alpha = .01$, determine if the linear model is appropriate.

b. Construct a 99% confidence interval for β_1.

6. Refer to the data in Problem 5.

a. Calculate r_s for the data.

b. Test using $\alpha = .01$ to determine if $\rho_s > 0$.

7. The accompanying data indicate the disposable income and the amount of money spent on food per week for families of four.

Disposable income, x (in thousands of dollars)	Food cost per week, y (in dollars)
30	106
36	109
27	81
20	77
25	83
24	50

a. By using $\alpha = .05$, determine if $\beta_1 > 0$.

b. Construct a 95% confidence interval for the average food cost per week for a family of four with a disposable income of $25,500.

c. For each additional $1000 of disposable income, approximate the increase in food costs per week.

8. A new medication has been developed to lower diastolic blood pressure. It was administered to six patients over a 2-week period, with the resulting data on diastolic blood pressure shown in the table.

Before treatment, x	107	120	92	127	114	105
After treatment, y	96	120	70	117	109	90

a. If a patient has a diastolic blood-presure reading of 100 before drug treatment, what is her predicted diastolic reading after treatment?

b. Construct a 95% prediction interval for the patient in part (a).

c. Construct a 95% confidence interval for the average diastolic reading after treatment for patients who had a reading of 100 before treatment.

9. To determine the effectiveness of a particular diet, the accompanying data were recorded for a random sample of eight dieters.

Number of weeks, x, on diet	Weight loss, y (in pounds)
1	8
4	22
3	17
2	15
3	19
1	10
2	11
4	20

a. Calculate r.

b. Calculate r_s.

c. Using $\alpha = .05$, determine if $\rho > 0$.

d. Using $\alpha = .05$, determine if $\rho_s > 0$.

e. Using $\alpha = .05$, determine if the linear model is appropriate.

f. Suppose a person who initially weighs 290 pounds and is 70 inches tall is on the diet for 1 year. What do you predict his weight to be at the end of 1 year of dieting? Explain.

10. A credit bureau compiled the tabled data relating annual family income to savings for six families.

Annual income, x (in thousands of dollars)	Annual savings, y (in hundreds of dollars)
12	6
15	11
9	2
22	24
16	12
36	36

a. Using $\alpha = .01$, determine if the linear model is appropriate.
b. For each increase of $1000 of income, what is the expected increase in savings?
c. How much in savings would you predict for a family whose annual income is $20,000?
d. Construct a 95% confidence interval for the average savings for a family having an income of $17,000.
e. Answer part (c) by constructing a 95% prediction interval for $y(20)$.

CHAPTER ACHIEVEMENT TEST

(20 points) **1.** Consider the following two sets of ratings for five brands of beer:

Brand	Rating 1	Rating 2
A	2	1
B	3	2
C	1	3
D	5	4
E	4	5

a. Find r_s.
b. Find μ_{r_s}, assuming $\rho_s = 0$.
c. Find σ_{r_s}.
d. Find the z value for r_s.
e. At $\alpha = .05$, test $H_0: \rho_s = 0$ against $H_1: \rho_s \neq 0$.

(20 points) **2.** A random sample of 20 salespeople took a test to measure assertiveness. The score of each salesperson was paired with his or her last 6-month gross sales figure. The correlation coefficient was found to be $r = .72$. Assuming test scores and gross sales are normally distributed, test at $\alpha = .05$ to determine if $\rho \neq 0$. In addition, provide the following information:
a. Value of the test statistic
b. Critical value(s)
c. Decision

(25 points) **3.** For the following data,

x	1	2	3	2
y	3	5	6	4

determine if the linear model is appropriate. Use $\alpha = .05$ and provide the following information:
a. Value of the t statistic
b. Critical value(s)

    c. Null hypothesis

    d. Decision

    e. Regression equation

(20 points)   **4.** Refer to the data in Problem 3.

    a. Determine a point estimate for b_0.

    b. Determine a point estimate for $E(y|3)$.

    c. Determine a point estimate for σ^2.

    d. Determine r^2 and interpret your result.

(15 points)   **5.** For the data in Problem 3, construct a 95% confidence interval for $E(y|2)$.

NONPARAMETRIC TESTS

Chapter Objectives

In this chapter you will learn:

- *what nonparametric methods are and under what circumstances they are usually used*

- *the advantages of using nonparametric tests*

- *the disadvantages of using nonparametric tests*

- *how to use and interpret the sign test*

- *how to use and interpret the rank-sum test*

- *how to use and interpret the Kruskal–Wallis test*

- *how to use and interpret the runs test*

Chapter Contents

Nonparametric procedures are used, for the most part, when assumptions underlying other tests cannot be satisfied and a data analysis is desired. Usually nonparametric methods assume no knowledge whatsoever about the shape of the sampled population. For this reason, they are sometimes referred to as **distribution-free methods**. Because the t and F tests require sampling from normal distributions, they are not distribution-free methods, but have traditionally been referred to as **parametric methods**.

As with parametric methods, nonparametric methods have many advantages and disadvantages. The chief disadvantages are that they are usually less efficient and not very sensitive for detecting real differences when they exist, particularly when sampling is from normal populations. They are less efficient because they waste information contained in a sample. For example, a few methods (such as Spearman's rank correlation coefficient) replace measurements with their corresponding ranks. Not being sensitive to real differences usually results in a higher probability for Type II errors and a reduction in statistical power. To offset some of this loss of power, larger samples frequently need to be collected.

In addition to being distribution-free, nonparametric tests enjoy certain advantages not enjoyed by their parametric counterparts. They are, on the whole, very quick and easy to carry out. Second, perhaps the most important advantage is that they can be used with qualitative data, such as ranks or categorical data. Third, nonparametric tests require fewer restrictive assumptions than their parametric counterparts. Usually nonparametric tests only require sampling from continuous populations that are symmetrical. Generally speaking, nonparametric tests perform nearly as well as their parametric counterparts on normal distributions, and often perform better than parametric methods on nonnormal distributions.

In this chapter we will study four nonparametric tests: the sign test, the rank-sum test, the Kruskal–Wallis test, and the runs test. All are founded on the basis of probability theory. We have previously studied two nonparametric tests, the chi-square test for analyzing categorical data and Spearman's rank correlation coefficient.

15.1 THE SIGN TEST (LARGE SAMPLES)

The **sign test** is a one-sample test that can be used to test a statistical hypothesis regarding the median of a continuous population. If $\tilde{\mu}$ denotes the population median and $\tilde{\mu} = 0$ is assumed to be true, then we would expect any finite random sample of values to have as many positive values as negative values, on the average. If a random sample has more positive values than negative values, there is some evidence to suggest that $\tilde{\mu} > 0$. For testing the null hypothesis $H_0: \tilde{\mu} \leq 0$ against the alternative hypothesis $H_1: \tilde{\mu} > 0$, the test statistic becomes the number of positive values contained in a sample. The null hypothesis is then rejected if a large number of positive values occur. The level of significance α determines how many positive values a sample must have in order for the null hypothesis to be rejected. If each positive value in the sample is identified as a success and each negative value as a failure, the number of positive values in a sample constitutes a binomial experiment with $p = .5$. Testing a null hypothesis of the form $H_0: \tilde{\mu} = 0$ using the sign test is equivalent to testing the null hypothesis $H_0: p = .5$, where p is the proportion of positive values (successes) in a binomial population. For sample sizes $n \geq 10$, the normal approximation to the binomial can be used to determine the critical values, while for $n < 10$, the binomial distribution should be used.

The sign test is also appropriate whenever paired data need to be analyzed to determine if two population means are equal. In this case, the null hypothesis $H_0: \mu_1 = \mu_2$ is equivalent to the null hypothesis $H_0: \mu_d = 0$, where d represents the distribution of difference scores. Each data pair in the sample is replaced by a "+" or "−" sign. The resulting sample of signs is used in the analysis. The test statistic becomes the number of "+" signs, and the statistic has a sampling distribution that is binomial. The normal approximation to the binomial is used to determine critical values whenever $n \geq 10$.

Recall that if x is the number of successes resulting from a binomial experiment, then

$$\mu_x = np$$

and

$$\sigma_x = \sqrt{np(1 - p)}$$

In addition, if $np \geq 5$ and $n(1 - p) \geq 5$, then a normal distribution can be used to approximate the binomial. The z value for x is

$$z = \frac{x - np}{\sqrt{np(1 - p)}} \tag{15-1}$$

If we assume that $H_0: p = .5$ is true, then (15-1) becomes

$$z = \frac{x - (n/2)}{\sqrt{n/4}}$$
$$= \frac{x - (n/2)}{\sqrt{n}/2}$$
$$= \frac{2x - n}{\sqrt{n}}$$

Thus, the test statistic for the sign test becomes

$$z = \frac{2x - n}{\sqrt{n}} \tag{15-2}$$

and (15-2) can be used whenever $np \geq 5$ and $n(1 - p) \geq 5$, or whenever $n(\frac{1}{2}) \geq 5$ since $p = .5$. This means that (15-2) can be used whenever $n \geq 10$.

As an illustration, the data in Table 15-1 represent the weights of 15 men who have been on a weight-reduction diet for 1 month. Let's use the sign test

Table 15-1 Diet Information

Male	Weight before	Weight after	Sign
1	210	195	+
2	175	162	+
3	187	179	+
4	189	185	+
5	198	199	−
6	205	200	+
7	198	193	+
8	178	178	0
9	164	162	+
10	176	169	+
11	192	188	+
12	187	184	+
13	210	198	+
14	178	172	+
15	205	206	−

and $\alpha = .01$ to test the hypothesis that the diet is effective in weight reduction. Suppose μ_B and μ_A represent the means of the population of before-diet weights and after-diet weights, respectively.

1. The null hypothesis is H_0: $\mu_B - \mu_A \leq 0$.
2. The alternative hypothesis is H_1: $\mu_B - \mu_A > 0$ (one-tailed test).
3. The level of significance is $\alpha = .01$.
4. The sampling distribution is binomial.

Note that the alternative hypothesis specifies the order in which we subtract. We let the random variable X denote the number of plus signs, since we are interested in positive "before-minus-after" differences (weight loss). The signs of the differences are recorded in the last column of Table 15-1. Since we are interested only in plus or minus signs (binomial experiment), we exclude the data for male number 8 from the analysis.

The number of signs is $n = 14$. The number of plus signs is $x = 12$ and hence the proportion of plus signs is

$$\hat{p} = \frac{x}{n} = \frac{12}{14} = .86$$

The value $\hat{p} = .86$ is an estimate of the true proportion of plus signs. If H_0: $\mu_B - \mu_A = 0$ is true, then the true proportion of plus signs in the population of signs is .5. Since we will consider only the resulting sample of 14 signs, the original hypotheses can be restated as:

H_0: $p \leq .5$

H_1: $p > .5$

If the diet is effective, then there will be more plus signs than minus signs and $p > .5$. Since $n = 14 > 10$, a normal distribution can be used. The z value for

$x = 12$ is

$$z = \frac{2(12) - 14}{\sqrt{14}} = 2.67$$

5. The critical value is $z_{.01} = 2.33$.

6. Decision: Since $z = 2.67 > z_{.01} = 2.33$, we reject $H_0: \mu_B - \mu_A = 0$ and conclude that the diet is effective in weight reduction.
7. Type of error possible: Type I; $\alpha = .01$.
8. p-value: p-value $= .5 - .4962 = .0038$.

The sign test also has the advantage of being applicable to data that are dichotomous, such as yes/no data. The following example illustrates the procedure for dichotomous data.

EXAMPLE 15-1 In a study to determine beer drinkers' preference for two new brands of beer, A and B, a random sample of 20 beer drinkers indicated 4 preferred brand A and 16 preferred brand B. Use the sign test at $\alpha = .05$ to test the hypothesis that brand B is preferred over brand A.

Solution Let p represent the proportion of all beer drinkers who prefer brand B. We proceed as follows:

1. The null hypothesis is $H_0: p \le .5$.
2. The alternative hypothesis is $H_1: p > .5$.
3. The level of significance is $\alpha = .05$.
4. The sampling distribution is binomial.

If x represents the number of beer drinkers who prefer brand B, then $x = 16$ and $n = 20$. Since $n = 20$, a normal distribution can be used to approximate the binomial distribution. By using (15-2), we obtain the value of the test statistic:

$$z = \frac{2x - n}{\sqrt{n}} = \frac{32 - 20}{\sqrt{20}} = 2.68$$

5. The critical value is $z_{.05} = 1.65$.
6. Decision: Since $z = 2.68 > z_{.05} = 1.65$, reject H_0 and conclude that brand B is preferred over brand A.
7. Type of error possible: Type I; $\alpha = .05$.
8. p-value: p-value $= .5 - .4963 = .0037$. ∎

The sign test can also be used to test null hypotheses of the form $H_0: \tilde{\mu} = \tilde{\mu}_0$, where $\tilde{\mu}$ is the population median and μ_0 is a fixed value. The following example illustrates the procedure.

EXAMPLE 15-2 The following data represent the number of minutes patients had to wait to see Doctor John on previous office visits:

22	30	31	40	37	25	29	14	30	17
23	32	20	40	28	26	33	25	34	21

Use the sign test with $\alpha = .05$ to test the doctor's claim that a patient will have to wait no more than 25 minutes before being seen by him.

Solution   Let $\tilde{\mu}$ represent the median amount of time that a patient waits to see the doctor.

1. H_0: $\tilde{\mu} \le 25$
2. H_1: $\tilde{\mu} > 25$ (one-tailed test)
3. $\alpha = .05$
4. Sampling distribution: Binomial.

A plus sign corresponds to a time greater than 25 minutes. If x represents the number of patients who have to wait longer than 25 minutes to see the doctor, then $x = 12$ and $n = 18$ (we discard the two measurements of 25). By using (15-2) we obtain the value of the test statistic:

$$z = \frac{2x - n}{\sqrt{n}} = \frac{24 - 18}{\sqrt{18}} = 1.41$$

5. Critical value: $z_{.05} = 1.65$.
6. Decision: We cannot reject H_0, since $z = 1.41 < z_{.05} = 1.65$. Hence, there is no statistical evidence to suggest that patients have to wait longer than 25 minutes to see the doctor.
7. Type of error possible: Type II, β is unknown.
8. p-value $= .5 - .4207 = .0793$.   ∎

Problem Set 15.1

A

1. The accompanying data were collected on 16 engineering students at the end of the semester at a certain college to determine if they did better in mathematics than physics. Use the sign test at $\alpha = .05$ to determine if the median math grade of engineering students is higher than the median physics grade.

Student	Math	Physics
1	97	92
2	78	80
3	85	82
4	84	83
5	92	96
6	33	32
7	62	65
8	80	72
9	80	80
10	82	92
11	96	89
12	85	80
13	50	74
14	100	99
15	82	72
16	80	64

2. The following nicotine contents (in milligrams) of 15 brand A cigarettes were obtained: 2.1, 4.0, 6.3, 5.4, 4.8, 3.7, 6.1, 2.3, 3.3, 6.4, 5.4, 2.5, 3.1, 4.1, and 4.7. At $\alpha = .05$, use the sign test to determine if the data indicate that $\tilde{\mu} < 4.5$.

3. Of 50 people randomly interviewed concerning a particular issue, 33 were in favor and 17 were against the issue. Use the sign test at the 5% level of significance to determine if the proportion of people in the population in favor of the issue is different from the proportion against the issue.

4. A random sample of teachers at a certain university yielded the following ages: 26, 49, 55, 55, 60, 61, 68, 57, 74, 75, 68, 68, 68, 67, 64, 64, and 63. Use the sign test at $\alpha = .05$ to determine if the median age is different from 64 years.

5. Two different brands of fertilizer were used to test the effects on peach tree yields. One season 16 peach trees were treated with fertilizer A, and the following season they were treated with fertilizer B. The number of good peaches (in bushels) for each tree is recorded in the table. Use the sign test at $\alpha = .05$ to determine if there is a difference in median peach tree yields for the two types of fertilizer.

Tree	First season	Second season
1	2.0	2.5
2	2.5	3.0
3	2.0	2.0
4	3.5	3.0
5	3.0	2.5
6	2.5	2.0
7	2.5	3.0
8	3.0	2.5
9	2.0	3.5
10	1.5	1.5
11	2.5	2.0
12	3.0	2.0
13	3.5	2.5
14	2.5	3.0
15	1.5	2.0
16	3.0	3.5

Person	Before	After
1	164	162
2	146	144
3	148	146
4	154	156
5	143	145
6	160	159
7	150	145
8	148	150
9	138	139
10	124	124
11	175	170
12	146	147
13	160	157
14	160	140
15	153	148

B

6. The data in the accompanying table indicate the systolic blood pressure measurements of fifteen subjects before and 1 hour following an exercise program. At $\alpha = .01$ does the exercise program help lower systolic blood pressure?

7. Answer Problem 6 if only the first six people are involved in the data analysis.

15.2 THE WILCOXON RANK-SUM TEST (LARGE SAMPLES) ————————

The **Wilcoxon rank-sum test** (also commonly known as the **Mann–Whitney U test**), a nonparametric test, is used to compare data from two continuous populations having the same shape and spread. It is the nonparametric alternative to the two-sample t test for independent samples and can be used when the normality assumption cannot be satisfied.

The rank-sum test involves the assignment of ranks to the data after the two samples have been combined and arranged in increasing order. To apply the rank-sum test, the data from the two samples are pooled together and ranked from smallest to largest. In case of ties, the tied ranks are assigned the average of the ranks had there been no ties. For example, if the fourth, fifth, and sixth observations are equal, a rank of 5 is assigned to each of the three identical observations. After we assign ranks to the data, we choose the smaller sample and find the sum of the ranks, denoted by W, for that sample. Either sample can be used to determine the sum of the ranks W if both samples are of the same size. If the sampled populations have unequal means, we would expect most of the lower ranks to be in the sample from the population with the smaller mean and we would expect most of the higher ranks to be in the sample from the population with the higher mean.

The rank-sum test is based on the test statistic U defined by

$$U = W - \frac{n_1(n_1 + 1)}{2} \tag{15-3}$$

where n_1 is the size of the smaller sample. The number $n_1(n_1 + 1)/2$ is the smallest value that W can assume, and the U statistic measures the distance between W and its smallest value. The U statistic is closely related to W; the values of U vary directly with the values of W. If the U test statistic is large, then W is large and the sample used to generate W should belong to the population with the larger mean. If we choose notation as that μ_1 is the mean of the population whose sample size is n_1, then for the value of U to be large means that $H_0: \mu_1 \leq \mu_2$ is rejected in favor of $H_1: \mu_1 > \mu_2$. And if the value for U is small, $H_0: \mu_1 \geq \mu_2$ is rejected in favor of $H_1: \mu_1 < \mu_2$.

If the samples come from continuous and identical populations and there are no ties in the ranks, then the sampling distribution of U has a mean and standard deviation given by

$$\mu_U = \frac{n_1 n_2}{2} \tag{15-4}$$

$$\sigma_U = \sqrt{\frac{n_1 n_2 (n_1 + n_2 + 1)}{12}} \tag{15-5}$$

where n_1 is the size of the smaller sample and n_2 is the size of the other sample. In addition, if both n_1 and n_2 are greater than 8, the sampling distribution of U is approximately normal. The z value for U is determined by

$$z = \frac{U - \mu_U}{\sigma_U}$$

If $n_1 \leq 8$, then special tables must be used to determine critical values for the U statistic. These tables are contained in most statistical handbooks.

The following two examples illustrate the use of the rank-sum test.

EXAMPLE 15-3   Two groups of students were taught statistics using different methods, A and B. A final examination was given to both groups at the end of the course. The results are given in Table 15-2. Use the rank-sum test at $\alpha = .05$ to determine if there is a difference in the average final examination scores for students taught by methods A and B.

Table 15-2 Data for Example 15-3

Method A	73	67	72	46	83	75	62	90	95	
Method B	71	47	68	87	77	92	65	86	79	57

Solution   The null hypothesis is $H_0: \mu_A - \mu_B = 0$ and the alternative hypothesis is $H_1: \mu_A - \mu_B \neq 0$.

The following table will be used to organize the ranking:

Data	Rank	Sample	Data	Rank	Sample
46	(1)	A	73	(10)	A
47	2	B	75	(11)	A
57	3	B	77	12	B
62	(4)	A	79	13	B
65	5	B	83	(14)	A
67	(6)	A	86	15	B
68	7	B	87	16	B
71	8	B	90	(17)	A
72	(9)	A	92	18	B
			95	(19)	A

Since sample A is the smaller sample, $n_1 = 9$ and $n_2 = 10$. The sum of the ranks for sample A (shown in parentheses in the table) is determined to be

$$W = 1 + 4 + 6 + 9 + 10 + 11 + 14 + 17 + 19 = 91$$

The U test statistic is found by using (15-3):

$$U = W - \frac{n_1(n_1 + 1)}{2} = 91 - \frac{(9)(10)}{2} = 91 - 45 = 46$$

The mean of the sampling distribution of U is found by using (15-4):

$$\mu_U = \frac{n_1 n_2}{2} = \frac{(9)(10)}{2} = 45$$

And the standard deviation of the sampling distribution of U is found by using (15-5):

$$\sigma_U = \sqrt{\frac{n_1 n_2 (n_1 + n_2 + 1)}{12}}$$
$$= \sqrt{\frac{(9)(10)(9 + 10 + 1)}{12}} = \sqrt{150} = 12.25$$

Since $n_1 > 8$ and $n_2 > 8$, the sampling distribution of U is approximately normal and the z statistic for U is

$$z = \frac{U - \mu_U}{\sigma_U} = \frac{46 - 45}{12.25} = .08$$

The critical values are $\pm z_{.025} = \pm 1.96$. Since $-1.96 < z = .08 < 1.96$, we cannot reject the null hypothesis. Therefore, there is no significant difference between the average final examination scores for students taught by the two methods.   ∎

It is interesting to note that the value $n_1(n_1 + 1)/2 = 45$ in Example 15-3 represents the smallest value that W could possibly be for sample A. This would occur if all the measurements of sample A were less than every measurement in sample B. In this case, W would become

$$W = 1 + 2 + 3 + \cdots + 9 = 45$$

the same value found by evaluating $n_1(n_1 + 1)/2$ for $n_1 = 9$. The value for U in this case would then be

$$U = W - \frac{n_1(n_1 + 1)}{2}$$

$$= 45 - 45 = 0$$

EXAMPLE 15-4   Patients with a certain disease are treated with two different drugs, and patients are evaluated in terms of complete recovery in days. A random sample of nine patients were treated with drug A and another random sample of nine patients were treated with drug B. The time in days to complete recovery is recorded for each patient in Table 15-3. Use the rank-sum test at the .05 level of significance to determine whether the mean recovery time for patients treated with drug B is less than the mean recovery time for patients treated with drug A.

Table 15-3 Data for Example 15-4

Drug A	13	10	12	14	14	15	16	16	17
Drug B	8	17	9	11	15	11	14	12	18

Solution   **1.** H_0: $\mu_B - \mu_A \geq 0$
2. H_1: $\mu_B - \mu_A < 0$
3. $\alpha = .05$
4. The following table is used for ranking the data:

Data	Rank	Sample	Data	Rank	Sample
8	(1)	B	14	10	A
9	(2)	B	14	(10)	B
10	3	A	15	12.5	A
11	(4.5)	B	15	(12.5)	B
11	(4.5)	B	16	14.5	A
12	6.5	A	16	14.5	A
12	(6.5)	B	17	16.5	A
13	8	A	17	(16.5)	B
14	10	A	18	(18)	B

Since both samples are of the same size (9), we choose sample B for finding the sum of the ranks W, since the alternative hypothesis is H_1: $\mu_B < \mu_A$ and the sum of the ranks corresponding to sample B should be less than the sum of the ranks corresponding to sample A. In this case, the left tail of the distribution should be chosen as the rejection region. [Had the size of sample A been smaller than the size of sample B, then sample A would have been chosen for finding the sum of the ranks, and the right tail of the U distribution would have been chosen as the rejection region.] The sum of the ranks corresponding to sample B (shown in parentheses in the table) is

$$W = 1 + 2 + 4.5 + 4.5 + 6.5 + 10 + 12.5 + 16.5 + 18 = 75.5$$

The value of the U statistic is found by using (15-3):

$$U = W - \frac{n_1(n_1 + 1)}{2} = 75.5 - \frac{(9)(10)}{2} = 30.5$$

The mean of the sampling distribution of U is given by (15-4):

$$\mu_U = \frac{n_1 n_2}{2} = \frac{(9)(9)}{2} = 40.5$$

The standard error of U is given by (15-5):

$$\sigma_U = \sqrt{\frac{n_1 n_2 (n_1 + n_2 + 1)}{12}}$$

$$= \sqrt{\frac{(9)(9)(19)}{12}} = \sqrt{128.25} = 11.32$$

Since $n_1 > 8$ and $n_2 > 8$, the sampling distribution of U is approximately normal and the value of the z statistic is found to be

$$z = \frac{U - \mu_U}{\sigma_U} = \frac{30.5 - 40.5}{11.32} = -.88$$

5. Critical value: $-z_{.05} = -1.65$.

6. Decision: Since $z = -.88 > -1.65$, we cannot reject H_0. Thus, there is no statistical evidence to suggest that the mean recovery time for patients treated with drug B is less than the mean recovery time for patients treated with drug A.

7. Type of error possible: Type II; β is unknown.

8. p-value $= .5 - .3106 = .1894$.   ■

It is interesting to note that had sample A been chosen in Example 15-4 to determine the sum of the ranks W, we would have had the following results:

$$W = 3 + 6.5 + 8 + 10 + 10 + 12.5 + 14.5 + 14.5 + 16.5 = 95.5$$

$$U = W - \frac{n_2(n_2 + 1)}{2} = 95.5 - \frac{(9)(10)}{2} = 50.5$$

$$z = \frac{U - \mu_U}{\sigma_U} = \frac{50.5 - 40.5}{11.32} = .88$$

The z value obtained here differs only in sign from the z value obtained in Example 15-4. In general, this relationship holds for the z values computed for the two U statistics corresponding to the two samples of data.

If W_1 denotes the sum of the ranks for sample A and W_2 denotes the sum of the ranks for sample B, then $W_1 + W_2$ can be found by using the formula

$$W_1 + W_2 = \frac{(n_1 + n_2)(n_1 + n_2 + 1)}{2}$$ (15-6)

As a consequence of (15-6), for Example 15-4 we have

$$W_1 + W_2 = \frac{(18)(19)}{2} = 171$$

Since the value of W_2 was found to be $W_2 = 75.5$, the value of W_1 could have been found easily by subtraction:

$$W_1 = 171 - W_2 = 95.5$$

which agrees with the result found above. The relationship (15-6) can be used as a check to ensure that the value of W has been calculated correctly.

Problem Set 15.2

A

1. The useful lives (in months) before failure of color cathode ray tubes made by two manufacturers, A and B, are as shown in the table. Use the rank-sum test and $\alpha = .05$ to determine if there is a difference in the average useful lifetimes of service for the tubes made by the two manufacturers.

A	42	35	50	41	45	39	47	49	40
B	52	51	46	57	44	58	54	53	43	48

2. The accompanying data represent the number of violent crimes committed in a certain city during a 9-week spring period and a 10-week fall period. Use the rank-sum test and $\alpha = .01$ to determine if there is, on the average, a difference in the weekly number of violent crimes committed for the two periods.

Spring	51	42	57	53	43	37	45	49	46
Fall	40	35	30	44	33	50	41	39	36	38

3. Suppose that the scores of trainees in a current training program are to be compared with the scores obtained from an independent sample of trainees from a previous program. The following results were obtained:

Previous: 7.5, 6.9, 6.7, 6.4, 6.2, 6.0, 5.9, 5.8, 5.6, 5.4, 5.3, 5.0, 4.9, 4.9, 4.3, 3.8, 3.0

Current: 6.7, 6.6, 6.3, 5.4, 5.2, 5.5, 4.9, 4.6, 4.5, 4.3, 3.9, 3.7, 3.2, 2.9, 2.3

By using the rank-sum test, determine whether the previous program produced higher average scores than the current program. Use $\alpha = .05$.

4. A study was made to investigate the effects of vitamin C on the common cold. A group of 20 students who had developed a cold were randomly assigned to two groups. Group A acted as the control group and received only a sugar tablet, while students in group B received 1 gram of vitamin C. The data listed in the table indicate the duration (in days) of cold symptoms for each subject. At $\alpha = .01$, test by using the rank-sum test to determine if the students in group B have a shorter average duration of cold symptoms than those in group A.

Group A	13	18	11	20	24	15	19	27	9
Group B	8	14	23	25	17	16	30	7	21	10	22

5. An algebra teacher teaches factoring by two different methods. To determine if method B is better than method A, two groups of algebra students were taught using the two different methods, one method for each group. A factoring examination was given to all students at the end of the factoring unit and the student scores for both methods are provided in the accompanying table. At the .05 level of significance, determine if the average score for students taught factoring by method B is greater than

the average score for students taught factoring by method A.

Method A	58	65	81	62	96	60	55	70	63	84
Method B	65	68	59	86	95	70	100	98	80	69

6. For Problem 4, use the *t* test at $\alpha = .01$ to determine if the students in group B have a shorter average duration of cold symptoms than students in group A. Compare your results with those obtained in Problem 4.

7. The data provided in the table represent the number of absences from a Math 101 class for two random samples of males and females.

Male	8	2	7	6	0	2	13	5	14	9	18
Female	3	20	1	11	5	4	2	11	10	3	1	13	21

a. At $\alpha = .05$ use the two-sample *t* test to determine if there is a difference in the average number of days missed for males and females.
b. At $\alpha = .05$ use the rank-sum test to determine if there is a difference in the average number of days missed for males and females.
c. Compare your results from parts (a) and (b). Which test should you use? Does it make a difference? Explain.

8. Two groups of adults were tested for critical-thinking ability. Group A had 11 women and group B consisted of 9 men. Both groups were administered the WG test to measure critical thinking. Their scores are given in the accompanying table.

Group A	72	66	39	101	90	86	48	109	118	64	73
Group B	148	97	83	75	33	67	98	133	70

a. By using the rank-sum test, determine if there is a difference in the average critical-thinking scores for the two groups. Use $\alpha = .01$ and determine the *p*-value.
b. Test to determine if the population variances are different. Use $\alpha = .10$.
c. By using the *t* test and the raw data, determine if there is a difference in the average scores for the two groups.

9. Consider the following data, which were obtained by taking random samples of size 15 from two normal populations with means 15 and 17 and standard deviations 2 and 2, respectively.

Sample A: 14.4, 12.1, 15.8, 12.0, 11.0, 16.6, 13.1, 13.8, 18.3, 13.3, 20.5, 14.2, 16.7, 11.0, 10.5

Sample B: 15.5, 16.9, 17.0, 19.5, 16.8, 18.8, 16.1, 11.9, 15.5, 15.1, 14.1, 14.6, 12.0, 18.3, 15.2

a. Use the *t* test at $\alpha = .05$ to determine if $\mu_A < \mu_B$.
b. Use the rank-sum test at $\alpha = .05$ to determine if $\mu_A < \mu_B$.
c. Compare your results in parts (a) and (b) with respect to sensitivity for detecting real differences. Would you expect the rank-sum test to be more sensitive than the *t* test? Why?

B
10. Suppose sample A of size $n_1 = 2$ is taken from population I and sample B of size $n_2 = 3$ is taken from population II. The accompanying table contains the ten possible arrangements for the ranks of the five pooled measurements, where *a* is a member of sample A and *b* is a member of sample B.

Rank					Statistic	
1	**2**	**3**	**4**	**5**	**W**	**U**
a	b	b	b	a	6	3
a	b	b	a	b		
a	a	b	b	b		
a	b	a	b	b		
b	a	b	a	b		
b	a	a	b	b		
b	b	b	a	a		
b	b	a	a	b		
b	a	b	b	a		
b	b	a	b	a		

a. For each of the ten possible arrangements, compute the values of *W* and *U* corresponding to sample A.
b. Construct a probability distribution table for *W*, the sum of the ranks for *A*.
c. Construct a probability distribution table for *U*.
d. Find μ_W and σ_W.
e. Find μ_U and σ_U. Compare these values with those obtained from the formulas contained in the text.
f. Can $\alpha = .05$ be used as a significance level for a rank-sum test with $n_1 = 2$ and $n_2 = 3$? Why? Specify the smallest level α that is possible for

such a test. [Your decision rule cannot be "never reject H_0."]

11. Refer to the ten arrangements listed in Problem 10.
    a. Compute the values of W and U corresponding to sample B.
    b. If W_1 and U_1 correspond to sample A and W_2 and U_2 correspond to sample B, how are W_1 and W_2 related? How are U_1 and U_2 related?

12. A variation of the rank-sum test involves the sampling distribution of W, instead of U. It can be shown that for $n_1 > 8$ and $n_2 > 8$, the sampling distribution of W is approximately normal, with

$$\mu_W = \frac{n_1(n_1 + n_2 + 1)}{2}$$

and

$$\sigma_W = \sqrt{\frac{n_1 n_2(n_1 + n_2 + 1)}{12}}$$

Use this result to do Problem 5.

13. By using the relationship $U = W - n_1(n_1 + 1)/2$, show that

$$\mu_W = \frac{n_1(n_1 + n_2 + 1)}{2} \qquad \text{and} \qquad \sigma_W = \sigma_U$$

14. If $U_1 = W_1 - n_1(n_1 + 1)/2$ and
$$U_2 = W_2 - n_2(n_2 + 1)/2,$$
show that the z value for U_1 is the negative of the z value for U_2.

15.3 THE KRUSKAL–WALLIS TEST

The **Kruskal–Wallis (KW) test** is a useful alternative when the normality assumption underlying a one-factor analysis of variance (ANOVA) cannot be satisfied. It is a nonparametric method useful for testing a null hypothesis H_0 that k independent samples are from identical continuous populations. It is a generalization ($k \geq 2$) of the rank-sum test studied in Section 15.2.

If the null hypothesis $H_0: \mu_1 = \mu_2 = \mu_3 = \cdots = \mu_k$ is assumed to be true and the N measurements from all the samples are ranked in ascending order, then each sample can be expected to have random ranks. If the data contain ties, ranks are assigned as in the rank-sum test. The tied values are assigned the average of the ranks occupying the tied positions. If H_0 is false, then some samples will have mostly small ranks and other samples will have mostly large ranks. If R_{ij} denotes the rank of x_{ij} and R_i and $\bar{R}_i$ denote the sum and average, respectively, of the ranks in the ith sample, then when H_0 is true the average value of R_{ij} is $(N + 1)/2$ and the average value of R_i is also $(N + 1)/2$. The KW test measures the extent to which the average ranks of the samples deviate from their average value of $(N + 1)/2$, much in the same fashion that MSB in ANOVA measures the extent to which each sample mean deviates about the grand mean. The test statistic H for the KW test is given by

$$H = \frac{12}{N(N + 1)} \cdot \sum n_i \left(\bar{R}_i - \frac{N + 1}{2} \right)^2 \tag{15-7}$$

where $N = \sum n_i = n_1 + n_2 + n_3 + \cdots + n_k$. A computational formula analogous to SSB for the test statistic H is given by

$$H = \frac{12}{N(N + 1)} \cdot \sum \left(\frac{R_i^2}{n_i} \right) - 3(N + 1) \tag{15-8}$$

If each value of n_i is greater than or equal to 5, then the sampling distribution of H is approximately a χ^2 distribution with df $= (k - 1)$. Consider the following examples.

EXAMPLE 15-5 The data in Table 15-4 represent the suicide rates (per 1000 population) for samples of cities in three sections of the United States. By using the KW test and the .05 level of significance, determine if the population mean suicide rates differ for the three sections of the country.

Table 15-4 Data for Example 15-5

Eastern	Central	Western
2.8	2.1	2.1
5.0	2.4	4.2
7.2	3.5	4.3
8.3	7.0	4.8
10.0	12.1	6.4
13.2	13.6	6.6
13.6	14.9	8.4
	15.6	8.9

Solution   **1.** We first rank the combined data from smallest to largest. The following table is used to organize the ranking:

Data	Rank	Group	Data	Rank	Group
2.1	1.5	C	7.2	13	E
2.1	1.5	W	8.3	14	E
2.4	3	C	8.4	15	W
2.8	4	E	8.9	16	W
3.5	5	C	10.0	17	E
4.2	6	W	12.1	18	C
4.3	7	W	13.2	19	E
4.8	8	W	13.6	20.5	E
5.0	9	E	13.6	20.5	C
6.4	10	W	14.9	22	C
6.6	11	W	15.6	23	C
7.0	12	C			

2. Next, we determine n_i and R_i for each area:

Eastern: $n_1 = 7, R_1 = 4 + 9 + 13 + 14 + 17 + 19 + 20.5 = 96.5$
Central: $n_2 = 8, R_2 = 1.5 + 3 + 5 + 12 + 18 + 20.5 + 22 + 23 = 105$
Western: $n_3 = 8, R_3 = 1.5 + 6 + 7 + 8 + 10 + 11 + 15 + 16 = 74.5$

The total number of observations is $N = n_1 + n_2 + n_3 = 23$.
As a check on our calculations, the following relation always holds:

$$\sum R_i = \frac{N(N + 1)}{2}$$

The sum of ranks R_i is

$$\sum R_i = 96.5 + 105 + 74.5 = 276$$

which is the same value as

$$\frac{N(N+1)}{2} = \frac{(23)(24)}{2} = 276 \quad (\text{check})$$

3. The value of the H test statistic is found by using (15-8):

$$H = \frac{12}{N(N+1)} \cdot \sum\left(\frac{R_i^2}{n_i}\right) - 3(N+1)$$

$$= \frac{12}{(23)(24)} \cdot \left[\frac{(96.5)^2}{7} + \frac{(105)^2}{8} + \frac{(74.5)^2}{8}\right] - 3(23+1) = 1.96$$

The sampling distribution of H is approximately a χ^2 distribution with df = $(k-1) = 3 - 1 = 2$. Note that n_1, n_2, and n_3 are all greater than or equal to 5.

4. The critical value is $\chi^2_{.05}(2) = 5.991$.

5. Decision: We cannot reject H_0: $\mu_1 = \mu_2 = \mu_3$. Hence, there is no statistical evidence that mean suicide rates for the three sections of the country differ.

6. Type of error possible: Type II, β is unknown.   ■

EXAMPLE 15-6   A study was done to compare the effectiveness of four different fertilizers on the growth of twenty plants of approximately the same size. Each fertilizer was used on five randomly chosen plants. The growth (in inches) was recorded for each plant after 3 weeks. The results shown in Table 15-5 were obtained. Determine using $\alpha = .05$ whether there is a significant difference between the fertilizers in terms of average growth.

Table 15-5 Data for Example 15-6

A	B	C	D
8.1	5.0	7.6	6.2
6.8	6.6	7.8	5.8
7.3	5.8	8.1	6.0
7.4	6.1	8.5	5.7
7.7	5.6	7.4	5.9

Solution   1. H_0: $\mu_A = \mu_B = \mu_C$

2. H_1:   At least two means are unequal.

3. $\alpha = .05$ (one-tailed test)

4. Sampling distribution: χ^2. In order to calculate the value of the H test statistic we shall proceed as follows.

First rank the combined data, as indicated in the following table:

Data	Rank	Group	Data	Rank	Group
5.0	1	B	6.8	11	A
5.6	2	B	7.3	12	A
5.7	3	D	7.4	13.5	A
5.8	4.5	D	7.4	13.5	C
5.8	4.5	B	7.6	15	C
5.9	6	D	7.7	16	A
6.0	7	D	7.8	17	C
6.1	8	B	8.1	18.5	C
6.2	9	D	8.1	18.5	A
6.6	10	B	8.5	20	C

The sample sizes are $n_1 = n_2 = n_3 = n_4 = 5$ and $N = 20$. The sum of the ranks for each group are $R_A = 71$, $R_B = 25.5$, $R_C = 84$, and $R_D = 29.5$. The value of the H test statistic is

$$H = \frac{12}{N(N + 1)} \cdot \sum \left(\frac{R_i^2}{n_i} \right) - 3(N + 1)$$

$$= \frac{12}{(20)(21)} \cdot \left[\frac{(71)^2}{5} + \frac{(25.5)^2}{5} + \frac{(84)^2}{5} + \frac{(29.5)^2}{5} \right] - 3(21)$$

$$= 14.81$$

5. Critical value: df $= k - 1 = 4 - 1 = 3$ and $\chi^2_{.05}(3) = 7.815$.

6. Decision: Since $H = 14.81 > 7.815$, we reject H_0 and conclude that there is a significant difference among the four fertilizers in terms of average growth. The means are judged unequal, but we do not know which means are different.

7. Type of error possible: Type I: $\alpha = .05$. ∎

Problem Set 15.3

A

In Problems 1–5 use the KW test to test the null hypothesis of equal population means. For each test use $\alpha = .05$.

1. Readability scores from Problem 1 of Problem Set 13.1:

Text A	50	49	54	57	58
Text B	48	57	61	55	60

2. Laboratory analysis results from Problem 3 of Problem Set 13.1:

Lab A   58.6   60.7   59.4   59.6   60.5

Lab B   61.6   64.8   62.8   59.2   60.4
Lab C   60.7   55.6   57.3   55.2   60.2

3. Typing scores from Problem 4 of Problem Set 13.1:

A	B	C	D
82	55	69	87
79	67	72	61
75	84	78	82
68	77	83	61
65	71	74	72

4. Test scores from Problem 5 of Problem Set 13.2:

Teacher A	Teacher B	Teacher C
75	90	17
91	80	81
83	50	55
45	93	70
82	53	61
75	87	43
68	76	89
62	58	73
47	82	73
95	98	93
38	78	58
79	64	81
	80	70
	33	
	79	

5. A test of basic reading skills was given to a random sample of 5 students in each of 4 schools. The results are as follows:

School A	School B	School C	School D
20	24	16	19
21	21	21	20
22	22	18	21
24	25	13	20
22	23	17	20

B

6. Consider the following three samples of data. The data were obtained by taking random samples of size 10 from three normal populations with means 10, 12, and 14 and standard deviations 2, 2, and 2, respectively.

A:	9.6	7.5	9.5	10.9	7.8	9.9	6.7	9.7	11.0	8.7
B:	12.9	10.8	10.5	7.8	12.2	11.4	10.4	10.5	9.0	12.7
C:	14.1	14.9	10.5	13.6	16.0	16.9	17.2	13.8	14.0	17.3

a. By using ANOVA and $\alpha = .05$, determine if the population means are unequal.

b. By using the Kruskal–Wallis test and $\alpha = .05$, determine if the population means are different.

c. Compare your results to parts (a) and (b). Would you expect the Kruskal–Wallis test to be more sensitive than ANOVA? Explain.

7. Prove that computational formula (15-8) is equivalent to formula (15-7).

15.4 A TEST FOR NONRANDOMNESS (LARGE SAMPLES) _____

A nonparametric test for determining the nonrandomness of data is the **runs test**. Suppose the following sequence consisting of M's and F's represents the successive occurrences of births of males (M) and females (F) in a particular hospital:

MMMMMMMMMM   FFFFFFFFFF   MMMMMMMMMM   FFFFFFFFFF

Would you question the randomness of occurrences? Probably so, but why? There are twenty of each symbol F and M occurring, so we have the same proportions of male births and female births. Undoubtedly, you question the order of occurrences rather than the frequency of occurrences. A random order should possess no pattern of occurrences. Thus, AABBAAABBBAAAABBBB is *probably* not a random sequence.

A **run** is defined to be a sequence of identical symbols that are followed or preceded by different symbols or none at all. For example, runs are underlined in the following sequences of two symbols:

1. $\underline{A}$ BBB $\underline{A}$ BBB $\underline{AA}$ B $\underline{A}$ B $\underline{AA}$,   9 runs

2. $\underline{+ + + +}$ $\underline{- - - -}$ $\underline{+ + +}$ $\underline{- - -}$ $\underline{+}$,   5 runs

For a sequence of five symbols of one kind and three symbols of another kind, the minimum number of runs is 2 and the maximum number of runs is 7. Why? We would not expect a sequence of two symbols to be random if it had too few or too many runs. For any random sequence of two symbols, we denote the number of runs by R, the number of one kind of symbol by n_1, and the number of the other kind of symbol by n_2.

The sampling distribution of R has a mean defined by

$$\mu_R = \frac{2n_1 n_2}{n_1 + n_2} + 1 \qquad\qquad (15\text{-}9)$$

and a standard deviation defined by

$$\sigma_R = \sqrt{\frac{2n_1 n_2 (2n_1 n_2 - n_1 - n_2)}{(n_1 + n_2)^2 (n_1 + n_2 - 1)}} \qquad\qquad (15\text{-}10)$$

If both n_1 and n_2 are greater than 10 or if either n_1 or n_2 is greater than 20, the sampling distribution of R is approximately normal. For situations where n_1 and n_2 are both less than or equal to 10, the sampling distribution of R is poorly approximated by a normal distribution and special tables must be used that contain the critical values. Such tables are provided in most handbooks of statistical tables. Consider the following practical application of the runs test.

A quality control study was conducted at a plant producing coil springs used in a certain type of camera. According to specifications, for each gram of pull, the spring should lengthen .01 mm. Since the springs are made to such exacting requirements, they are relatively expensive. Table 15-6 displays the order in which 50 springs were produced, as well as their elongations for 1 gram of force.

After examining the chart, we suspect that something is wrong, since the spring elongation measurements appear to be decreasing over time. The runs test for nonrandomness can help us verify our suspicion that the production process is in trouble or the instrument used for testing is defective. If changes in elongation are recorded "$+$" or "$-$" ("$+$" for increase and "$-$" for decrease) as the springs are produced on the assembly line, the following sequence of changes results:

$$- - + - + - + - + - + - + + - + - - + - + - + - + - - + - - + - + - -$$

There are $R = 31$ runs. The number of "$+$" symbols is $n_1 = 16$ and the number of "$-$" symbols is $n_2 = 21$.

We can test to determine if this sequence is nonrandom. The statistical hypotheses are:

H_0:   The sequence is random.

H_1:   The sequence is not random.

Table 15-6 Quality Control Chart

Order	Elongation (in .01 mm)	Order	Elongation (in .01 mm)
1	1.3	26	1.2
2	1.3	27	1.0
3	1.2	28	.9
4	1.2	29	1.0
5	1.1	30	.9
6	1.2	31	1.0
7	1.1	32	.9
8	1.3	33	1.0
9	1.1	34	1.0
10	1.1	35	.9
11	1.2	36	.9
12	1.0	37	.9
13	1.1	38	.9
14	1.0	39	1.0
15	1.0	40	.9
16	1.1	41	.8
17	1.1	42	1.0
18	1.0	43	.8
19	1.1	44	.8
20	1.1	45	.7
21	1.1	46	.9
22	1.0	47	.8
23	1.1	48	.9
24	1.2	49	.8
25	1.0	50	.6

1. The mean of the sampling distribution of R is

$$\mu_R = \frac{2n_1 n_2}{n_1 + n_2} + 1$$

$$= \frac{(2)(16)(21)}{16 + 21} + 1 = 19.16$$

2. The standard error of R is

$$\sigma = \sqrt{\frac{2n_1 n_2 (2n_1 n_2 - n_1 - n_2)}{(n_1 + n_2)^2 (n_1 + n_2 - 1)}}$$

$$= \sqrt{\frac{(2)(16)(21)(635)}{(37)^2 (36)}} = 2.94$$

3. The z value for $R = 31$ is

$$z = \frac{R - \mu_R}{\sigma_R} = \frac{31 - 19.16}{2.94} = 4.03$$

4. For a 5% level of significance, the critical values are $\pm z_{.025} = \pm 1.96$. We can reject H_0 and conclude that the sequence is not random, since

4.03 > 1.96. The sign test in Section 15.1 can be used to verify the downward trend in elongation as the springs come off the assembly line.

To aid in analyzing the data, a frequency table (Table 15-7) and line chart (Figure 15-1) were constructed. By examining the line chart, we see that it appears to be approximately symmetric and centered close to the specification point of .01 mm. However, by examining the original data (Figure 15-2), we

Table 15-7 Frequency Table for Elongations

Elongation (in .01 mm)	Frequency
.6	1
.7	1
.8	5
.9	10
1.0	13
1.1	11
1.2	6
1.3	3

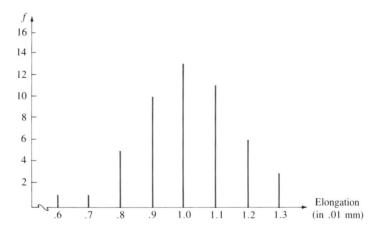

FIGURE 15-1 Frequency graph for elongations

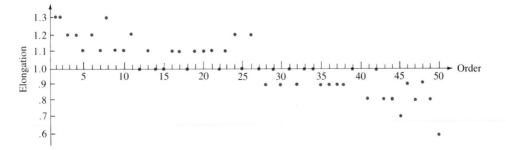

FIGURE 15-2 Original data for elongations

can detect a general drift downward as the springs are taken off the assembly line.

EXAMPLE 15-7   The runs test is sometimes used to test a sequence of stock price changes for evidence of nonrandomness. The symbol "+" denotes a price increase and "−" denotes a decrease. Test the following sequence of stock price changes for non-randomness by using the .05 level of significance:

$$- - - + + + - - + - - + - + - - - + + - - + - + - - + - + +$$

Solution   Here $R = 18$, $n_1 = 17$, and $n_2 = 13$. Thus, the mean of the sampling distribution of R is determined by using (15-9):

$$\mu_R = \frac{2n_1 n_2}{n_1 + n_2} + 1$$

$$= \frac{(2)(17)(13)}{30} + 1 = 15.73$$

The standard error of R is found by using (15-10):

$$\sigma_R = \sqrt{\frac{2n_1 n_2 (2n_1 n_2 - n_1 - n_2)}{(n_1 + n_2)^2 (n_1 + n_2 - 1)}}$$

$$= \sqrt{\frac{(2)(17)(13)[(2)(17)(13) - 17 - 13]}{(17 + 13)^2 (17 + 13 - 1)}} = 2.64$$

The z value for $R = 18$ is

$$z = \frac{R - \mu_R}{\sigma_R} = \frac{18 - 15.73}{2.64} = .86$$

To test for nonrandomness using $\alpha = .05$, we perform the following steps:

1. H_0: The sequence is random.
2. H_1: The sequence is not random.
3. $\alpha = .05$ (two-tailed test, too many runs or too few runs)
4. Sampling distribution: Distribution of the number of runs, R. The test statistic is the z value for R. In the above solution, the z value for $R = 18$ was found to be $z = .86$.
5. Critical values: $\pm z_{.025} = \pm 1.96$.
6. Decision: Since $z = .86 < 1.96 = z_{.025}$, we cannot reject H_0. Thus, the number of runs is neither too small nor too large for us to conclude that the sequence of stock prices is not random.   ∎

EXAMPLE 15-8   Last semester a 2:00 P.M. statistics class met every Monday, Wednesday, and Friday at a certain college for 22 weeks. The number of absences was recorded each Friday. The number of absences, in the order that they occurred, are written from left to right as follows:

12	1	8	7	3	10	5	7	15	12	9
18	12	17	1	7	18	6	14	5	2	11

By using $\alpha = .05$, do the data indicate that Friday absences are not random?

Solution   We will test for nonrandomness by using the median value. By ranking the data, we find the median value is 8.5. We will compare each number in the original sequence to $\tilde{x} = 8.5$ and indicate that a number is below the median by using a "b" and that a number is above the median by using an "a." The resulting sequence of a's and b's is given by

$$a\ b\ b\ b\ b\ a\ b\ b\ a\ a\ a\ a\ a\ a\ b\ b\ a\ b\ a\ b\ b\ a$$

Let n_1 denote the number of a's and n_2 denote the number of b's. Then we have $n_1 = 11$ and $n_2 = 11$. The number of runs is $R = 11$. The sampling distribution of R is approximately normal, since both n_1 and n_2 are greater than 10. The mean of the sampling distribution of R is

$$\mu_R = \frac{2n_1 n_2}{n_1 + n_2} + 1$$

$$= \frac{(2)(11)(11)}{11 + 11} + 1 = 12$$

The standard error of R is

$$\sigma_R = \sqrt{\frac{2n_1 n_2 (2n_1 n_2 - n_1 - n_2)}{(n_1 + n_2)^2 (n_1 + n_2 - 1)}}$$

$$= \sqrt{\frac{(2)(11)(11)[(2)(11)(11) - 11 - 11]}{(11 + 11)^2 (11 + 11 - 1)}}$$

$$= 2.29$$

The z value for $R = 11$ runs is

$$z = \frac{R - \mu_R}{\sigma_R} = \frac{11 - 12}{2.29} = -.44$$

To test for nonrandomness, we have:
1. H_0:   The sequence of a's and b's is random.
2. H_1:   The sequence of a's and b's is not random.
3. $\alpha = .05$ (two-tailed test)
4. Sampling distribution: Distribution of the number of runs, R. The value of the test statistic is $z = -.44$.
5. Critical values: $\pm z_{.025} = \pm 1.96$.
6. Decision: We cannot reject H_0, since $-1.96 < -.44 < 1.96$. Hence, we have no statistical evidence to suggest the sequence is nonrandom.
7. Type of error possible: Type II; β is unknown.
8. p-value: $2P(z < -.44) = 2(.33) = .66$.   ■

Note that the runs test for nonrandomness is always a nondirectional test. There can either be too many runs or too few runs to indicate nonrandom behavior of a sequence of symbols.

The runs test can be applied to numerical data as well as nominal data (such as symbols). Number sequences can be tested for nonrandomness by noting whether each number is above or below the median. Thus, number sequences generated by hand-held calculators or computers can be tested for randomness by noting the number of runs above and below 4.5, the median of the integers from 0 to 9. For example, the sequence of digits

$$1\ 9\ 7\ 6\ 5\ 4\ 3\ 7\ 4\ 1\ 7\ 8\ 6\ 5\ 3\ 2\ 1$$

has $R = 7$ runs. We could also test this sequence for nonrandomness by noting the number of runs of odd-numbered digits and even-numbered digits. For the given example, we would have $R = 11$ runs if runs were determined by using odd and even digits:

$$1\ 9\ 7\ 6\ 5\ 4\ 3\ 7\ 4\ 1\ 7\ 8\ 6\ 5\ 3\ 2\ 1$$

Problem Set 15.4

A

1. Find $R, n_1, n_2, \mu_R, \sigma_R,$ and z for each of the following sequences:
a. M M F M F F M F M F F M M M F M M F F F M F M
b. T F F T F T F F T F T T T F F F F T F T T F T T
c. $+ - + + - - - + + - + - + - - + - - + - + + -$
d. a a b b b a a a b a b b b b a a b a b b a a b a

2. In order to determine whether speeders and non-speeders occur nonrandomly on a certain stretch of highway, a state policeman monitors speeds using radar. Each time a vehicle exceeds the 55-mph speed limit, he writes F and each time a vehicle goes the speed limit or less, he writes S. He recorded the following results:

SSFFSFFFFSFSFFSSFFFFFSSFSSSFFSFFFFSFSF

Test for nonrandomness at $\alpha = .05$.

3. The daily high temperature in a Maryland town was recorded one winter for a 25-day period. Each day was classified as above normal (A) or below normal (B). The following sequence was obtained:

A A B B B A B A A A B B B B A A B
A B B B B A A A

Determine using $\alpha = .01$ if the daily high temperatures follow a pattern of nonrandomness.

4. At $\alpha = .05$ determine if the first 50 digits in the random number table (Table 8-1 in Section 8.1) are in a nonrandom order by using the median.

5. The following is a calculation of π that has been carried out to 200 decimal places:

$\pi = 3.14159$	16535	89793	23846	26433
83279	50288	41971	69399	37510
58209	74944	59230	78164	06286
20899	86280	34825	34211	70679
82148	08651	32823	06647	09384
46095	50582	23172	53594	08128
48111	74502	84102	70193	85211
05559	64462	29489	54930	38196

Test for nonrandomness of the digits with $\alpha = .05$ by:
a. Using the median
b. Letting E represent an even digit and O represent an odd digit

6. The following are down times (in minutes) during which a college's computer system was not functional during a 1-year period: 50, 47, 40, 36, 33, 45, 49, 34, 37, 45, 39, 30, 36, 44, 32, 29, 34, 25, 30, 37, 40, 37, 33, 22, 30, 41, 34, 40, 38, 39, 43, 46, 22, 29, 32, 25, 33, 34, 38, 49, 55, 52, 46, 41, 32, 43, 24, 42, 53, and 39. Use the method of runs below and above the median and the 5% level of significance to test for nonrandomness.

7. The total number of televisions (in thousands) produced by a certain firm during the years 1965–1985 was 95, 115, 140, 146, 114, 145, 137, 103, 155, 138, 117, 133, 120, 107, 105, 117, 138, 150, 120, 112, and 163. By using the 5% level of significance, determine if there is a significant trend in the data.

B

8. Would it make any sense to construct a confidence interval for R, the number of runs? Explain.

9. This problem develops the logic of the runs test.
   a. How many different arrangements (sequences) are possible for $n_1 = 3$ symbols of one kind and $n_2 = 3$ symbols of another kind? List all the possibilities.
   b. By using the list compiled in part (a), calculate the number of runs R for each sequence and construct a frequency table for the number of runs.
   c. Construct a vertical line graph for the sampling distribution of R.
   d. For the sampling distribution of R, find the mean and standard deviation using the frequency table constructed for part (b).
   e. Verify your answers to part (d) by using the formulas listed in the text for finding the mean and standard deviation of the sampling distribution of R.
   f. If the level of significance is 40%, find the critical values for the runs test, i.e., find the values of a and b such that $P(R < a \text{ or } R > b) = .4$.

CHAPTER SUMMARY

In this chapter we presented four popular nonparametric tests: the sign test, the rank-sum test, the Kruskal–Wallis test, and the runs test. These tests do not require any special assumptions about the sampled populations and commonly use ranks instead of the raw data. The sign test is the nonparametric counterpart of the two-sample t test for paired data. The rank-sum test is the nonparametric counterpart of the paired-data t test. The Kruskal–Wallis test is used instead of ANOVA when the assumptions for ANOVA cannot be met. Finally, we learned that the runs test is a test for randomness.

The nonparametric tests discussed in this textbook are summarized in Table 15-8 on page 564.

C H A P T E R   R E V I E W

IMPORTANT TERMS

For each of the following terms, provide a definition in your own words. Then check your responses against the definitions given in the chapter.

distribution-free methods

Kruskal–Wallis (KW) test

Mann-Whitney μ test

nonparametric procedures

parametric methods

Wilcoxon rank-sum test

runs test

sign test

IMPORTANT SYMBOLS

U, Wilcoxon U statistic

W, sum of ranks

r, number of plus signs

H, Kruskal–Wallis test statistic

R_{ij}, rank associated with x_{ij}

$\bar{R}_i$, average of the ranks in the ith sample

R_i, sum of the ranks in the ith sample

R, number of runs

Table 15-8 Summary of Nonparametric Tests Covered in Text

Test	Assumptions	Sampling distribution	Test statistic	Parametric counterpart
Spearman's rank correlation	$n \geq 10$	z	$z = r_s \sqrt{n-1}$	t test for ρ
Chi-square **a.** independence **b.** goodness-of-fit **c.** multinomial parameters **d.** populattion proportions **e.** homogeneity	**a.** at least 80% of E's ≥ 5 **b.** no E's < 1	χ^2	$\chi^2 = \sum \dfrac{(O_{ij} - E_{ij})^2}{E_{ij}}$	None
Runs test	$n_1 > 10,\ n_2 > 10$	z	$z = \dfrac{R - \mu_R}{\sigma_R}$ $\mu_R = \dfrac{2n_1 n_2}{n_1 + n_2} + 1$ $\sigma_R = \sqrt{\dfrac{2n_1 n_2 (2n_1 n_2 - n_1 - n_2)}{(n_1 + n_2)^2 (n_1 + n_2 - 1)}}$	None
Rank-sum	**a.** independent samples **b.** populations have same continuous distributions **c.** $n_1 > 8,\ n_2 > 8$	z	$z = \dfrac{U - \mu_U}{\sigma_U}$ $U = W - \dfrac{n_1(n_1 + 1)}{2}$ $\mu_U = \dfrac{n_1 n_2}{2}$ $\sigma_U = \sqrt{\dfrac{n_1 n_2 (n_1 + n_2 + 1)}{12}}$	t test (independent samples)
Sign test	$n \geq 10$	z	$z = \dfrac{2r - n}{\sqrt{n}}$	t test (dependent samples)
Kruskal–Wallis	**a.** independent samples **b.** populations have same continuous distributions **c.** $n_i \geq 5$	χ^2	$H = \dfrac{12}{N(N+1)} \cdot \sum \left(\dfrac{R_i^2}{n_i}\right) - 3(N+1)$	ANOVA

IMPORTANT FACTS AND FORMULAS

1. $z = \dfrac{2r - n}{\sqrt{n}}$, the z value for r

2. $U = W - \dfrac{n_1(n_1 + 1)}{2}$, the U statistic

3. $\mu_U = \dfrac{n_1 n_2}{2}$, the mean of U

4. $\sigma_U = \sqrt{\dfrac{n_1 n_2 (n_1 + n_2 + 1)}{12}}$, the standard deviation of U

5. $H = \dfrac{12}{N(N + 1)} \cdot \sum n_i \left(\bar{R}_i - \dfrac{N + 1}{2} \right)^2$, the H statistic

6. $H = \dfrac{12}{N(N + 1)} \cdot \sum \left(\dfrac{R_i^2}{n_i} \right) - 3(N + 1)$, computational formula

7. $\mu_R = \dfrac{2n_1 n_2}{n_1 + n_2} + 1$, the mean of R

8. $\sigma_R = \sqrt{\dfrac{2n_1 n_2 (2n_1 n_2 - n_1 - n_2)}{(n_1 + n_2)^2 (n_1 + n_2 - 1)}}$, the standard deviation of R

REVIEW PROBLEMS

1. Two different types of brake lining were tested for difference in wear. Two independent samples were tested with the results listed in the table (in thousands of miles). Use the rank-sum test to test for a difference in average brake lining wear at the .05 level of significance.

Brand A	42	58	64	40	47	50	62
Brand B	48	40	30	44	54	38	32

Brand A	54	42	38	66	52
Brand B	42	40	52	50	34

2. In order to evaluate two methods of teaching German, students were randomly assigned to two groups. Students in one group were taught by method A and students in the other group were taught by method B. At the end of the instructional units, both groups were given the same achievement test. The scores for the two groups are given in the accompanying table. At the .05 level of significance determine if method B produces higher average results than method A using the rank-sum test.

Group A	84	86	91	93	84	88	69	74	81	82
Group B	90	88	92	94	84	85	92	89	80	

3. Eleven workers performed a task using two different methods. The completion times (in minutes) for each task are given in the table. Use the sign test to test for a difference in average completion times for the two methods. Use $\alpha = .01$.

Worker	Method A	Method B
1	15.2	14.5
2	14.6	14.8
3	14.2	13.8
4	15.6	15.6
5	14.9	15.3
6	15.2	14.3
7	15.6	15.5
8	15.0	15.0
9	16.2	15.6
10	15.7	15.2
11	15.6	14.8

4. An auto manufacturing plant hopes to institute a new employee incentive plan. To evaluate the new plan, five employees will be under the incentive plan for an experimental period. Their work outputs before and after implementing the new plan are recorded in the table. At $\alpha = .05$, use the sign test to determine if the new incentive plan results in greater average output.

Employee	Output before	Output after
A	20	23
B	17	19
C	23	24
D	20	23
E	21	23

5. The military conducted a study to determine the accuracy of two types of rifles. A random sample

of equally proficient soldiers was chosen to test the accuracy of the rifles. The soldiers were divided into two groups; each group shot only one type of rifle. The results are provided in the table. (A high score indicates a more accurate rifle.) Use the rank-sum test at the .01 level of significance to determine if, on the average, rifle B is more accurate than rifle A.

Rifle A	88	84	88	90	86	92	88	89	87
Rifle B	90	94	92	90	88	86	91	87	90

6. The following sequence represents the order in which the last 30 babies were born at a local hospital (the symbol M represents a male and the symbol F represents a female):

F F M F M M F F M F M M M F M
F F M M F M F M M F M M F F M

Test at $\alpha = .05$ to determine if the births occur in nonrandom order.

7. For Problem 2, use the Kruskal–Wallis test to determine if methods A and B produce different average results. Use $\alpha = .05$.

8. The following sequence of M's and P's shows the order in which cars passed the Mason–Dixon Line on U.S. Route 40. A car with a Maryland license plate is denoted by M and a car with a Pennsylvania license plate is denoted by P.

M P P M P M M P M M M P P M P
M P P M P M M P M

Test for nonrandomness using $\alpha = .05$.

9. The accompanying table shows the miles per gallon of gasoline a driver obtained with eighteen tanks of three different kinds of gasoline. Use the Kruskal–Wallis test to determine if there is a difference in the true average mileage-per-gallon ratings for the three kinds of gasoline. Use $\alpha = .01$.

Gasoline A	Gasoline B	Gasoline C
27	22	16
15	15	27
29	19	19
24	29	29
21	20	21
26	18	
28		

CHAPTER ACHIEVEMENT TEST

(24 points)   **1.** For the following sequence of symbols, let n_1 denote the number of A's and n_2 denote the number of B's; test at the .05 level of significance for nonrandomness. In addition, provide the following information:

A B A B B A B B A B B A A A B A B A B B A A B A A

a. R **b.** μ_R **c.** σ_R **d.** z value for R **e.** Critical value   **f.** Decision

(32 points)   **2.** Consider the following paired data:

Pair	1	2	3	4	5	6	7	8	9	10	11
Sample 1	11.2	10.6	10.2	11.6	10.9	11.2	11.6	11.0	12.2	11.7	11.6
Sample 2	10.5	10.8	9.8	11.1	11.3	10.3	11.5	11.0	11.6	11.2	10.8

Use the sign test to test $H_0: \mu_1 - \mu_2 = 0$ against $H_1: \mu_1 - \mu_2 \neq 0$ at the .01 level of significance and provide the following information:
a. Find the value of the test statistic.
b. Find the critical values.
c. Find the p-value for the test.
d. Construct a 95% confidence interval for $(\mu_1 - \mu_2)$.

(20 points)   **3.** In order to evaluate two methods of teaching statistics, students were randomly assigned to two groups. Students in one group were taught by method A and students

in the other group were taught by method B. At the end of instruction, both groups were given the same achievement test. The scores for the two groups follow:

Group A: 89 91 96 98 89 93 87 90 94
Group B: 95 93 97 99 89 90 97 88 92 98

At the .01 level of significance use the rank-sum test to determine if method B produces higher average results than method A.

(24 points) **4.** For Problem 3, use the Kruskal–Wallis test to determine if methods A and B produce different average results. Use $\alpha = .05$.

SUMMATION NOTATION AND RULES

Throughout the text the abbreviated notation $\sum X$ is used to mean the sum of the X measurements. This abbreviated notation does not make it explicitly clear what or how many measurements are being added. To take care of this problem, we use the more formal summation notation:

$$\sum_{i=1}^{n} X_i = X_1 + X_2 + X_3 + \cdots + X_n$$

where i is called the **index of summation** and 1 and n are called the **limits of summation**.

Suppose X_i and Y_i have the values shown in the table and consider the following examples:

i	1	2	3	4	5
X_i	2	6	8	3	4
Y_i	1	3	4	0	2

a. $\displaystyle\sum_{i=1}^{4} X_i = X_1 + X_2 + X_3 + X_4$

$\qquad = 2 + 6 + 8 + 3 = 19$

b. $\displaystyle\sum_{i=2}^{5} X_i = X_2 + X_3 + X_4 + X_5$

$\qquad = 6 + 8 + 3 + 4 = 21$

c. $\displaystyle\sum_{i=1}^{3} 3X_i = 3X_1 + 3X_2 + 3X_3$

$\qquad = 6 + 18 + 24 = 48$

d. $\displaystyle\sum_{i=2}^{3} X_i^2 = X_2^2 + X_3^2$

$\qquad = 6^2 + 8^2 = 36 + 64 = 100$

e. $\displaystyle\sum_{i=1}^{3} (X_i^2 + Y_i^2) = (X_1^2 + Y_1^2) + (X_2^2 + Y_2^2) + (X_3^2 + Y_3^2)$

$\qquad = (2^2 + 1^2) + (6^2 + 3^2) + (8^2 + 4^2)$

$\qquad = (4 + 1) + (36 + 9) + (64 + 16) = 5 + 45 + 80 = 130$

f. $\displaystyle\sum_{i=2}^{4} X_i Y_i^2 = X_2 Y_2^2 + X_3 Y_3^2 + X_4 Y_4^2$

$$= (6)(3^2) + (8)(4^2) + (3)(0^2)$$

$$= (6)(9) + (8)(16) + (3)(0) = 54 + 128 + 0 = 182$$

g. $\displaystyle\sum_{i=2}^{3} (2X_i + 3Y_i) = (2X_1 + 3Y_1) + (2X_2 + 3Y_2)$

$$= [(2)(6) + (3)(3)] + [(2)(8) + (3)(4)]$$

$$= (12 + 9) + (16 + 12) = 21 + 28 = 49$$

h. $\displaystyle\sum_{i=1}^{3} X_{i+1} = X_{1+1} + X_{2+1} + X_{3+1}$

$$= X_2 + X_3 + X_4 = 6 + 8 + 3 = 17$$

There are two basic rules for summations. They are as follows:

Rule 1: $\displaystyle\sum_{i=1}^{n} (X_i + Y_i) = \sum_{i=1}^{n} X_i + \sum_{i=1}^{n} Y_i$

Rule 2: $\displaystyle\sum_{i=1}^{n} KX_i = K \sum_{i=1}^{n} X_i,$ where K is a constant.

Rule 1 can be verified by noting that

$$\sum_{i=1}^{n} (X_i + Y_i) = (X_1 + Y_1) + (X_2 + Y_2) + (X_3 + Y_3) + \cdots + (X_n + Y_n)$$

$$= (X_1 + X_2 + X_3 + \cdots + X_n) + (Y_1 + Y_2 + Y_3 + \cdots + Y_n)$$

$$= \sum_{i=1}^{n} X_i + \sum_{i=1}^{n} Y_i$$

Similarly, Rule 2 can be verified by noting that

$$\sum_{i=1}^{n} KX_i = KX_1 + KX_2 + KX_3 + \cdots + KX_n$$

$$= K(X_1 + X_2 + X_3 + \cdots + X_n)$$

$$= K \sum_{i=1}^{n} X_i$$

Suppose for each value of the index i, $X_i = 1$. Then as a consequence of Rule 2, we have

$$\sum_{i=1}^{n} K = \sum_{i=1}^{n} (K)(1) = \sum_{i=1}^{n} (KX_i) = K \sum_{i=1}^{n} X_i$$

$$= K(1 + 1 + \cdots + 1) = (K)(n)$$
$$\uparrow$$
$$n \text{ 1s}$$

As a result, we have the following rule:

Rule 3: $\displaystyle\sum_{i=1}^{n} K = nK$

Frequently, we have occasion to use double summation notation. In such expressions we could first expand the expression on the index j. For example, to expand the expression

$$\sum_{i=1}^{3} \sum_{j=1}^{2} X_{ij}$$

we first evaluate the expression $\sum_{j=1}^{2} X_{ij}$:

$$\sum_{i=1}^{3} \sum_{j=1}^{2} X_{ij} = \sum_{i=1}^{3} (X_{i1} + X_{i2}) = (X_{11} + X_{12}) + (X_{21} + X_{22}) + (X_{31} + X_{32})$$

Since addition is a commutative operation, we could have expanded the above expression on the index j first, as shown here:

$$\sum_{i=1}^{3} \left(\sum_{j=1}^{2} \right) X_{ij} = \sum_{j=1}^{2} X_{1j} + \sum_{j=1}^{2} X_{2j} + \sum_{j=1}^{2} X_{3j}$$
$$= (X_{11} + X_{12}) + (X_{21} + X_{22}) + (X_{31} + X_{32})$$

This is the same result as we found above.

Problem Set Appendix A

A

1. Write each of the following without the summation symbol:

a. $\sum_{i=1}^{4} X_i$ **b.** $\sum_{i=2}^{4} (X_i^2 + 2)$

c. $\sum_{i=1}^{4} (X_i^2 + Y_i)$ **d.** $\sum_{i=1}^{3} X_i^2 Y_i$

e. $\sum_{i=1}^{5} f_i X_i^2$ **f.** $\sum_{i=1}^{2} \sum_{j=1}^{3} X_i Y_j^2$

2. Write the following expressions using summation notation:

a. $X_1 Y_1 + X_2 Y_2 + X_3 Y_3$
b. $X_1 Y_2 + X_2 Y_3 + X_3 Y_4 + X_4 Y_5$
c. $X_2^2 + X_3^2 + X_4^2$
d. $X_1^2 + 2X_2^2 + 3X_3^2 + 4X_4^2$
e. $(2X_3 + 3) + (2X_4 + 3) + (2X_5 + 3) + (2X_6 + 3)$

3. Suppose X_i and Y_i take on the following values:

i	1	2	3	4	5
X_i	3	4	5	2	1
Y_i	1	3	0	2	4

Find the value of each of the following expressions:

a. $\sum_{i=1}^{5} (X_i + Y_i)$ **b.** $\sum_{i=2}^{5} (3X_i - Y_i)$

c. $\left(\sum_{i=1}^{3} X_i \right)^2$ **d.** $\sum_{i=1}^{3} X_i^2$

e. $\sum_{i=1}^{3} X_i Y_i$ **f.** $\left(\sum_{i=1}^{2} X_i \right) \left(\sum_{i=1}^{2} Y_i \right)$

g. $\sum_{i=1}^{4} (X_i^2 - Y_i^2)$ **h.** $\sum_{i=1}^{5} (X_i - 3)$

i. $\sum_{i=1}^{5} (Y_i - 2)$ **j.** $\sum_{i=1}^{5} (X_i - 3)(Y_i - 2)$

4. Suppose X_i and Y_j take on the following values.

i	1	2	3	4	5
X_i	2	3	1	2	3

j	1	2	3	4
Y_j	1	3	1	2

Find the values of the following expressions:

a. $\sum_{i=1}^{3} \sum_{j=1}^{2} (X_i + Y_j)$ **b.** $\sum_{i=1}^{3} \sum_{j=1}^{3} X_i Y_j$

c. $\sum_{i=2}^{4} \sum_{j=3}^{4} X_i Y_j^2$ **d.** $\sum_{j=1}^{4} \sum_{i=1}^{5} 3X_i$

e. $\sum_{i=1}^{2} X_i^2 + \sum_{j=1}^{3} Y_j^2$ **f.** $\sum_{i=1}^{5} \sum_{j=1}^{4} 2$

B

5. Prove that

$$\sum_{i=1}^{n} i = \frac{n(n+1)}{2}$$

6. Prove that

$$\sum_{i=1}^{n} i^2 = \frac{n(n+1)(2n+1)}{6}$$

7. Prove that

$$\sum_{i=a}^{b} f(i) = \sum_{i=a-c}^{b-c} f(i+c) \text{ where } a > c \text{ and } b > c$$

and f is a function of i.

8. Prove that

$$\sum_{i=a}^{c} f(i) = \sum_{i=a}^{b} f(i) + \sum_{i=b+1}^{c} f(i) \quad \text{where } a \leq b < c$$

9. Express each of the following using summation notation:
  a. $3 + 5 + 7 + 9 + \cdots + 23$
  b. $9 + 16 + 25 + 36 + \cdots + 400$
  c. $1 + 6 + 15 + 28 + \cdots + 190$
  d. $12 + 20 + 30 + 42 + \cdots + 90$

STATISTICAL TABLES

Table 1 Binomial Distribution Tables

n	x	.01	.05	.1	.2	.3	.4	.5	.6	.7	.8	.9	.95	.99	x
1	0	.990	.950	.900	.800	.700	.600	.500	.400	.300	.200	.100	.050	.010	0
	1	.010	.050	.100	.200	.300	.400	.500	.600	.700	.800	.900	.950	.990	1
2	0	.980	.903	.810	.640	.490	.360	.250	.160	.090	.040	.010	.003	.000	0
	1	.020	.095	.180	.320	.420	.480	.500	.480	.420	.320	.180	.095	.020	1
	2	.000	.003	.010	.040	.090	.160	.250	.360	.490	.640	.810	.903	.980	2
3	0	.970	.857	.729	.512	.343	.216	.125	.064	.027	.008	.001	.000	.000	0
	1	.029	.135	.243	.384	.441	.432	.375	.288	.189	.096	.027	.007	.000	1
	2	.000	.007	.027	.096	.189	.288	.375	.432	.441	.384	.243	.135	.029	2
	3	.000	.000	.001	.008	.027	.064	.125	.216	.343	.512	.729	.857	.970	3
4	0	.961	.815	.656	.410	.240	.130	.063	.026	.008	.002	.000	.000	.000	0
	1	.039	.171	.292	.410	.412	.346	.250	.154	.076	.026	.004	.000	.000	1
	2	.001	.014	.049	.154	.265	.346	.375	.346	.265	.154	.049	.014	.001	2
	3	.000	.000	.004	.026	.076	.154	.250	.346	.412	.410	.292	.171	.039	3
	4	.000	.000	.000	.002	.008	.026	.063	.130	.240	.410	.656	.815	.961	4
5	0	.951	.774	.590	.328	.168	.078	.031	.010	.002	.000	.000	.000	.000	0
	1	.048	.204	.328	.410	.360	.259	.156	.077	.028	.006	.000	.000	.000	1
	2	.001	.021	.073	.205	.309	.346	.313	.230	.132	.051	.008	.001	.000	2
	3	.000	.001	.008	.051	.132	.230	.313	.346	.309	.205	.073	.021	.001	3
	4	.000	.000	.000	.006	.028	.077	.156	.259	.360	.410	.328	.204	.048	4
	5	.000	.000	.000	.000	.002	.010	.031	.078	.168	.328	.590	.774	.951	5
6	0	.941	.735	.531	.262	.118	.047	.016	.004	.001	.000	.000	.000	.000	0
	1	.057	.232	.354	.393	.303	.187	.094	.037	.010	.002	.000	.000	.000	1
	2	.001	.031	.098	.246	.324	.311	.234	.138	.060	.015	.001	.000	.000	2
	3	.000	.002	.015	.082	.185	.276	.313	.276	.185	.082	.015	.002	.000	3
	4	.000	.000	.001	.015	.060	.138	.234	.311	.324	.246	.098	.031	.001	4
	5	.000	.000	.000	.002	.010	.037	.094	.187	.303	.393	.354	.232	.057	5
	6	.000	.000	.000	.000	.001	.004	.016	.047	.118	.262	.531	.735	.941	6
7	0	.932	.698	.478	.210	.082	.028	.008	.002	.000	.000	.000	.000	.000	0
	1	.066	.257	.372	.367	.247	.131	.055	.017	.004	.000	.000	.000	.000	1
	2	.002	.041	.124	.275	.318	.261	.164	.077	.025	.004	.000	.000	.000	2
	3	.000	.004	.023	.115	.227	.290	.273	.194	.097	.029	.003	.000	.000	3
	4	.000	.000	.003	.029	.097	.194	.273	.290	.227	.115	.023	.004	.000	4

Binomial Distribution Tables (*continued*)

n	x	.01	.05	.1	.2	.3	.4	.5	.6	.7	.8	.9	.95	.99	x
	5	.000	.000	.000	.004	.025	.077	.164	.261	.318	.275	.124	.041	.002	5
	6	.000	.000	.000	.000	.004	.017	.055	.131	.247	.367	.372	.257	.066	6
	7	.000	.000	.000	.000	.000	.002	.008	.028	.082	.210	.478	.698	.932	7
8	0	.923	.663	.430	.168	.058	.017	.004	.001	.000	.000	.000	.000	.000	0
	1	.075	.279	.383	.336	.198	.090	.031	.008	.001	.000	.000	.000	.000	1
	2	.003	.051	.149	.294	.296	.209	.109	.041	.010	.001	.000	.000	.000	2
	3	.000	.005	.033	.147	.254	.279	.219	.124	.047	.009	.000	.000	.000	3
	4	.000	.000	.005	.046	.136	.232	.273	.232	.136	.046	.005	.000	.000	4
	5	.000	.000	.000	.009	.047	.124	.219	.279	.254	.147	.033	.005	.000	5
	6	.000	.000	.000	.001	.010	.041	.109	.209	.296	.294	.149	.051	.003	6
	7	.000	.000	.000	.000	.001	.008	.031	.090	.198	.336	.383	.279	.075	7
	8	.000	.000	.000	.000	.000	.001	.004	.017	.058	.168	.430	.663	.923	8
9	0	.914	.630	.387	.134	.040	.010	.002	.000	.000	.000	.000	.000	.000	0
	1	.083	.299	.387	.302	.156	.060	.018	.004	.000	.000	.000	.000	.000	1
	2	.003	.063	.172	.302	.267	.161	.070	.021	.004	.000	.000	.000	.000	2
	3	.000	.008	.045	.176	.267	.251	.164	.074	.021	.003	.000	.000	.000	3
	4	.000	.001	.007	.066	.172	.251	.246	.167	.074	.017	.001	.000	.000	4
	5	.000	.000	.001	.017	.074	.167	.246	.251	.172	.066	.007	.001	.000	5
	6	.000	.000	.000	.003	.021	.074	.164	.251	.267	.176	.045	.008	.000	6
	7	.000	.000	.000	.000	.004	.021	.070	.161	.267	.302	.172	.063	.003	7
	8	.000	.000	.000	.000	.000	.004	.018	.060	.156	.302	.387	.299	.083	8
	9	.000	.000	.000	.000	.000	.000	.002	.010	.040	.134	.387	.630	.914	9
10	0	.904	.599	.349	.107	.028	.006	.001	.000	.000	.000	.000	.000	.000	0
	1	.091	.315	.387	.268	.121	.040	.010	.002	.000	.000	.000	.000	.000	1
	2	.004	.075	.194	.302	.233	.121	.044	.011	.001	.000	.000	.000	.000	2
	3	.000	.010	.057	.201	.267	.215	.117	.042	.009	.001	.000	.000	.000	3
	4	.000	.001	.011	.088	.200	.251	.205	.111	.037	.006	.000	.000	.000	4
	5	.000	.000	.001	.026	.103	.201	.246	.201	.103	.026	.001	.000	.000	5
	6	.000	.000	.000	.006	.037	.111	.205	.251	.200	.088	.011	.001	.000	6
	7	.000	.000	.000	.001	.009	.042	.117	.215	.267	.201	.057	.010	.000	7
	8	.000	.000	.000	.000	.001	.011	.044	.121	.233	.302	.194	.075	.004	8
	9	.000	.000	.000	.000	.000	.002	.010	.040	.121	.268	.387	.315	.091	9
	10	.000	.000	.000	.000	.000	.000	.001	.006	.028	.107	.349	.599	.904	10
11	0	.895	.569	.314	.086	.020	.004	.000	.000	.000	.000	.000	.000	.000	0
	1	.099	.329	.384	.236	.093	.027	.005	.001	.000	.000	.000	.000	.000	1
	2	.005	.087	.213	.295	.200	.089	.027	.005	.001	.000	.000	.000	.000	2
	3	.000	.014	.071	.221	.257	.177	.081	.023	.004	.000	.000	.000	.000	3
	4	.000	.001	.016	.111	.220	.236	.161	.070	.017	.002	.000	.000	.000	4
	5	.000	.000	.002	.039	.132	.221	.226	.147	.057	.010	.000	.000	.000	5
	6	.000	.000	.000	.010	.057	.147	.226	.221	.132	.039	.002	.000	.000	6
	7	.000	.000	.000	.002	.017	.070	.161	.236	.220	.111	.016	.001	.000	7
	8	.000	.000	.000	.000	.004	.023	.081	.177	.257	.221	.071	.014	.000	8
	9	.000	.000	.000	.000	.001	.005	.027	.089	.200	.295	.213	.087	.005	9
	10	.000	.000	.000	.000	.000	.001	.005	.027	.093	.236	.384	.329	.099	10
	11	.000	.000	.000	.000	.000	.000	.000	.004	.020	.086	.314	.569	.895	11

Binomial Distribution Tables (*continued*)

n	x	.01	.05	.1	.2	.3	.4	.5	.6	.7	.8	.9	.95	.99	x
12	0	.886	.540	.282	.069	.014	.002	.000	.000	.000	.000	.000	.000	.000	0
	1	.107	.341	.377	.206	.071	.017	.003	.000	.000	.000	.000	.000	.000	1
	2	.006	.099	.230	.283	.168	.064	.016	.002	.000	.000	.000	.000	.000	2
	3	.000	.017	.085	.236	.240	.142	.054	.012	.001	.000	.000	.000	.000	3
	4	.000	.002	.021	.133	.231	.213	.121	.042	.008	.001	.000	.000	.000	4
	5	.000	.000	.004	.053	.158	.227	.193	.101	.029	.003	.000	.000	.000	5
	6	.000	.000	.000	.016	.079	.177	.226	.177	.079	.016	.000	.000	.000	6
	7	.000	.000	.000	.003	.029	.101	.193	.227	.158	.053	.004	.000	.000	7
	8	.000	.000	.000	.001	.008	.042	.121	.213	.231	.133	.021	.002	.000	8
	9	.000	.000	.000	.000	.001	.012	.054	.142	.240	.236	.085	.017	.000	9
	10	.000	.000	.000	.000	.000	.002	.016	.064	.168	.283	.230	.099	.006	10
	11	.000	.000	.000	.000	.000	.000	.003	.017	.071	.206	.377	.341	.107	11
	12	.000	.000	.000	.000	.000	.000	.000	.002	.014	.069	.282	.540	.886	12
13	0	.878	.513	.254	.055	.010	.001	.000	.000	.000	.000	.000	.000	.000	0
	1	.115	.351	.367	.179	.054	.011	.002	.000	.000	.000	.000	.000	.000	1
	2	.007	.111	.245	.268	.139	.045	.010	.001	.000	.000	.000	.000	.000	2
	3	.000	.021	.100	.246	.218	.111	.035	.006	.001	.000	.000	.000	.000	3
	4	.000	.003	.028	.154	.234	.184	.087	.024	.003	.000	.000	.000	.000	4
	5	.000	.000	.006	.069	.180	.221	.157	.066	.014	.001	.000	.000	.000	5
	6	.000	.000	.001	.023	.103	.197	.209	.131	.044	.006	.000	.000	.000	6
	7	.000	.000	.000	.006	.044	.131	.209	.197	.103	.023	.001	.000	.000	7
	8	.000	.000	.000	.001	.014	.066	.157	.221	.180	.069	.006	.000	.000	8
	9	.000	.000	.000	.000	.003	.024	.087	.184	.234	.154	.028	.003	.000	9
	10	.000	.000	.000	.000	.001	.006	.035	.111	.218	.246	.100	.021	.000	10
	11	.000	.000	.000	.000	.000	.001	.010	.045	.139	.268	.245	.111	.007	11
	12	.000	.000	.000	.000	.000	.000	.002	.011	.054	.179	.367	.351	.115	12
	13	.000	.000	.000	.000	.000	.000	.000	.001	.010	.055	.254	.513	.878	13
14	0	.869	.488	.229	.044	.007	.001	.000	.000	.000	.000	.000	.000	.000	0
	1	.123	.359	.356	.154	.041	.007	.001	.000	.000	.000	.000	.000	.000	1
	2	.008	.123	.257	.250	.113	.032	.006	.001	.000	.000	.000	.000	.000	2
	3	.000	.026	.114	.250	.194	.085	.022	.003	.000	.000	.000	.000	.000	3
	4	.000	.004	.035	.172	.229	.155	.061	.014	.001	.000	.000	.000	.000	4
	5	.000	.000	.008	.086	.196	.207	.122	.041	.007	.000	.000	.000	.000	5
	6	.000	.000	.001	.032	.126	.207	.183	.092	.023	.002	.000	.000	.000	6
	7	.000	.000	.000	.009	.062	.157	.209	.157	.062	.009	.000	.000	.000	7
	8	.000	.000	.000	.002	.023	.092	.183	.207	.126	.032	.001	.000	.000	8
	9	.000	.000	.000	.000	.007	.041	.122	.207	.196	.086	.008	.000	.000	9
	10	.000	.000	.000	.000	.001	.014	.061	.155	.229	.172	.035	.004	.000	10
	11	.000	.000	.000	.000	.000	.003	.022	.085	.194	.250	.114	.026	.000	11
	12	.000	.000	.000	.000	.000	.001	.006	.032	.113	.250	.257	.123	.008	12
	13	.000	.000	.000	.000	.000	.000	.001	.007	.041	.154	.356	.359	.123	13
	14	.000	.000	.000	.000	.000	.000	.000	.001	.007	.044	.229	.488	.869	14

Binomial Distribution Tables (*continued*)

n	x	.01	.05	.1	.2	.3	.4	.5	.6	.7	.8	.9	.95	.99	x
15	0	.860	.463	.206	.035	.005	.000	.000	.000	.000	.000	.000	.000	.000	0
	1	.130	.366	.343	.132	.031	.005	.000	.000	.000	.000	.000	.000	.000	1
	2	.009	.135	.267	.231	.092	.022	.003	.000	.000	.000	.000	.000	.000	2
	3	.000	.031	.129	.250	.170	.063	.014	.002	.000	.000	.000	.000	.000	3
	4	.000	.005	.043	.188	.219	.127	.042	.007	.001	.000	.000	.000	.000	4
	5	.000	.001	.010	.103	.206	.186	.092	.024	.003	.000	.000	.000	.000	5
	6	.000	.000	.002	.043	.147	.207	.153	.061	.012	.001	.000	.000	.000	6
	7	.000	.000	.000	.014	.081	.177	.196	.118	.035	.003	.000	.000	.000	7
	8	.000	.000	.000	.003	.035	.118	.196	.177	.081	.014	.000	.000	.000	8
	9	.000	.000	.000	.001	.012	.061	.153	.207	.147	.043	.002	.000	.000	9
	10	.000	.000	.000	.000	.003	.024	.092	.186	.206	.103	.010	.001	.000	10
	11	.000	.000	.000	.000	.001	.007	.042	.127	.219	.188	.043	.005	.000	11
	12	.000	.000	.000	.000	.000	.002	.014	.063	.170	.250	.129	.031	.000	12
	13	.000	.000	.000	.000	.000	.000	.003	.022	.092	.231	.267	.135	.009	13
	14	.000	.000	.000	.000	.000	.000	.000	.005	.031	.132	.343	.366	.130	14
	15	.000	.000	.000	.000	.000	.000	.000	.000	.005	.035	.206	.463	.860	15
20	0	.818	.358	.122	.012	.001	.000	.000	.000	.000	.000	.000	.000	.000	0
	1	.165	.377	.270	.058	.007	.000	.000	.000	.000	.000	.000	.000	.000	1
	2	.016	.189	.285	.137	.028	.003	.000	.000	.000	.000	.000	.000	.000	2
	3	.001	.060	.190	.205	.072	.012	.001	.000	.000	.000	.000	.000	.000	3
	4	.000	.013	.090	.218	.130	.035	.005	.000	.000	.000	.000	.000	.000	4
	5	.000	.002	.032	.175	.179	.075	.015	.001	.000	.000	.000	.000	.000	5
	6	.000	.000	.009	.109	.192	.124	.037	.005	.000	.000	.000	.000	.000	6
	7	.000	.000	.002	.055	.164	.166	.074	.015	.001	.000	.000	.000	.000	7
	8	.000	.000	.000	.022	.114	.180	.120	.035	.004	.000	.000	.000	.000	8
	9	.000	.000	.000	.007	.065	.160	.160	.071	.012	.000	.000	.000	.000	9
	10	.000	.000	.000	.002	.031	.117	.176	.117	.031	.002	.000	.000	.000	10
	11	.000	.000	.000	.000	.012	.071	.160	.160	.065	.007	.000	.000	.000	11
	12	.000	.000	.000	.000	.004	.035	.120	.180	.114	.022	.000	.000	.000	12
	13	.000	.000	.000	.000	.001	.015	.074	.166	.164	.055	.002	.000	.000	13
	14	.000	.000	.000	.000	.000	.005	.037	.124	.192	.109	.009	.000	.000	14
	15	.000	.000	.000	.000	.000	.001	.015	.075	.179	.175	.032	.002	.000	15
	16	.000	.000	.000	.000	.000	.000	.005	.035	.130	.218	.090	.013	.000	16
	17	.000	.000	.000	.000	.000	.000	.001	.012	.072	.205	.190	.060	.001	17
	18	.000	.000	.000	.000	.000	.000	.000	.003	.028	.137	.285	.189	.016	18
	19	.000	.000	.000	.000	.000	.000	.000	.000	.007	.058	.270	.377	.165	19
	20	.000	.000	.000	.000	.000	.000	.000	.000	.001	.012	.122	.358	.818	20

Binomial Distribution Tables (*continued*)

n	x	.01	.05	.1	.2	.3	.4	.5	.6	.7	.8	.9	.95	.99	x
25	0	.778	.277	.072	.004	.000	.000	.000	.000	.000	.000	.000	.000	.000	0
	1	.196	.365	.199	.024	.001	.000	.000	.000	.000	.000	.000	.000	.000	1
	2	.024	.231	.266	.071	.007	.000	.000	.000	.000	.000	.000	.000	.000	2
	3	.002	.093	.226	.136	.024	.002	.000	.000	.000	.000	.000	.000	.000	3
	4	.000	.027	.138	.187	.057	.007	.000	.000	.000	.000	.000	.000	.000	4
	5	.000	.006	.065	.196	.103	.020	.002	.000	.000	.000	.000	.000	.000	5
	6	.000	.001	.024	.163	.147	.044	.005	.000	.000	.000	.000	.000	.000	6
	7	.000	.000	.007	.111	.171	.080	.014	.001	.000	.000	.000	.000	.000	7
	8	.000	.000	.002	.062	.165	.120	.032	.003	.000	.000	.000	.000	.000	8
	9	.000	.000	.000	.029	.134	.151	.061	.009	.000	.000	.000	.000	.000	9
	10	.000	.000	.000	.012	.092	.161	.097	.021	.001	.000	.000	.000	.000	10
	11	.000	.000	.000	.004	.054	.147	.133	.043	.004	.000	.000	.000	.000	11
	12	.000	.000	.000	.001	.027	.114	.155	.076	.011	.000	.000	.000	.000	12
	13	.000	.000	.000	.000	.011	.076	.155	.114	.027	.001	.000	.000	.000	13
	14	.000	.000	.000	.000	.004	.043	.133	.147	.054	.004	.000	.000	.000	14
	15	.000	.000	.000	.000	.001	.021	.097	.161	.092	.012	.000	.000	.000	15
	16	.000	.000	.000	.000	.000	.009	.061	.151	.134	.029	.000	.000	.000	16
	17	.000	.000	.000	.000	.000	.003	.032	.120	.165	.062	.002	.000	.000	17
	18	.000	.000	.000	.000	.000	.001	.014	.080	.171	.111	.007	.000	.000	18
	19	.000	.000	.000	.000	.000	.000	.005	.044	.147	.163	.024	.001	.000	19
	20	.000	.000	.000	.000	.000	.000	.002	.020	.103	.196	.065	.006	.000	20
	21	.000	.000	.000	.000	.000	.000	.000	.007	.057	.187	.138	.027	.000	21
	22	.000	.000	.000	.000	.000	.000	.000	.002	.024	.136	.226	.093	.002	22
	23	.000	.000	.000	.000	.000	.000	.000	.000	.007	.071	.266	.231	.024	23
	24	.000	.000	.000	.000	.000	.000	.000	.000	.001	.024	.199	.365	.196	24
	25	.000	.000	.000	.000	.000	.000	.000	.000	.000	.004	.072	.277	.778	25

Table 2 Critical Values $\chi^2_\alpha(df)$ for the Chi-Square Distributions

α = shaded area

$\chi^2_\alpha(df)$ = critical value

df	.995	.99	.975	.95	.90	.10	.05	.025	.01	.005
1	0.000	0.000	0.001	0.004	0.016	2.706	3.841	5.024	6.635	7.879
2	0.010	0.020	0.051	0.103	0.211	4.605	5.991	7.378	9.210	10.597
3	0.072	0.115	0.216	0.352	0.584	6.251	7.815	9.348	11.345	12.838
4	0.207	0.297	0.484	0.711	1.064	7.779	9.488	11.143	13.277	14.860
5	0.412	0.554	0.831	1.145	1.610	9.236	11.070	12.832	15.086	16.750
6	0.676	0.872	1.237	1.635	2.204	10.645	12.592	14.449	16.812	18.548
7	0.989	1.239	1.690	2.167	2.833	12.017	14.067	16.013	18.475	20.278
8	1.344	1.646	2.180	2.733	3.490	13.362	15.507	17.535	20.090	21.955
9	1.735	2.088	2.700	3.325	4.168	14.684	16.919	19.023	21.666	23.589
10	2.156	2.558	3.247	3.940	4.865	15.987	18.307	20.483	23.209	25.188
11	2.603	3.053	3.816	4.575	5.578	17.275	19.675	21.920	24.725	26.757
12	3.074	3.571	4.404	5.226	6.304	18.549	21.026	23.337	26.217	28.300
13	3.565	4.107	5.009	5.892	7.042	19.812	22.362	24.736	27.688	29.819
14	4.075	4.660	5.629	6.571	7.790	21.064	23.685	26.119	29.141	31.319
15	4.601	5.229	6.262	7.261	8.547	22.307	24.996	27.488	30.578	32.801
16	5.142	5.812	6.908	7.962	9.312	23.542	26.296	28.845	32.000	34.267
17	5.697	6.408	7.564	8.672	10.085	24.769	27.587	30.191	33.409	35.718
18	6.265	7.015	8.231	9.390	10.865	25.989	28.869	31.526	34.805	37.156
19	6.844	7.633	8.907	10.117	11.651	27.204	30.144	32.852	36.191	38.582
20	7.434	8.260	9.591	10.851	12.443	28.412	31.410	34.170	37.566	39.997
21	8.034	8.897	10.283	11.591	13.240	29.615	32.670	35.479	38.932	41.401
22	8.643	9.542	10.982	12.338	14.042	30.813	33.924	36.781	40.289	42.796
23	9.260	10.196	11.688	13.090	14.848	32.007	35.172	38.076	41.638	44.181
24	9.886	10.856	12.401	13.848	15.659	33.196	36.415	39.364	42.980	45.558
25	10.520	11.524	13.120	14.611	16.473	34.382	37.652	40.646	44.314	46.928
26	11.160	12.198	13.844	15.379	17.292	35.563	38.885	41.923	45.642	48.290
27	11.808	12.879	14.573	16.151	18.114	36.741	40.113	43.194	46.963	49.645
28	12.461	13.565	15.308	16.928	18.939	37.916	41.337	44.461	48.278	50.993
29	13.121	14.256	16.047	17.708	19.768	39.088	42.557	45.772	49.588	52.336
30	13.787	14.954	16.791	18.493	20.599	40.256	43.773	46.979	50.892	53.672
31	14.458	15.655	17.539	19.281	21.434	41.422	44.985	48.232	52.190	55.003
32	15.134	16.362	18.291	20.072	22.271	42.585	46.194	49.480	53.486	56.328
33	15.815	17.074	19.047	20.867	23.110	43.745	47.400	50.725	54.776	57.649
34	16.501	17.789	19.806	21.664	23.952	44.903	48.602	51.966	56.061	58.964
35	17.192	18.509	20.569	22.465	24.797	46.059	49.802	53.203	57.340	60.275
36	17.887	19.233	21.336	23.269	25.643	47.212	50.998	54.437	58.619	61.581
37	18.586	19.960	22.106	24.075	26.492	48.363	52.192	55.668	59.892	62.883
38	19.289	20.691	22.878	24.884	27.343	49.513	53.384	56.896	61.162	64.181
39	19.996	21.426	23.654	25.695	28.196	50.660	54.572	58.120	62.428	65.476
40	20.706	22.164	24.433	26.509	29.050	51.805	55.758	59.342	63.691	66.766

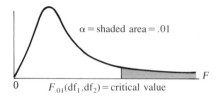

$\alpha = $ shaded area $= .01$

$F_{.01}(df_1, df_2) = $ critical value

Table 3a Critical Values $F_{.01}(df_1, df_2)$ for the F Distributions*

df_2	1	2	3	4	5	6	7	8	9
					df_1				
1	4052	5000	5403	5625	5764	5859	5928	5982	6022
2	98.5	99.0	99.2	99.3	99.3	99.4	99.4	99.4	99.4
3	34.1	30.3	29.5	28.7	28.2	27.9	27.7	27.5	27.3
4	21.2	18.0	16.7	16.0	15.5	15.2	15.0	14.8	14.7
5	16.3	13.3	12.1	11.4	11.0	10.7	10.5	10.3	10.2
6	13.7	10.9	9.78	9.15	8.75	8.47	8.26	8.10	7.98
7	12.2	9.55	8.45	7.85	7.46	7.19	6.99	6.84	6.72
8	11.3	8.65	7.59	7.01	6.63	6.37	6.18	6.03	5.91
9	10.6	8.02	6.99	6.42	6.06	5.80	5.61	5.47	5.35
10	10.0	7.56	6.55	5.99	5.64	5.39	3.20	5.06	4.94
11	9.65	7.21	6.22	5.67	5.32	5.07	4.89	4.74	4.63
12	9.33	6.93	5.95	5.41	5.06	4.82	4.64	4.50	4.39
13	9.07	6.70	5.74	5.21	4.86	4.62	4.44	4.30	4.19
14	8.86	6.51	5.56	5.04	4.70	4.46	4.28	4.14	4.03
15	8.68	6.36	5.42	4.89	4.56	4.32	4.14	4.00	3.89
16	8.53	6.23	5.29	4.77	4.44	4.20	4.03	3.89	3.78
17	8.40	6.11	5.18	4.67	4.34	4.10	3.93	3.79	3.68
18	8.29	6.01	5.09	4.58	4.25	4.01	3.84	3.71	3.60
19	8.18	5.93	5.01	4.50	4.17	3.94	3.77	3.63	3.52
20	8.10	5.85	4.94	4.43	4.10	3.87	3.70	3.56	3.46
21	8.02	5.78	4.87	4.37	4.04	3.81	3.64	3.51	3.40
22	7.95	5.72	4.82	4.31	3.99	3.76	3.59	3.45	3.35
23	7.88	5.66	4.76	4.26	3.94	3.71	3.54	3.41	3.30
24	7.82	5.61	4.72	4.22	3.90	3.67	3.50	3.36	3.26
25	7.77	5.57	4.68	4.18	3.86	3.63	3.46	3.32	3.22
26	7.72	5.53	4.64	4.14	3.82	3.59	3.42	3.29	3.18
27	7.68	5.49	4.60	4.11	3.78	3.56	3.39	3.26	3.15
28	7.64	5.45	4.57	4.07	3.75	3.53	3.36	3.23	3.12
29	7.60	5.42	4.54	4.04	3.73	3.50	3.33	3.20	3.09
30	7.56	5.39	4.51	4.02	3.70	3.47	3.30	3.17	3.07
40	7.31	5.18	4.31	3.83	3.51	3.29	3.12	2.99	2.89
60	7.08	4.98	4.13	3.65	3.34	3.12	2.95	2.82	2.72
125	6.84	4.78	3.94	3.47	3.17	2.95	2.79	2.65	2.56

* df_1 (associated with the numerator)

Table 3a (*continued*)

	df$_1$								
df$_2$	10	12	14	16	20	30	40	50	100
1	6056	6106	6142	6169	6208	6258	6286	6302	6334
2	99.4	99.4	99.4	99.4	99.5	99.5	99.5	99.5	99.5
3	27.2	27.1	26.9	26.8	26.7	26.5	26.4	26.3	26.2
4	14.5	14.4	14.2	14.2	14.0	13.8	13.7	13.7	13.6
5	10.1	9.89	9.77	9.68	9.55	9.38	9.29	9.24	9.13
6	7.87	7.72	7.60	7.52	7.39	7.23	7.14	7.09	6.99
7	6.62	6.47	6.35	6.27	6.15	5.98	5.90	5.85	5.75
8	5.82	5.67	5.56	5.48	5.36	5.20	5.11	5.06	4.96
9	5.26	5.11	5.00	4.92	4.80	4.64	4.56	4.51	4.41
10	4.85	4.71	4.60	4.52	4.41	4.25	4.17	4.12	4.01
11	4.54	4.40	4.29	4.21	4.10	3.94	3.86	3.80	3.70
12	4.30	4.16	4.05	3.98	3.86	3.70	3.61	3.56	3.46
13	4.10	3.96	3.85	3.78	3.67	3.51	3.42	3.37	3.27
14	3.94	3.80	3.70	3.62	3.51	3.34	3.26	3.21	3.11
15	3.80	3.67	3.56	3.48	3.36	3.20	3.12	3.07	2.97
16	3.69	3.55	3.45	3.37	3.25	3.10	3.01	2.96	2.86
17	3.59	3.45	3.35	3.27	3.16	3.00	2.92	2.86	2.76
18	3.51	3.37	3.27	3.19	3.07	2.91	2.83	2.78	2.68
19	3.43	3.30	3.19	3.12	3.00	2.84	2.76	2.70	2.60
20	3.37	3.23	3.13	3.05	2.94	2.77	2.69	2.63	2.53
21	3.31	3.17	3.07	2.99	2.88	2.72	2.63	2.58	2.47
22	3.26	3.12	3.02	2.94	2.83	2.67	2.58	2.53	2.42
23	3.21	3.07	2.97	2.89	2.78	2.62	2.53	2.48	2.37
24	3.17	3.03	2.93	2.85	2.74	2.58	2.49	2.44	2.33
25	3.13	2.99	2.89	2.81	2.70	2.54	2.45	2.40	2.29
26	3.09	2.96	2.86	2.77	2.66	2.50	2.41	2.36	2.25
27	3.06	2.93	2.83	2.74	2.63	2.47	2.38	2.33	2.21
28	3.03	2.90	2.80	2.71	2.60	2.44	2.35	2.30	2.18
29	3.00	2.87	2.77	2.68	2.57	2.41	2.32	2.27	2.15
30	2.98	2.84	2.74	2.66	2.55	2.38	2.29	2.24	2.13
40	2.80	2.66	2.56	2.49	2.37	2.20	2.11	2.05	1.94
60	2.63	2.50	2.40	2.32	2.20	2.03	1.93	1.87	1.74
125	2.47	2.33	2.23	2.15	2.03	1.85	1.75	1.68	1.54

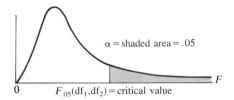

α = shaded area = .05

$F_{.05}(df_1, df_2)$ = critical value

Table 3b Critical Values $F_{.05}(df_1, df_2)$ for the F Distributions*

df$_2$	1	2	3	4	5	6	7	8	9
1	161	200	216	225	230	234	237	239	241
2	18.5	19.0	19.2	19.3	19.3	19.4	19.4	19.4	19.4
3	10.1	9.55	9.28	9.12	9.01	8.94	8.89	8.85	8.81
4	7.71	6.94	6.59	6.39	6.29	6.16	6.09	6.04	6.00
5	6.61	5.79	5.41	5.19	5.05	4.95	4.88	4.82	4.77
6	5.99	5.14	4.76	4.53	4.39	4.28	4.21	4.15	4.10
7	5.59	4.74	4.35	4.12	3.97	3.87	3.79	3.73	3.68
8	5.32	4.46	4.07	3.84	3.69	3.58	3.50	3.44	3.39
9	5.12	4.26	3.86	3.63	3.48	3.37	3.29	3.23	3.18
10	4.96	4.10	3.71	3.48	3.33	3.22	3.14	3.07	3.02
11	4.84	3.98	3.59	3.36	3.20	3.09	3.01	2.95	2.90
12	4.75	3.89	3.49	3.26	3.11	3.00	2.91	2.85	2.80
13	4.67	3.81	3.41	3.18	3.03	2.92	2.83	2.77	2.71
14	4.60	3.74	3.34	3.11	2.96	2.85	2.76	2.70	2.65
15	4.54	3.68	3.29	3.06	2.90	2.79	2.71	2.64	2.59
16	4.49	3.63	3.24	3.01	2.85	2.74	2.66	2.59	2.54
17	4.45	3.59	3.20	2.96	2.81	2.70	2.61	2.55	2.49
18	4.41	3.55	3.16	2.93	2.77	2.66	2.58	2.51	2.46
19	4.38	3.52	3.13	2.90	2.74	2.63	2.54	2.48	2.42
20	4.35	3.49	3.10	2.87	2.71	2.60	2.51	2.45	2.39
21	4.32	3.47	3.07	2.84	2.68	2.57	2.49	2.42	2.37
22	4.30	3.44	3.05	2.82	2.66	2.55	2.46	2.40	2.34
23	4.28	3.42	3.03	2.80	2.64	2.53	2.44	2.37	2.32
24	4.26	3.40	3.01	2.78	2.62	2.51	2.42	2.36	2.30
25	4.24	3.39	2.99	2.76	2.60	2.49	2.40	2.34	2.28
26	4.23	3.37	2.98	2.74	2.59	2.47	2.39	2.32	2.27
27	4.21	3.35	2.96	2.73	2.57	2.46	2.37	2.31	2.25
28	4.20	3.34	2.95	2.71	2.56	2.45	2.36	2.29	2.24
29	4.18	3.33	2.93	2.70	2.55	2.43	2.35	2.28	2.22
30	4.17	3.32	2.92	2.69	2.53	2.42	2.33	2.27	2.21
40	4.08	3.23	2.84	2.61	2.45	2.34	2.25	2.18	2.12
60	4.00	3.15	2.76	2.53	2.37	2.25	2.17	2.10	2.04
125	3.92	3.07	2.68	2.44	2.29	2.17	2.08	2.01	1.95

* df$_1$ (associated with the numerator)

Table 3b (*continued*)

df$_2$	df$_1$								
	10	12	14	16	20	30	40	50	100
1	242	244	245	246	248	250	251	252	253
2	19.4	19.4	19.4	19.4	19.4	19.5	19.5	19.5	19.5
3	8.78	8.74	8.71	8.69	8.66	8.62	8.60	8.58	8.56
4	5.96	5.91	5.87	5.84	5.80	5.74	5.71	5.70	5.66
5	4.74	4.68	4.64	4.60	4.56	4.50	4.46	4.44	4.40
6	4.06	4.00	3.96	3.92	3.87	3.81	3.77	3.75	3.71
7	3.63	3.57	3.52	3.49	3.44	3.38	3.34	3.32	3.28
8	3.34	3.28	3.23	3.20	3.15	3.08	3.05	3.03	2.98
9	3.13	3.07	3.02	2.98	2.93	2.86	2.82	2.80	2.76
10	2.97	2.91	2.86	2.82	2.77	2.70	2.67	2.64	2.59
11	2.86	2.79	2.74	2.70	2.65	2.57	2.53	2.50	2.45
12	2.76	2.69	2.64	2.60	2.54	2.46	2.42	2.40	2.35
13	2.67	2.60	2.55	2.51	2.46	2.38	2.34	2.32	2.26
14	2.60	2.53	2.48	2.44	2.39	2.31	2.27	2.24	2.19
15	2.55	2.48	2.43	2.39	2.33	2.25	2.21	2.18	2.12
16	2.49	2.42	2.37	2.33	2.28	2.20	2.16	2.13	2.07
17	2.45	2.38	2.33	2.29	2.23	2.15	2.11	2.08	2.02
18	2.41	2.34	2.29	2.25	2.19	2.11	2.07	2.04	1.98
19	2.38	2.31	2.26	2.21	2.15	2.07	2.02	2.00	1.94
20	2.35	2.28	2.23	2.18	2.12	2.04	1.99	1.96	1.90
21	2.32	2.25	2.20	2.15	2.09	2.00	1.96	1.93	1.87
22	2.30	2.23	2.18	2.13	2.07	1.98	1.93	1.91	1.84
23	2.28	2.20	2.14	2.10	2.04	1.96	1.91	1.88	1.82
24	2.26	2.18	2.13	2.09	2.02	1.94	1.89	1.86	1.80
25	2.24	2.16	2.11	2.06	2.00	1.92	1.87	1.84	1.77
26	2.22	2.15	2.10	2.05	1.99	1.90	1.85	1.82	1.76
27	2.20	2.13	2.08	2.03	1.97	1.88	1.84	1.80	1.74
28	2.19	2.12	2.06	2.02	1.96	1.87	1.81	1.78	1.72
29	2.18	2.10	2.05	2.00	1.94	1.85	1.80	1.77	1.71
30	2.16	2.09	2.04	1.99	1.93	1.84	1.79	1.76	1.69
40	2.07	2.00	1.95	1.90	1.84	1.74	1.69	1.66	1.59
60	1.99	1.92	1.86	1.81	1.75	1.65	1.59	1.56	1.48
125	1.90	1.83	1.77	1.72	1.65	1.55	1.49	1.45	1.36

C

ANSWERS TO SELECTED PROBLEMS

Note. Your answers may sometimes vary slightly from those listed here depending on the number of decimal places carried in subcomputations. Do not be overly concerned with minor differences between your answers and those given here. Many of these numerical answers were obtained with a hand-held calculator or a computer.

CHAPTER 1

Problem Set 1.2

1. a. The cause of each accident in the state of Maryland for the month of June
b. The cause of each accident in the counties served by the five highway patrol offices for the month of June
c. The percentage of accidents where alcohol is a contributing factor for the counties served by the five highway patrol offices can serve as an estimate of the percentage of accidents in which alcohol is a factor for the state of Maryland.

3. a. The outcomes (cured or not) of all arthritis patients who take cod liver oil and the outcomes (cured or not) of all arthritis patients not taking cod liver oil
b. The outcomes (cured or not) of the group of 50 arthritis patients taking cod liver oil and the outcomes (cured or not) of the group of 50 arthritis patients not taking cod liver oil
c. The percentages of those cured in the two samples can be used to estimate the percentages that are cured in the two populations. Or, the difference of the sample percentages that are cured could be used to estimate the difference in the population percentages that are cured.

5. a. The relationship (increase or decrease) between the number of helper and T-cells and stress and mood for every person
b. The relationship (increase or decrease) between the number of helper and T-cells and stress and mood for each of the 36 people involved in the study

Problem Set 1.3

2. a. All responses from the 1500 registered voters
b. 500 responses from the sampled voters
c. Number of votes for Mr. Jackson in sample
d. Number of votes for Mr. Jackson in population
e. He will not intensify his campaign efforts.

3. a. Descriptive   **b.** Descriptive
   c. Inferential   **d.** Inferential

5. a. The collection of textbook costs for the 1200 students
b. The textbook costs for the 25 polled students
c. 1200, the size of the population; $135, the average textbook cost for the population
d. 25, the size of the sample; $152.25, the average textbook cost for the sample
e. A mistake had been made or the sample is not representative of the population.

CHAPTER 2

Problem Set 2.1

1. a. Qualitative   **b.** Qualitative
   c. Qualitative   **d.** Qualitative
   e. Qualitative   **f.** Quantitative

g. Quantitative   **h.** Qualitative
i. Quantitative   **j.** Qualitative
3. a. Nominal   **b.** Ordinal
c. Ordinal   **d.** Nominal
e. Nominal   **f.** Ordinal
g. Nominal

5. Class	f	**Cum. f**
28–38	4	4
39–49	6	10
50–60	3	13
61–71	4	17
72–82	4	21
83–93	4	25

Problem Set 2.2

1. a.

Class	f	**Relative f**
1–4	14	.175
5–8	18	.225
9–12	12	.150
13–16	16	.200
17–20	20	.250

b.

Class	f	**Cumulative f**
1–4	14	14
5–8	18	32
9–12	12	44
13–16	16	60
17–20	20	80

c.

Class	f	**Cum. rel. f**
1–4	14	.175
5–8	18	.400
9–12	12	.550
13–16	16	.750
17–20	20	1.000

3.

Class	**Frequency**
28–38	4
39–49	6
50–60	3
61–71	4
72–82	4
83–93	4

7.

Class	f
115.9–120.8	10
120.9–125.8	6
125.9–130.8	5
130.9–135.8	3
135.9–140.8	1

9.

X	f	**Rel. f**
2	1	.033
3	4	.133
4	8	.267
5	6	.200
6	4	.133
7	2	.067
8	2	.067
9	1	.033
10	1	.033
11	1	.033

11. a. 5.45   **b.** 27.5   **c.** 23.08   **d.** 15.38
e. 30.77   **f.** 30.77
12. a. 44.44   **b.** 77.78   **c.** 11.43   **d.** 25
13. a. 64   **b.** 20   **c.** $w = 5.26$

Class	f
9.80–15.05	9
15.06–20.31	5
20.32–25.57	5
25.58–30.83	4
30.84–36.09	2

15. a. 5
b. 6.95, 11.95, 16.95, 21.95, 26.95
c. 4.45–9.45

Problem Set 2.3

1. a. Grade distribution

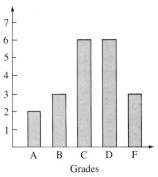

b. Grade distribution

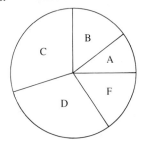

2. *f*

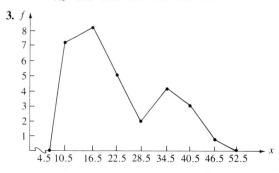

3. *f*

4.
0	8	9								
1	0	1	1	2	3	4	5	6	6	6	7	7
2	1	4	4	4	5	8				
3	1	2	3	4	7					
4	1	2	3	8						

7. Cum *f*

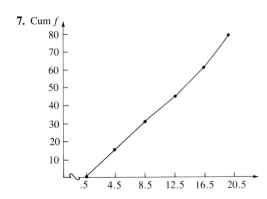

9.

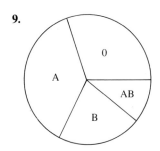

11.
2	2
3	4
4	6 6 6
5	1 4 5 7
6	1 3 5 6 7 9
7	1 1 4
8	4
9	0

13.
1*a*	
1*b*	9
2*a*	0 1 1 2 3
2*b*	5 7 7 7 7 7 7 7 9 9
3*a*	0 1 1 2 4 4
3*b*	5 5 5 8
4*a*	1 4
4*b*	7
5*a*	1

Chapter 2 Review

1. a.

Class	f
148–150	6
151–153	5
154–156	6
157–159	6
160–162	5
163–165	7
166–168	3
169–171	7
172–174	3
175–177	2

b.

```
14 | 8 9
15 | 0 0 0 0 1 1 2 3 3 4 4 5 5 6 6 7 7 7 8 8 8
16 | 0 1 1 2 2 3 3 3 3 4 4 5 7 7 7 9 9 9
17 | 0 0 0 1 2 4 4 5 6
```

2. a. Quantitative   **b.** Qualitative
c. Quantitative   **d.** Qualitative
e. Qualitative   **f.** Quantitative
g. Qualitative   **h.** Qualitative
i. Quantitative   **j.** Qualitative
k. Quantitative

4. a. 10   **b.** 4.5   **c.** 3.3   **d.** 3.56   **e.** .07

5. 11

6. a. 14   **b.** 57.7   **c.** 44

7. b.

Class	f
120–136	1
137–153	2
154–170	3
171–187	3
188–204	4
205–221	5
222–238	5
239–255	3
256–272	2
273–289	1

9.

```
5b | 9
6a | 4 4 4 3 3 4
6b | 6 5 8 9 5 8 8 6 7 9 7 7 9 9 6 7 6 7
7a | 1 0 1 2 0 0 0
7b |
8a | 0
```

Chapter 2 Achievement Test

1. X   f

X	f
0	1
1	3
2	2
3	5
4	3
5	5
6	1
7	3
8	5
9	4
10	2
11	5
12	4
13	1
14	1
15	2
16	0
17	2
18	1

2.
0a	0	1	1	1	2	2	3	3	3	3	3	4	4	4				
0b	5	5	5	5	5	6	7	7	7	8	8	8	8	8	9	9	9	9
1a	0	0	1	1	1	1	1	2	2	2	2	3	4					
1b	5	5	7	7	8													

3.

X	f	Rel. f	Cum. rel. f
0	1	.02	.02
1	3	.06	.08
2	2	.04	.12
3	5	.10	.22
4	3	.06	.28
5	5	.10	.38
6	1	.02	.40
7	3	.06	.46
8	5	.10	.56
9	4	.08	.64
10	2	.04	.68
11	5	.10	.78
12	4	.08	.86
13	1	.02	.88
14	1	.02	.90
15	2	.04	.94
16	0	.00	.94
17	2	.04	.98
18	1	.02	1.00

4.

Class	f
0–3	11
4–7	12
8–11	16
12–15	8
16–19	3

5. Cum. rel. f

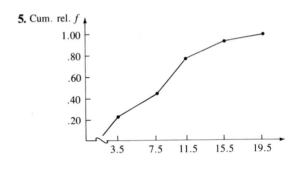

6. f

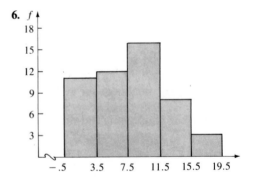

7. 11   **8.** 8

9. a. 41.67   **b.** 25   **c.** 36.36   **d.** 54.55
   e. 27.27

CHAPTER 3

Problem Set 3.1

1. The sample mean; $\bar{x} = 4.1$

2. $\bar{x} = 4.1$; $\tilde{x} = 4.5$; mode = 5; midrange = 4; the data are negatively skewed.

3. $800,000

4. a. $5500   **b.** $9300   **c.** $7000

   d. $17750   **e.** Positively skewed   **f.** Median

5. 21   **6. a.** 11   **b.** 30

7. $\bar{x} = 24.5$; $\tilde{x} = 23.65$; mode $= 22.5$;
midrange $= 22.5$

8. a. 3.825   **b.** $P_{40} = 3.51$, $P_{75} = 4.33$
   c. Negatively skewed

10. Mode   **11.** Mode   **12.** 113.55

16. *Hint:* Use $\bar{x}_g$.

19. *Hint:* Determine x so that $(x/7)(6) = 4.5$.

20. *Hint:* First determine the measures for a smaller
sample, e.g., 4, 6, 8, 10.

21. *Hint:* Use the geometric mean.

Problem Set 3.2

1. Zero

2. No. The standard deviation is smaller than the
variance only when the standard deviation is
greater than 1.

4. $R = 4$; $s^2 = 2$; $s = 1.41$

5. $R = 9$; $\sigma^2 = 5.56$; $\sigma = 2.36$

6. a. $\bar{x} = 1.67$; $s^2 = 13.60$; $s = 3.69$

7. No; SS < 0.   **8.** 68.75   **9.** Yes.

11. All of the data values must be identical.

12. a. Mean of car A $= .94$; mean of car B $= 1.15$
   b. Variance for car A $= .0055$;
     variance for car B $= .04$
   c. A
   d. A

13. 10.71

14. Enter any number and press the variance key.
If the display shows an error, then it is computing
sample variance. If the display shows 0, then it
is computing population variance.

15. 7851.49   **17.** 5.76

19. Yes; 16 is more than 3 standard deviations from
the mean.

22. $c = \bar{x}$ **24.** 3

28. *Hint:* After expanding $(x - \bar{x})^2$, use rules 1 and
2 in Appendix A.

30. No. As a consequence of Problem 29, the
variance of the new set is c^2 times the variance
of the old set. Since $\sqrt{c^2} = |c|$, the standard
deviation of the new set is c times the standard
deviation of the old set only if $c \geq 0$.

31. $\frac{10}{9}$

34. No. If the mean is negative and the coefficient
of variation expresses the standard deviation

as a percentage of the mean, then the standard
deviation would be negative, a contradiction, since
the standard deviation is always nonnegative.

35. .99

36. *Hint:* Try the measures of central tendency.

38. *Hint:* Two possibilities are $\{3, 3, 3\}$ and $\{-1, 1\}$.

Problem Set 3.3

1.

x	80	65	60	11.45	47	92
z	2.2	1.2	.87	-2.37	0	3

2. a. 12   **b.** 5.66   **d.** $\mu = 0$; $\sigma = 1$

3. Sue's z score $= .357$; Mary's z score $= 1$; Mary

4. Dave's z score $= -2.93$; Rick's z score $= -2.75$;
Dave

5. (e)   **6.** Dick

Chapter 3 Review

1. a. $\mu = 5$; $\sigma^2 = 4.67$; $\sigma = 2.16$; $\tilde{\mu} = 5$; midrange $= 5$;
   no mode
   c. $\mu = 4$; $\sigma^2 = 10.8$; $\sigma = 3.29$; $\tilde{\mu} = 4$;
    midrange $= 4.5$; no mode

2. a. $\bar{x} = 3.75$; $\tilde{x} = 3$; mode $= 2$; $R = 5$; $s^2 = 5.58$;
   $s = 2.4$
   b. $\bar{x} = 5.2$; $\tilde{x} = 4$; mode $= 4$; $R = 8$; $s^2 = 10.7$;
   $s = 3.27$
   c. $\bar{x} = 2.4$; $\tilde{x} = 1$; mode $= 0, 1$; $R = 10$; $s^2 = 18.3$;
   $s = 4.28$

3. a. 3.5   **b.** -1.88 **d.** -23.75 **e.** 1

4. $\bar{x} = 1.9$; $s^2 = 1.21$; $s = 1.10$; $\tilde{x} = 2$; mode $= 3$

5. $\bar{x} = 5.95$; $s^2 = 2.37$; $s = 1.54$; mode $= 6$; $\tilde{x} = 6$

6. a. $\bar{x} = 12$; $s = 5.66$
   b. Decrease; 20 is closer to the mean
   c. Decrease; 8 is closer to the mean

7. a. Sample
   b. $\bar{x} = 85$; $s = 8.94$
   c. $z = -1.57$; $z = 1.12$

8. a. 4   **b.** Professor   **c.** Republican
   d. B   **e.** Slow

9. a. 7.26   **b.** 3.20   **c.** .86

10. a. 1.06   **b.** .96   **c.** 31

11. $\bar{x} = 32.73$; $s = 6.37$ **12.** $\bar{x} = 8.51$; $s = 1.41$

Chapter 3 Achievement Test

1. a. Mode   **b.** C   **c.** 29   **d.** 34.48
   e. 72.41

2. a. 9   **b.** 8   **c.** 8   **d.** 7.5   **e.** 11.5
   f. 3.39   **g.** .59

3. a. 4.7   **b.** 2.19

4. a. x is the mean.
   b. x is 2 standard deviations above the mean.
   c. x is 1 standard deviation below the mean.
5. (c)   **6.** 4

CHAPTER 4

Problem Set 4.1

2. b. $\frac{1}{3}$ **c.** $\frac{3}{7}$
3. a. $y = 4x - 10$ **c.** $y = -(\frac{2}{3})x + \frac{8}{3}$
4. a. $y = (\frac{1}{2})x$ **b.** $y = -2x$
   c. $y = -2x + 10$ **d.** $y = (-\frac{5}{2})x + 17$
   e. $y = 2$ **f.** $x = -3$
5. a. $a = -2; b = \frac{1}{2}$ **c.** $a = -4; b = -3$
   e. $a = -\frac{1}{2}, b = -\frac{1}{2}$
6. $12

Problem Set 4.2

1. a. 50.83   **b.** 22.83   **c.** 17.83
2. a. Positive   **b.** None   **c.** None
   d. Negative   **e.** Positive   **f.** None
   g. Positive   **h.** Positive
3. b. Positive   **c.** 1   **5. c.** .53   **8.** $-.73$
9. .49   **10. b.** 0   **14.** 56.91
21. a. $\hat{y} = 4.0269615 - .0065461x$
   b. $-.98$ **c.** 1999

Problem Set 4.3

2. $\hat{y} = 3.1378 + .3508x$; SSE = 16.577
3. $\hat{y} = -2 + x$; SSE = 0
5. $\hat{y} = 33.7506 + .5339x$; SSE = 1234.928
7. a. $\hat{y}' = 1.3x'$ **b.** 6.5
8. $\hat{y} = 160.1 + 8.1x$; $\hat{y}(6) = 208.7$
9. $\hat{y} = 355.7122x - 1.849x$; $\hat{y}(65) = 235.51$

Chapter 4 Review

1. b. .93   **c.** $\hat{y} = -.7 + 3.9x$ **d.** 12,950
   e. 3900
2. b. $-.9139$ **c.** $\hat{y} = 100.8 - 5.3x$ **d.** 84.9
   e. -5.3 points
4. a. .904   **b.** $\hat{y} = -6.3 + 5.4x$ **d.** 13.95
6. a. $\hat{y} = 9.139 - .139x$ **b.** 7.89 billion dollars
7. b. .98   **c.** $\hat{y} = 2.8589 + 5.0753x$ **e.** 2.86
10. $\hat{y} = -3.208 + .0789x$; $\hat{y}(300) = 20.46$

Chapter 4 Achievement Test

1. a. 13.1   **b.** Increase   **c.** 1.4
   d. No; the number of push-ups cannot be negative.
2. a. 61.33   **b.** 2151.5
   c. 153   **d.** 2.495
   e. -22.97 **f.** $\hat{y} = -22.97 + 2.495x$

g. .4212   **h.** 1769.77
i. 139.205   **j.** 1.39
3. a. No   **b.** Inflation
4. Since $\hat{y} = a + bx$ and $a = \bar{y} - b\bar{x}$, $\hat{y} = (\bar{y} - b\bar{x}) + bx$. Therefore, $\hat{y} = \bar{y} + b(x - \bar{x})$. Substituting $\bar{y}$ for $\hat{y}$ and $\bar{x}$ for x, we have $\bar{y} = \bar{y}$. Hence, the point $(\bar{x}, \bar{y})$ is on the regression line.
5. 6

CHAPTER 5

Problem Set 5.1

1. a. Two of many possibilities are
   $S_1 = \{\text{girl, boy}\}$ and
   $S_2 = \{\text{student 1, not student 1}\}$.
   b. $S_1 = \{1, 2, 3, 4, 5, 6, 7, 8, 9, 10\}$
   $S_2 = \{\text{odd, even}\}$
2. a. It will be hot or it will rain.
   b. It will be hot and it will rain.
   c. It will not be hot and it will rain.
   d. It will not be hot or it will not rain.
   e. It will not be hot or it will rain.
   f. It will not be hot and it will not rain.
   g. It will not be hot and it will rain.
   h. It will not be hot and or it will not rain.
4. a. 9   **b.** 18   **c.** 27   **d.** 9   **e.** 18
   f. 14
5. a. $S = \{$TTT, HHH, HTH,THH, HHT, THT, TTH, HTT$\}$ **b.** 4   **c.** 6   **d.** 4   **e.** 4   **f.** 2
7. a. 1, 2   **b.** 1, 2, 3, 4, 5, 6
   c. 4, 6   **d.** 8
   e. 1, 3, 4, 6, 7, 8   **f.** 1
   g. 1, 3, 7   **h.** 4, 6, 7, 8
   i. 1, 2, 4, 7   **j.** 7, 8
9. a. {NYY, YNY, YYN, YYY}
   b. {NNN, NYN, NNY, NYY}
   c. {NNN, YNN, NYN, NNY}
   d. {NNN, NYN, NNY,NYY, YNY, YYN, YYY}
   e. {NYY}
   f. {NNN, NYN, NNY}

Problem Set 5.2

1. a. 0   **b.** 1   **c.** .01   **d.** .99
3. a. $\frac{3}{10}$ **b.** $\frac{1}{5}$ **c.** $\frac{1}{2}$ **5.** $\frac{1}{50}$
6. a. $\frac{9}{19}$ **b.** 10:9   **c.** $\frac{9}{19}$ **d.** 9:10
   e. $\frac{18}{19}$ **f.** $\frac{1}{38}$ **g.** 1:18   **h.** 18:1
7. c. .05   **d.** .30
9. a. $\frac{17}{30}$ **b.** $\frac{13}{30}$ **c.** $\frac{1}{6}$ **d.** $\frac{1}{10}$ **e.** $\frac{3}{20}$
11. a. He is a prime suspect, since
   $P(\text{ridge count} \geq 220) \simeq .07$.
14. The probability is approximately .79.

16. a. 684

18. *Hint:*   Consider all the possibilities.

Problem Set 5.3

1. 30   **3.** 720   **5.** 720   **7.** 70   **9.** 120

11. 2^{10}

Problem Set 5.4

1. a. .247   **b.** .26   **c.** .47

3. a. .21   **b.** .29   **c.** .57   **d.** .43

5. $\frac{1}{24}$ **7.** .2   **9.** .4; .6   **11. a.** .2   **b.** .5

Problem Set 5.5

2. a. 0   **b.** $\frac{1}{3}$ **c.** 1

3. a. $\frac{3}{10}$ **b.** $\frac{7}{10}$ **c.** $\frac{4}{5}$ **d.** $\frac{1}{2}$

4. a. $\frac{1}{3}$ **b.** $\frac{2}{3}$

5. a. $\frac{4}{17}$ **b.** $\frac{1}{17}$ **c.** $\frac{25}{51}$ **d.** $\frac{11}{51}$ **e.** $\frac{11}{663}$

6. a. $\frac{1}{3}$ **b.** $\frac{2}{15}$ **c.** $\frac{2}{3}$ **d.** $\frac{8}{15}$

7. a. .66   **b.** $\frac{22}{53}$ **c.** $\frac{22}{35}$ **8.** .122; .271

9. a. .527   **b.** $\frac{6}{7}$ **c.** $\frac{1}{7}$ **d.** .049

     e. $\frac{466}{951}$ **f.** .042   **g.** $\frac{42}{527}$

11. a. $\frac{5}{8}$ **b.** $\frac{1}{4}$ **c.** $\frac{4}{21}$ **d.** $\frac{3}{32}$ **e.** $\frac{11}{32}$

12. a. Yes   **b.** No   **c.** No

13. No. See Problem 12, part (g).

16. $P(E \text{ and } \bar{F}) = P(E) - P(E \text{ and } F)$

Problem Set 5.6

1. a. $\frac{1}{4}$ **b.** $\frac{1}{16}$ **c.** $\frac{1}{2}$ **d.** $\frac{3}{13}$ **e.** .308

3. a. No   **b.** No   **5.** Yes

6. a. $\frac{1}{8}$ **b.** $\frac{1}{8}$ **c.** $\frac{3}{8}$ **d.** $\frac{7}{8}$ **e.** $\frac{7}{8}$

8. a. .5   **b.** 0   **9.** No   **14.** 0   **15.** No

Problem Set 5.7

1. a. .23   **b.** .26   **c.** .60

     d. 1.28   **e.** 1.84

2. a. .30   **b.** .70   **c.** .57

     d. .95   **e.** 2.4   **f.** 1.66

3. $\mu = 1.5$; $\sigma = .866$

5.

x	# of days	P(x)
0	44	.044
1	87	.087
2	128	.128
3	234	.234
4	297	.297
5	155	.155
6	30	.030
7	25	.025

7. $E(w) = -\$0.26$; $\sigma^2 = 150.74$

9. b. $\frac{4}{55}$ **c.** $\frac{6}{55}$ **d.** $\frac{9}{11}$ **e.** $\frac{9}{55}$ **f.** $\frac{3}{55}$

10. 2.5   **12.** \$3220

Chapter 5 Review

1. a. .43   **b.** .82   **c.** .13

3. a. .2   **b.** .91

4. a. .76   **b.** .46   **c.** .14

     d. .27   **e.** .86

5. c. 3.5   **d.** 1.71

7. $\mu = \$800$; $\sigma = \$3429.29$ **9. b.** $\mu = \frac{4}{3}$; $\sigma = \frac{2}{3}$

10. a. \$11.30/share   **b.** \$1.30 gain   **c.** 13

     d. 1.51   **e.** 1.51

Chapter 5 Achievement Test

1. 720

2. a. $\frac{6}{11}$ **b.** $\frac{14}{33}$ **c.** $\frac{8}{33}$ **d.** $\frac{8}{11}$

     e. $\frac{4}{7}$ **f.** $\frac{3}{10}$ **g.** No   **h.** No

3. a. $S = \{0M, 1M, 2M\}$

     b. 1

     c. 3 (not counting order), 6 (counting order)

     d. .3

     e. .6

4. a. .33   **b.** .62   **c.** .29   **d.** .91

5. $E(w) = \$.36$ **6.** $\mu = .5$, $\sigma = .61$

CHAPTER 6

Problem Set 6.1

1. a. A binomial experiment for which we have 8 trials and are interested in 4 successes

3. Binomial with $p = .3$, $n = 10$, and $0 \le x \le 10$ (integer values only)

4. Not binomial

5. Binomial with $p = .5$, $n = 10$, and $0 \le x \le 10$ (integer values only)

7. Not binomial

8. a. 120   **b.** 1

     d. 1   **e.** 330

     f. 31,824   **g.** 792

     i. 300   **j.** 1536

10. *Hint:*   Examine the first several rows in Pascal's triangle.

Problem Set 6.2

1. a. .001   **b.** .209   **c.** .595   **d.** .009

3. a. .236   **b.** .154   **c.** .179

5. a. .313   **b.** .657   **c.** .234   **d.** .016

7. a. .107   **b.** .296   **c.** .263

     d. .000   **e.** .666

9. a. .000   **b.** .878   **c.** .026   **d.** .201

11. a. .073   **b.** .195   **c.** .433   **d.** .567

Problem Set 6.3

1. **a.** $\mu = 6$; $\sigma^2 = 2.4$; $\sigma = 1.55$
   c. $\mu = 13.5$; $\sigma^2 = 1.35$; $\sigma = 1.16$
   e. $\mu = 17$; $\sigma^2 = 2.55$; $\sigma = 1.60$
   g. $\mu = 2.7$; $\sigma^2 = 1.49$; $\sigma = 1.22$
   h. $\mu = 2.4$; $\sigma^2 = .96$; $\sigma = .98$

3. $\mu = 2.9$; $\sigma^2 = 1.5$ 5. $\mu = 6.8$; $\sigma = 1.48$

9. **b.** The two standard deviations are equal.

11. *Hint:* Complete the square for the quadratic expression.

Chapter 6 Review

1. **a.**

x	0	1	2	3	4	5
$P(x)$	.116	.312	.336	.181	.049	.005

   c. 1.75   **d.** 1.07   **e.** .994
2. **a.** .219   **b.** .004   **c.** .219   **d.** 4   **e.** 2
3. **a.** .250   **b.** .944   **c.** .526   **4.** .041
5. 1.53   **6.** 50   **7.** $\mu = 8.5$; $\sigma = 1.13$
9. .111   **10.** .874

Chapter 6 Achievement Test

1. **a.** .358   **b.** .735   **c.** .000   **d.** .003
2. **a.** .026   **b.** .677   **c.** 1.000   **d.** .503
3. **a.** 2   **b.** 1.6

4. **a.**

x	0	1	2	3	4	5
$P(x)$	.237	.396	.264	.088	.015	.001

   c. $\mu = 1.25$, $\sigma = .97$

CHAPTER 7

Problem Set 7.1

1. **a.** .16   **b.** .68   **c.** .815   **d.** .025
   e. .0235   **f.** .1585   **g.** .025   **h.** .475
2. **a.** .5000   **b.** .9270   **c.** .0933   **d.** .0215
   e. .0026   **f.** .00011   **g.** 1.000
3. **a.** .3907   **b.** .4904   **c.** .9538   **d.** .0248
   e. .1063   **f.** .9904   **g.** .0015
4. **a.** 1.645   **b.** 1.96   **c.** 1.28   **d.** 2.58
   e. -1.04 **f.** -1.78 **g.** .72

Problem Set 7.2

1. **a.** .0062   **b.** .9876   **c.** .0013   **d.** 1.000
2. **a.** 448   **b.** 500   **c.** 567
   d. 433   **e.** 552   **f.** 448
3. **a.** .9544   **b.** .1587   **c.** .9861
   d. .6915   **e.** .3551
5. .1587   **7.** 73.33 inches   **9.** .62%
10. **a.** 39.8   **b.** 38.25   **c.** 17.1

12. 8 years and 8 months
13. **a.** 84   **d.** 2

Problem Set 7.3

1. **a.** .5438   **b.** .2148   **c.** .4404   **d.** 0^+
3. **a.** .9732   **b.** .7397   **c.** .0069   **d.** .7257
5. **a.** .6563   **b.** .6578   **c.** .6578
7. .0582   **9.** .024

Chapter 7 Review

1. **a.** .8710   **b.** .1075   **c.** .9625
   d. .0162   **e.** .9429   **f.** .571
   g. .0308   **h.** .1242   **i.** .1036
2. **a.** 1.04   **b.** $-.25$ **c.** .25
   d. .84   **e.** $-.52$
3. **a.** .0808   **b.** .9452   **c.** .2119
   d. .9861   **e.** .7333   **f.** .3446
4. **a.** 59.9   **b.** 69.6   **c.** 44.8
   d. 55.2   **e.** 44.8   **f.** 52.5   **5.** 273
7. **a.** .2743   **b.** .0548   **c.** .9594   **d.** .1069
9. 64.12 mph
12. **a.** .0313   **b.** .6615   **c.** 2877   **d.** .9738

Chapter 7 Achievement Test

1. **a.** .6950   **b.** .8742   **c.** 2095
   d. .8729   **e.** .4052
2. **a.** .52   **b.** $-.67$ **c.** $-.25$ **d.** 1.28
3. **a.** .0099   **b.** .9082   **c.** .0376   **d.** .0437
4. **a.** 16.16   **b.** 20.75   **c.** 22.52   **d.** 22.31
5. **a.** 44.02   **b.** 38.5   **6.** .62%   **7.** .8925

CHAPTER 8

Problem Set 8.1

1. 13:Robert Moon, 04:Ed Doe, 01:Mike Able, 20:Maud Tuck, 12:James Lum, 17:Rick Quest, 07:Pete Gum, 02:Mary Baker, 10:Helen Jewel, 18:Bart Rat

3. We cannot tell, because we do not know how the numbers were selected.

4.

Digit	f
0	18
1	27
2	34
3	35
4	24
5	21
6	27
7	15
8	27
9	22

We would expect each digit to occur 25 times.

5.

Digit	Frequency	Expected frequency	Sampling error
0	18	25	−7
1	27	25	2
2	34	25	9
3	35	25	10
4	24	25	−1
5	21	25	−4
6	27	25	2
7	15	25	−10
8	27	25	2
9	22	25	−3
			0

6. a. Teacher 057, teacher 013, teacher 044, teacher 193, teacher 112
b. 057: teacher 57, 629: teacher 29, 013: teacher 13, 843: teacher 43, 840: teacher 40

7.

Digit	Frequency	Expected frequency	Sampling error
1	1	2	−1
2	3	2	1
3	2	2	0
4	1	2	−1
5	0	2	−2
6	5	2	3
			0

9. a. No; the sample is chosen on the basis of a physical trait.
b. No; the sample would be biased toward gym students.
c. No, especially if the number of students enrolled is greater than 10,000. Two different students can have different SS numbers but have the same last four digits of their SS numbers equal.
d. No; the sample would be biased toward sports fans.
e. No; the names are in alphabetical order and they all have phones.

Problem Set 8.2

1. a. The sampling distribution of the median is the distribution of all possible values of the median computed from samples of the same size.

b. The sampling distribution of the variance is the distribution of all possible values of the variance computed from samples of the same size.
c. The sampling distribution of the range is the distribution of all possible values of the range computed from samples of the same size.

2. None, since the mode of a sample does not always exist.

3. a. Either formula (8-2) or formula (8-3)
b. Formula (8-3)
c. Either formula (8-2) or formula (8-3)

5. It changes from .75 to .5, and gets smaller and approaches 0 as the sample size becomes large.

7. $\mu_x = 3.5$; $\sigma_{\bar{x}} = 2.09$ **8.** $\mu_{\bar{x}} = 3.5$; $\sigma_{\bar{x}} = 1.71$

11. a.

x	P(x)
30	.01
33	.08
36	.26
39	.40
42	.25

b. 38.4   **c.** 2.81

Problem Set 8.3

1. a. .0212   **b.** .7912   **c.** .2709   **d.** .8438
e. .9925

3. .005   **5.** .95   **7.** .045

8. The value of n does not exist. The limiting t distribution is the standard normal distribution.

9. −2.14   **10.** .975

11. a. .0808   **b.** .8791   **c.** .9599   **d.** .0179
13. a. .8531   **b.** .6448   **c.** .9826
14. a. 2.718   **b.** 1.796
15. a. 32.58   **b.** 33.43   **c.** 31.72
d. 32.28   **e.** 30.77

16. *Hint:*   Use the t distribution to determine the probability of obtaining a sample mean at least as large as the one given by the sample.

Problem Set 8.4

1. $\mu_{\bar{x}} = 6$; $\sigma_{\bar{x}} = .77$;
the sampling distribution of the mean is approximately normal.

3. a. $\mu = 4$; $\sigma = 2.24$
b. $\mu_{\bar{x}} = 4$; $\sigma_{\bar{x}} = 1.29$

5. a. .9049   **b.** .8078   **c.** .9120   **d.** .3300
8. a. .1190   **b.** .7620   **c.** .1190
10. a. .8854   **b.** .0918   **c.** .7486
11. a. $\mu_{\bar{x}} = 75$, $\sigma_{\bar{x}} = 1.58$
b. 238
c. 291

Problem Set 8.5

1. a. .8340   b. .8506   c. .6255
3. a. .2877   b. .4246   c. .9871
4. a. .60   b. .045
5. 95.44
7. a. .0245   b. .3409   c. .6827
8. .018
9. a. .0194   b. (.72, .78)
10. .0188

Chapter 8 Review

1. a. Yes
   b. No; all possible samples of size 5 are not available using this method.
4. a. 4   b. 1.58
5. a. 4   b. 1.29

7. a.

Median	f	Sampling error	f(sampling error)
1	10	$1 - 3 = -2$	-20
2	22	$2 - 3 = -1$	-22
3	22	$3 - 3 = 0$	0
6	10	$6 - 3 = 3$	30
	64		-12

c. $-.1875$
d. 2.8125
e. 1.5297
f. 1.5297

8. a, b.

Sample	p
3, 4, 5, 6	.75
2, 4, 5, 6	.50
2, 3, 5, 6	.50
2, 3, 4, 6	.50
2, 3, 4, 5	.75
1, 4, 5, 6	.75
1, 3, 5, 6	.75
1, 3, 4, 6	.75
1, 2, 5, 6	.50
1, 3, 4, 5	1.00
1, 2, 4, 6	.50
1, 2, 4, 5	.75
1, 2, 3, 6	.50
1, 2, 3, 5	.75
1, 2, 3, 4	.75

c. $\frac{2}{3}$
d. .15
9. a. .9554
   b. .8187
   c. .9977
13. a. .3108
    b. .0703
    c. .5793
14. 232.62

Chapter 8 Achievement Test

1. -5 3. 1.5
4. a. 2.53
   b. 9
   c. 2.11
5. a. .7492
   b. .5
   c. 40.3
6. a. 0.1056
   b. 0.4649
7. -3.16 8. .65
9. a. 4
   b. .71

CHAPTER 9

Problem Set 9.1

1. a. 79.94%
   b. 85.02%
   c. 99.02%
   d. 99.42%
2. a. 1.88
   b. 1.65
   c. 1.56
3. 2.145
5. 2.462
6. .5513
7. 6.15
9. .5061
10. 77.38%
11. .7132

Problem Set 9.2

1. a. (27.56, 29.04); $w = 1.48$
   b. (27.42, 29.18); $w = 1.76$
   c. (27.14, 29.46); $w = 2.32$
   d. (26.96, 29.64); $w = 2.68$
3. (69.59, 71.21)
5. (26.19, 30.01)
7. (147.38, 156.62)

8. (155.41, 171.59)

9. a. 1.80

    b. .30

    c. .85

11. (.98, 1.04)

12. (74.38, 131.12)

17. $\mu > 68.08$

Problem Set 9.3

1. a. (.21, .29)

    b. (.22, .28)

    c. (.58, .62)

    d. (.34, .46)

3. a. .037

    b. (.47, .55)

4. (.017, .063)

5. (.024, .056)

6. (.65, .95)

7. (.47, .79)

9. (.46, .60)

10. (.57, .63)

15. (.53, .65)

17. (.72, .78)

18. (.35, .61)

19. (.14, .26)

Problem Set 9.4

1. 2090; 2401

2. 4096

3. 6670

4. 16, 641

5. 8487

7. 9604

9. 466

10. 711

11. 25

15. *Hint:*   Complete the square.

Problem Set 9.5

1. a. 34.267; 5.142

    b. 39.364; 12.401

    c. 20.483; 3.247

    d. 40.113; 16.151

2. a. A 90% confidence interval for σ^2 is (.015, .039); a 90% confidence interval for σ is (.12, .20).

    c. A 90% confidence interval for σ^2 is (.72, 1.57); a 90% confidence interval for σ is (.85, 1.25).

3. a. 1.61

    b. (1.37, 1.98)

4. 6.087

5. A 95% confidence interval for σ^2 is (.020, .11); a 95% confidence interval for σ is (.14, .34).

7. .99   **8.** .01

9. (1.07, 5.14)   **11.** (.0018, .0055)

Chapter 9 Review

1. a. (.55, .57)

    b. $n = 9466$

3. 666

4. (.00091, .0029)

5. (.346; .444); $E = .049$

7. (.089, .50)

8. a. .75

    b. (1.55, 3.05)

    c. 4532

9. (57.76, 66.24)

11. (3.15, 6.63)

12. a. 1.35

    b. (25.85, 28.55)

    c. 565

13. a. .028

    b. (7.992, 8.048)

15. a. 2.42

    b. (232.68, 237.52)

    c. 188

Chapter 9 Achievement Test

1. a. 1.74

    b. 42.980

2. a. 4.26

    b. -9.62

    c. 540

3. (16.47, 17.93)

4. (.12, .22)

5. (.97, 2.42)

6. 547   **7.** 80

8. a. .60

    b. .85

    c. .80

    d. .95

    e. .98

CHAPTER 10

Problem Set 10.1

1. a. *B*

    b. *A*

    c. *B*

    d. *B*

2. (a) and (d)

3. a. H_0: The effectiveness of the new drug is less than or equal to the old drug.
H_1: The effectiveness of the new drug is greater than that of the old drug. Decision: reject H_0; Type I error

b. H_0: School 1 is as effective as school 2.
H_1: School 1 is not as effective as school 2. Decision: do not reject H_0; good decision

c. H_0: The proportion of repairs for RCAs $\leq$ the proportion of repairs for Zeniths.
H_1: The proportion of repairs for RCAs > the proportion of repairs for Zeniths. Decision: reject H_0; good decision

d. H_0: A is as efficient as B.
H_1: A is not as efficient as B. Decision: do not reject H_0; Type II error

4. a. Two-tailed
c. Two-tailed
d. One-tailed

7. a. When smoking is harmful to your health, deciding that smoking is harmful.
b. When smoking is harmful to your health, deciding that smoking is not harmful.
c. Deciding that smoking is harmful when it is harmful or deciding that smoking is not harmful when it is not harmful are good decisions.
A Type II is the more serious error.

8. a. Type I: Deciding that the discovery method is not better than the expository method when in reality the opposite is true;
Type II: Deciding that the discovery method is better than the expository method when in reality the opposite is true.
b. Type I: Deciding that more than 2% of machines are defective when in reality at most 2% are defective;
Type II: Deciding that at most 2% of machines are defective when in reality more than 2% are defective.
c. Type I: Deciding that less than 95% of Americans are against war when in reality at least 95% are against war;
Type II: Deciding that at least 95% of Americans are against war when in reality less than 95% are against war.

9. a. False
b. False
c. True
d. False
e. True

11. Not in general. Since $0 \leq \alpha \leq 1$ and $0 \leq \beta \leq 1$, the value of $\alpha + \beta$ need not equal 1; it can equal any value between 0 and 2, inclusive.

12. b. (1) Incorrect; Type II
(2) Correct
(3) Incorrect; Type I
(4) Correct

13. No

14. No

15. No

Problem Set 10.2

1. $C = 479.38$; all $\bar{x} < 479.38$
3. $C = 3,140.25$; all $\bar{x} > 3,140.25$
4. $C = 149.07$; all $\bar{x} < 149.07$
5. a. H_0: $\mu \geq 500$
b. H_1: $\mu < 500$
c. 480
d. .05
e. Cannot reject H_0
f. Type II

7. a. H_0: $\mu \leq 30000$
b. H_1: $\mu > 30000$
c. 30,200
d. .05
e. Reject H_0
f. Type I

9. Yes

10. a. H_0: $\mu = 10.4$
b. H_1: $\mu \neq 10.4$
c. $t = -4.34$
d. .05
e. ± 2.306
f. Reject H_0
g. Type I

11. a. No
b. No
c. Yes
d. Yes
e. No

14. *Hint:* What would you use as the null value?

Problem Set 10.3

1. a. H_0: $\mu \geq 95$
b. H_1: $\mu < 95$, one-tailed test
c. $\alpha = .05$
d. Distribution of the mean
e. $z = -1.94$
f. $-z_{.05} = -1.65$
g. Reject H_0
h. Type I, $\alpha = .05$
i. p-value $= .5 - .4738 = .0262$

3. a. H_0: $\mu \geq 10$
b. H_1: $\mu < 10$, one-tailed test
c. $\alpha = .01$

d. t distribution
e. $t = -2.15$
f. $-t_{.01}(14) = -2.624$
g. Cannot reject H_0
h. Type II, β is unknown
i. $.01 < p\text{-value} < .025$

5. a. $H_0: \mu = 70$
b. $H_1: \mu \neq 70$, two-tailed test
c. $\alpha = .05$
d. t distribution
e. $t = .67$
f. $t_{.025}(9) = 2.262$
g. Do not reject H_0
h. Type II, β is unknown
i. $p\text{-value} > .10$

6. a. $H_0: \mu \geq .25$
b. $H_1: \mu < .25$, one-tailed test
c. $\alpha = .10$
d. Sampling distribution of the mean
e. $z = -1.18$
f. $-z_{.10} = -1.28$
g. Do not reject H_0
h. Type II error, β is unknown
i. $p\text{-value} = .119$

7. a. $H_0: \mu \geq .84$
b. $H_1: \mu < .84$, one-tailed test
c. $\alpha = .05$
d. Sampling distribution of the mean
e. $z = -.22$
f. $-z_{.05} = -1.65$
g. Do not reject H_0
h. Type II error, β is unknown
i. $p\text{-value} = .4129$

9. a. $H_0: \mu = 14$
b. $H_1: \mu \neq 14$, two-tailed test
c. $\alpha = .05$
d. t distribution
e. $t = .70$
f. $t_{.025} = 2.365$
g. Do not reject H_0
h. Type II error, β is unknown
i. $p\text{-value} > .10$

10. $(62.93, 83.07)$ **11.** $(13.29, 15.31)$

13. $(4.86, 4.94)$

14. *Hint:* Find $\beta = P(\bar{x} < 36.57$ whenever $\mu = 37)$.

15. $.7224$

Problem Set 10.4

1. a. $z = -2.24$, $\pm z_{.005} = -2.58$; do not reject H_0
c. $z = -2.89$, $-z_{.01} = -2.33$; reject H_0

2. a. $\chi^2 = 13.5$, $\chi^2_{.95}(27) = 16.151$; reject H_0
c. $\chi^2 = 28.8$, $\chi^2_{.025}(36) = 54.437$,
$\chi^2_{.975}(36) = 21.336$; do not reject H_0

3. $H_0: p \leq .20$, $H_1: p > .20$; $z = 2.80$,
$z_{.01} = 2.33$, $p\text{-value} = .0026$;
reject H_0

5. $H_0: p \geq .55$, $H_1: p < .55$; $z = -1.28$,
$-z_{.05} = -1.65$, $p\text{-value} = .0778$; do not reject H_0

7. $H_0: p = .60$, $H_1: p \neq .60$; $z = -1.63$,
$\pm z_{.025} = \pm 1.96$, $p\text{-value} = .1032$; do not
reject H_0

8. $H_0: \sigma \geq 1.5$, $H_1: \sigma < 1.5$; $\chi^2 = 8.727$,
$\chi^2_{.95}(10) = 3.940$; do not reject H_0

9. $H_0: \sigma^2 = 121$, $H_1: \sigma^2 \neq 121$; $\chi^2 = 18.577$,
$\chi^2_{.005}(19) = 38.580$, $\chi^2_{.995}(19) = 6.843$; do not
reject H_0

11. $H_0: \sigma^2 \leq 5$, $H_1: \sigma^2 > 5$; $\chi^2 = 12.54$,
$\chi^2_{.05}(11) = 19.675$; do not reject H_0

12. $H_0: \sigma^2 \leq .0001$, $H_1: \sigma^2 > .0001$; $\chi^2 = 54$,
$\chi^2_{.01}(24) = 42.980$; reject H_0

13. $(.42, .62)$

Chapter 10 Review

1. $H_0: \mu \leq 24$, $H_1: \mu > 24$; $z = 3.94$, $z_{.05} = 1.65$,
decision: reject H_0; $p\text{-value} = .0005$

3. $H_0: p \geq .62$, $H_1: p < .62$; $z = -1.84$,
$-z_{.05} = -1.65$, decision: reject H_0; $p\text{-value} = .0329$

4. $H_0: \mu \geq 7$, $H_1: \mu < 7$; $t = -3.27$,
$-t_{.01}(14) = -2.624$, decision: reject H_0

5. $H_0: \mu = 140$, $H_1: \mu \neq 140$; $t = 1.38$,
$\pm t_{.025}(6) = \pm 2.447$, do not reject H_0

7. $H_0: \sigma^2 \leq .05$, $H_1: \sigma^2 > .05$; $\chi^2 = 33.6$,
$\chi^2_{.05}(24) = 36.415$, decision: do not reject H_0

9. $H_0: p \leq .6$, $H_1: p > .6$; $z = 2.91$, $z_{.05} = 1.65$,
decision: reject H_0

10. $H_0: \sigma = 9.8$, $H_1: \sigma \neq 9.8$; $\chi^2 = 6.61$,
$\chi^2_{.975}(9) = 2.700$, $\chi^2_{.025}(9) = 19.023$, decision:
do not reject H_0

11. $H_0: \mu = 16$, $H_1: \mu \neq 16$; $t = -1.11$,
$\pm t_{.005}(5) = \pm 4.032$, decision: do not reject H_0

Chapter 10 Achievement Test

1. $H_0: \mu = 71$, $H_1: \mu \neq 71$
a. $t = 1.15$ **b.** $\pm t_{.025}(11) = \pm 2.201$
c. Do not reject H_0. **d.** $p\text{-value} > .20$

2. $H_0: p \geq .85$, $H_1: p < .85$
a. $z = -1.08$ **b.** $-z_{.01} = -2.33$
c. Do not reject H_0. **d.** $p\text{-value} = .1401$

3. a. 23.75
b. 30.144
c. Do not reject H_0.

4. $H_0: \mu \geq 60{,}000$, $H_1: \mu < 60{,}000$; $z = -2.44$,
$-z_{.05} = -1.65$; decision: reject H_0; thus, the data
indicate that $\mu < 60{,}000$.

5. H_0: $\sigma \geq 1200$, H_1: $\sigma < 1200$; $\chi^2 = 22.12$,
$\chi^2_{.99}(34) = 17.789$; decision: do not reject H_0

CHAPTER 11

Problem Set 11.1

1. Dependent
2. a. Randomly assign subjects to the two groups, eight per group.
  b. For each twin pair, assign one to group A and one to group B.
3. Independent
5. Independent
6. Use the same subjects for a period of 6 months. Replicate the experiment twice, reversing the order of treatment.
8. Dependent

Problem Set 11.2

1. .0793
2. .2502
3. a. $z = -1.44$, $\pm z_{.005} = \pm 2.58$; do not reject H_0.
  b. $(-.16, .56)$. We cannot say there is a difference.
5. $z = 2.50$, $z_{.01} = 2.33$; reject H_0, p-value $= .0062$
7. $z = 2.67$, $z_{.05} = 1.65$; p-value $= .0038$, reject H_0
9. $z = 1.93$, $z_{.05} = 1.65$; reject H_0, p-value $= .0268$
10. a. $z = 1.98$, $\pm z_{.005} = \pm 2.58$; cannot reject H_0
  b. $(-.423, 3.223)$

Problem Set 11.3

1. a. .2
  b. .061
  c. $z = -.66$
2. a. $z = .53$, $\pm z_{.025} = \pm 1.96$; p-value $= .5962$, do not reject H_0
  b. $(-.018, .058)$
3. $z = 1.30$, $\pm z_{.025} = \pm 1.96$; p-value $= .1936$ do not reject H_0
5. $(-.145, .185)$
6. $z = .24$, $\pm z_{.025} = \pm 1.96$; do not reject H_0
7. $(-.048, .064)$
8. $(-.11, .14)$
10. $z = .67$, $z_{.025} = 1.96$; p-value $= .5028$, do not reject H_0
11. $z = 2.09$, $\pm z_{.025} = 1.96$; reject H_0; a 95% confidence interval is $(.0053, .1421)$.

Problem Set 11.4

1. $F = 1.21$, df $= (19, 14)$
2. a. .490
  b. .392

c. .337
d. .392
e. .219
f. .206
3. a. $(1.09, 5.13)$
  b. $(.17, 1.08)$
4. a. $(1.044, 2.26)$
  b. $(.41, 1.041)$
5. a. $F = 1.70$, $F_{.05}(25, 15) = 2.33$; do not reject H_0
  c. $F = 4.27$, $F_{.05}(10, 25) = 2.24$; reject H_0
6. a. $F = 1.65$, df $= (30, 24)$, $F_{.01}(30, 24) = 2.58$; do not reject H_0
  c. $F = 3.81$, df $= (9, 11)$, $F_{.01}(9, 11) = 4.63$; do not reject H_0
7. a. $F = 1.78$, df $= (20, 30)$, $F_{.05}(20, 30) = 1.93$; do not reject H_0
  b. $(.53, 1.04)$
8. a. $(.28, 2.84)$
  b. $F = 1.12$, df $= (9, 9)$, $F_{.05}(9, 9) = 3.18$; do not reject H_0
9. $F = 1.78$, df $= (20, 19)$, $F_{.05}(20, 19) = 2.15$; do not reject H_0
10. a. Yes
  b. $(.36, .81)$
  c. $(1.23, 2.78)$

Problem Set 11.5

1. a. 1.22
  b. 3
2. $t = 3.71$, $-t_{.05}(18) = 1.734$; reject H_0
3. $t = -1.20$; df $= 10 - 1 = 9$
4. $t = 3.04$, $t_{.05}(5) = 2.015$; we can conclude that the program was effective.
5. $t = 3.03$, $t_{.025}(5) = 2.571$; we can conclude that there is a difference in tread wear.
6. $(.048, .586)$
7. a. $F = 1.4826$, $F_{.05}(12, 17) = 2.38$; do not reject H_0
  b. $t = -4.15$, $-t_{.01}(29) = -2.462$; we can conclude that drug A is associated with lower mean cholesterol than drug B.
9. a. $t = 1.50$, $\pm t_{.005}(21) = \pm 2.831$; do not reject H_0
  b. $(-14.42, 4.42)$
10. a. $t = 1.41$, $\pm t_{.025}(29) = \pm 2.045$; do not reject H_0
  b. $(-.31, 1.71)$
11. $t = 3.13$, $t_{.05}(13) = 1.771$; we can conclude that the diet is effective.
13. a. $(-6.73, .73)$
  b. $t = -1.68$, $t_{.025}(20) = -2.086$; do not reject H_0

15. a. $(-3.32, 10.32)$
   b. $t = 1.13$, $t_{.025}(11) = 2.201$; do not reject H_0

Chapter 11 Review

1. a. $z = 2.58$, $z_{.025} = \pm 1.96$; reject H_0,
   p-value $= .0098$
   b. $(2.01, 14.83)$
3. $(-.13, .20)$
4. $t = 2.38$, $\pm t_{.025}(10) = \pm 2.228$; reject H_0
5. $z = 4.82$, $\pm z_{.025} = \pm 1.96$; reject H_0
6. $(.07, .17)$
7. $z = 1.29$, $z_{.01} = 2.33$; do not reject H_0,
   p-value $= .0985$
8. $F = 2.25$, $F_{.01}(30, 14) = 3.34$; do not reject H_0
9. $(.82, 2.48)$
10. a. $z = 3.64$, $\pm z_{.025} = \pm 1.96$; reject H_0
   b. $(.65, 2.15)$
12. $(.009, .051)$
13. a. $t = 1.74$, $t_{.025}(11) = 2.201$; cannot reject H_0,
   $.05 < p$-value $< .10$
   b. $(-2.21, 19.05)$
15. a. $t = .96$, $\pm t_{.005}(10) = \pm 3.169$; cannot reject H_0
   b. $(-.37, .69)$
17. $(1.16, 3.24)$; no
18. $t = 1.33$, $t_{.01}(11) = 2.718$; cannot reject H_0
19. $z = 3.87$, $\pm z_{.025} = \pm 1.96$; reject H_0
21. $F = 2.09$, $F_{.05}(49, 39) \simeq 1.66$; reject H_0; yes, the test
   should not be used.
22. $z = 3.16$, $\pm z_{.025} = \pm 1.96$; reject H_0

Chapter 11 Achievement Test

1. a. $F = 1.33$, $F_{.05}(30, 30) = 1.84$; do not reject H_0
   b. $z = 2.83$, $z_{.05} = 1.65$; reject H_0
   c. $(1.29, 4.91)$
2. a. $\hat{p} = .56$
   b. $z = .32$
   c. $z_{.05} = 1.65$; do not reject H_0
   d. $(-.16, .22)$
3. a. $(.72, 2.44)$
   b. $(.64, 1.18)$
4. a. 2.83
   b. 2.12
   c. $z = 2.47$; reject H_0
   d. p-value $= .0068$
   e. $(2.33, 11.67)$
5. a. 4.31
   b. $F = 1.33$, $F_{.05}(14, 14) = 2.48$; cannot reject the
   hypothesis of equal variances
   c. $t = 1.97$, $t_{.05}(28) = 1.701$; reject H_0
   d. $(.42, 5.78)$
   e. $(-5.78, -.42)$

CHAPTER 12

Problem Set 12.1

1. a. df $= 5$; $\chi^2_{.05}(5) = 11.070$
   b. df $= 20$; $\chi^2_{.01}(20) = 37.566$
   c. df $= 18$; $\chi^2_{.05}(18) = 28.869$
   d. df $= 9$; $\chi^2_{.01}(9) = 21.666$
2. a. $z = 1.12$, $\pm z_{.025} = \pm 1.96$; do not reject H_0
   b. $\chi^2 = 1.25$, $\chi^2_{.05}(1) = 3.841$; do not reject H_0
   c. $1.12^2 \simeq 1.25$
4. $\chi^2 = 4.19$, df $= 2$, $\chi^2_{.05}(2) = 5.991$; do not reject H_0
5. $\chi^2 = 64.79$, $\chi^2_{.01}(3) = 11.345$; we can conclude that
   the proportions differ.
7. $\chi^2 = 2.12$, $\chi^2_{.05}(3) = 7.815$; we cannot conclude that
   the proportions differ.
9. $\chi^2 = .42$, $\chi^2_{.05}(3) = 7.815$; do not reject H_0

Problem Set 12.2

1. $\chi^2 = 1.22$, $\chi^2_{.01}(4) = 13.277$; do not reject H_0
3. $\chi^2 = 26.6$, $\chi^2_{.05}(9) = 16.919$; we can conclude that
   the frequencies differ.
5. $\chi^2 = 32.6$, $\chi^2_{.05}(4) = 9.488$; we can conclude that
   the color preferences differ.
6. $\chi^2 = 158.52$, $\chi^2_{.05}(4) = 9.488$; we can conclude that
   the grades do not follow a normal distribution.
7. $\chi^2 = 11.98$, $\chi^2_{.05}(3) = 7.815$; we can conclude that
   the results not follow the specified binomial
   distribution.
8. $\chi^2 = 7.04$, $\chi^2_{.05}(5) = 11.070$; do not reject H_0
9. $\chi^2 = 6.8$, $\chi^2_{.05}(9) = 16.919$; do not reject H_0. The
   digits may occur with equal frequency.

Problem Set 12.3

1. $\chi^2 = .15$, $\chi^2_{.05}(2) = 5.991$; do not reject H_0
3. $\chi^2 = 16.81$, $\chi^2_{.05}(4) = 9.488$; we can conclude that
   interest and ability are related.
5. $\chi^2 = 19.33$, $\chi^2_{.01}(6) = 16.812$; we can conclude that
   social status and high school program are
   dependent.
7. $\chi^2 = 30.15$, $\chi^2_{.05}(10) = 18.307$; we can conclude
   that cigarette smoking and systolic blood pressure
   are related.
9. $\chi^2 = 12.36$, $\chi^2_{.05}(2) = 5.991$; we can conclude that
   student responses are related to grade level.

Chapter 12 Review

1. $\chi^2 = 9.5$, $\chi^2_{.05}(4) = 9.488$; we can conclude that
   there are differences in the preferences.
3. $\chi^2 = 8.82$, $\chi^2_{.05}(1) = 3.841$; we can conclude that
   the efficacy of the drug differs from 85%.
5. $\chi^2 = 7.6$, $\chi^2_{.05}(5) = 11.070$; there is no evidence
   that the die is biased.

7. $\chi^2 = .94$, $\chi_{.05}^2(1) = 3.841$; we cannot conclude that the percentages differ.

9. $\chi^2 = 5.33$, $\chi_{.05}^2(2) = 5.991$; we cannot conclude that the percentages have changed.

11. $\chi^2 = 1.38$, $\chi_{.01}^2(3) = 11.345$; we cannot conclude that the proportions differ.

13. $\chi^2 = 35.25$, $\chi_{.05}^2(2) = 5.991$; we can conclude that social class and newspaper subscriptions are related.

15. $\chi^2 = 157.39$, $\chi_{.05}^2(2) = 5.991$; the data indicate that the categories are not homogeneous.

Chapter 12 Achievement Test

1. $z = 3.87$

2. a. $p_A \neq .9$ **b.** 129.60
  c. 6.635   **d.** Reject H_0

3. a. 8.63   **b.** 13.277   **c.** Cannot reject H_0

4. a. 15.87   **b.** 5.991
  c. Reject H_0; the variables are dependent.

CHAPTER 13

Problem Set 13.1

1. a. $t = .8866$, $t_{.025}(8) = 2.306$, df = 8; do not reject H_0
  b. $F = .786$, $F_{.05}(1, 8) = 5.32$; do not reject H_0
  c. $t^2 = (.8866)^2 = .786$

3. $F = 4.92$, $F_{.05}(2, 12) = 3.89$; the laboratories produce different mean results.

5. $F = 5.93$, $F_{.05}(3, 12) = 3.49$; there is a difference in the mean scores for the four schools.

7. $F = 2.29$, $F_{.05}(2, 57) \simeq 3.23$; there is no evidence that the mean participation times differ.

9. $F = 3.7382$, $F_{.01}(2, 15) = 6.36$; there is no evidence that the three diets produce different mean blood pressure measurements.

Problem Set 13.2

1. SST = 188.9, SSB = 16.9, SSW = 172.0, MSB = 16.9, MSW = 21.5, $F = .79$

3. SST = 1419.8, SSB = 52.2, SSW = 1367.6, MSB = 17.4, MSW = 85.48, $F = .20$

5. SST = 13,658.38, SSB = 335.54, SSW = 13,322.84, MSB = 167.77, MSW = 360.08, $F = .47$

Problem Set 13.3

1. $F = 27.62$, $F_{.05}(3, 32) \simeq 2.92$; reject H_0; CD = 30.61, A $\neq$ D, A $\neq$ C, B $\neq$ D, B $\neq$ C; $\hat{\omega}^2 = .69$

3. $F = 9.50$, $F_{.05}(2, 9) = 4.26$; CD = 9.07, DB2 $\neq$ DB3; $\hat{\omega}^2 = .59$

5. $F = 22.83$, $F_{.05}(2, 18) = 3.55$; reject H_0 A versus B: CD = 3.46, A $\neq$ B;

B versus C: CD = 3.61;
A versus C: CD = 3.72, A $\neq$ C; $\hat{\omega}^2 = .68$

Chapter 13 Review

1. $F = 4.45$, $F_{.05}(2, 27) = 3.35$; reject H_0

3. $F = 6.48$, $F_{.05}(3, 8) = 4.07$; reject H_0

5. $F = 1.835$, $F_{.05}(2, 10) = 4.10$; cannot reject H_0

Chapter 13 Achievement Test

1. a. 30.0   **b.** 7.5   **c.** 22.5   **d.** 2
  e. 6   **f.** 3.75   **g.** 3.75   **h.** 3.75
  i. 3.75   **j.** 1

2.

Source	SS	df	MS	F
Between	7.5	2	3.75	1
Within	22.5	6	3.75	
Total	30.0	8		

3. 0   **4.** We cannot reject H_0.

CHAPTER 14

Problem Set 14.1

1. a. $y = 3 + 4x + \varepsilon$
  b. 11.5
  c. -5

2. a. 4   **b.** -8 **c.** 4   **d.** 4

3. a. 83.2712
  b. 28.18
  c. 3.82
  d. 96.91
  e. 96.91
  f. 116, the mean of y in both cases

5. a. 57.56   **b.** 4.50
  c. 170.44   **d.** 219.45
  e. 187.97

Problem Set 14.2

1. a. $F = 13.280$, $F_{.05}(1, 5) = 6.61$; the linear model is appropriate; $\hat{y} = 9.0256 + .3020x$
  b. (.089, .52)

3. a. $F = 277,634.15$, $F_{.05}(1, 3) = 10.1$; the linear model is appropriate; $\hat{y} = 69.7085 + 203.758x$
  b. (879.98, 889.50)
  c. (1686.97, 1712.57)
  d. (1688.43, 1711.11)
  e. 203.76

5. $t = 3.64$, $\pm t_{.025}(5) = \pm 2.571$; the linear model is appropriate.

7. $t = 526.91$, $\pm t_{.025}(3) = \pm 3.182$; the linear model is appropriate.

Problem Set 14.3

1. $t = 2.137$, $\pm t_{.025}(10) = \pm 2.228$; there is no evidence that ρ is different from 0.
2. $t = 1.638$, $t_{.005}(10) = 3.169$; cannot reject H_0. No, do not use bone sizes to predict body sizes.
4. $t = 7.309$, $t_{.05}(28) = 1.701$; reject H_0
5. $t = -2.843$, $t_{.05}(18) = 1.734$; reject H_0
7. **a.** .842
   b. $z = 2.526$, $z_{.01} = 2.33$; we can conclude that the correlation coefficient is greater than 0.
9. **a.** $-.6675$
   b. $t = -2.835$, $\pm t_{.025}(10) = \pm 2.228$; reject H_0
11. .45

Chapter 14 Review

1. $t = 1.06$, $\pm t_{.025}(4) = \pm 2.776$; cannot reject H_0
2. **a.** .0564   **b.** $z = 1.69$, $\pm z_{.025} = \pm 1.96$; cannot reject H_0
3. **a.** .6   **b.** $z = 1.8$, $z_{.05} = 1.65$; reject H_0
5. **a.** $F = 58.10$, $F_{.01}(1, 3) = 34.1$; reject H_0; the linear model is appropriate; $\hat{y} = -4.9824 + 3.1985x$
   b. (.75, 5.65)
7. **a.** $t = 2.25$, $t_{.05}(4) = 2.132$; reject H_0
   b. 58.96, 98.03
   c. 2.92
9. **a.** .96
   b. .97
   c. $t = 8.093$, $t_{.05}(6) = 1.943$; reject H_0
   d. $z = 2.57$, $z_{.05} = 1.65$; reject H_0
   e. The linear model is appropriate by part (d). β equals 0 if and only if ρ equals 0.
   f. $\hat{y}(52) = 218.2$. Thus, at the end of the year, the person would weigh 71.8 pounds—a very unlikely weight with his height. The value $x = 52$ should not be used since it is not near the given fixed values of x.

Chapter 14 Achievement Test

1. **a.** .60
   b. 0
   c. .5
   d. 1.20
   e. H_0: $\rho = 0$, H_1: $\rho \neq 0$; $z = 1.20$, $\pm z_{.025} = \pm 1.96$; cannot reject H_0
2. **a.** $t = 4.4$
   b. $\pm t_{.025}18 = \pm 2.101$
   c. Cannot reject H_0
3. **a.** 4.24
   b. ± 4.30
   c. $\beta = 0$
   d. The linear model is not appropriate.
   e. $\hat{y} = 1.5 + 1.5x$

4. **a.** 1.5   **b.** 6   **c.** .25   **d.** .90
5. (3.42, 5.58)

CHAPTER 15

Problem Set 15.1

1. $z = 1.29$, $z_{.05} = 1.65$; do not reject H_0
3. $z = 2.26$, $\pm z_{.025} = \pm 1.96$; reject H_0
5. $z = 0$, $\pm z_{.025} = \pm 1.96$; do not reject H_0
7. Use the binomial with $n = 6$ and $p = .5$. $P(x \geq 4) = .344$. Since $\alpha < .344$, we have no evidence that the exercise program helps to lower blood pressure.

Problem Set 15.2

1. $U = 13$, $z = -2.61$, $\pm z_{.025} = \pm 1.96$; there is a difference in the useful lives.
3. $U = 79.5$, $z = -1.81$, $-z_{.05} = -1.65$; there is a difference.
5. $U = 29$, $z = -1.59$, $-z_{.05} = -1.65$; we have no evidence that method B is better.
7. **a.** $t = .17$, $\pm t_{.025}(22) = \pm 2.074$; there is no evidence that there is a difference.
   b. $U = 72$, $z = .29$, $\pm z_{.025} = \pm 1.96$; there is no evidence that there is a difference.
9. **a.** $-t = 1.71$, $-t_{.05}(28) = -1.701$; we can conclude that $\mu_A < \mu_B$.
   b. $U = 69$, $z = -1.80$, $-z_{.05} = -1.65$; we can conclude that $\mu_A < \mu_B$.
   c. No. For the rank-sum test the p-value = .005, and for the t test the p-value $\simeq$ .05.

Problem Set 15.3

1. $H = .70$, $\chi^2_{.05}(1) = 3.841$; do not reject H_0
3. $H = .54$, $\chi^2_{.05}(3) = 7.815$; do not reject H_0
5. $H = 12.79$, $\chi^2_{.05}(3) = 7.815$; reject H_0

Problem Set 15.4

1. **a.** $R = 15$, $n_1 = 12$, $n_2 = 11$; $\mu = 12.4783$, $\sigma = 2.3381$, $z = 1.079$
   c. $R = 16$, $n_1 = 11$, $n_2 = 12$; $\mu = 12.478$, $\sigma = 2.337$, $z = 1.507$
3. $z = -1.02$, $\pm z_{.005} = \pm 2.58$; the sequence may be random.
5. **a.** The test is based on the median score of 4.5. $z = 2.01$, $\pm z_{.025} = \pm 1.96$; the sequence is not random.
   b. The test is based on even/odd digits. $z = -1.23$, $\pm z_{.025} = \pm 1.96$; the sequence may be random.
7. The median is 120. $z = .73$, $\pm z_{.025} = \pm 1.96$; there is no evidence that there is a significant trend.
8. (4.71, 17.30)

9. a. 20

b. Runs	f
2	2
3	4
4	8
5	4
6	2

d. $\mu = 4$, $\sigma = 1.10$
e. $n_1 = 3$, $n_2 = 3$; $\mu = 4$, $\sigma = 1.2$
f. $a = 2$, $b = 6$

Chapter 15 Review

1. $U = 109$, $z = 2.14$, $\pm z_{.025} = \pm 1.96$; reject H_0
3. $z = 1.67$, $z_{.005} = 2.58$; cannot reject H_0
5. $U = 24.5$, $z = -1.81$, $-z_{.01} = -2.33$; cannot reject H_0
7. $H = 2.54$, $\chi^2_{.05}(1) = 3.841$; cannot reject H_0
9. $H = 1.56$, $\chi^2_{.01}(2) = 9.210$; cannot reject H_0

Chapter 15 Achievement Test

1. a. 17   **b.** 13.48   **c.** 2.44   **d.** 1.44
   e. ± 1.96 **f.** Cannot reject H_0
2. a. $z = 1.90$ **b.** -2.58, 2.58
   c. .0574   **d.** $(.47, 1)$

3. $U = 32.5$, $z = -1.02$, $-z_{.01} = -2.33$; cannot reject H_0
4. $H = 1.13$, $\chi^2_{.05}(1) = 3.841$; cannot reject H_0

Problem Set Appendix A

1. a. $X_1 + X_2 + X_3 + X_4$
   b. $(X_2^2 + 2) + (X_3^2 + 2) + (X_4^2 + 2)$
   c. $(X_1^2 + Y_1) + (X_2^2 + Y_2) + (X_3^2 + Y_3) + (X_4^2 + Y_4)$
   d. $X_1^2 Y_1 + X_2^2 Y_2 + X_3^2 Y_3$
   e. $f_1 X_1^2 + f_2 X_2^2 + f_3 X_3^2 + f_4 X_4^2 + f_5 X_5^2$
   f. $X_1 Y_1^2 + X_1 Y_2^2 + X_1 Y_3^2 + X_2 Y_1^2 + X_2 Y_2^2 + X_3 Y_3^2$

2. a. $\sum\limits_{i=1}^{3} X_i Y_i$

   b. $\sum\limits_{i=1}^{4} X_i Y_{i+1}$

   c. $\sum\limits_{i=2}^{4} X_i^2$

   d. $\sum\limits_{i=1}^{4} i X_i^2$

   e. $\sum\limits_{i=3}^{6} (2X_i + 3)$

3. a. 25   **b.** 27   **c.** 144   **d.** 50   **e.** 15
   f. 28   **g.** 40   **h.** 0   **i.** 0   **j.** -7
4. a. 24   **b.** 30   **c.** 30
   d. 132   **e.** 24   **f.** 40

9. a. $\sum\limits_{i=1}^{11} (2i + 1)$ **b.** $\sum\limits_{i=3}^{20} i^2$

INDEX

t Distributions

df	.10	.05	.025	.01	.005
1	3.078	6.314	12.706	31.821	63.657
2	1.886	2.920	4.303	6.965	9.925
3	1.638	2.353	3.182	4.541	5.841
4	1.533	2.132	2.776	3.747	4.604
5	1.476	2.015	2.571	3.365	4.032
6	1.440	1.943	2.447	3.143	3.707
7	1.415	1.895	2.365	2.998	3.499
8	1.397	1.860	2.306	2.896	3.355
9	1.383	1.833	2.262	2.821	3.250
10	1.372	1.812	2.228	2.764	3.169
11	1.363	1.796	2.201	2.718	3.106
12	1.356	1.782	2.179	2.681	3.055
13	1.350	1.771	2.160	2.650	3.012
14	1.345	1.761	2.145	2.624	2.977
15	1.341	1.753	2.131	2.602	2.947
16	1.337	1.746	2.120	2.583	2.921
17	1.333	1.740	2.110	2.567	2.898
18	1.330	1.734	2.101	2.552	2.878
19	1.328	1.729	2.093	2.539	2.861
20	1.325	1.725	2.086	2.528	2.845